International Textbooks in

MECHANICAL ENGINEERING

Consulting Editor

EDWARD F. OBERT

Professor of Mechanical Engineering

University of Wisconsin

Mechanical Systems Analysis

Mechanical Systems Analysis

ALI SEIREG
Professor of Mechanical Engineering
University of Wisconsin

INTERNATIONAL TEXTBOOK COMPANY
Scranton, Pennsylvania

Library of Congress Catalog Card Number: 77-86868

Standard Book Number 7002 2246 4

Then said a teacher, speak to us of teaching.

And he said:

No man can reveal to you aught but that which already lies half asleep in the dawning of your knowledge.

The teacher who walks in the shadow of the temple, among his followers, gives not of his wisdom but rather of his faith and his lovingness.

If he is indeed wise he does not bid you enter the house of his wisdom, but rather leads you to the threshold of your own mind.

The astronomer may speak to you of his understanding of space but he cannot give you his understanding.

The musician may sing to you the rhythm which is in all space, but he cannot give you the ear which arrests the rhythm nor the voice that echoes it.

And he who is versed in the science of numbers can tell of the regions of weight and measure but he cannot conduct you thither.

For the vision of one man lends not its wings to another man.

And even as each one of you stands alone in God's knowledge, so must each one be alone in his knowledge of God and in his understanding of the earth.

—Kahlil Gibran, *The Prophet*

Preface

Mechanical systems analysis has log been an important part of the mathematical, physical and engineering disciplines. A system in the engineering sense is an entity which can be modeled as a transformation from an input to an output. Problems in system analysis can therefore be classified as:

1. Determining the output if the input and system are known.
2. Determining the input if the output and system are known.
3. Determining the system if the input and output are known.

The special class of systems discussed in this text are those which can be modeled by means of Newtonian mechanics. The objective is to give in a compact form a treatment, from the standpoint of the mechanical engineering student which exposes him to the broad field of mechanical systems and enables him to develop mathematical models and analogs which easily lend themselves to analytical treatment.

The book aims to achieve the following objectives:

(a) To emphasize problem formulation as an important analytical Step worthy of special attention. This is a logical result of the development of high speed digital and analog computers which caused this shift of emphasis. The purpose is to develop in the reader the ability to describe a complex mechanical system or element in terms of an idealized mathematical model. The model and the boundary conditions should be simple enough to lend themselves to analysis and yet accurate enough to provide acceptable representation of the actual systems.

(b) To give a general introduction to the basic principles governing the design of mechanical systems operating under forcing functions which may be known deterministically or probabilistically. The text provides a general treatment of mechanical system dynamics and the behavior of system materials under different types of stress regimes.

(c) Nonlinear systems are treated along the way without undue restraint. This unorthodox approach is also a result of the computer age.

In an effort to familiarize the reader with as many techniques of solution as possible, the depth of coverage of some of these methods has to be sacrificed. Those who would like to master any particular technique will have to refer to more specialized texts. The emphasis is placed on the numerical and graphical methods of solution, especially the phase-plane technique.

This book has grown out of lectures to senior and early graduate students in the Colleges of Engineering of Marquette University and the University of Wisconsin. The lectures have also been given as a part of a professional development program with the purpose of familiarizing the practicing engineer with useful techniques applicable to the analysis of dynamic systems. Throughout the book an effort has been made to avoid mathematical methods and notations which may distract the reader from the physical aspects of the problems.

In order to ensure completeness, the material presented in this book exceeds what can be covered in a single course. It is thus up to the instructor to select the topics he judges are best suited to the student's background and structure of the curriculum. A typical coverage for a senior or early graduate course is as follows:

1. Introduction (2 lectures).
   Historical Introduction.
   Review of Mechanics and Kinematics.
2. Mathematical Models and Analogs (10 lectures).
   Stability of equilibrium. Elastically supported particles. Elastically connected systems of particles. Elastically supported rigid bodies. The solid and fluid continuum. Influence of friction. The disturbance. Electrical and acoustical analogy.
3. Linear and nonlinear systems subjected to general disturbances. The generalized phase-plane method. Response spectra—Lagrange's method. (13 lectures.)
4. The propagation of the disturbance. Wave motion. Reflection of waves. Boundary effects. Exact solution of the differential equation. Ritz method. Impact of spheres. Sound. (6 lectures.)
5. Statistical concepts and definitions. Statistical system dynamics. Response of single degree-of-freedom systems to random excitations. Systems with two degrees of freedom. Continuous systems. (4 lectures.)
6. Experimental methods—scale models (3 lectures).
7. Mechanism of fatigue failure. Factors influencing fatigue failure. Analysis of fatigue data. Stress-life diagrams. Statistical nature of fatigue. Endurance limits for different stress cycles. Cumulative fatigue damage. Plastic strain-life criterion. Design consideration. Probabilistic fatigue criteria due to random loading. Failure under impulsive loads. (3 lectures.)

In view of the growing interest of the mechanical engineer in problems of medicine and biology, two sections are included in Chapters 14 and 15 on biological systems and biological materials.

Chapter 16, although brief, is intended as an introduction to the application of programming methods to the design and synthesis of mechanical systems. As the concluding chapter, it helps put in the proper prospective the role of analysis in the total engineering activity. A book on mechanical systems design is currently in preparation by the author.

My association with the Falk Corporation during the last two decades shows in the practical accent detectable throughout the book. To the many engineers at Falk, especially Mr. E. J. Wellauer, I would like to express my appreciation for this long association.

I would also like to thank many of my colleagues who showed interest in this manuscript. I especially mention Professors E. F. Obert (University of Wisconsin), W. G. Murphy and E. J. Weiter (Marquette University). Professor Obert gave on many occasions valuable advise and suggestions. A special acknowledgement is due to Professor Thomas J. Higgins (University of Wisconsin) for his generous offer to check the entire manuscript and page proofs which proved invaluable. I am grateful to Professor Higgins for his help with this onerous task.

To my former students I would like to extend my appreciation for their comments on the class notes on which this book is based. Particular recognition is due to L. K. Howard, Electronics Division of General Motors for his contribution to the section on the approximate normal mode, B. Hamad (University of Khartoum) who contributed to Chapter 16 and R. A. Schunck (Falk Corporation) who contributed to the stability analysis by means of the phase-plane.

This book would not have been possible without the cooperation and understanding of my wife Shirley and daughters Mirette and Pamela.

ALI SEIREG

Madison, Wisconsin
August, 1969

Contents

Historical Introduction

From earliest times certain concepts about the relationship between force and motion were developed. Although vague and primitive, these concepts were passed on from generation to generation and used in everyday living. Levers were utilized in raising heavy objects long before there were any laws of mechanics to explain why the objects were heavy or how the levers were able to overcome their weight. Drums and other forms of sound-producing devices were used many centuries before any effort was made to analyze their vibration.

The first attempt toward understanding and formulating these concepts was made by the philosophers of Athens and Alexandria in the few centuries before the birth of Christ. Among the most distinguished men of this era were Aristotle (384–322 B.C.), Euclid (c. 300 B.C.), Archimedes (287–212 B.C.), and Hero of Alexandria (1st cent. A.D.). The main contributions to dynamics during this period are the correct formulation of the laws of buoyancy and the analysis of the lever.

The Aristotelian concepts of mechanics, based on philosophical interpretations without experimental evidence, dominated scientific thought during this period and were responsible for the lack of progress in the field for many centuries. During the Renaissance the contributions of Leonardo da Vinci (1452–1519) to the development of mechanical systems and his keen interest in the analysis of their behavior are noteworthy. It was not until Galileo Galilei (1564–1642) that Aristotelian mechanics was disputed and the foundations laid for Newtonian mechanics as we know it today. Galileo's famous experiments with falling bodies and his astronomical observations, together with the laws of Johannes Kepler (1571–1603) on the motion of the planets, marked the dawn of a new era that led to Newton's celebrated laws. It was Sir Isaac Newton (1642–1727) who introduced the concept of force and mass and unquestionably formulated the basic laws of mechanics. Leonhard Euler (1707–1783) extended Newton's equations of motion from the single particle to the rigid body. Galileo also significantly contributed to the field of vibrations and demonstrated for the first time the association between vibration and sound. He discovered the relationship between the length of the pendulum and its frequency. He also observed resonance or "sympathetic vibrations" of two bodies oscillating at the same frequency and connected by some energy-transferring medium.

Robert Hooke (1635–1703) established the relation between force and deformation in elastic bodies which led to better understanding of the mechanics of continuous media. He illustrated the reason for the isochronal motion of a spring or extended string and the uniform sound developed by those strings whose vibrations are rapid enough to produce an audible sound. Charles Augustin de Coulomb (1736–1806) analyzed the torsional pendulum and determined the torsional rigidity of wires by observing the oscillation of the suspended mass. He found that the period of oscillation is independent of the angle of twist and showed that the principal cause of damping these vibrations is some imperfection in the material, not the air resistance.

Although as early as the sixth century B.C. Pythagoras observed that shorter strings fastened at the ends emitted higher notes than longer strings, it was Joseph Louis Lagrange (1736–1813) who correctly calculated the frequency of vibration of strings. He was also able to predict the approximate harmonic frequencies of closed and open pipes. One of the main contributions of Lagrange is using the scalar *work* and *energy* rather than the vector quantities *force* and *acceleration* in analyzing the motion of mechanical systems. His well-known equation is widely used in many fields of physics and engineering.

It is due to Sir William Rowan Hamilton (1805–1865) and Karl Gustav Jacobi (1804–1851) that variational methods were used in dynamics. Hamilton's principle, as well as "the principle of least action," provides a more general and fundamental treatment of mechanics. It is this formulation which proved useful in quantum mechanics. Hamilton's principle may be expressed by stating that if the configuration of the system is given at two instants, then the value of the time integral of the sum of the potential and kinetic energies is stationary (that is, less or greater) for the path actually described in the natural motion than in any other path. This may be expressed more freely as: Nature tends to equalize the mean potential and kinetic energies during any motion.

Albert Einstein in 1905 introduced the *theory of relativity*, which gives results identical to those obtained by Newtonian mechanics except when the speed of the bodies approach the speed of light. This relativistic mechanics is also capable of describing the microscopic motions of the nucleus where classical mechanics becomes inadequate.

The subject of oscillating systems is a branch of mechanics which attracted the interest of physicists and mathematicians for many years. Mechanical systems are for the most part vibratory, and the analysis of their behavior relies heavily on vibration theory. Among the noted contributions to the literature in this area are those of Félix Savart (1791–1841), Thomas Seebeck (1770–1831), Jean Biot (1774–1862), Rudolf Koenig

(1832–1901), Hermann Helmholtz (1821–1894), August Kundt (1839–1894), Ernst Chladni (1756–1827), and many others. The most significant contribution to the subject until the early part of the twentieth century is due to John William Strutt, Lord Rayleigh (1842–1919). His comprehensive book, *Theory of Sound*, first published in 1877, still retains a position of importance in the literature of the field. It presented the first connected exposition of acoustics and vibration which up to that time could be found only in scattered publications.

The analysis of dynamic systems, which was considered as part of the mathematical and physical disciplines, became of considerable interest to the engineer toward the end of the nineteenth century. Many problems with practical significance arose with the development of high-speed machinery and alternating current. The civil engineer also found the application of dynamic theory very useful in the design of structures subjected to wind, waves, earthquakes, and suddenly applied or moving loads. The works of Frahm on torsional systems and Foppl on the whirling of shafts carrying disks are among the early significant publications in these areas. Also among the important pioneering contributions are those of Stodola on the dynamic behavior of rotating machinery and Kryloff on the lateral vibrations of bridges under traveling loads.

Another area that attracted considerable interest is impact. Newton founded the theory of impact by formulating the laws of motion (conservation of momentum) as well as the restitution notation. One of the most discussed problems in impact is the propagation of longitudinal waves in long prismatic bars. Bernard de St. Venant (1797–1886) correctly stated the condition for termination of their impact and gave the stress and velocity for any given point in the two colliding bars during impact. According to St. Venant's theory, the stress waves generated in colliding bodies determine the pulse shape. Heinrich Hertz (1857–1894) treated the impact of two bodies as a statical problem, giving no consideration to the dissipation of energy. According to Hertz, the stresses and strains in the region of contact are the important factors in determining the impact. The theory of St. Venant relates to impacts between the flat end faces of long rods, whereas the Hertz theory applies when the time of transit of an elastic wave through the colliding bodies is small compared to the time of contact, as in the case of colliding small spheres.

In the last two decades engineering requirements in aerospace and other applications have necessitated the adoption of techniques of random-process analysis to mechanical systems. These techniques, which have their origin in the theory of probability and statistics, became essential in the analysis of many mechanical systems subjected to random excitations.

There have also been many significant contributions in recent years to the general theory of dynamic systems. These developments include concepts fundamental to problems of vibrations, control, communication, information processing, economic forecasting, and the like, as well as analytical methods in specialized problem areas that show promise of relating to the general class of problems. Although no unifying system theory presently exists, it is anticipated, from these related fields, that a theory of physical systems will emerge that can accept certain processes as inputs and generate from them certain other processes as outputs. Many of the tools of modern system theory are rigorous mathematical developments and, for the most part, are beyond the scope of this introductory book. However, any modern development which may be of interest to the reader will be referred to in the text.

Advances in techniques of system analysis can be effectively utilized by the designer of mechanical systems only when corresponding advances are made in the areas of fracture and failure of materials under dynamic load regimes. Modern literature is rich with developments in this area and contains many useful studies on the behavior of metals, nonmetals, synthetics, and composite materials. A brief account of some of these developments is included in a chapter on material behavior under dynamic loading.

REFERENCES

Clagett, M. *Science of Mechanics in the Middle Ages*, U. of Wisconsin Press, Madison, Wisc., 1959.

Dugas, R. *A History of Mechanics*, Central Book, New York, 1955.

Fox, Jerome (ed.), *Proceedings of the Symposium on Systems Theory*, Wiley, Interscience, New York, 1965.

Goodman, L. E., and W. Warner. *Dynamics*, Wadsworth, Belmont, Calif., 1963.

Strutt, John, Lord Rayleigh, *Theory of Sound*, Dover, New York, 1945.

Timoshenko, S., *History of Strength of Materials*, McGraw-Hill, New York, 1953.

Webster, A. G. *The Dynamics of Particles*, Dover, New York, 1959.

part I

Problem Formulation

In this sense rational mechanics will be the science of motions resulting from any forces whatsoever, and the forces required to produce any motions accurately proposed and demonstrated.

Isaac Newton (1642–1727)

chapter 1

Mechanics—The Analysis of Motion and Deformation

1-1. DEFINITION

Mechanics is the science dealing with force and motion. Its objective is to describe all motions occurring in nature in the simplest manner and to relate them to the forces producing them on the basis of the simplest possible laws. This chapter gives a brief review of some of the laws governing the behavior of particles, rigid bodies, and elastic elements.

1-2. FREE BODY OR SYSTEM ISOLATION

In order to study and analyze a portion of the universe, the portion selected is imagined to be detached from its surroundings. This portion can be a planet, a rocket, the crankshaft of an engine, or an atom. The effects exerted by the surroundings should be considered as external actions on that portion when treated as a free body. The system should then be idealized in order to make a solution possible through the laws of mechanics. This idealization is usually based on certain assumptions which permit the translation of the physical system into a mathematical model.

Mechanical systems can be idealized into one or a combination of the following categories.

The Particle

A particle is an object whose mass and not its size is of significance in the analysis of its motion. A particle can move only in a translatory fashion. This motion is represented by the change in position of the center of mass, and the distribution of mass is of no consequence.

In the general sense, a particle has three degrees of freedom. The number of degrees of freedom that a system possesses is the number of independent coordinates which must be specified in order to define its motion in space. In the case of a particle, these coordinates are conveniently represented by the displacements x, y, z in the direction of three fixed orthogonal axes X, Y, Z defining an inertial reference.

The body shown in Fig. 1-1, for example, is dynamically treated as a particle, since the line of action of all the forces (F_1, F_2, F_3) acting on

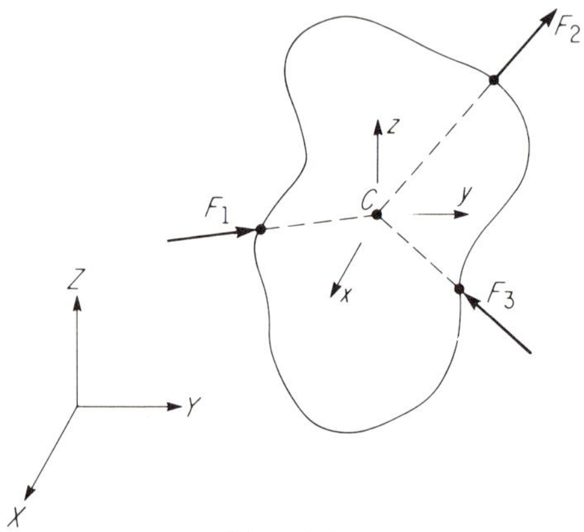

FIG. 1-1

it passes by its center of mass. The body will therefore translate without rotation. The motion of all points in the body is the same as that of its center of mass, and the body dimensions will have no effect on this motion.

The Rigid Body

A rigid body is an object whose size and mass distribution are of primary concern in the analysis of its dynamic behavior. All points in a rigid body are also assumed to remain at fixed distances from each other during its motion.

Any displacement of a rigid body may be reduced to a succession of translation and rotation. Such body will have, in general, six degrees of freedom. Three of them can be conveniently considered to represent the general translation (x, y, z) of the center of mass and the other three to represent the rotation $(\theta_x, \theta_y, \theta_z)$ of the body about the three reference axes as shown in Fig. 1-2.

The Deformable Body or Continuous Medium

In the previous definition of rigid bodies, the assumption is made that the particles in the bodies remain at constant distances from each other in spite of the external forcing effects. This assumption can be tolerated when the motion of the body as a whole is the principal interest.

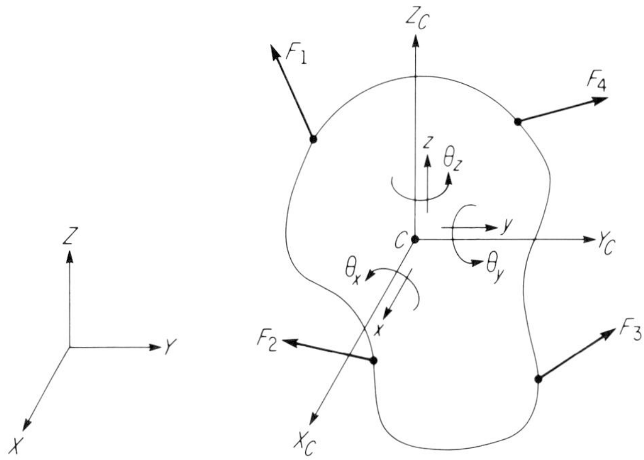

FIG. 1-2

However, this assumption is an obvious idealization, and it is well known that all bodies deform to a certain extent when external effects are applied to them. The relative displacements between the different particles result from the load distribution within the body and are mutually dependent. A finite continuum will have infinite degrees of freedom, since it is composed of infinite particles which can move relative to each other.

1-3. NEWTON'S LAWS

The laws of mechanics used to analyze the motion of the previous systems are based on Newton's laws, which can be stated as follows:

1. A body continues in the state of rest or of uniform motion in a straight line unless it is compelled to change that state by forces impressed upon it.
2. The change of motion is proportional to the motive force impressed, and is in the direction of the straight line in which the force is impressed.
3. To every action there is always a reaction equal in magnitude and opposite in direction.
4. Two particles are attracted toward each other along a line connecting them with a force whose magnitude is directly proportional to the product of the masses and inversely proportional to the square of the distance between them.

1-4. KINEMATICS

Kinematics is the geometry of motion. The kinematics of a particle deals with the motion of a point in space, while that of rigid bodies and continuous media takes into consideration changes in position as well as orientation. The position of a point in space is known when its three rectangular cartesian coordinates with respect to a fixed system of axes are given. Its motion is completely described if its coordinates (x, y, z) are known at all instants of time. These coordinates may be given as

$$x = f_1(t)$$
$$y = f_2(t)$$
$$z = f_3(t)$$

where f_1, f_2, f_3 are continuous functions of time t, since in no actual motion does a moving point disappear in one position to reappear after a very small interval of time in a new position at a finite distance from the old. These functions are also assumed to have definite derivatives for every value of time. The first and second derivatives give the components of the velocity (v_x, v_y, v_z) and acceleration (a_x, a_y, a_z) in the direction of the reference axes. This is expressed analytically as

$$v_x = \frac{dx}{dt}, \qquad a_x = \frac{d^2x}{dt^2}$$

$$v_y = \frac{dy}{dt}, \qquad a_y = \frac{d^2y}{dt^2}$$

$$v_z = \frac{dz}{dt}, \qquad a_z = \frac{d^2z}{dt^2}$$

The distance S between any two points (x_1, y_1, z_1) and (x_2, y_2, z_2) in space is shown in Fig. 1-3 to be

$$S = \sqrt{(x_2 - x_1)^2 + (y_2 - y_1)^2 + (z_2 - z_1)^2} \tag{1-1}$$

The direction of the vector connecting (x_1, y_1, z_1) to (x_2, y_2, z_2) is defined by its orientation with respect to the reference axes. This is generally given as the cosines of the angles $(\theta_{sx}, \theta_{sy}, \theta_{sz})$ made with the X-, Y-, and Z-axes respectively. Accordingly,

$$l = \cos \theta_{sx} = \frac{x_2 - x_1}{S} \tag{1-2a}$$

$$m = \cos \theta_{sy} = \frac{y_2 - y_1}{S} \tag{1-2b}$$

$$n = \cos \theta_{sz} = \frac{z_2 - z_1}{S} \tag{1-2c}$$

where l, m, n are known as the *directional cosines* of the vector under consideration. It can be concluded from Eqs. 1-1 and 1-2 that the following relation exists between the directional cosines of any line in space:

$$l^2 + m^2 + n^2 = 1 \tag{1-3}$$

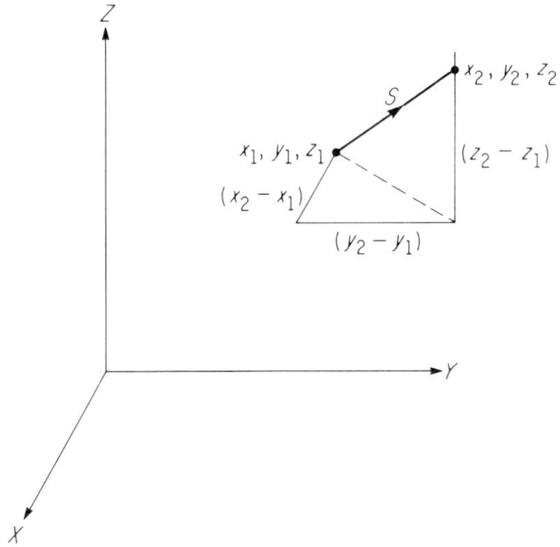

FIG. 1-3

The rotation of a rigid body with respect to a fixed frame of reference can be defined by the change in the orientation of any fixed line in the body with respect to the three reference axes. This orientation is given at any instant by

$$\theta_x = f_4(t)$$
$$\theta_y = f_5(t)$$
$$\theta_z = f_6(t)$$

where $\theta_x, \theta_y, \theta_z$ are the angles in radians between any body-fixed line and the three axes X, Y, Z respectively. The functions f_4, f_5, f_6 are also continuous and have definite derivatives. The components of the angular velocity $(\omega_x, \omega_y, \omega_z)$ and the angular acceleration $(\alpha_x, \alpha_y, \alpha_z)$ are therefore

$$\omega_x = \frac{d\theta_x}{dt}, \qquad \alpha_x = \frac{d^2\theta_x}{dt^2}$$

$$\omega_y = \frac{d\theta_y}{dt}, \qquad \alpha_y = \frac{d^2\theta_y}{dt^2}$$

$$\omega_z = \frac{d\theta_z}{dt}, \qquad \alpha_z = \frac{d^2\theta_z}{dt^2}$$

1-5. PARTICLE DYNAMICS

Equations of Motion

Since a particle is in effect a mass point, its motion in a fixed frame of reference (Fig. 1-4) is completely described by direct application of New-

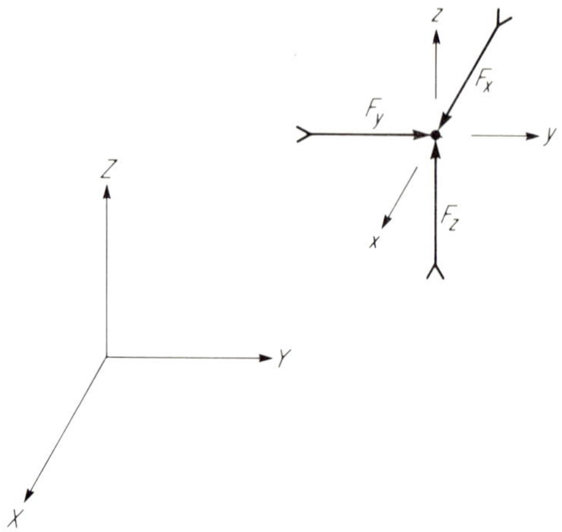

FIG. 1-4

ton's second law:

$$m\frac{d^2x}{dt^2} = m\frac{dv_x}{dt} = F_x \tag{1-4a}$$

$$m\frac{d^2y}{dt^2} = m\frac{dv_y}{dt} = F_y \tag{1-4b}$$

$$m\frac{d^2z}{dt^2} = m\frac{dv_z}{dt} = F_z \tag{1-4c}$$

where m = mass of particle
x, y, z = components of displacement
F_x, F_y, F_z = forces acting on the particle

Work and Energy

The previous equations can be used to derive the work and energy equations. Equation 1-4a for example can be rewritten as

$$m\frac{dv_x}{dx}\frac{dx}{dt} = F_x$$

or

$$mv_x dv_x = F_x\,dx$$

and by integrating this equation between an initial position i and a final position f, we get

$$\frac{1}{2}m(V_x)_f^2 - \frac{1}{2}m(V_x)_i^2 = F_x[x_f - x_i] \tag{1-5a}$$

Similarly,

$$\frac{1}{2}m(V_y)_f^2 - \frac{1}{2}m(V_y)_i^2 = F_y[y_f - y_i] \tag{1-5b}$$

and

$$\frac{1}{2}m(V_z)_f^2 - \frac{1}{2}m(V_z)_i^2 = F_z[z_f - z_i] \tag{1-5c}$$

Since $\frac{1}{2}mV^2$ represents the kinetic energy of the particle at any instant, these equations illustrate the relationship between work and energy: *The work done by all forces acting on a particle in any direction equals the change in the kinetic energy of the particle in that direction.* Notice that work and energy are scalar quantities which make this formulation more advantageous, in many situations, than the differential equations of motion.

Impulse and Momentum

Newton's second law can also be rewritten to derive the relation between impulse and momentum. Using Eq. 1-4a as an example, we have

$$m\frac{dv_x}{dt} = F_x$$

or

$$m\,dv_x = F_x\,dt$$

and by integrating between an initial condition i and a final condition f we get

$$m(V_x)_f - m(V_x)_i = \int_{t_i}^{t_f} F_x\,dt \tag{1-6a}$$

Similarly,

$$m(V_y)_f - m(V_y)_i = \int_{t_i}^{t_f} F_y\,dt \tag{1-6b}$$

and

$$m(V_z)_f - m(V_z)_i = \int_{t_i}^{t_f} F_z\,dt \tag{1-6c}$$

The left-hand side of these equations represents the change in momentum and the right-hand side is the impulse in the particular direction during the given interval.

In the case of an impact between two bodies as shown in Fig. 1-5, we have

$$m_1(V_1)_f - m_1(V_1)_i = \int_{t_i}^{t_f} R\,dt$$

$$m_2(V_2)_f - m_2(V_2)_i = \int_{t_i}^{t_f} -R\,dt$$

where R is the reaction between the two bodies in the line of impact at any time during the impact.

By adding and arranging the two equations, we get

$$m_1(V_1)_f + m_2(V_2)_f = m_1(V_1)_i + m_2(V_2)_i \tag{1-7}$$

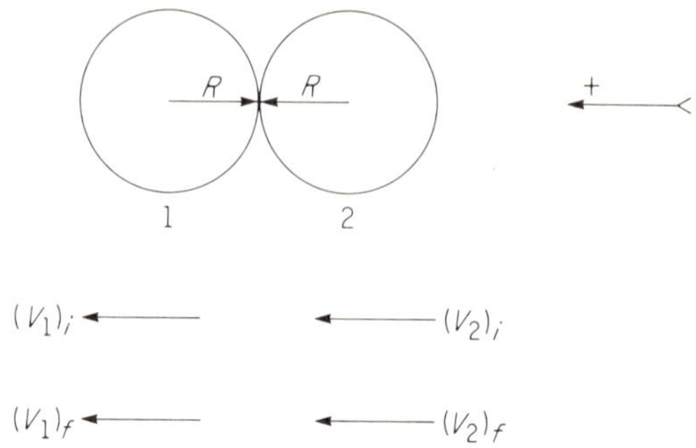

FIG. 1-5

which shows that the total momentum after impact is equal to the total momentum before impact. This is known as the *principle of conservation of momentum*. It should be noted here that the momentum equation is a vector one. Values of the different velocities should be substituted in the equation with the proper sign with respect to a reference positive direction.

Newton also introduced the restitution notation to define a coefficient of restitution e as *the ratio between the velocity of separation and the velocity of approach*:

$$e = -\frac{(V_2)_f - (V_1)_f}{(V_2)_i - (V_1)_i} \tag{1-8}$$

This coefficient is an empirical factor which depends on the materials in impact, their geometry, and the velocity of approach. Its value ranges between 1.0 for elastic impact and zero for plastic impact. It can be seen from Eq. 1-8 that in the latter case the two impacting bodies do not separate after impact.

The simultaneous solution of Eqs. 1-7 and 1-8 gives the following expressions for the velocities after impact for the two bodies:

$$(V_1)_f = \frac{(V_1)_i\left(1 - \dfrac{m_2}{m_1}e\right) + \dfrac{m_2}{m_1}(V_2)_i(1 + e)}{\left(1 + \dfrac{m_2}{m_1}\right)}$$

$$(V_2)_f = \frac{(V_1)_i(1 + e) + (V_2)_i\left(\dfrac{m_2}{m_1} - e\right)}{\left(1 + \dfrac{m_2}{m_1}\right)}$$

The energy dissipated in the impact can be calculated from

$$\text{Kinetic energy loss} = \frac{1 - e^2}{2}\,\frac{m_1 m_2}{m_1 + m_2}\,[(V_1)_i - (V_2)_i]^2$$

1-6. RIGID-BODY DYNAMICS

Equations of Motion

The rigid-body motion (Fig. 1-2) can be described generally by the following six equations:

$$F_x = m \frac{d^2 x}{dt^2} \tag{1-9a}$$

$$F_y = m \frac{d^2 y}{dt^2} \tag{1-9b}$$

$$F_z = m \frac{d^2 z}{dt^2} \tag{1-9c}$$

which govern the translation of the center of mass, and

$$M_x = I_{x_c x_c}\left(\frac{d^2\theta_x}{dt^2}\right) + \left(\frac{d\theta_z}{dt} \cdot \frac{d\theta_y}{dt}\right)(I_{z_c z_c} - I_{y_c y_c}) \tag{1-9d}$$

$$M_y = I_{y_c y_c}\left(\frac{d^2\theta_y}{dt^2}\right) + \left(\frac{d\theta_x}{dt} \cdot \frac{d\theta_z}{dt}\right)(I_{x_c x_c} - I_{z_c z_c}) \tag{1-9e}$$

$$M_z = I_{z_c z_c}\left(\frac{d^2\theta_z}{dt^2}\right) + \left(\frac{d\theta_y}{dt} \cdot \frac{d\theta_x}{dt}\right)(I_{y_c y_c} - I_{x_c x_c}) \tag{1-9f}$$

which govern the rotation of the body as a whole. In these equations

m = mass of the body
$I_{x_c x_c}, I_{y_c y_c}, I_{z_c z_c}$ = principal moments of inertia at the center of mass (X_c, Y_c, Z_c are body-fixed axes)
x, y, z = displacements of the center of mass in the reference frame X, Y, Z
$\theta_x, \theta_y, \theta_z$ = angular rotation of the body about reference axes
F_x, F_y, F_z = forces acting on the body in the X-, Y-, and Z-directions
M_x, M_y, M_z = the moments of all forces acting on the body about the center of mass in the three reference directions

The first three equations can be easily recognized as Newton's second law. The last three equations are known as *Euler's equations* and represent an extension of Newton's law to account for the rotation of the rigid body about its center of mass. Euler's equations are also applicable when the rotation is about a fixed point in space. The reader interested in the proof of these equations should consult any of the texts on the dynamics of rigid bodies.

Work and Energy

Since the motion of rigid bodies combines a translation of the center of mass and a rotation about it, the kinetic energy (K.E.) for a rigid body should include both the energy of displacement as well as the energy of rotation. It is defined by

$$\text{K.E.} = \frac{1}{2} mV^2 + \frac{1}{2} I_{x_c x_c}\left(\frac{d\theta_x}{dt}\right)^2 + \frac{1}{2} I_{y_c y_c}\left(\frac{d\theta_y}{dt}\right)^2 + \frac{1}{2} I_{z_c z_c}\left(\frac{d\theta_z}{dt}\right)^2 \tag{1-10}$$

where

$$V = \left[\left(\frac{dx}{dt}\right)^2 + \left(\frac{dy}{dt}\right)^2 + \left(\frac{dz}{dt}\right)^2\right]^{1/2}$$

= magnitude of the velocity of the center of mass

The work and energy equation can therefore be written as

Change in kinetic energy = Work done by all forces and moments acting on the body during a particular motion

Impulse and Momentum

A rigid body may be subjected to linear impulse due to the resultant force at the center of mass, as well as to angular impulse due to the moment of all forces acting on it about the center of mass. It will accordingly undergo a change in both its linear and angular momentum. The linear momentum is obtained by the product of the mass and the velocity V of the center of mass. The angular momentum about each principal axis at the center of mass is obtained by multiplying the moment of inertia by the angular velocity about that axis. This is also true for the angular momentum about a fixed axis of rotation as well as for a principal axis at a fixed center of rotation.

1-7. DEFORMABLE BODIES OR CONTINUUM

Deformable bodies, by definition, exhibit relative displacements between their particles when subjected to external forces. These relative displacements can be recoverable (*elastic*) after withdrawal of the external effects, or permanent and irrecoverable (*plastic*). Most engineering materials exhibit both characteristics and are called *viscoelastic*. The theories of elasticity, viscoelasticity, plasticity, and fluids discuss the relationships between the external effects and the resulting displacements. These theories make it possible to investigate the distribution of body deformations (or strains) and reactive body forces (or stresses) in the continuum. If these deformations or forces are time-dependent, the discussions of system behavior are generally included in the theory of vibrations and impact. It should be noted here that although rigorous methods are available to the engineer for analysis of the continuum, application of such methods can be tedious in complex problems. We shall consider in this text only the motion of simple cases of deformable bodies which are of interest to the engineer. The vibration of strings, beams, membranes, plates, and fluids in pipes is discussed in Chapter 3. The following gives a brief summary of some useful relationships between force, displacement, and stress for examples of these elements.

The String

The string is the most elementary of all structures, since it is capable only of transmitting a tensile force normal to its cross section. This is evaluated from the conditions for equilibrium when the configuration of the string is known. The general expressions for stress and deformation at any section can be written as

$$\text{Stress} = \frac{T_x}{A_x}$$

where T_x and A_x are the transmitted force and the cross-sectional area respectively. Also,

$$\text{Deformation} = \int_0^L \frac{T_x}{A_x E_x}\, dx$$

where E_x is the modulus of elasticity. Common examples of strings are illustrated in the accompanying table.

The Beam

A beam is an elastic element capable of transmitting a general force and a general moment. The general force can be resolved into two components, one normal to the cross section (compressive or tensile force) and the other in the plane of the section (shearing force). The general moment can similarly be resolved into a bending moment (normal to the section) and a twisting moment (in the plane of the section). General expressions for evaluation of stress and deflection are given in the following:

Normal Force

$$\text{Maximum normal stress} = \sigma_x = k_{N_x}\left(\frac{N_x}{A_x}\right)$$

where N_x and A_x are the normal force and cross-sectional area respectively and k_{N_x} is a factor accounting for the distribution of stress over the

Description	*Governing Equations*
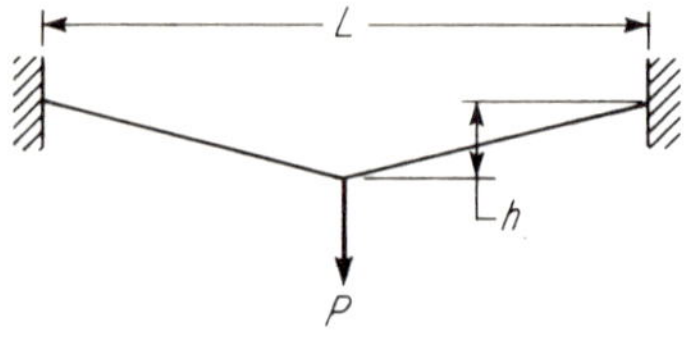 String with concentrated load at the center	$T = \dfrac{PL}{4h}$
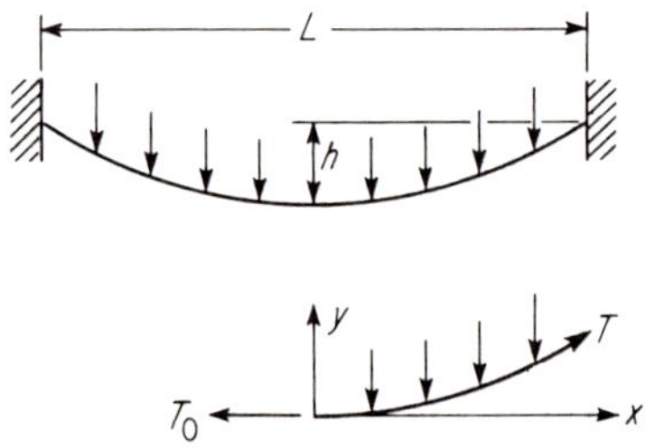 String with uniform load intensity w along the span	$y = \dfrac{wx^2}{2T_0}$ $T_0 = \dfrac{wL^2}{8h}$ $T = w\sqrt{x^2 + \dfrac{L^4}{64h^2}}$
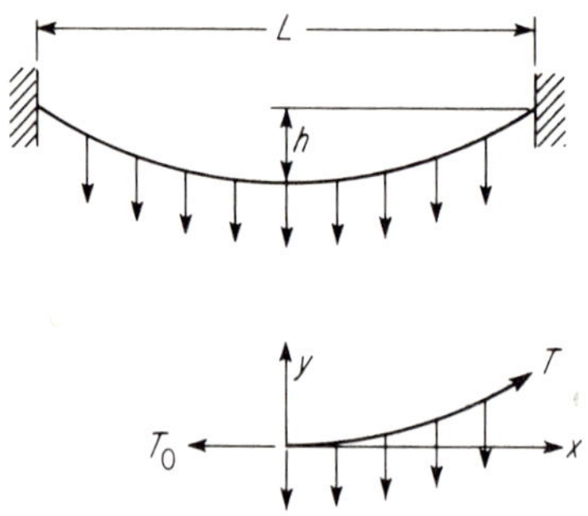 String with uniform load p per unit length of string	$y = \dfrac{T_0}{p}\left(\cosh \dfrac{px}{T_0} - 1\right)$ $T = T_0 \cosh \dfrac{px}{T_0}$

cross section

$$\text{Elongation} = \int_0^L \frac{N_x}{A_x E_x}\,dx$$

where E_x is the modulus of elasticity.

Shearing Force

$$\text{Maximum shear stress} = \tau_x = k_{S_x}\left(\frac{S_x}{A_x}\right)$$

where S_x is the shearing force on the cross section and k_{S_x} is a factor accounting for the distribution of shear stress which can be evaluated from the geometry of the section. For a rectangular cross section $k_{S_x} = 1.5$ and for a circular cross section $k_{S_x} = 1.33$.

$$\text{Shear deflection} = \int_0^L \frac{\tau_x}{G_x}\,dx$$

where G_x is the shear modulus of the material.

Bending Moment

The expressions for calculation of stress and deformation in any cross section of a beam due to a bending moment are

$$\text{Bending stress} = \sigma_b = \frac{M_b}{I}C$$

$$\text{Radius of beam curvature} = R = \frac{EI}{M_b}$$

$$\text{Slope of elastic curve} = \int \frac{M_b}{EI}\,dx = \frac{dy}{dx}$$

$$\text{Deflection} = y = \iint\left(\frac{M_b}{EI}\,dx\right)dx$$

where

M_b = bending moment
I = moment of inertia of the cross section about the centroidal axis
C = distance from the centroidal axis
E = modulus of elasticity.

The accompanying table summarizes results for common cases of uniform beams subjected to transverse loads.

Condition	*Equation of Elastic Curve*
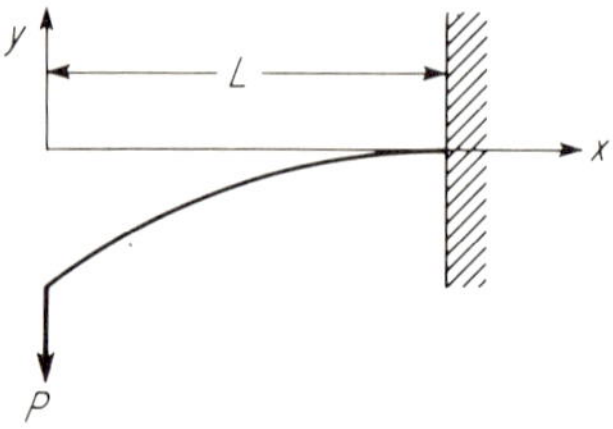	$y = -\frac{PL^3}{EI}\left[\frac{1}{3} - \frac{1}{2}\frac{x}{L} + \frac{1}{6}\left(\frac{x}{L}\right)^3\right]$
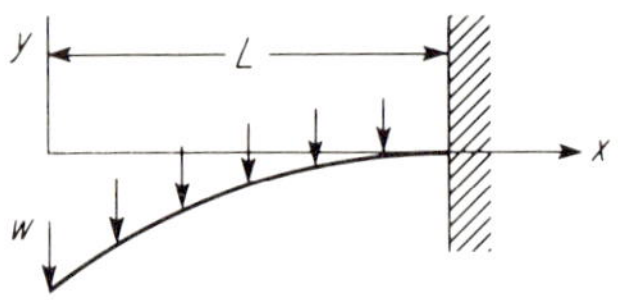 Total load $P = wL$	$y = -\frac{PL^3}{EI}\left[\frac{1}{24}\left(\frac{x}{L}\right)^4 - \frac{1}{6}\frac{x}{L} + \frac{1}{8}\right]$
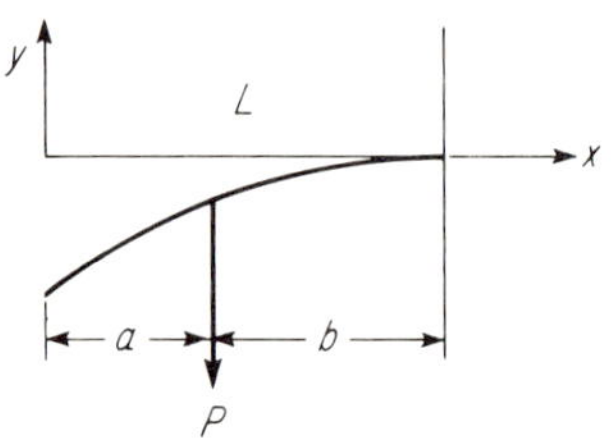	$x \leq a$ $y = \frac{PL^3}{EI}\left[\left(1 - \frac{a}{L}\right)^2\left(\frac{1}{2}\frac{x}{L} - \frac{1}{3} - \frac{1}{6}\frac{a}{L}\right)\right]$ $x \geq a$ $y = \frac{PL^3}{EI}\left[\frac{1}{6}\left(1 - \frac{a}{L}\right)^3 - \frac{1}{6}\left(\frac{x}{L} - \frac{a}{L}\right)^3 - \frac{1}{2}\left(1 - \frac{a}{L}\right)^2\left(1 - \frac{x}{L}\right)\right]$
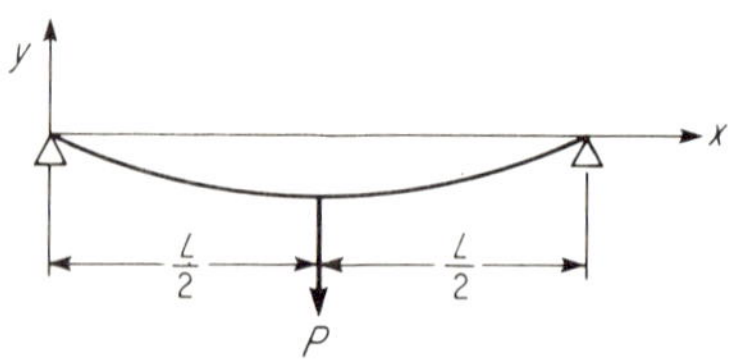	$x \leq \frac{L}{2}$ $y = -\frac{PL^3}{EI}\left[\frac{1}{16}\frac{x}{L} - \frac{1}{12}\left(\frac{x}{L}\right)^3\right]$

Condition	Equation of Elastic Curve
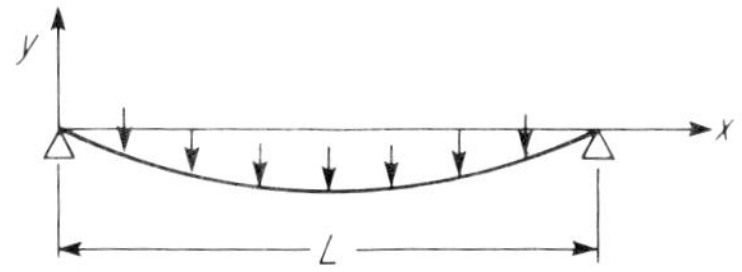 Total load $P = wL$	$y = -\frac{PL^3}{EI}\left[\frac{1}{24}\frac{x}{L} - \frac{1}{12}\left(\frac{x}{L}\right)^3 + \frac{1}{24}\left(\frac{x}{L}\right)^4\right]$
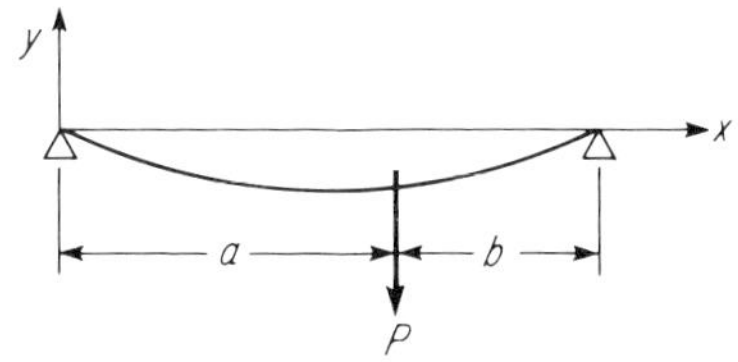	$x \leq a$ $y = \frac{PL^3}{EI}\left[\frac{1}{6}\frac{b}{L}\left(\frac{x}{L}\right)^3 - \frac{1}{6}\left(\frac{b}{L}\right)\left(\frac{x}{L}\right)\left(1 - \left(\frac{b}{L}\right)^2\right)\right]$ $x \geq a$ $y = \frac{PL^3}{EI}\left[\frac{1}{6}\left(\frac{b}{L}\right)\left(\frac{x}{L}\right)^3 - \frac{1}{6}\left(\frac{x}{L} - \frac{a}{L}\right)^3 - \frac{1}{6}\frac{b}{L}\frac{x}{L}\left(1 - \left(\frac{b}{L}\right)^2\right)\right]$
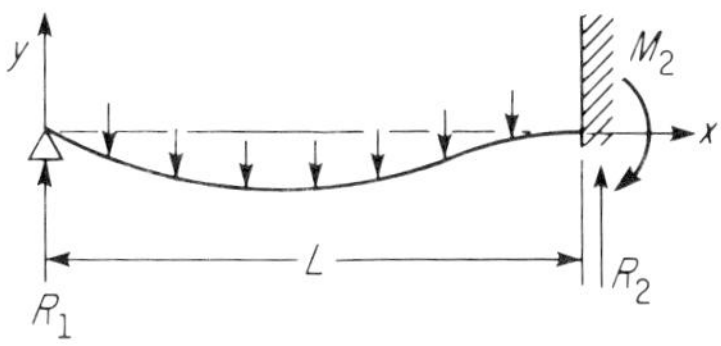 $P = wL$	$y = \frac{PL^3}{EI}\frac{1}{48}\left[\left(\frac{x}{L}\right)^3 - 2\left(\frac{x}{L}\right)^4 - \left(\frac{x}{L}\right)\right]$ $R_1 = \frac{3}{8}p$ $R_2 = \frac{5}{8}p$ $M_2 = \frac{1}{8}PL$

Condition	*Equation of Elastic Curve*
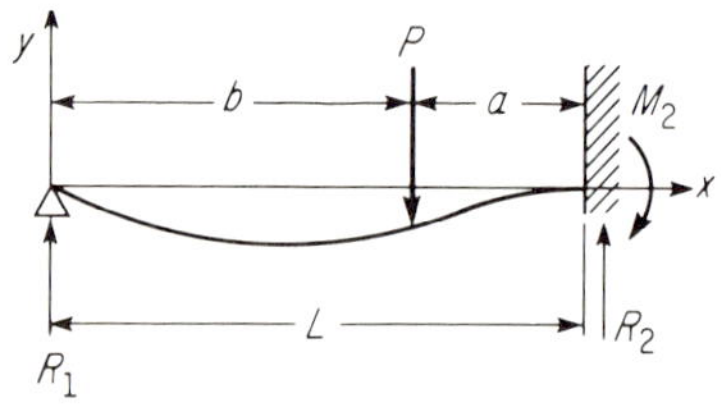	$0 \le x \le b$ $y = \frac{PL^3}{EI}\left[\left\{\frac{1}{4}\left(\frac{a}{L}\right)^2 - \frac{1}{12}\left(\frac{a}{L}\right)^3\right\} \times \left\{\left(\frac{x}{L}\right)^3 - 3\left(\frac{x}{L}\right)\right\} + \frac{1}{2}\left(\frac{a}{L}\right)^2\left(\frac{x}{L}\right)\right]$ $b \le x \le L$ $y = \frac{PL^3}{EI}\left[\left\{\frac{1}{4}\left(\frac{a}{L}\right)^2 - \frac{1}{12}\left(\frac{a}{L}\right)^3\right\} \times \left\{\left(\frac{x}{L}\right)^3 - 3\left(\frac{x}{L}\right)\right\} + \left\{\frac{1}{2}\left(\frac{a}{L}\right)^2\left(\frac{x}{L}\right) - \frac{1}{6}\left(\frac{x-b}{L}\right)^3\right\}\right]$ $R_1 = \frac{P}{2}\left[3\left(\frac{a}{L}\right)^2 - \left(\frac{a}{L}\right)^3\right]$ $M_2 = \frac{PL}{2}\left[\left(\frac{a}{L}\right)^3 - 3\left(\frac{a}{L}\right)^2 + 2\left(\frac{a}{L}\right)\right]$
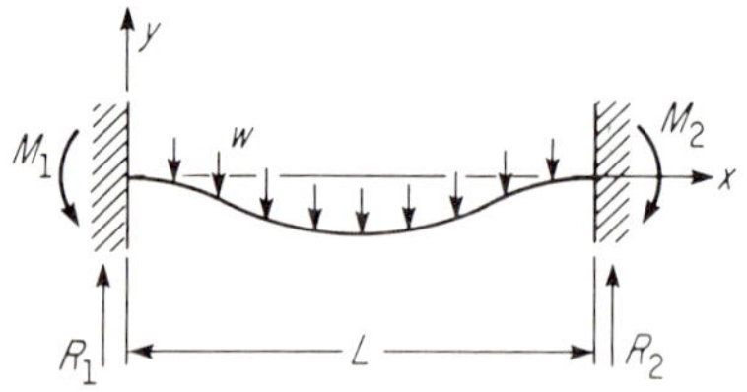$P = wL$	$y = \frac{PL^3}{EI}\frac{1}{24}\left[2\left(\frac{x}{L}\right)^3 - \left(\frac{x}{L}\right)^2 - \left(\frac{x}{L}\right)^4\right]$ $R_1 = \frac{P}{2} = R_2$ $M_1 = \frac{PL}{12} = M_2$

Condition	*Equation of Elastic Curve*
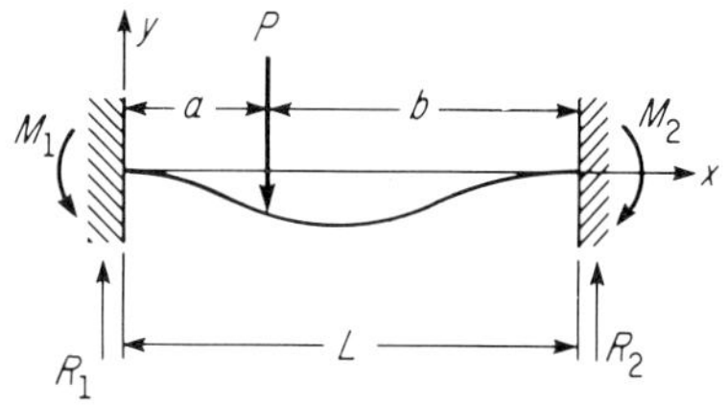	$0 \le x \le a$ $y = -\frac{PL^3}{EI}\left[\frac{1}{2}\left(\frac{x}{L}\right)^3\left(\frac{b}{L}\right)^2\left(\frac{a}{L}\right) + \frac{1}{6}\left(\frac{x}{L}\right)^3\left(\frac{b}{L}\right)^3 - \frac{1}{2}\left(\frac{b}{L}\right)^2\left(\frac{x}{L}\right)^2\left(\frac{a}{L}\right)\right]$ $a \le x \le L$ $y = -\frac{PL^3}{EI}\frac{1}{6}\left(\frac{a}{L}\right)^2\left(1 - \frac{x}{L}\right)^2 \times \left[\left(\frac{3b + a}{L}\right)\left(1 - \frac{x}{L}\right) - 3\left(\frac{b}{L}\right)\right]$ $R_1 = P\left(\frac{b}{L}\right)^2\left(\frac{3a + b}{L}\right)$ $R_2 = P\left(\frac{a}{L}\right)^2\left(\frac{3b + a}{L}\right)$ $M_1 = PL\left(\frac{a}{L}\right)\left(\frac{b}{L}\right)^2$ $M_2 = PL\left(\frac{a}{L}\right)^2\left(\frac{b}{L}\right)$

It can be easily seen that the general deflection equation takes the form

$$y = k_b \frac{PL^3}{EI}$$

where P = total transverse load on the beam
L = beam length
k_b = constant depending on the load distribution, the beam-support condition and the location where the deflection is calculated

The following table gives the bending stiffness of common cases of beams.

Condition	*Spring Constant*
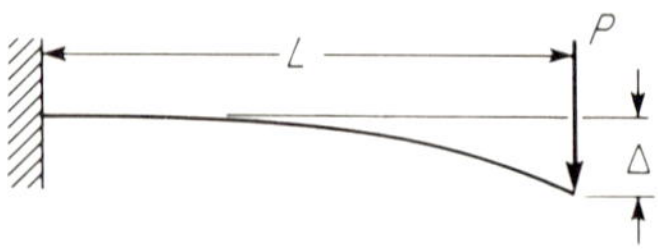	$\dfrac{P}{\Delta} = \dfrac{3EI}{L^3}$
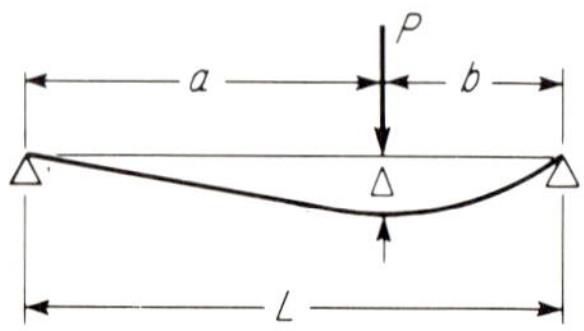	$\dfrac{P}{\Delta} = \dfrac{3EI}{L^3\left(\dfrac{a}{L}\right)^2\left(\dfrac{b}{L}\right)^2}$
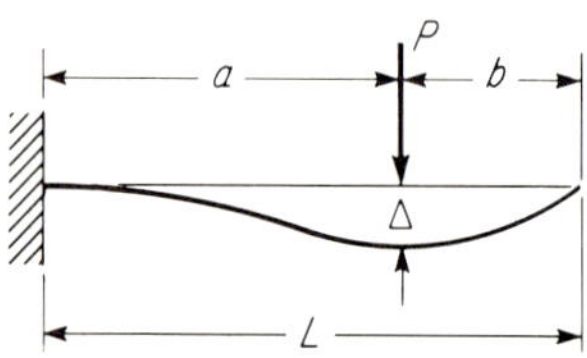	$\dfrac{P}{\Delta} = \dfrac{96EI}{L^3\left[5\left(\dfrac{b}{L}\right)^3 - 3\left(\dfrac{b}{L}\right)\right]}$
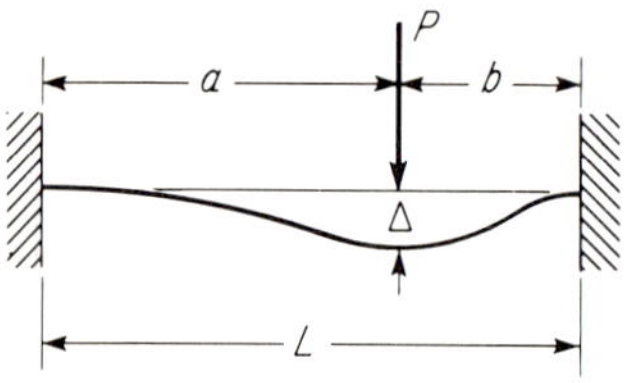	$\dfrac{P}{\Delta} = \dfrac{3EI}{L^3\left(\dfrac{a}{L}\right)^3\left(\dfrac{b}{L}\right)^3}$
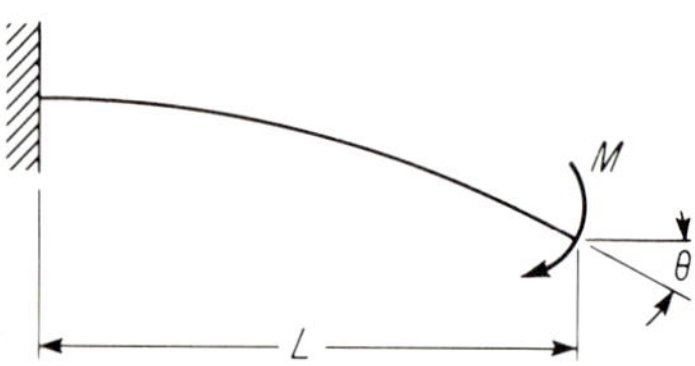	$\dfrac{M}{\theta} = \dfrac{EI}{L}$
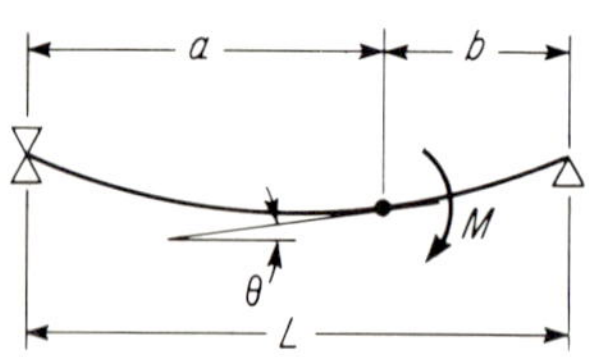	$\dfrac{M}{\theta} = \dfrac{3EI}{L\left[1 - 3\left(\dfrac{a}{L}\right)\left(\dfrac{b}{L}\right)\right]}$

Condition	Spring Constant
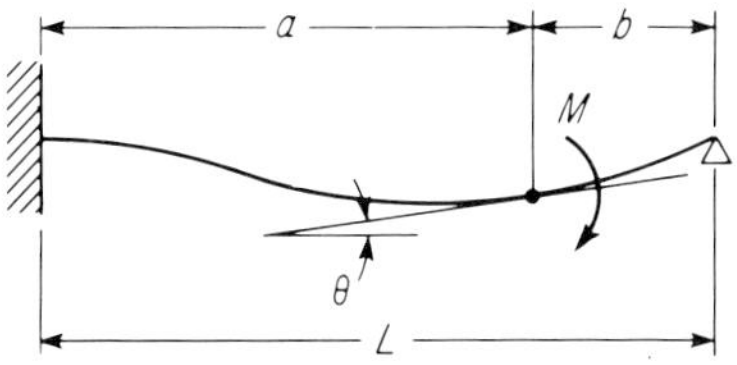	$\frac{M}{\theta} = \frac{EI}{L\left[\left(\frac{b}{L}\right) - \frac{3}{4}\left(\frac{b}{L}\right)^2\left(1 - \frac{a}{L}\right)\left(1 - \frac{b}{L}\right)\right]}$
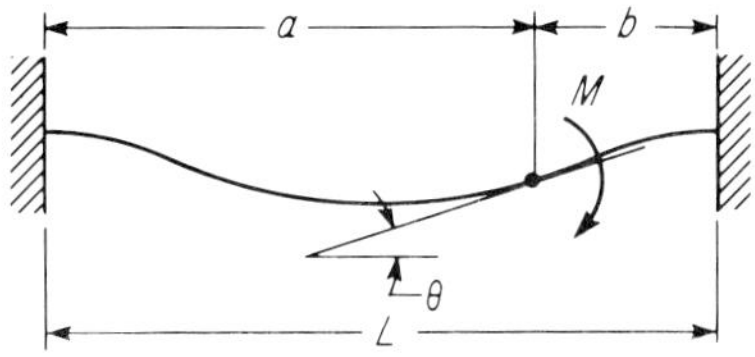	$\frac{M}{\theta} = \frac{EI}{L\left[\left(\frac{a}{L}\right)\left(\frac{b}{L}\right) - 3\left(\frac{a}{L}\right)^2\left(\frac{b}{L}\right)^2\right]}$
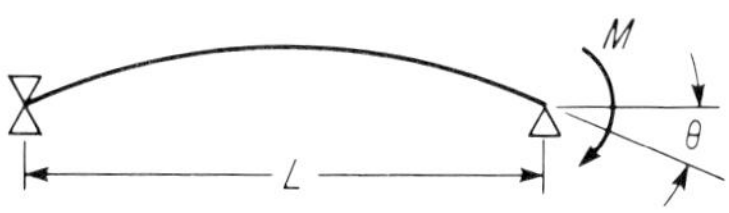	$\frac{M}{\theta} = \frac{3EI}{L}$
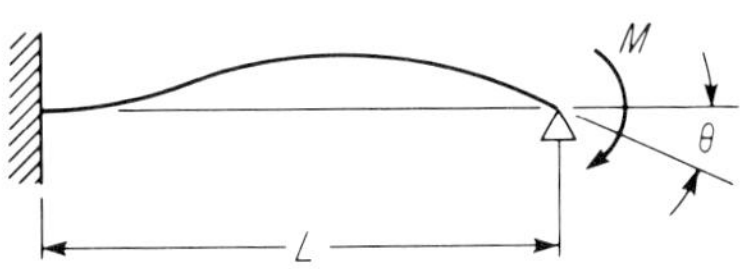	$\frac{M}{\theta} = \frac{4EI}{L}$

Twisting Moment

The stress in this case is a shear stress and the deformation is an angle of twist. The general equations for these quantities resulting from a twisting moment M_t are

$$\text{Maximum shear stress} = \frac{M_t}{Z_t}$$

$$\text{Angle of twist per unit length} = \frac{M_t}{k_t G}$$

where Z_t, k_t = moduli of the cross-sectional area
G = shear modulus for the material

Expressions of Z_t and k_t for some common cross sections are tabulated in the following.

Cross Section	Z_t	k_t
Circular 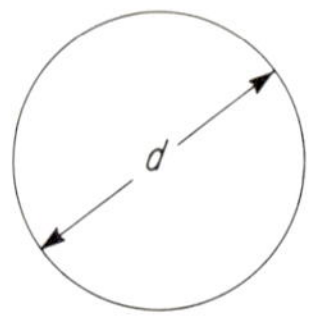	$\frac{\pi}{16} d^3$	$\frac{\pi}{32} d^4$
Rectangular 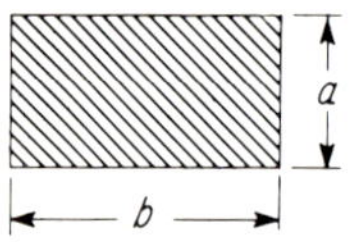 $\frac{b}{a}$ = 1.0 2.0 5.0 ∞	$0.208ab^2$ $0.246ab^2$ $0.291ab^2$ $0.333ab^2$	$0.141ab^3$ $0.229ab^3$ $0.291ab^3$ $0.333ab^3$
Equilateral triangle 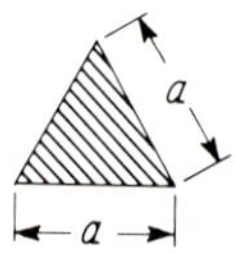		$0.02165a^4$
Tubular 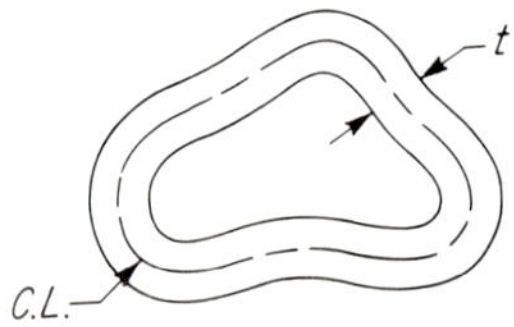 A_c = area bounded by center line C = length of center line t = wall thickness	$2A_ct$	$\frac{(2A_c)^2t}{C}$

Curved Beams

The following table gives a summary of maximum deflections δ and bending moments M for some curved beams which are commonly used as springs and are therefore of particular interest in this text. In all the considered cases it is assumed that the thickness of the bar is small compared to the radius of curvature.

Membrane

A membrane is a two dimensional elastic element capable of supporting load by tensile stresses only. Considering the circular membrane of Fig. 1-6 the stress σ at any radius r due to a concentrated force P at the

Condition	*Governing Equations*
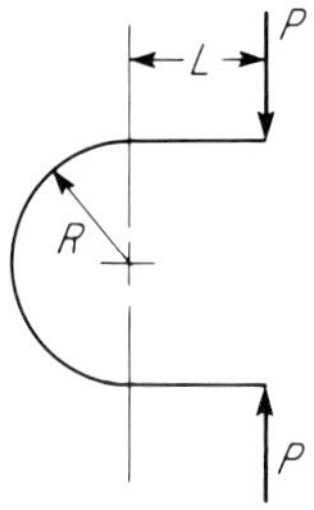 $K = \dfrac{L}{R}$	$\delta = \dfrac{PR^3}{EI}\left[\dfrac{2}{3}K^3 + \pi K^2 + 4K + \dfrac{\pi}{2}\right]$ $M = PR\,[1 + K]$
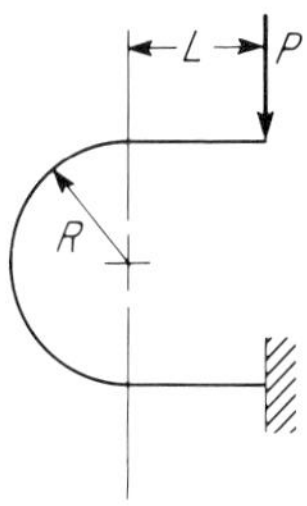 $K = \dfrac{L}{R}$	$\delta = \dfrac{PR^3}{EI}\left[\dfrac{2}{3}K^3 + \pi K^2 + 4K + \dfrac{\pi}{2}\right]$ $M = PR\,[1 + K]$
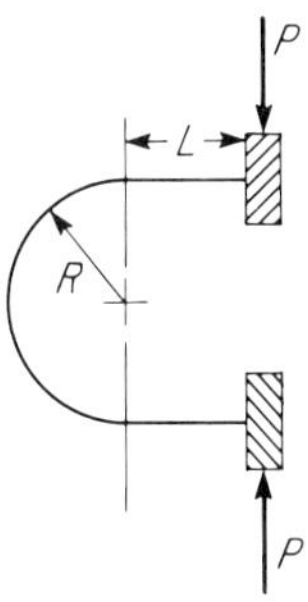 $K = \dfrac{L}{R}$	$\delta = \dfrac{PR^3}{EI}\left[\dfrac{2}{3}K^3 + K^2 - 0.4297 + (2 - K^2) \times \left(\dfrac{K^2 + 2K + 1.1416}{\pi + 2K}\right)\right]$ $M = PR\left[\dfrac{1.1416 + 2K + K^2}{3.1416 + 2K}\right]$
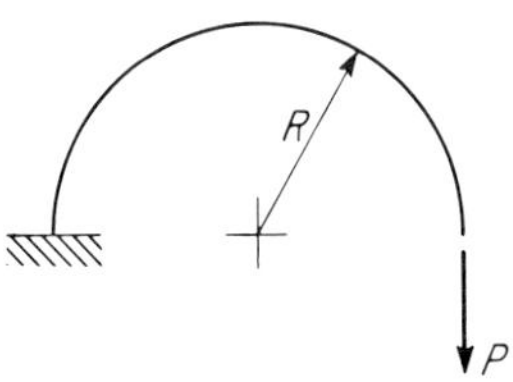	$\delta = \dfrac{2PR^3}{EI}$ $M = 2PR$

Condition	*Governing Equations*
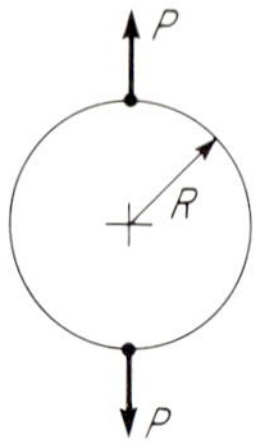	$\delta = 0.149 \dfrac{PR^3}{EI}$ $M = 0.318PR$
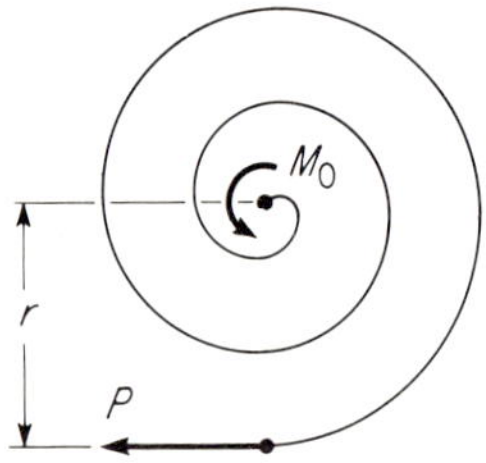 S = total length of the spiral	ϕ = angle of rotation of one end of the spring relative to the other during winding $= \dfrac{PrS}{EI} = \dfrac{M_0 S}{EI}$ $M = 2Pr = 2M_0$
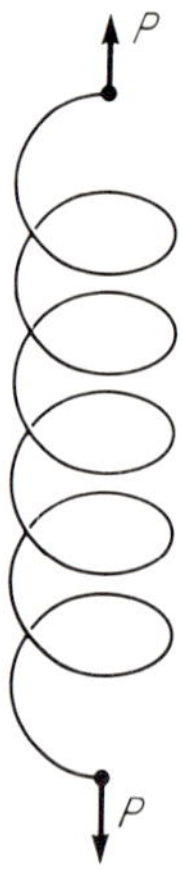 Helical spring N = number of coils P = axial load R = mean radius d = wire diameter C = spring index $= \dfrac{2R}{d}$ K_s = stress correction factor $= 1 + \dfrac{0.615}{C}$	$\tau_{\max}$ = maximum shearing stress $= K_s \dfrac{16PR}{\pi d^3}$ $= K_s \dfrac{8PC}{\pi d^2}$ $= K_s \dfrac{2PC^3}{\pi R^2}$ δ = deflection $= \dfrac{64R^3N}{d^4G} P$ $= \dfrac{8C^3NP}{dG}$ $= \left(\dfrac{4NC^4}{RG}\right) P$

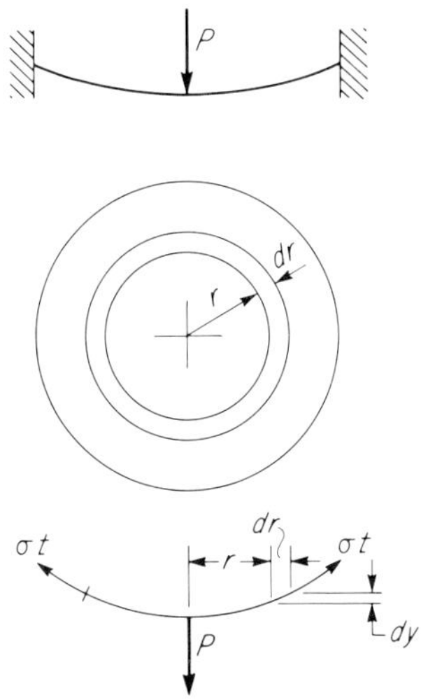

FIG. 1-6

center can be calculated from the equilibrium condition as

$$P = 2\pi rt\sigma \sin\theta$$
$$= 2\pi rt\sigma \frac{dy}{dr}$$
$$\text{Strain} = \frac{\sigma}{E}$$

Change in length

$$(\text{of the small element } dr) = \frac{\sigma}{E} dr$$
$$= \sqrt{(dr)^2 + (dy)^2} - dr$$
$$= dr\left[\sqrt{1 + \left(\frac{dy}{dr}\right)^2} - 1\right]$$

For small values of $\dfrac{dy}{dr}$,

$$\sqrt{1 + \left(\frac{dy}{dr}\right)^2} \cong 1 + \frac{1}{2}\left(\frac{dy}{dr}\right)^2$$

accordingly,

$$\frac{\sigma}{E} dr \cong \frac{1}{2}\left(\frac{dy}{dr}\right)^2 dr$$

from which

$$\frac{dy}{dr} = \sqrt{\frac{2\sigma}{E}}$$

and by substituting in the equilibrium equation we obtain

$$P = 2\pi t \sigma r \sqrt{\frac{\sigma}{E}}$$

which gives the relationship between force and stress in this case.

The Plate

The expressions for the maximum bending stresses and deflections for plates subjected to transverse loads can be generalized as

$$\sigma_{\max} = k\frac{P}{h^2} \quad \text{or} \quad \sigma_{\max} = k\frac{qa^2}{h^2}$$

$$\omega_{\max} = k_1\frac{Pa^2}{Eh^3} \quad \text{or} \quad \omega_{\max} = k_1\frac{qa^4}{Eh^3}$$

depending on whether the applied load P is concentrated along the edge or distributed over the surface as a pressure q. The factors k and k_1 depend on the geometry of the plate, the support conditions, and Poisson's ratio. The adjoining table gives some values of these constants calculated for common cases of plates using a Poisson's ratio equal to 0.3. The dimension a is a characteristic dimension of the plate, and h is the thickness of the plate.

Condition	a/b	k	k_1
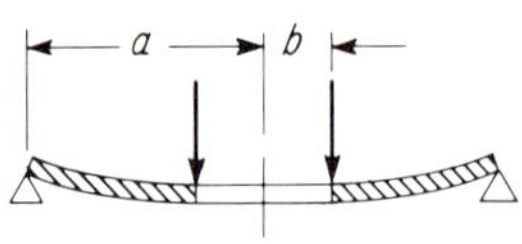	1.25	1.1	0.341
	1.5	1.26	0.519
	2.0	1.48	0.672
	3.0	1.88	0.734
	4.0	2.17	0.724
	5.0	2.34	0.704
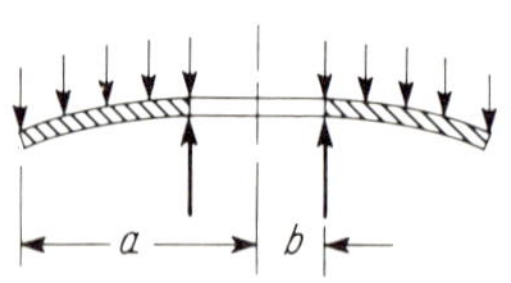	1.25	0.66	0.202
	1.5	1.19	0.491
	2.0	2.04	0.902
	3.0	3.34	1.220
	4.0	4.3	1.3
	5.0	5.1	1.31

Condition	a/b	k	k_1
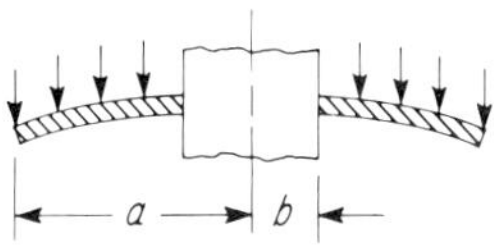	1.25	0.135	0.00231
	1.5	0.410	0.0183
	2.0	1.04	0.0938
	3.0	2.15	0.293
	4.0	2.99	0.448
	5.0	3.69	0.564
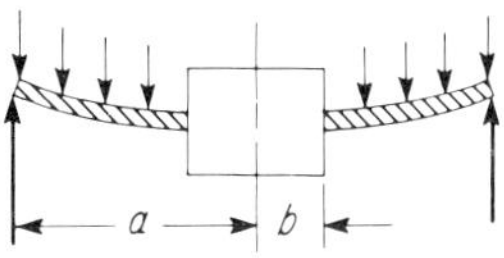	1.25	0.122	0.00343
	1.5	0.336	0.0313
	2.0	0.74	0.125
	3.0	1.21	0.291
	4.0	1.45	0.417
	5.0	1.59	0.492
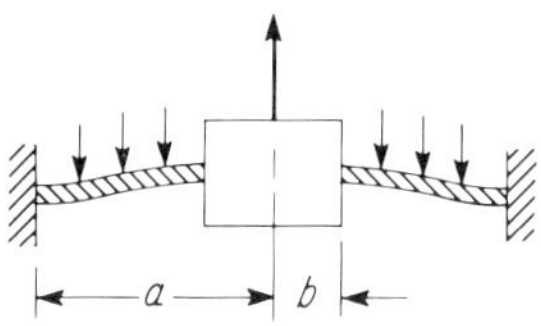	1.25	0.090	0.00077
	1.5	0.273	0.0062
	2.0	0.71	0.0329
	3.0	1.54	0.110
	4.0	2.23	0.179
	5.0	2.80	0.234
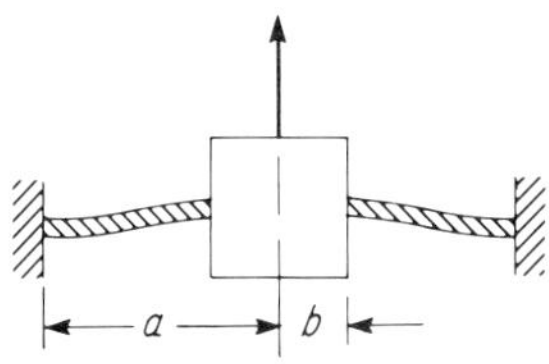	1.25	0.115	0.00129
	1.5	0.220	0.0064
	2.0	0.405	0.0237
	3.0	0.703	0.062
	4.0	0.933	0.092
	5.0	1.13	0.114
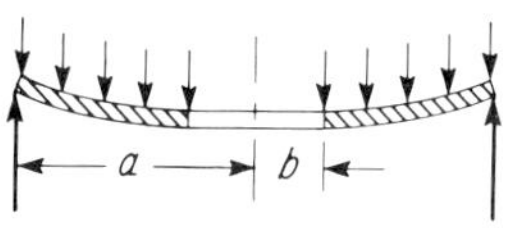	1.25	0.592	0.184
	1.5	0.976	0.414
	2.0	1.44	0.664
	3.0	1.880	0.824
	4.0	2.08	0.830
	5.0	2.19	0.813
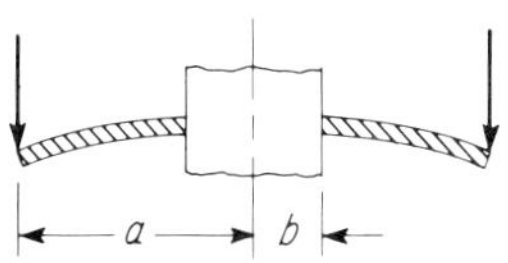	1.25	0.227	0.00510
	1.5	0.428	0.0249
	2.0	0.753	0.0877
	3.0	1.205	0.209
	4.0	1.514	0.293
	5.0	1.745	0.350

Condition	a/b	k	k_1
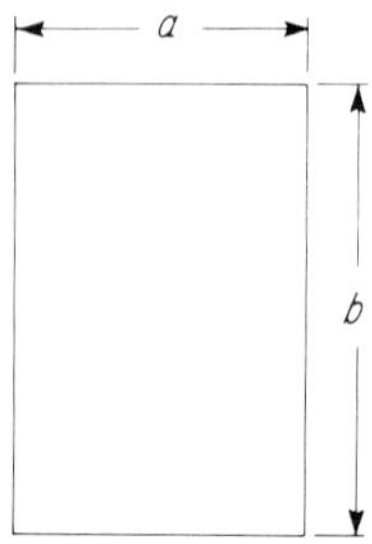 Plate supported at the edges with uniformly distributed load q	0 0.2 0.5 0.625 0.77 1.0	0.75 0.7476 0.6102 0.5172 0.4158 0.3118	0.1422 0.1416 0.1106 0.0906 0.0697 0.0443
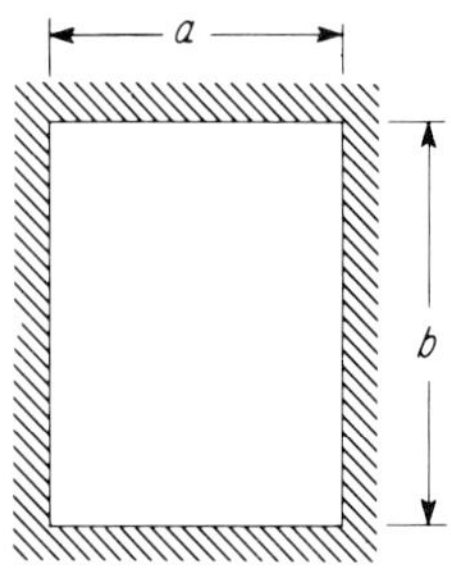 Plate with built-in edges and uniformly distributed load q	0 0.5 0.666 0.8 1.0	0.5 0.4974 0.4542 0.3990 0.3078	0.0284 0.0277 0.0240 0.0199 0.0138
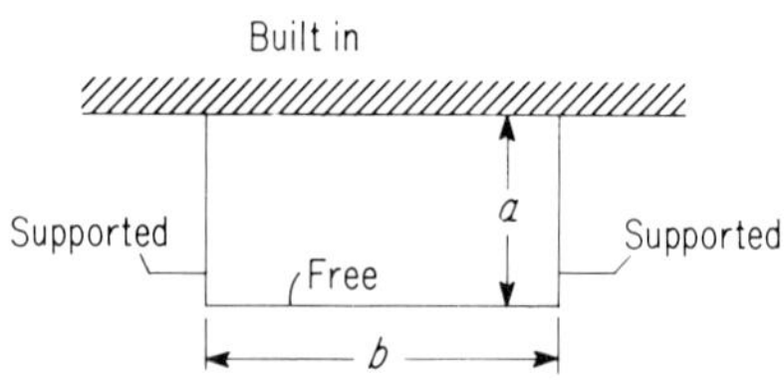 Plate with two opposite sides supported, third side built in, and fourth side free with uniformly distributed load q	1 1.5 2 3 ∞	0.714 1.362 1.914 2.568 3.0	0.123 0.366 0.635 1.03 1.37

Condition	a/b	k	k_1
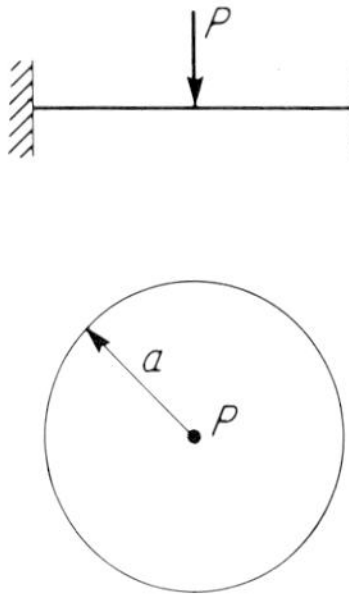 All clamped edges			0.262
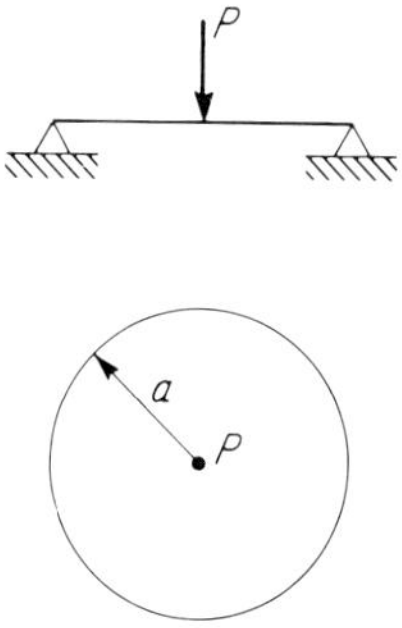 All simply supported edges			0.865
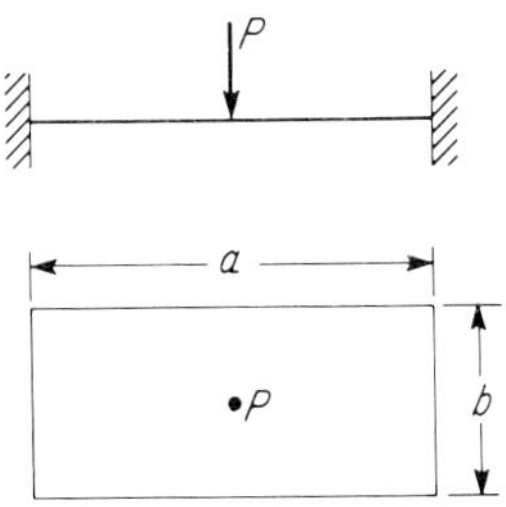 All clamped edges	4 2 1		0.0045 0.0186 0.057

Condition	a/b	k	k_1
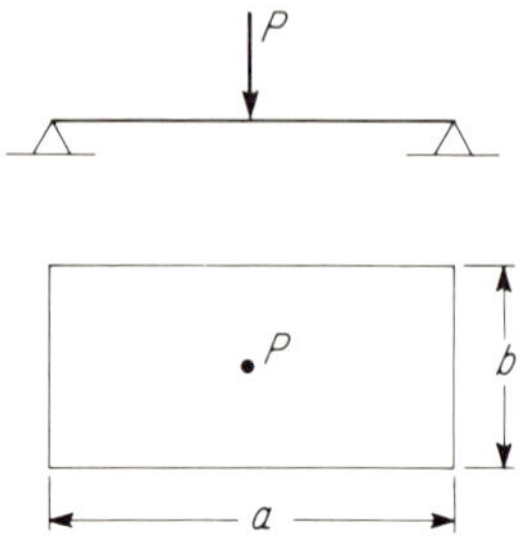 All simply supported edges	4 2 1		0.0115 0.0446 0.126
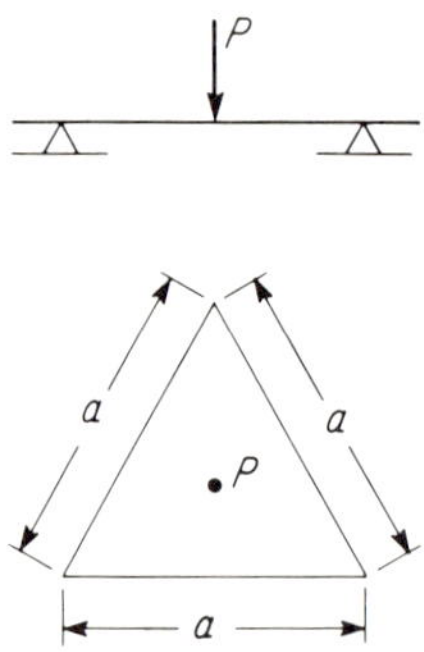 All simply supported edges			0.0624

Elastic Bodies in Contact

If two spherical bodies are pressed against each other by a force P, the surface of contact is approximately circular with radius a. The maximum pressure $p_{\max}$ occurs at the center of the area of contact and the centers of the two spheres approach one another by a distance Z. These quantities can be calculated by the following equations.

$$p_{\max} = 0.616 \sqrt[3]{\frac{P}{R_e^2} E_e^2}$$

$$a = 0.88 \sqrt[3]{\frac{PR_e}{E_e}}$$

and

$$Z = \frac{a^2}{R_e}$$

where

$$\frac{1}{R_e} = \frac{1}{R_1} + \frac{1}{R_2}$$

R_1, R_2 = radii of the two spheres (positive when convex and negative when concave)

$$\frac{1}{E_e} = \frac{1}{E_1} + \frac{1}{E_2}$$

E_1, E_2 = moduli of elasticity for the two materials.

If two cylindrical bodies are contacting along a generatix and pressed together by a force P' per unit length, the width of the contact area b, the maximum pressure p_{max} and the approach Z of the axes are given by

$$b = 2.15\sqrt{\frac{P'}{E_e}R_e}$$

$$p_{max} = 0.59\sqrt{\frac{P'}{R_e}E_e}$$

$$Z = \frac{b^2}{4}\left(\frac{1}{R_e}\right)$$

where

$$\frac{1}{R_e} = \frac{1}{R_1} + \frac{1}{R_2}$$

R_1, R_2 = radii of cylinders (positive when convex and negative when concave)

$$\frac{1}{E_e} = \frac{1}{E_1} + \frac{1}{E_2}$$

E_1, E_2 = moduli of elasticity for the two materials

Beams on Elastic Foundation

The general equation describing the elastic curve of the beam is

$$EI\frac{d^4y}{dx^4} = -ky$$

where k = foundation reaction per unit length (called the *modulus of the foundation*)

E = modulus of elasticity of beam material

I = moment of inertia of the beam

Using the notation

$$\beta = \sqrt[4]{\frac{k}{4EI}}$$

the general solution for beam deflection can be represented by

$$y = e^{\beta x}(A\cos\beta x + \beta\sin\beta x) + e^{-\beta x}(C\cos\beta x + D\sin\beta x)$$

where A, B, C, and D are integration constants which must be determined from known conditions.

For relatively short beams with length smaller than $0.6/\beta$ the beam can be considered as rigid, since the deflection due to bending is negligible compared to the deflection of the foundation. In this case the deflection will be constant and is given by

$$\delta = \frac{P}{kL}$$

and the maximum bending moment $= PL/4$.

For relatively long beams with length greater than $5/\beta$ the deflection will have a waveform with gradually diminishing amplitudes. The general solution can be found in texts on advanced strength of materials. (See, for example, Ref. 5, S. P. Timoshenko, *Advanced Strength of Materials*, McGraw-Hill, 1950.)

The following table gives expressions for deflection, slope and bending moment for long beams loaded at the center and at the end.

Condition	*Governing Equation*
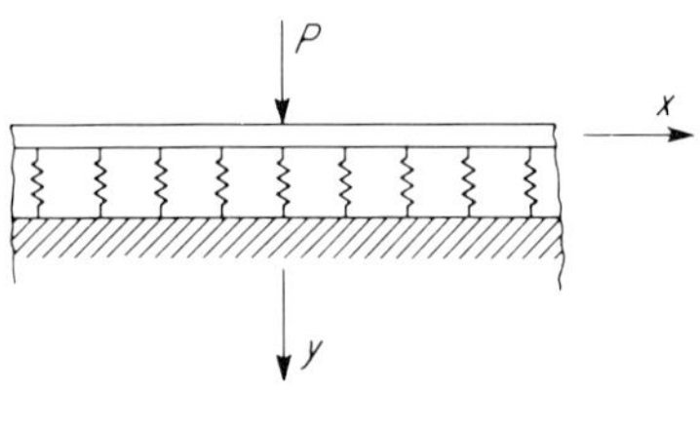	$y_{max} = \frac{P\beta}{2k}$ at $x = 0$ $\theta_{max} = -\frac{P\beta^3}{k}(0.3223)$ at $x = \frac{0.8}{\beta}$ $M_{max} = \frac{P}{4\beta}$ at $x = 0$
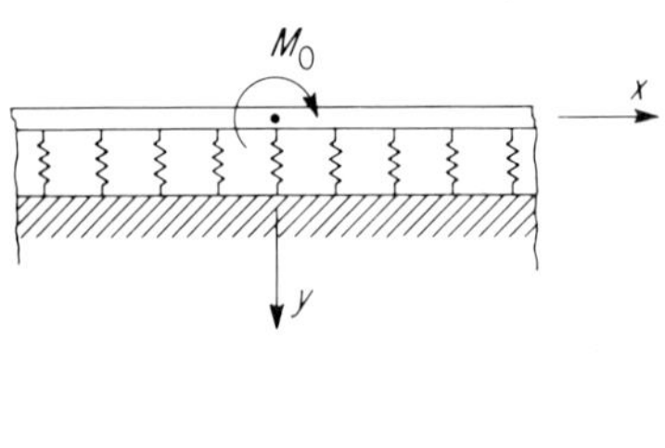	$y_{max} = \frac{M_0\beta^2}{k}(0.3223)$ at $x = \frac{0.8}{\beta}$ $\theta_{max} = \frac{M_0\beta^3}{k}$ at $x = 0$ $M_{max} = \frac{M_0}{2}$ at $x = 0$

Description	*Governing Equations*
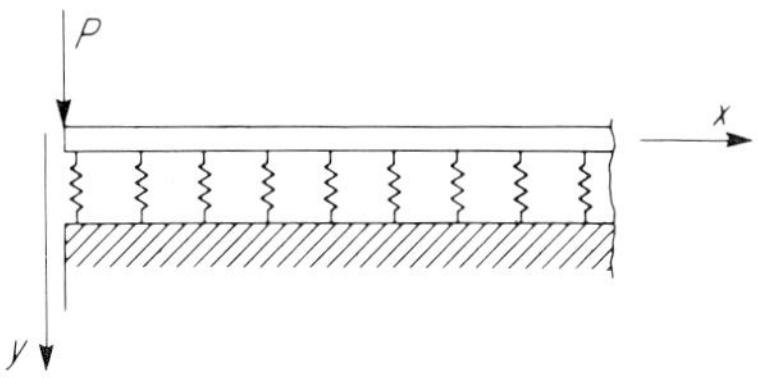	$y_{\max} = \dfrac{P}{2EI\beta^3}$ at $x = 0$ $\theta_{\max} = \dfrac{P}{2EI\beta^2}$ at $x = 0$
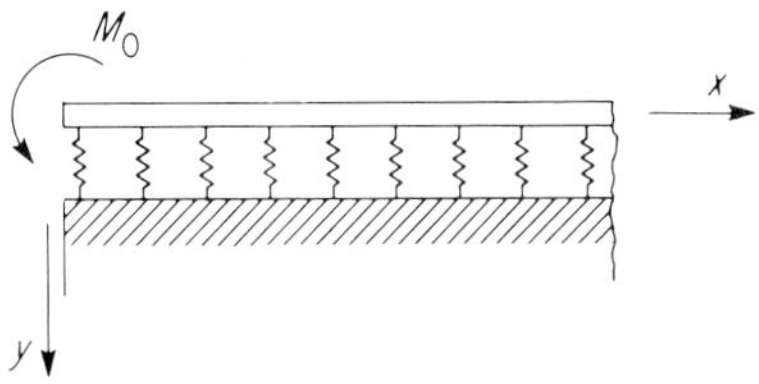	$y_{\max} = \dfrac{M_0}{2EI\beta^2}$ at $x = 0$ $\theta_{\max} = \dfrac{M_0}{EI\beta}$ at $x = 0$

1-8. SUMMARY

This chapter provides a brief review of some basic relationships for the analysis of particles, rigid bodies, and deformable elements. A good comprehension of these fundamental aspects of mechanics is essential to the modeling of mechanical systems

PROBLEMS

1-1. A space probe weighing 100 lb is moving at a constant velocity of 1,000 mph in a line with directional cosines $l = 0.2$, $m = 0.3$ in the inertial reference XYZ. A thrust $F = 100$ lb with directional cosines $l = 0.1$, $m = 0.4$ is applied

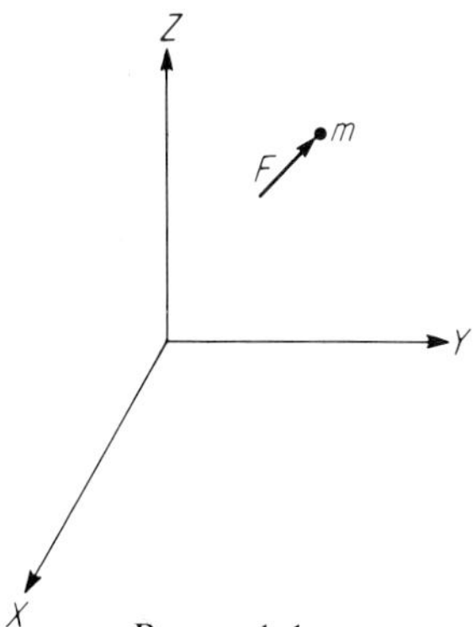

PROB. 1-1

for 10 sec. Calculate the final velocity of the rocket. Neglect all gravitational forces.

1-2. A wagon weighing 3220 lb starts from rest at point A and travels due to its own weight until it is stopped by a spring as shown in the sketch. Neglecting any friction or air resistance, calculate (a) the reaction between the wagon and the track at point *B*, (b) the radius *R* necessary for zero reaction between the wagon and the track at point *C* just before it reaches the spring, (c) the spring constant *K* required to stop the wagon after a 2 ft compression in the spring.

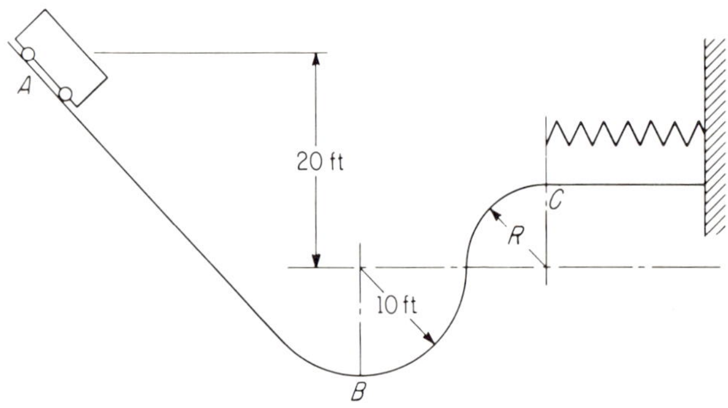

PROB. 1-2

1-3. A small rocket-propelled test vehicle weighs 200 lb and travels along a track with a coefficient of friction = 0.2 until it impacts a spring at *C* and produces a maximum compression of 5 ft. If the rocket exerts a constant thrust of 250 lb from *A* to *B* where it is shut off, calculate the stiffness *k* of the spring. Assume that the vehicle started from rest at point *A*.

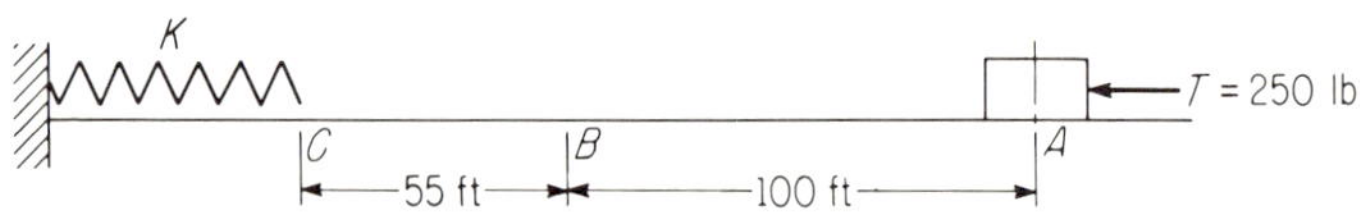

PROB. 1-3

1-4. An experimental rocket weighs 32,200 lb and has a thrust of 100,000 lb. It travels on a circular path from point *A* where it fires to point *B* where the fuel ends and the rocket goes on a trajectory as shown. Assuming that the thrust remains constant and neglecting the air resistance, calculate (a) the velocity of the rocket at point *B*, (b) the maximum range *R* and maximum height *h* of the trajectory.

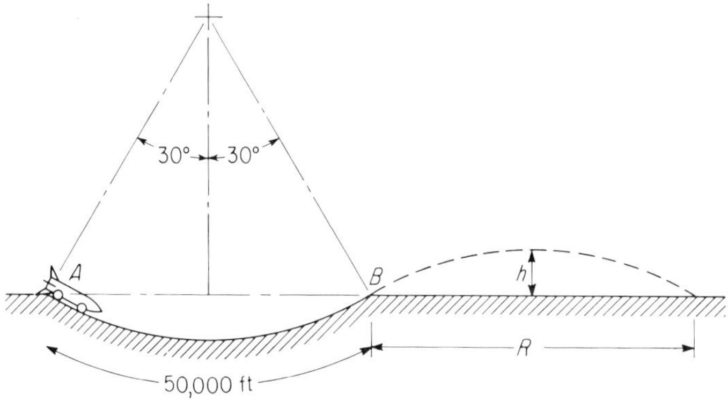

PROB. 1-4

1-5. An air gun utilizes a source of constant pressure p_0 lb/in.^{2} If a rod 1 in. in diameter and weighing W lb is loaded as shown, calculate the velocity of the rod as it leaves the muzzle. If the shown spring is to stop the rod after a 1-ft compression, calculate the stiffness of the spring. Neglect friction.

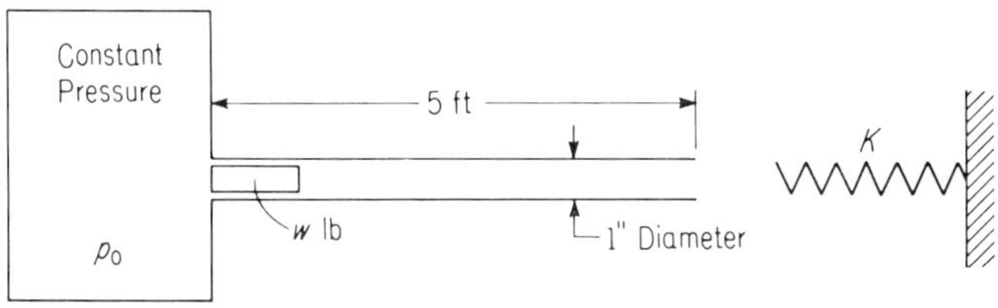

PROB. 1-5

1-6. A wooden block weighing 2 lb is suspended by a string 1 ft long and capable of carrying a maximum load of 50 lb. The block is fired at with a $\frac{1}{4}$-lb bullet. Calculate the bullet velocity required to break the string. Assume plastic impact ($e = 0$).

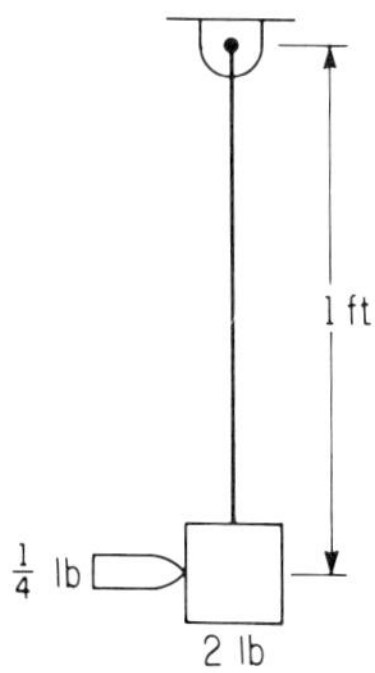

PROB. 1-6

1-7. A pendulum ball is released from rest at a 60° angle from the vertical to strike the ball of another pendulum as shown. Neglecting the mass of the cable and air resistance, find the maximum angles from the vertical for both pendulums after impact. The coefficient of restitution is 0.8.

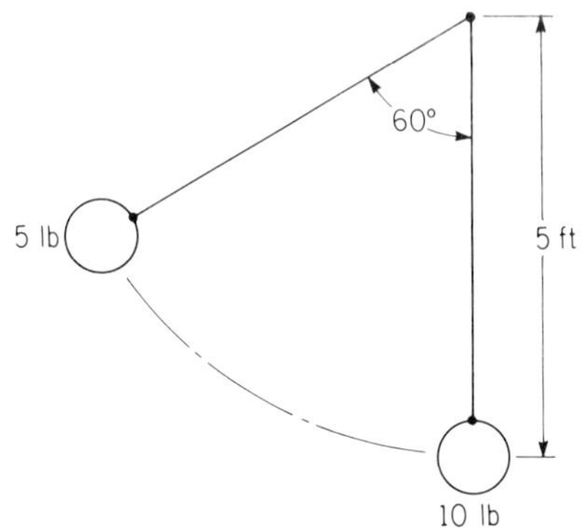

PROB. 1-7

1-8. Neglecting the weight of the cable, find the speed of the 100-lb weight after it has moved 5 ft along the incline from a position of rest. The coefficient of friction along the incline is 0.3.

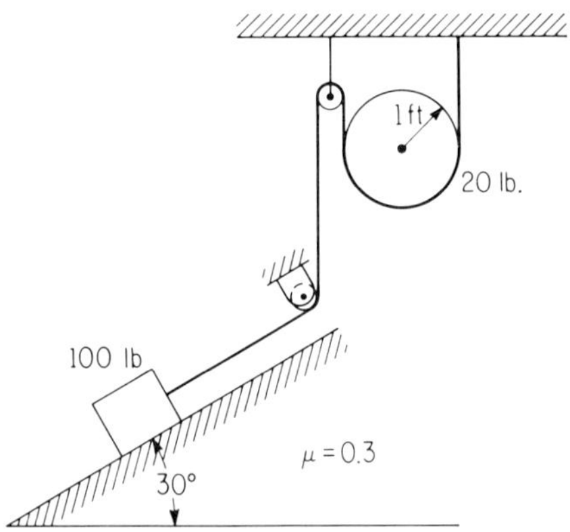

PROB. 1-8

1-9. A 15-lb weight is suspended from a lightweight cord wrapped around a uniform cylinder of radius 2 ft and weight 100 lb. Find the angular velocity of the cylinder after it has started from rest and rolled without slipping a distance of 10 ft.

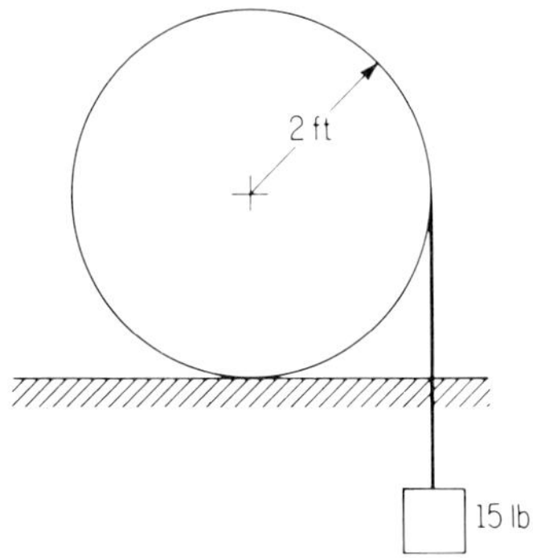

PROB. 1-9

1-10. A sphere of radius R and mass m is pushed along a horizontal plane by a horizontal force P at a distance h above the plane. The coefficient of kinetic friction between the sphere and the surface is μ. Find (a) the acceleration of the center of mass and the angular acceleration of the sphere if sliding occurs, (b) the frictional force F for a pure rolling condition.

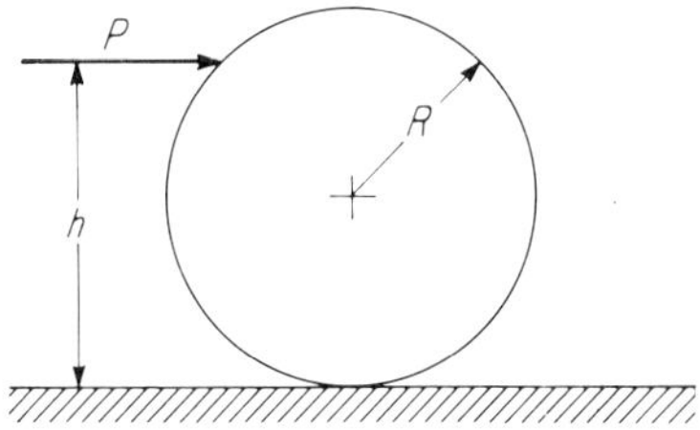

PROB. 1-10

1-11. The given wheel is free to rotate about its center of gravity O. Find the reaction at the pin at O and the angular acceleration α at the time when a force $P = 100$ lb is applied as shown. Given:

W = weight of the wheel = 32.2 lb
K_0 = radius of gyration of the wheel about O
= $3\sqrt{2}$ in.
r = radius of the wheel = 6 in.

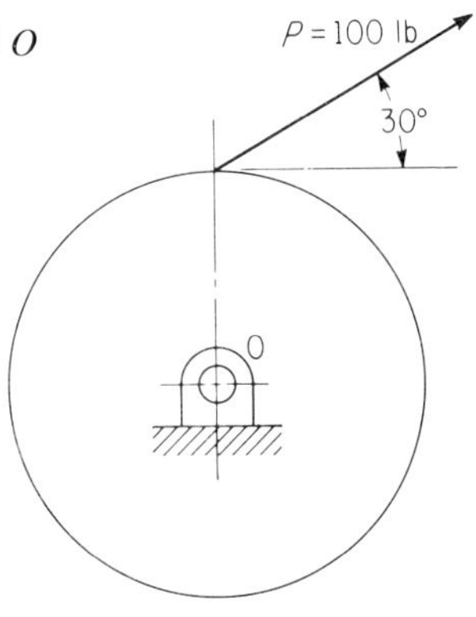

PROB. 1-11

1-12. A flywheel of radius R having a moment of inertia I about the axis of rotation has an angular velocity ω rad/sec. At time $t = 0$ a brake is applied with a normal braking force P as shown. The brake coefficient of friction is μ. Find (a) the time required to stop the flywheel, (b) the number of revolutions occurring during this time.

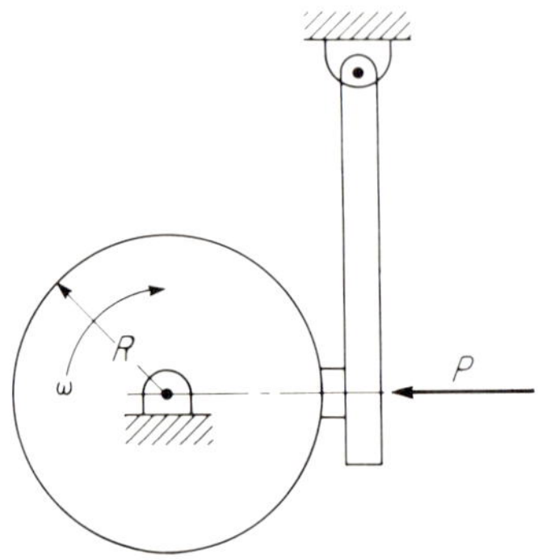

PROB. 1-12

1-13. The shown mass is moved so that it extends the spring 4 in. from its unextended position. If the coefficient of friction between the mass and the supporting surface is 0.3, find (a) the velocity of the mass as it returns to the undeformed position of the spring, (b) the amount of spring compression when the mass reaches zero velocity.

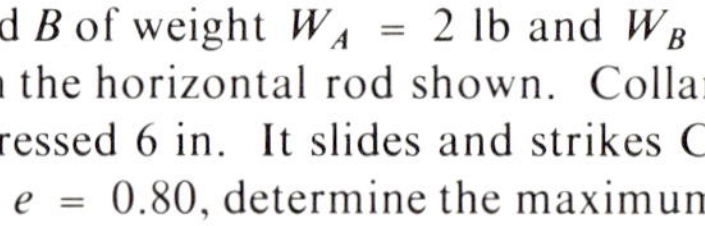

PROB. 1-13

1-14. Collars A and B of weight $W_A = 2$ lb and $W_B = 3$ lb respectively may slide without friction on the horizontal rod shown. Collar A is released from rest when spring C is compressed 6 in. It slides and strikes Collar B. Assuming the coefficient of restitution $e = 0.80$, determine the maximum deflection of spring D.

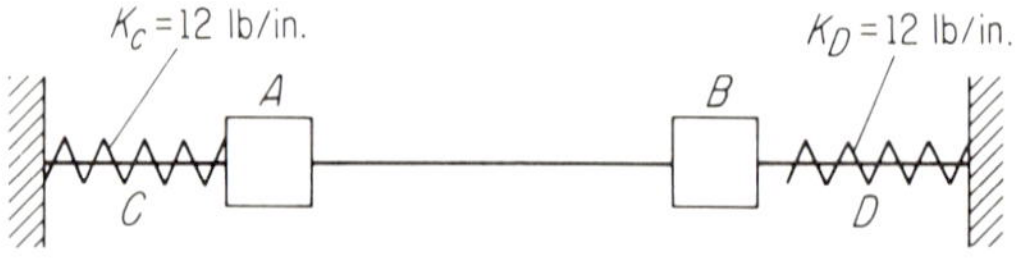

PROB. 1-14

1-15. A $^3/_4$-oz bullet is fired with a velocity of 1800 ft/sec into a 10-lb block. Knowing that the coefficient of friction between the block and the floor is 0.3, determine (a) through what distance the block will slide, (b) the percentage of the initial kinetic energy lost in friction between the block and the floor.

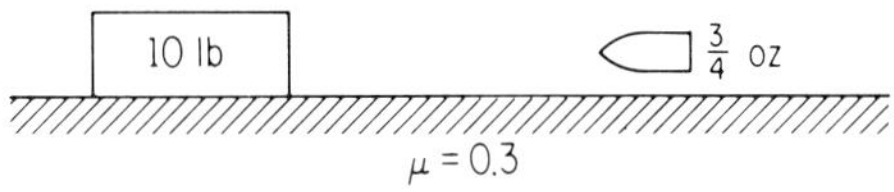

PROB. 1-15

1-16. Calculate the angular acceleration of the rod shown in the sketch. The weight of the rod is W_3 and the length is L. Neglect the weight of pulleys A and B.

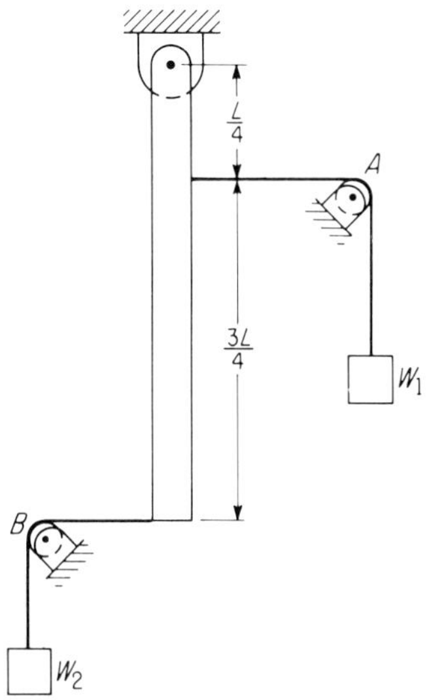

PROB. 1-16

1-17. A 966-lb flywheel rotates in a horizontal plane. Its mass center is 0.25 ft from the vertical axis of rotation. Find the bearing reactions for the position shown when the angular velocity is 180 rpm (clockwise when looking down) and is increasing in magnitude at the rate of 100 rad/sec^2.

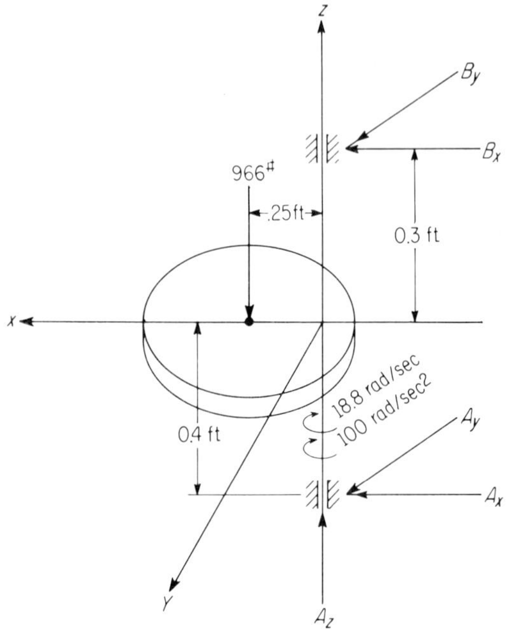

PROB. 1-17

1-18. A cylinder with radius 1 ft, height 3 ft, and weight 32.2 lb can rotate about an axis kk with directional cosines $l_k = 0.5$, $m_k = 0.5$, and $n_k = 0.707$. If a torque T is applied to the axis kk, calculate the angular velocity of the cylinder after 20 sec. (*Hint*: X_c, Y_c, Z_c are principal axes at center of mass)

$$I_{kk} = l_k^2 I_{x_c x_c} + m_k^2 I_{y_c y_c} + n_k^2 I_{z_c z_c}$$

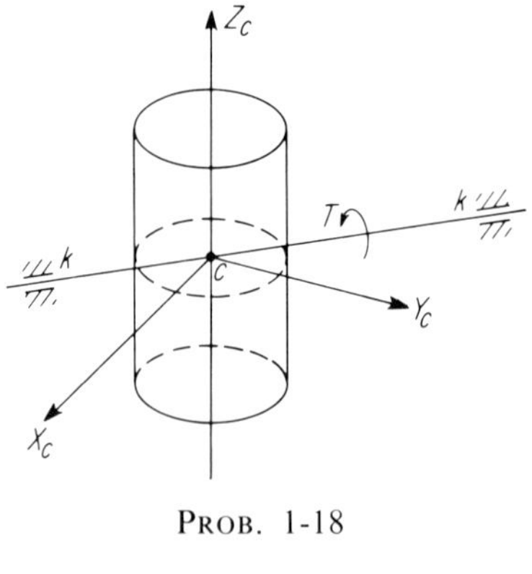

PROB. 1-18

chapter 2

Kinematics of Oscillations

2-1. INTRODUCTION

In many dynamic situations the motion of a particle, rigid body, or of a continuum is such that the moving elements do not depart indefinitely from their original position and repeatedly pass through it during the course of the motion. This type of motion is called an *oscillation*. In cases where the time interval taken by the vibrating object between any two successive repetitions of a particular position is the same during the motion, the vibration is called *periodic*. When this time interval is variable and unpredictable, the vibration is called *random*. The simplest type of periodic oscillations is the *simple harmonic motion*. In this type of motion the displacement of the object from its original configuration follows a sinusoidal function of time.

The vibratory motion is a purely geometrical or kinematic concept and is discussed as such in this chapter.

2-2. THE SIMPLE HARMONIC MOTION

Geometrical Representation

A simple harmonic motion is conveniently represented by the projection—in any particular direction—of a point moving on a circle with a constant angular velocity. Taking for example a point N as shown in Fig. 2-1, which moves on a circle of radius A with constant angular velocity ω, the motion of its projection P on a fixed axis X will be simple harmonic. This simple harmonic motion will have a range equal to the diameter of the circle. This range is equally divided by the projection O of the center C of the circle. The point P will therefore move within $\pm A$ from O which is called the *center of oscillation*. The distance A is known as the *amplitude* of the motion. It can be easily seen from Fig. 2-1 that P will repeat its motion whenever N completes a full revolution. The motion of P between two successive repetitions of a particular position is called a *cycle*. The time taken to complete one cycle is called the *period* of the oscillation τ and is calculated from

$$\omega\tau = 2\pi$$

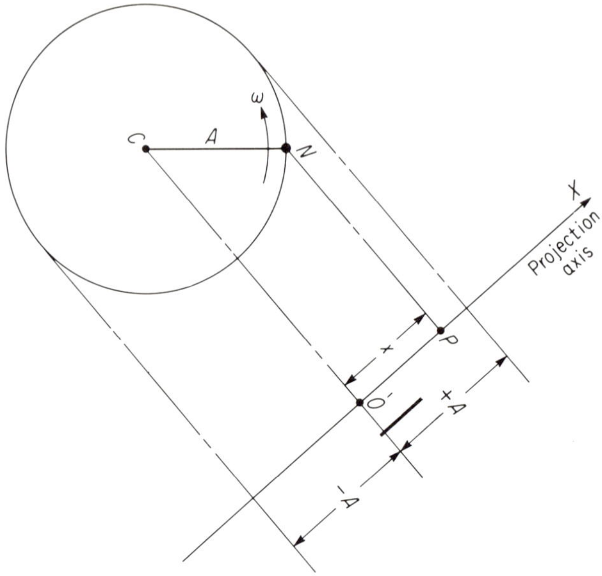

FIG. 2-1

or

$$\tau = \frac{2\pi}{\omega} \tag{2-1}$$

where ω is the circular angular velocity of the motion.

The number of cycles performed in a unit of time can therefore be calculated from

$$f = \frac{1}{\tau} \text{ cycles per unit of time} \tag{2-2}$$

and is known as the *frequency* of the oscillation. Substituting Eq. 2-1 in Eq. 2-2 we get

$$f = \frac{\omega}{2\pi} \tag{2-3}$$

which shows that ω can also be used as a measure of frequency. Since one cycle in a simple harmonic motion corresponds to 2π radians, ω will indicate the frequency in radians per unit of time.

The position of point P at any time t is defined by the circular angle (ωt) measured from the initial position of point N on the circle.

If the time measured from the instant at which P leaves the center of motion in the positive direction, the vibration will be as shown in Fig. 2-2a. The analytical expression for the motion will be

$$x = A \sin \omega t$$

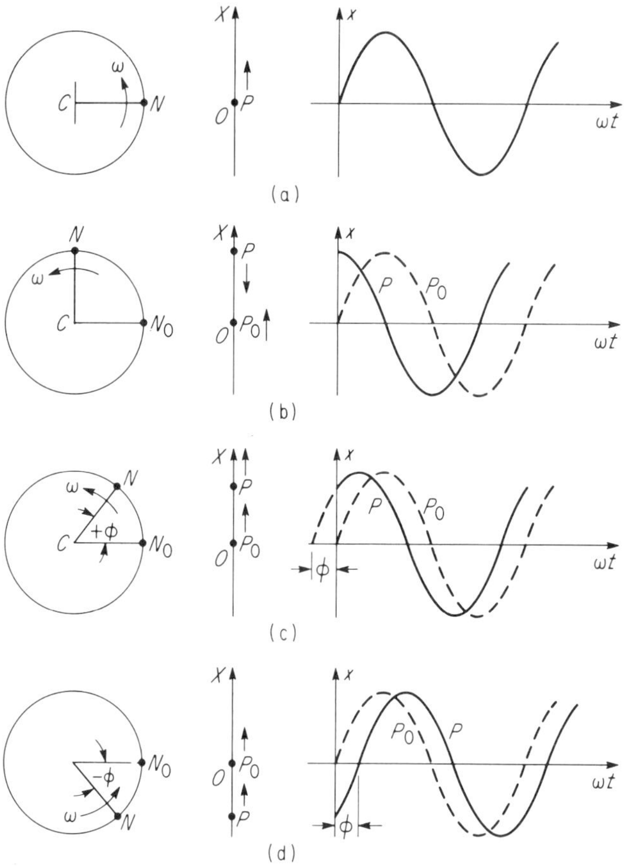

FIG. 2-2

Similarly, if the time is measured from the instant at which P leaves the peak of its displacement, Fig. 2-2b, the equation of motion will be

$$x = A \cos \omega t$$

If the time is measured from the instant where P is in a position corresponding to an angular displacement ϕ from the center of oscillation, the graphical representation of the motion will be as shown in Fig. 2-2c and its equation will take the form

$$x = A \sin (\omega t + \phi)$$

In this case P will be leading the point P_0 (which starts at O at the same time) by a circular angle ϕ.

On the other hand, if the time measurement starts when P is as shown in Fig. 2-2d, its motion will lag behind that of P_0 (which starts at the center of oscillation) by the angle ϕ. The motion in this case is represented by

$$x = A \sin(\omega t - \phi)$$

The angle ϕ is usually referred to as the *phase angle* and defines the initial position of the vibrating point.

Vector Representation

The previous discussion shows that a simple harmonic motion can be completely described, as shown in Fig. 2-3 by the component of a rotating

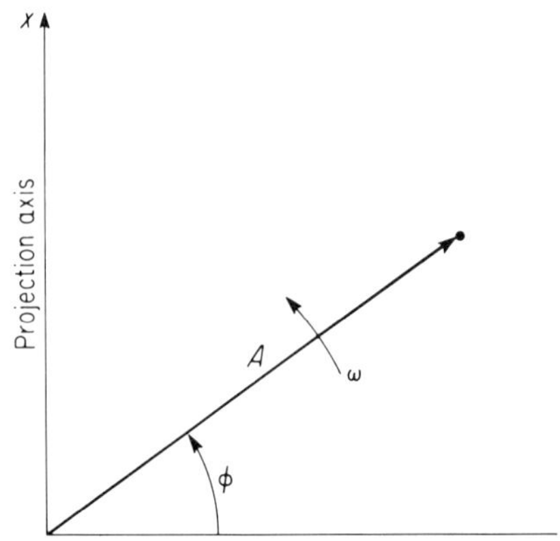

FIG. 2-3

vector in a fixed direction. This vector has a constant magnitude representing the amplitude of the vibration and rotates with a constant angular velocity which corresponds to the frequency of the motion. The initial position of the vector from the axis perpendicular to the projection axis represents the phase angle ϕ of the motion.

This vector representation is very useful in analysis of vibratory motions with different phase relations.

Mathematical Representation

Analytically, a motion is classified as simple harmonic when the acceleration at any time is proportional in magnitude and opposite in sign to the displacement. Considering an object in a rectilinear motion with coordinate x, it will undergo a simple harmonic motion if

$$\frac{d^2x}{dt^2} = -\omega^2 x \tag{2-4}$$

where ω^2 is the constant of proportionality. It may be easily verified that the solution for the differential equation describing the simple harmonic motion is

$$x = a \sin \omega t + b \cos \omega t \tag{2-5}$$

from which

$$x = A \sin (\omega t + \phi) \tag{2-6}$$

where (a, b), (A, ϕ) are interrelated integration constants representing the characteristics of the oscillation. These constants are evaluated from the initial or boundary conditions of the motion. By examination of Eq. 2-6 it can be found that ω is the frequency of the motion while A and ϕ are its amplitude and phase angle respectively. Differentiating Eq. 2-6 with respect to time we obtain the expressions for the velocity and acceleration as

$$\begin{aligned} v &= \frac{dx}{dt} \\ &= \omega A \cos (\omega t + \phi) \\ &= \omega A \sin \left(\omega t + \phi + \frac{\pi}{2}\right) \end{aligned} \tag{2-7}$$

and

$$\begin{aligned} a &= \frac{d^2x}{dt^2} \\ &= -\omega^2 A \sin(\omega t + \phi) \\ &= \omega^2 A \sin (\omega t + \phi + \pi) \end{aligned} \tag{2-8}$$

These expressions show that both the velocity and acceleration can also be represented by vectors rotating with the angular frequency ω. These vectors (Fig. 2-4) have amplitudes (ωA), $(\omega^2 A)$, and phase angles $(\pi/2 + \phi)$, $(\pi + \phi)$ respectively. It is easy to note, from Eqs. 2-7, 2-8, and Fig. 2-4, that for a simple harmonic motion:

1. The vectors representing the velocity and acceleration lead the displacement vector by $\pi/2$ and π radians respectively.
2. The velocity is zero at the two end positions and reaches its maximum value ωA at the center of oscillation.
3. The velocity changes sign when it reaches any of the two end points.
4. The acceleration is proportional to the displacement at all times. Its value is zero at the center of the motion and reaches its maximum value $\omega^2 A$ at the two end points.
5. The direction of the acceleration is always toward the center of the oscillation irrespective of the position during the motion.

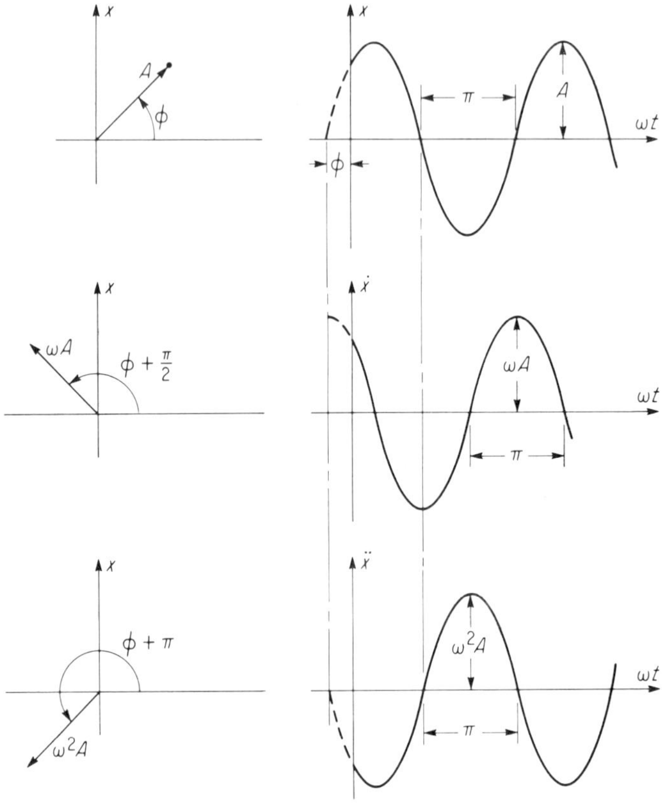

FIG. 2-4

2-3. SUPERPOSITION OF SIMPLE HARMONIC MOTIONS—THE PERIODIC MOTION

A point in three-dimensional space can simultaneously perform several simple harmonic motions. The resultant motion will be periodic. On the other hand, any periodic motion can be analyzed into simple harmonic components.

The present section deals with the superposition of simple harmonic motions. The harmonic analysis of periodic vibrations is discussed in the following section.

Superposition of Two Harmonic Motions in the Same Direction (Rectilinear Periodic Motion)

If we assume that a point in a state of rectilinear motion is simultaneously undergoing two harmonic oscillations:

$$x_1 = A_1 \sin (\omega_1 t + \phi_1)$$

and

$$x_2 = A_2 \sin(\omega_2 t + \phi_2)$$

the coordinate x of the point at any time is obtained from the superposition of the two oscillations:

$$\begin{aligned} x &= x_1 + x_2 \\ &= A_1 \sin(\omega_1 t + \phi_1) + A_2 \sin(\omega_2 t + \phi_2) \end{aligned}$$

We shall discuss in the following examples some simple forms of this equation as well as the general case.

Case 1

The simplest case of superposition is when the two harmonics have the same frequency and phase angles. In this case,

$$\omega_1 = \omega_2 = \omega \qquad \text{and} \qquad \phi_1 = \phi_2 = \phi$$

The resultant motion is

$$x = (A_1 + A_2) \sin(\omega t + \phi)$$

which represents a simple harmonic motion having the same frequency and phase, with an amplitude which equals the algebraic sum of the amplitudes of the harmonic components. This is represented vectorially in Fig. 2-5.

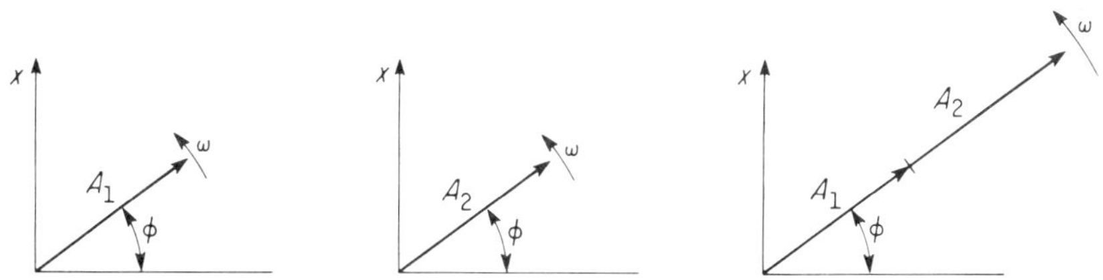

FIG. 2-5

Case 2

In this case when the two harmonic components have the same frequency but otherwise have different magnitudes and phase angles:

$$\omega_1 = \omega_2 = \omega$$

we can see from Fig. 2-6 that the resultant motion will be a simple harmonic represented by

$$x = R \sin(\omega t + \phi) \tag{2-9a}$$

where

$$R = \sqrt{A_1^2 + A_2^2 + 2A_1 A_2 \cos(\phi_2 - \phi_1)} \tag{2-9b}$$

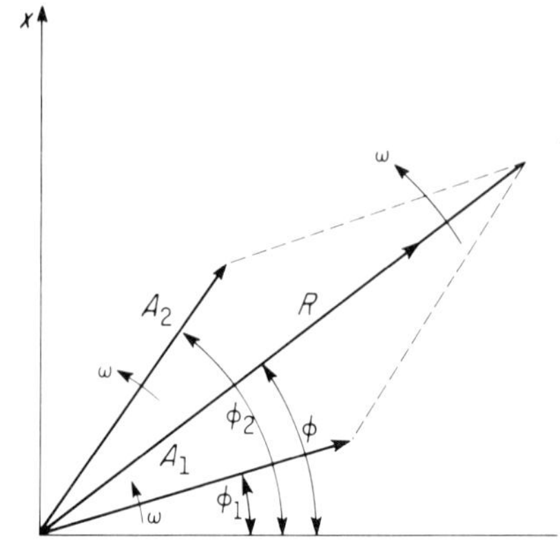

FIG. 2-6

and

$$\phi = \arctan \frac{A_1 \sin \phi_1 + A_2 \sin \phi_2}{A_1 \cos \phi_1 + A_2 \cos \phi_2} \tag{2-9c}$$

Case 3

We consider here the general case where the amplitudes, the frequencies and the phase angles are different. Referring to Fig. 2-7, the

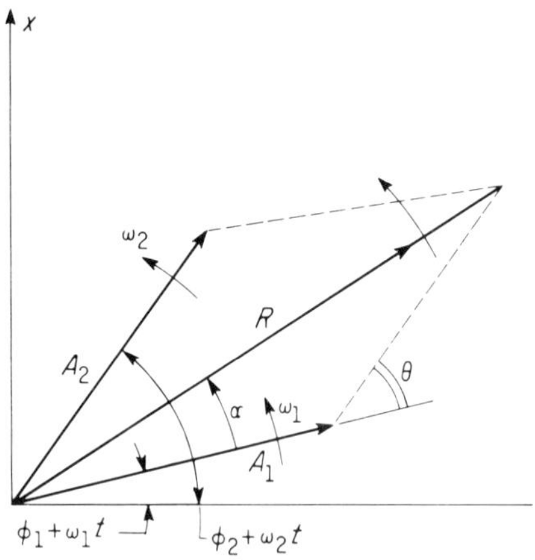

FIG. 2-7

trigonometric analysis gives the following expression for the resultant motion:

$$x = R \sin (\omega_1 t + \phi_1 + \alpha) \tag{2-10a}$$

where

$$R = \sqrt{A_1^2 + A_2^2 + 2A_1A_2\cos\theta} \tag{2-10b}$$

$$\alpha = \arctan\frac{A_2\sin\theta}{A_1 + A_2\cos\theta} \tag{2-10c}$$

$$\theta = (\omega_2 - \omega_1)t + (\phi_2 - \phi_1) \tag{2-10d}$$

It should be noted here that in cases where $\omega_1 \neq \omega_2$ the combined motion, although periodic, is no longer simple harmonic.

For the special case when $\phi_1 = \phi_2 = 0$ and $A_1 = A_2$, Eq. 2-10 will give

$$\theta = (\omega_2 - \omega_1)t$$

$$\alpha = \arctan\frac{\sin\theta}{1 + \cos\theta} = \frac{\theta}{2}$$

$$R = A\sqrt{2(1 + \cos\theta)} = 2A\cos\frac{\theta}{2}$$

from which

$$x = 2A\cos\left(\frac{\omega_2 - \omega_1}{2}\right)t\sin\left(\frac{\omega_1 + \omega_2}{2}\right)t$$

When ω_2 is approximately equal to ω_1, or in other words when $(\omega_2 - \omega_1)$ is much smaller than $(\omega_2 + \omega_1)$, the resultant motion can be looked upon as a vibration with a frequency $\left(\frac{\omega_1 + \omega_2}{2}\right)$ whose amplitude fluctuates between $2A$ and zero with a period of $\frac{2\pi}{(\omega_2 - \omega_1)/2} = \frac{4\pi}{\omega_2 - \omega_1}$ seconds. This type of motion is known as a *beat vibration* or *amplitude-modulated vibration*, and is illustrated in Fig. 2-8.

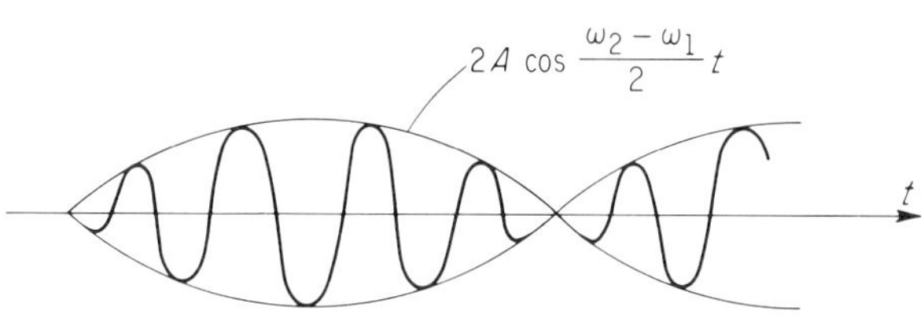

FIG. 2-8

Superposition of Two Harmonic Motions in Two Perpendicular Directions (Plane Periodic Motion)

We consider here the resultant motion of a point undergoing two harmonic oscillations in the two perpendicular directions X and Y respectively. Taking

$$x = A_x \sin(\omega_x t + \phi_x)$$

and

$$y = A_y \sin(\omega_y t + \phi_y)$$

The combined motion, in general, will be periodic in the X-Y plane. The following are simple examples of this plane periodic motion.

Case 1

When

$$\omega_x = \omega_y = \omega$$

and

$$\phi_x = \phi_y = 0$$

the combined motion (Fig. 2-9) will be simple harmonic in a direction

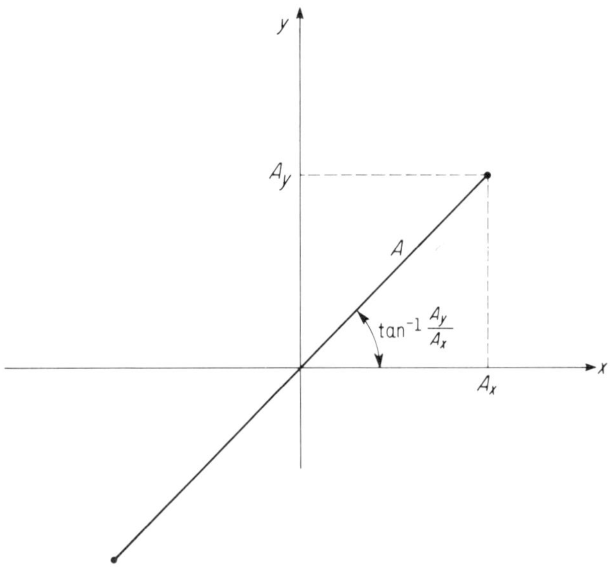

FIG. 2-9

defined by

$$y = \frac{A_y}{A_x} x$$

The amplitude of the motion will be

$$A = \sqrt{A_x^2 + A_y^2}$$

and its frequency will be the same as the frequency ω of the harmonic components.

Case 2

In this case we assume that the two harmonic components have the same frequency ω and are $\pi/2$ radians out of phase. Taking for example

$$x = A_x \sin \omega t$$

$$y = A_y \sin \left(\omega t + \frac{\pi}{2}\right)$$

the combined motion will be elliptical, as shown in Fig. 2-10, and will have a period $2\pi/\omega$.

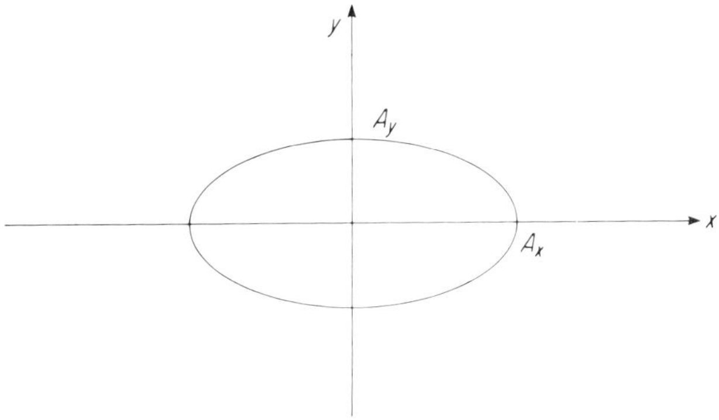

FIG. 2-10

For the special case when $A_x = A_y$ the resultant motion will be circular and is usually referred to as a *circular harmonic motion.* An example of this type of motion is that of water particles on the surface of a pond as shown in Fig. 2-11. Instead of traveling with the waves, the

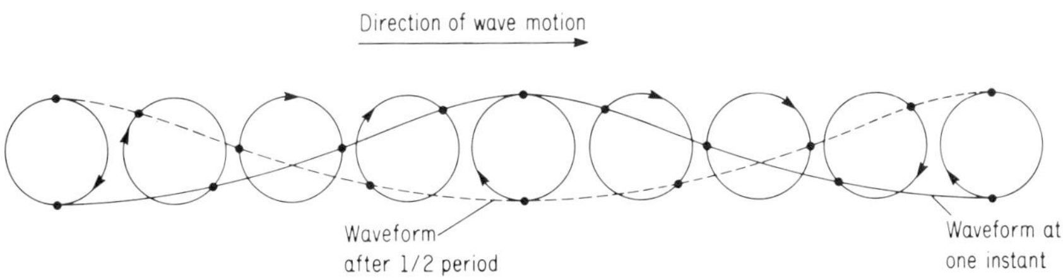

FIG. 2-11

particles themselves move in nearly stationery circular orbits. These circular orbits have diameters equal to the depth of the wave from crest to trough. All particles complete one orbit in the same period taken by the surface wave itself to advance one wavelength.

Case 3

In this case we assume

$$A_x = A_y = A$$
$$\omega_y = 2\omega_x$$
$$\phi_x = \phi_y = 0$$

The resultant motion will be as shown in Fig. 2-12 and will have a period equal to $2\pi/\omega_x$.

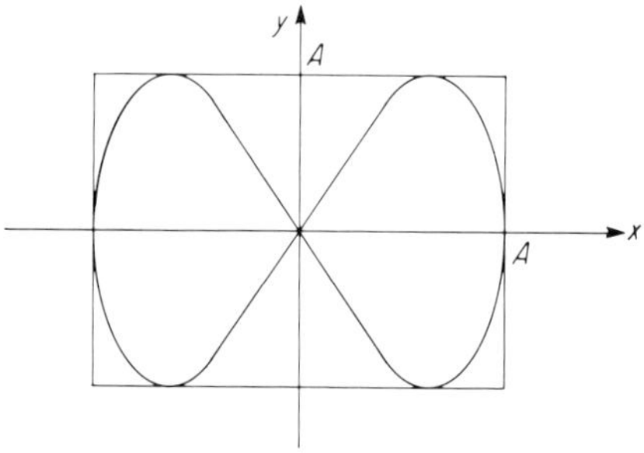

FIG. 2-12

Case 4

When the two harmonic components in the previous example are $\pi/2$ radians out of phase, the resultant periodic motion is as shown in

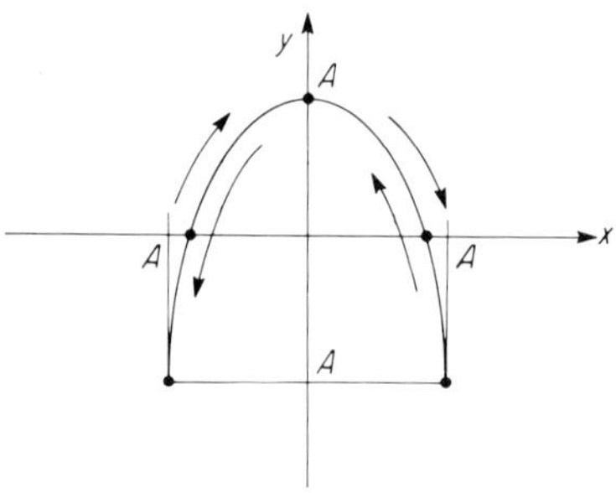

FIG. 2-13

Fig. 2-13. The motion parameters in this case are

$$A_x = A_y = A$$
$$\omega_y = 2\omega_x$$
$$\phi_x = 0$$
$$\phi_y = \frac{\pi}{2}$$

Superposition of Harmonic Motions in Three Perpendicular Directions (Space Periodic Motion)

The general case in this category is represented by a point which is simultaneously undergoing three harmonic motions in the X-, Y-, and Z-directions. The coordinates of such a point are therefore defined by

$$x = A_x \sin(\omega_x t + \phi_x)$$
$$y = A_y \sin(\omega_y t + \phi_y)$$
$$z = A_z \sin(\omega_z t + \phi_z)$$

The resultant motion will be periodic and is usually referred to as *space periodic motion*. This motion can be simple harmonic in the special case when

$$\phi_x = \phi_y = \phi_z = \phi$$
$$\omega_x = \omega_y = \omega_z = \omega$$

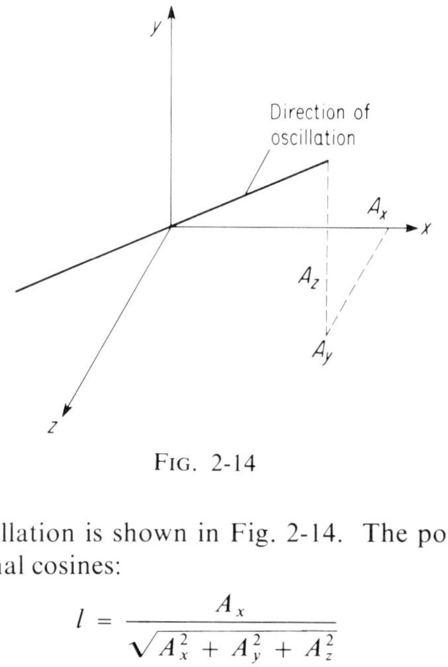

FIG. 2-14

This particular oscillation is shown in Fig. 2-14. The point will move in a line with directional cosines:

$$l = \frac{A_x}{\sqrt{A_x^2 + A_y^2 + A_z^2}}$$

$$m = \frac{A_y}{\sqrt{A_x^2 + A_y^2 + A_z^2}}$$

$$n = \frac{A_z}{\sqrt{A_x^2 + A_y^2 + A_z^2}}$$

The frequency of the oscillation will be ω and its amplitude will be $\sqrt{A_x^2 + A_y^2 + A_z^2}$.

Frequency of Periodic Motion

When a periodic motion (rectilinear, plane, or space) is defined by its harmonic components, its period is calculated from the minimum number of full periods of the lowest frequency component that can accommodate an integer number of all the other harmonics. Analytically, this is written as

$$\tau = a\tau_1 = b\tau_2 = c\tau_3 = \ldots, \text{etc.} \tag{2-11}$$

where

$$\begin{aligned} \tau &= \text{period of the resultant motion} \\ \tau_1, \tau_2, \tau_3, \ldots, &= \text{periods of the harmonic components} \\ a, b, c, \ldots, &= \text{smallest integers which satisfy Eq. 2-11.} \end{aligned}$$

As an example, the period of the resultant motion of the harmonics

$$\begin{aligned} x &= A_x \sin(0.5t + \phi_x) \\ y &= A_y \sin(3.5t + \phi_y) \\ z &= A_z \sin(5t + \phi_z) \end{aligned}$$

is calculated from Eq. 2-11 as

$$\tau = a\frac{2\pi}{0.5} = b\frac{2\pi}{3.5} = c\frac{2\pi}{5}$$

It can be seen that

$$\begin{aligned} a &= 1 \\ b &= 7 \\ c &= 10 \end{aligned}$$

The period of the resultant vibration is therefore

$$\tau = 4\pi \text{ sec}$$

and the frequency of the periodic motion is

$$\omega = \frac{2\pi}{\tau} = 0.5 \text{ rad/sec}$$

2-4. THE ANALYSIS OF PERIODIC MOTION

When a periodic motion is described by a time function, it can be expressed as a mean value and an infinite series of harmonics by utilizing a mathematical technique known as *Fourier analysis.* In the case of a general rectilinear periodic motion

$$x(t) = f(t)$$

with a period τ and a frequency $\omega = 2\pi/\tau$, the Fourier expansion is written as

$$x(t) = \bar{x} + \sum_{n=1}^{\infty} A_n \sin(n\omega t + \phi_n) \tag{2-12}$$

where $\bar{x}$ is the mean value of the function $f(t)$. A_n, $n\omega t$, ϕ_n are the amplitude, the frequency, and phase angle of the nth harmonic respectively. Equation 2-12 can be rewritten in the form of sine and cosine series as

$$x(t) = \bar{x} + \sum_{n=1}^{\infty} B_n \sin n\omega t + \sum_{n=1}^{\infty} C_n \cos n\omega t \tag{2-13}$$

where

$$A_n = \sqrt{B_n^2 + C_n^2}$$

$$\phi_n = \arctan \frac{C_n}{B_n}$$

B_n = amplitude of nth sine harmonic
C_n = amplitude of nth cosine harmonic

By means of the Fourier theory (see for example Ref. 4 for details of this theory) the values $\bar{x}$, B_n, and C_n are evaluated from

$$\bar{x} = \frac{1}{\tau} \int_0^{\tau} f(t)\, dt \tag{2-14}$$

$$B_n = \frac{2}{\tau} \int_0^{\tau} f(t) \sin n\omega t\, dt \tag{2-15}$$

$$C_n = \frac{2}{\tau} \int_0^{\tau} f(t) \cos n\omega t\, dt \tag{2-16}$$

The proof of Eq. 2-14 is self-evident, since the mean value of the function is equal to its average over one period. Equations 2-15 and 2-16 are derived from multiplying both sides of Eq. 2-13 by $\sin n\omega t$ and $\cos n\omega t$ respectively, and by performing the integration over one period τ. The following relations are utilized in evaluating B_n and C_n:

$$\int_0^{2\pi} \sin(n\omega t) d(n\omega t) = 0$$

$$\int_0^{2\pi} \cos(n\omega t) d(n\omega t) = 0$$

$$\int_0^{2\pi} \sin^2(n\omega t) d(n\omega t) = \pi$$

$$\int_0^{2\pi} \cos^2 (n\omega t) d(n\omega t) = \pi$$

$$\int_0^{2\pi} \sin (n\omega t) \cos (n\omega t) d(n\omega t) = 0$$

$$\int_0^{2\pi} \sin (n\omega t) \cos (m\omega t) d(n\omega t) = 0 \qquad n \neq m$$

Example

As an illustrative example for the application of Fourier analysis to expanding a periodic function, we consider the function represented in Fig. 2-15. The function has a period τ as shown in the graph and the

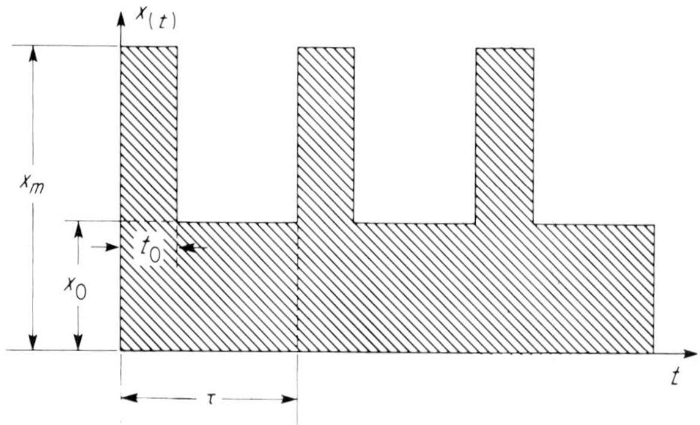

FIG. 2-15

cycle can be described mathematically by

$$x(t) = x_m \qquad 0 < t < t_0$$

$$x(t) = x_0 \qquad t_0 < t < \tau$$

By substitution in Eqs. 2-14, 2-15, and 2-16 we obtain

$$\bar{x} = \frac{1}{\tau}\left[\int_0^{t_0} x_m \, dt + \int_{t_0}^{\tau} x_0 \, dt\right]$$

$$= \frac{x_m t_0 + x_0 \tau - x_0 t_0}{\tau}$$

$$= x_0 + (x_m - x_0)\frac{t_0}{\tau}$$

The coefficient B_n is

$$B_n = \frac{2}{\tau}\left[\int_0^{t_0} x_m \sin n\omega t\, dt + \int_{t_0}^{\tau} x_0 \sin n\omega t\, dt\right]$$

$$= \frac{2}{n\tau\omega}[x_m(1 - \cos n\omega t_0) + x_0(\cos n\omega t_0 - \cos n\omega\tau)]$$

Since $\tau = \dfrac{2\pi}{\omega}$, therefore

$$B_n = \frac{1}{n\pi}[x_m(1 - \cos n\omega t_0) + x_0(\cos n\omega t_0 - 1)]$$

$$= \frac{1}{n\pi}(1 - \cos n\omega t_0)(x_m - x_0)$$

and the coefficient C_n is

$$C_n = \frac{2}{\tau}\left[\int_0^{t_0} x_m \cos n\omega t\, dt + \int_{t_0}^{\tau} x_0 \cos n\omega t\, dt\right]$$

$$= \frac{2}{n\tau\omega}[x_m \sin n\omega t_0 + x_0(\sin n\omega\tau - \sin n\omega t_0)]$$

Since $\tau = \dfrac{2\pi}{\omega}$, therefore

$$C_n = \frac{1}{n\pi}[x_m \sin n\omega t_0 + x_0(\sin 2n\pi - \sin n\omega t_0)]$$

$$= \frac{1}{n\pi}(x_m - x_0)\sin n\omega t_0$$

The values of B_n, and C_n which represent the amplitudes of the different harmonic components can therefore be computed for any value of n and substituted in Eq. 2-13 to give the Fourier series for $x(t)$. It should be noted here that in many situations the periodic functions are defined by their numerical values or by experimental curves. In such cases $\overline{x}$, B_n, and C_n are evaluated from Eqs. 2-14, 2-15, and 2-16 by numerical integration. If the experimental periodic function is available in the form of an electrical signal, the harmonic components can be evaluated experimentally by means of an apparatus known as the *harmonic analyzer* (see Part VI).

2-5. RANDOM VIBRATION

An oscillation is called *random* when it does not follow any particular pattern. Although the vibrating point alternates between positions above and below a central position, it does not repeat its motion in any regular or predictable pattern, as shown in Fig. 2-16. The frequency of passage

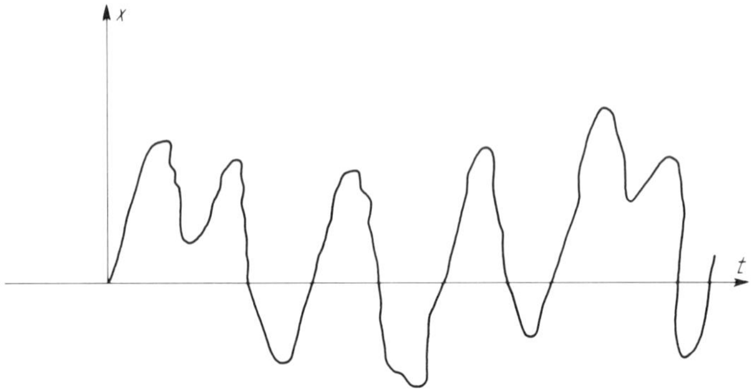

FIG. 2-16

through any particular position is a matter of probability and can only be estimated by the laws of statistics. This type of oscillation is discussed in Part V.

2-6. SUMMARY

The kinematics or geometry of the vibratory motion is discussed in this chapter without regard to the forces that produce this motion or the systems where it occurs. This aspect of the problem is discussed in the following chapter.

PROBLEMS

2-1. Two electrical signals are fed into an X-Y plotter and are represented by

$$E_x = 10 \sin 5t \text{ volts}$$

$$E_y = 5 \sin\left(3t + \frac{\pi}{2}\right) \text{ volts}$$

If the scale is set such as 1 in. in the X-direction represents 4 volts and 1 in. in the Y-direction represents 2 volts, what would be the expected plot? How long does it take the plotter to complete one cycle?

2-2. If the two signals in the previous problem are fed into the same channel of a recorder where the scale is set at 5 volts per inch, what will be the shape of the plot? If the chart has a speed of 1 in. per sec, what chart length will be required for one complete cycle?

2-3. A surface wave in a pond is traveling at a speed of 3 ft/sec. Calculate the angular velocity of the surface particles if the wavelength is 1 ft.

2-4. Calculate the period as well as the peak displacement for the rectilinear motion having the following components:

$$x_1 = 10 \sin 50t$$

$$x_2 = 10 \sin 45t$$

If the resultant periodic motion is assumed as an amplitude-modulated oscillation, calculate the beat frequency.

2-5. A particle starts a simple harmonic motion with a velocity $v = 15$ ft per sec. The period of its oscillation is 5 sec. Calculate the amplitude of the motion. At what distance is the center of this oscillation from the starting point?

2-6. An arm OA is rotating in the X-Y plane about a fixed axis at O with an angular velocity of 4 rad/sec. The arm carries a disk at A which also rotates in X-Y plane and has an angular velocity of 5 rad/sec with respect to the arm. Write the expression for the motion of a point P on the circumference of the disk and find its period. Take $OA = 4$ in. and the radius of the disk $= 1$ in.

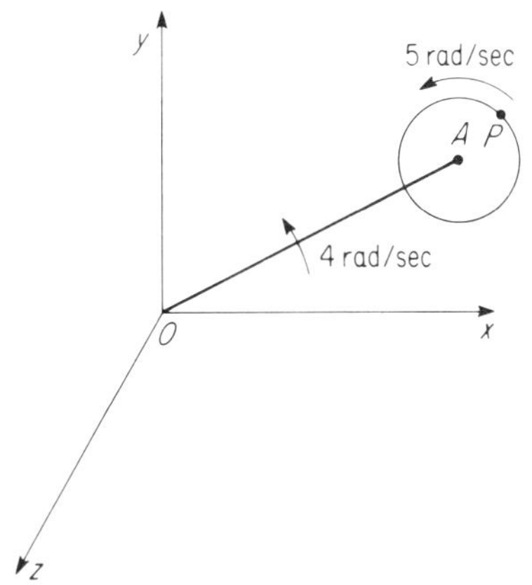

PROB. 2-6

2-7. The disk in the previous problem rotates about A in the AOZ plane while the arm OA rotates about O in the X-Y plane as shown. Write the mathematical expression for the motion of P with respect to the fixed plane XYZ and calculate the period of the motion.

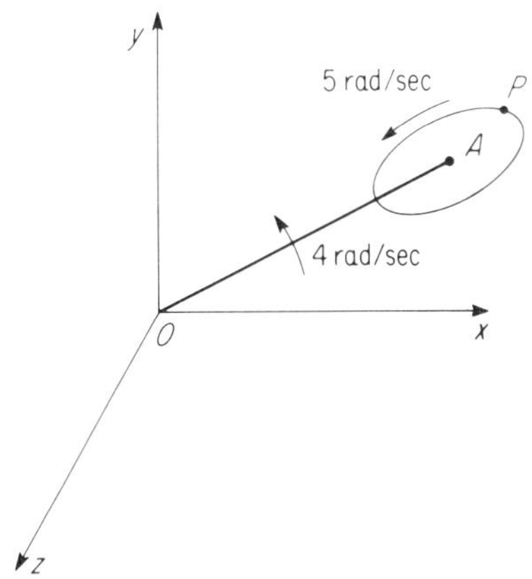

PROB. 2-7

2-8. A particle P performs a simple harmonic motion on a rod AB fixed to a horizontal platform as shown. The center of oscillation of P lies on the axis of rotation of the platform. The amplitude of its oscillation on the rod is 5 in. and its frequency is 10 rad/sec. The angular velocity of the platform is 7 rad/sec. Describe the motion of P in a fixed frame of reference and calculate its period.

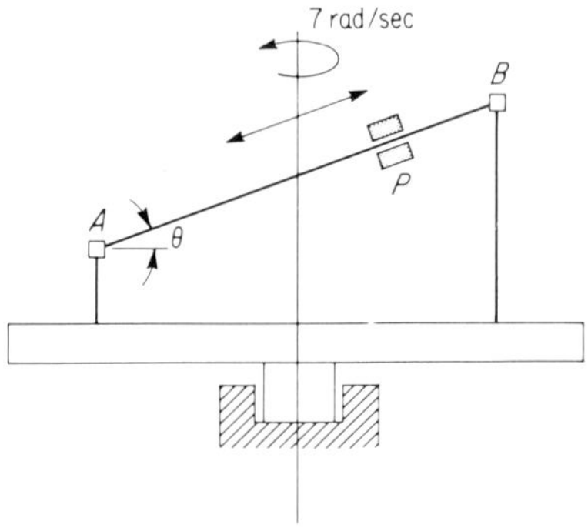

Prob. 2-8

2-9. Calculate the first two harmonic components of the square wave shown.

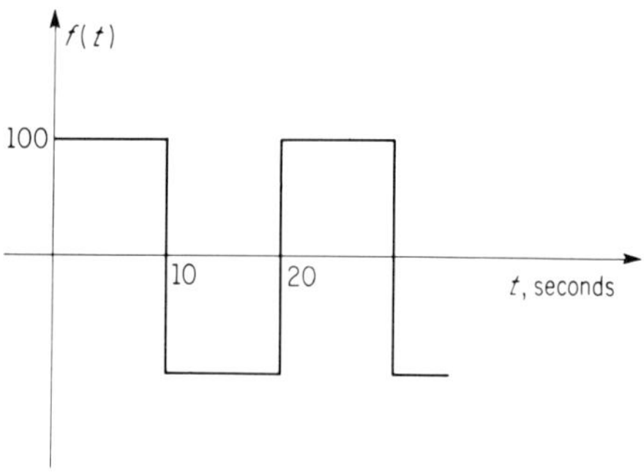

Prob. 2-9

2-10. Calculate the amplitude, frequency, and phase angle of the fundamental harmonic component (with the same frequency as the periodic function) of the sawtooth function shown in the sketch.

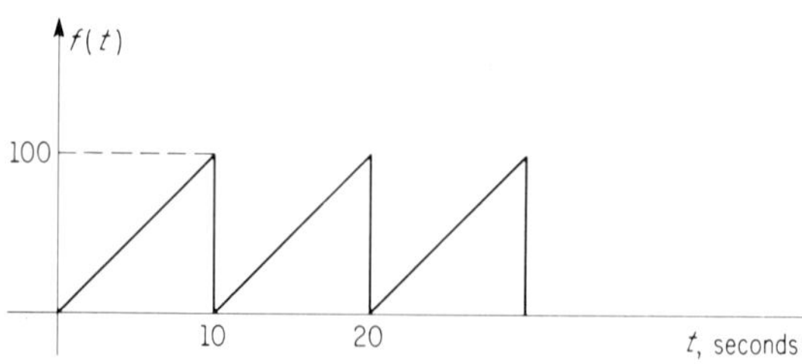

Prob. 2-10

2-11. An idealized crankshaft torque cycle due to one cylinder of an internal combustion engine is given in terms of the crank angles θ as shown in the sketch. Calculate the amplitude of the third-order harmonic component (i.e., harmonic component with three times the frequency of the periodic function).

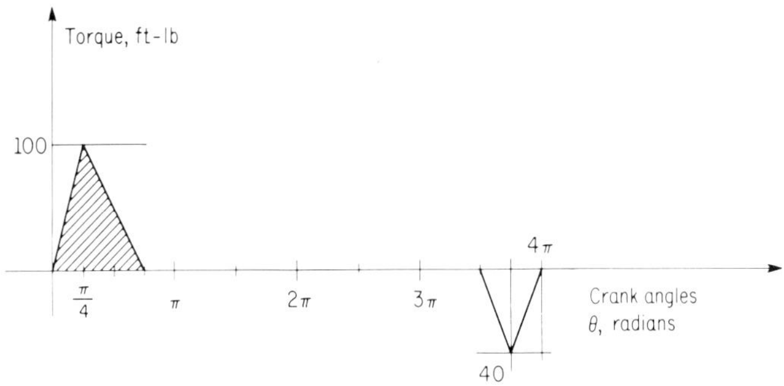

PROB. 2-11

chapter 3

Vibratory Systems Mathematical Models and Analogs

3-1. INTRODUCTION

It has been shown in Chapter 1 that a mechanical system can be idealized into one or a combination of the following categories: a particle, a system of particles, a rigid body, a system of rigid bodies, and a deformable body or continuum. According to Newton's first law, such a system will remain at rest or in a state of uniform motion until it is compelled to change by forces imposed upon it.

Vibratory systems are no exception to other dynamic systems. They remain at rest until they are disturbed by an external influence. One of the requirements of an oscillating system is that its equilibrium should be stable. Another requirement is that an external disturbance produces a force which tends to restore the body to its original position. Under the influence of this restoring force and in the absence of considerable friction, the body will accelerate toward its initial position. Because of its inertia it will continue to move beyond this position and will therefore create a restoring force in the opposite direction. This will decelerate the body, eventually cause it to stop, and then accelerate it again toward the initial position. The system is therefore in a state of vibration. If no frictional forces exist, the system will continue to oscillate indefinitely, repeating its motion at a constant time interval which is the *period of oscillation*. This discussion shows that a system must satisfy four basic requirements in order to vibrate:

1. Stability of its equilibrium position.
2. The system should have mass or inertia and must be influenced by a restoring force.
3. The frictional forces should be of a relatively small magnitude such as not to overshadow the restoring and inertia effects.
4. Subjection to a disturbance.

3-2. STABLE EQUILIBRIUM

The systems illustrated in Fig. 3-1 show the importance of the stability of the equilibrium position. The equilibrium in cases a and b is stable,

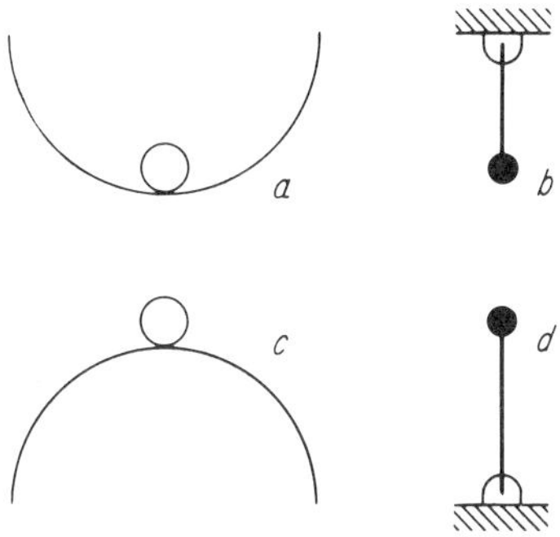

FIG. 3-1

and if the system is subjected to a small disturbance it will vibrate about its equilibrium position. On the other hand, cases c and d are unstable and a small disturbance will move the system away from the equilibrium position.

3-3. THE VIBRATORY SYSTEM

In describing the vibratory system we shall be primarily concerned with the masses and the restoring forces in the systems. Although the disturbing and the frictional forces are part of the vibratory system, they are considered as external influences and are discussed separately in the following sections. The systems will be mathematically represented by their differential equations of motion, and no attempt will be made toward a solution at this stage.

Elastically Supported Particle

Let us consider the general motion of a particle having a mass m and supported by a spring as shown in Fig. 3-2. The particle is assumed to be free to move in any direction. The spring is considered hinged at both ends and is therefore capable of resisting only axial displacements. The elastic constant of the spring is k and its directional cosines with respect to a fixed frame XYZ are l, m, and n respectively. The origin of the fixed frame is taken at the equilibrium position of the particle. Now if the particle is displaced a small distance x along the x-axis, it will cause an extension (lx) in the spring as shown. The corresponding restoring force

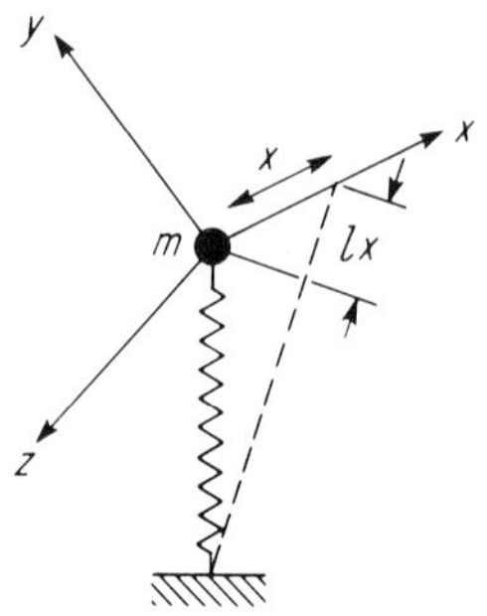

FIG. 3-2

will be (klx). Since the particle displacement is relatively small, the components of this force in the three reference directions will be

$$\begin{aligned} F_x &= -(klx)l \\ F_y &= -(klx)m \\ F_z &= -(klx)n \end{aligned}$$

Similarly, if the particle is displaced a small distance y in Y-direction, the restoring force will have the following components in the X-, Y-, and Z-directions:

$$\begin{aligned} F_x &= -(kmy)l \\ F_y &= -(kmy)m \\ F_z &= -(kmy)n \end{aligned}$$

Also for a small displacement z we get

$$\begin{aligned} F_x &= -(knz)l \\ F_y &= -(knz)m \\ F_z &= -(knz)n \end{aligned}$$

Therefore the equations of motion for a particle undergoing displacements (x, y, z) can be written as

$$m\ddot{x} = F_x = -(kl^2)x - (klm)y - (kln)z \tag{3-1a}$$

$$m\ddot{y} = F_y = -(kml)x - (km^2)y - (kmn)z \tag{3-1b}$$

$$m\ddot{z} = F_z = -(knl)x - (knm)y - (kn^2)z \tag{3-1c}$$

where $\ddot{x}$, $\ddot{y}$, $\ddot{z}$ are the accelerations in the reference directions and kl^2x, $kmlx$, $knlx$, etc., are the restoring-force components.

This analysis can be easily extended to describe the general case of a particle suspended by N springs as shown in Fig. 3-3. Taking the elastic constant for each spring as k_i and its directional cosines as l_i, m_i, n_i where i is 1, 2, 3, . . . , N, the equations of motion for a small displacement

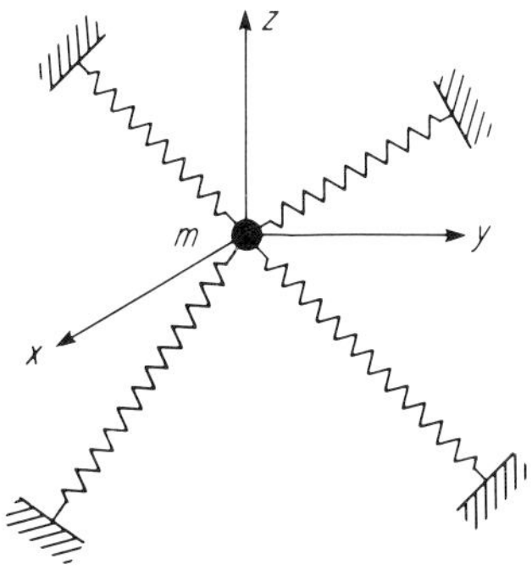

FIG. 3-3

x, y, z, will be

$$m\ddot{x} = \Sigma F_x = -\left[\sum_{i=1}^{N} l_i^2 k_i\right] x - \left[\sum_{i=1}^{N} l_i m_i k_i\right] y - \left[\sum_{i=1}^{N} l_i n_i k_i\right] z \qquad \text{(3-2a)}$$

$$m\ddot{y} = \Sigma F_y = -\left[\sum_{i=1}^{N} m_i l_i k_i\right] x - \left[\sum_{i=1}^{N} m_i^2 k_i\right] y - \left[\sum_{i=1}^{N} m_i n_i k_i\right] z \qquad \text{(3-2b)}$$

$$m\ddot{z} = \Sigma F_z = -\left[\sum_{i=1}^{N} n_i l_i k_i\right] x - \left[\sum_{i=1}^{N} n_i m_i k_i\right] y - \left[\sum_{i=1}^{N} n_i^2 k_i\right] z \qquad \text{(3-2c)}$$

In many practical problems, the particle may be restrained to move in a fixed direction. If we assume, for example, that the particle in Fig. 3-4 is

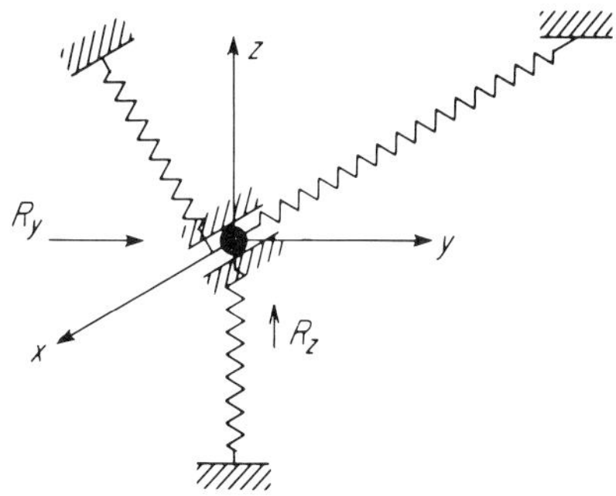

FIG. 3-4

restricted to move in the x-direction by means of frictionless guides, then

$$y = \ddot{y} = 0$$

and

$$z = \ddot{z} = 0$$

Equations 3-2 will be in this case

$$m\ddot{x} = -\left[\sum_{i=1}^{N} k_i l_i^2\right] x$$

$$0 = -\left[\sum_{i=1}^{N} m_i l_i k_i\right] x + R_y$$

$$0 = -\left[\sum_{i=1}^{N} n_i l_i k_i\right] x + R_z$$

and the particle will have a single degree of freedom represented by the first equation. The second and third equations give the support reactions at any instant.

Example 1

A particle is suspended by three springs with stiffness 10 lb/in., 20 lb/in., and 15 lb/in. When at rest the directional cosines of these springs with respect to a stationary (or inertial) frame of reference are (0.5, 0.5, 0.707), (0, 0.8, 0.6) and (1, 0, 0) respectively. The particle is displaced a small distance d inches in the direction of the negative z-axis and then released. Write the equations of motion.

Solution: From Eqs. 3-2 the differential equations of motion are

$$m\ddot{x} = -[(0.5)^2(10) + (0) + (1)^2(15)]x - [2.5]y - [3.535]z$$
$$m\ddot{y} = -[(0.5)^2(10) + (0.8)^2(20) + (0)]y - [13.135]z - [2.5]x$$
$$m\ddot{z} = -[(0.707)^2(10) + (0.6)^2(20) + (0)]z - [3.535]x - [13.135]y$$

and the initial conditions are

at $t = 0$

$$x = 0 \qquad \dot{x} = 0$$
$$y = 0 \qquad \dot{y} = 0$$
$$z = -d \qquad \dot{z} = 0$$

Example 2

If the particle in the previous example is only free to move in the y-direction, the equations of motion become

$$0 = -[2.5]y + R_x$$
$$m\ddot{y} = -[3.78]y$$
$$0 = -[13.135]y + R_z$$

which define the displacement y as well as the support reactions R_x and R_z.

Elastically Connected System of Particles

Each particle in a system of particles, Fig. 3-5, has three degrees of freedom. The total number of degrees of freedom is therefore equal to $3N$, where N is the number of particles in the system.

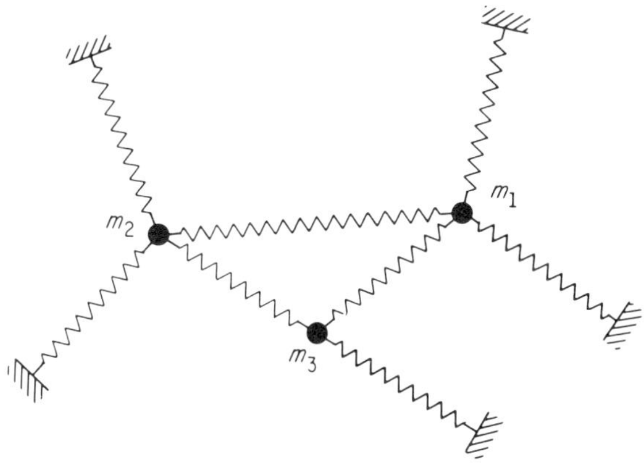

FIG. 3-5

Three equations of motion can be written for each particle. The simultaneous solution of these equations which satisfies the initial conditions of the motion gives the complete description of the system vibration.

When two particles are connected as shown in Fig. 3-6 by a spring

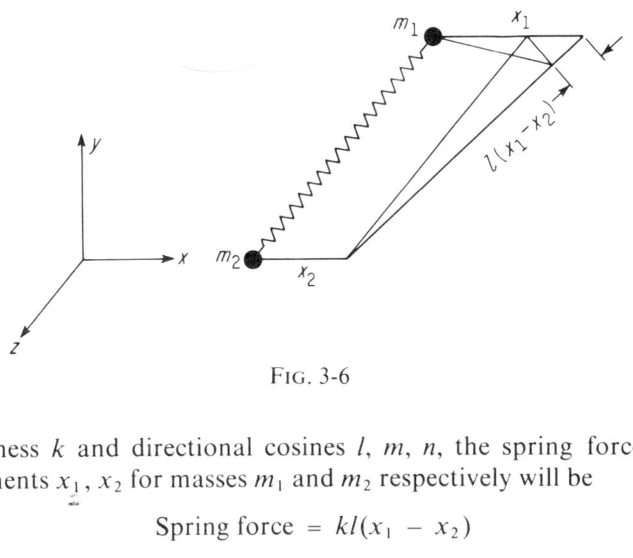

FIG. 3-6

with stiffness k and directional cosines l, m, n, the spring force due to displacements x_1, x_2 for masses m_1 and m_2 respectively will be

$$\text{Spring force} = kl(x_1 - x_2)$$

The components of this force in the three reference directions are therefore

$$F_x = \pm kl^2(x_1 - x_2) \quad (3\text{-}3a)$$
$$F_y = \pm klm(x_1 - x_2) \quad (3\text{-}3b)$$
$$F_z = \pm kln(x_1 - x_2) \quad (3\text{-}3c)$$

where the lower and upper signs correspond to the forces on particles 1 and 2 respectively.

Similarly, for small displacements y_1, y_2 the force components will be

$$F_x = \pm kml(y_1 - y_2) \quad (3\text{-}3d)$$
$$F_y = \pm km^2(y_1 - y_2) \quad (3\text{-}3e)$$
$$F_z = \pm kmn(z_1 - z_2) \quad (3\text{-}3f)$$

and for displacements z_1, z_2 we have

$$F_x = \pm knl(z_1 - z_2) \quad (3\text{-}3g)$$
$$F_y = \pm knm(z_1 - z_2) \quad (3\text{-}3h)$$
$$F_z = \pm kn^2(z_1 - z_2) \quad (3\text{-}3i)$$

The reader is now prepared to apply Newton's second law for any elastically connected system of particles.

Example 3

Write the equations for studying the small vibration of the given system, Fig. 3-7.

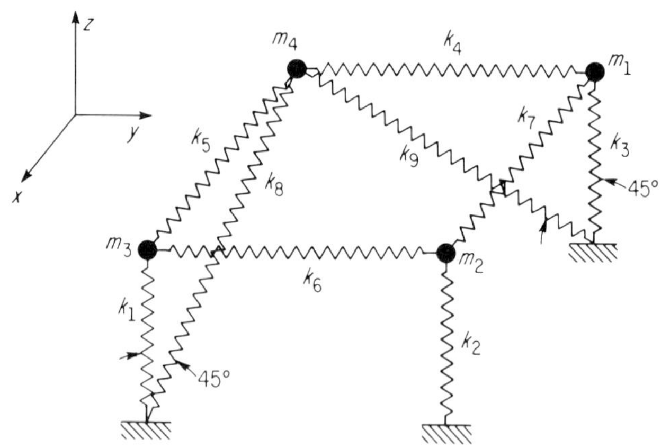

FIG. 3-7

Solution: The equations of motion for each mass can be found by application of Newton's second law and utilizing Eqs. 3-2 and 3-3.

For m_1

$$m_1\ddot{x}_1 = -k_7(x_1 - x_2)$$
$$m_1\ddot{y}_1 = -k_4(y_1 - y_4)$$
$$m_1\ddot{z}_1 = -k_3z_1$$

For m_2

$$m_2\ddot{x}_2 = -k_7(x_2 - x_1)$$
$$m_2\ddot{y}_2 = -k_6(y_2 - y_3)$$
$$m_2\ddot{z}_2 = -k_2z_2$$

For m_3

$$m_3\ddot{x}_3 = -k_5(x_3 - y_4)$$
$$m_3\ddot{y}_3 = -k_6(y_3 - y_2)$$
$$m_3\ddot{z}_3 = -k_1z_3$$

For m_4

$$m_4\ddot{x}_4 = -k_5(x_4 - x_3) - (0.707)^2k_8x_4 + (0.707)^2k_8z_4$$
$$m_4\ddot{y}_4 = -k_4(y_4 - y_1) - (0.707)^2k_9y_4 + (0.707)^2k_9z_4$$
$$m_4\ddot{z}_4 = -(0.707)^2k_8z_4 - (0.707)^2k_9z_4 + (0.707)^2k_8x_4 + (0.707)^2k_9y_4$$

The Elastically Supported Rigid Body

A mass which is elastically supported in space has, in the general case, six degrees of freedom. Three of these define the displacement of the center of mass (x, y, z) and the other three describe the rotation of the body about the center of mass $(\theta_x, \theta_y, \theta_z)$. Figure 3-8 illustrates such system where (X, Y, Z) is the fixed frame of reference.

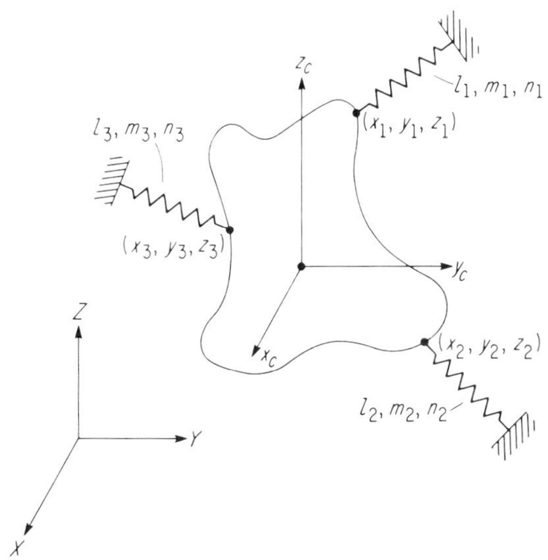

FIG. 3-8

As shown in Chapter 1, the differential equations describing the motion of such a mass are the Newton and Euler equations given below.

$$m\ddot{x} = F_x \tag{3-4a}$$
$$m\ddot{y} = F_y \tag{3-4b}$$
$$m\ddot{z} = F_z \tag{3-4c}$$

which describe the translation X, Y, Z of the center of mass C,

$$I_{x_c x_c}\ddot{\theta}_x + \dot{\theta}_z\dot{\theta}_y(I_{z_c z_c} - I_{y_c y_c}) =' M_x \tag{3-4d}$$
$$I_{y_c y_c}\ddot{\theta}_y + \dot{\theta}_x\dot{\theta}_z(I_{x_c x_c} - I_{z_c z_c}) = M_y \tag{3-4e}$$
$$I_{z_c z_c}\ddot{\theta}_z + \dot{\theta}_y\dot{\theta}_x(I_{y_c y_c} - I_{x_c x_c}) = M_z \tag{3-4f}$$

which define the rotation of the body. (x_c, y_c, z_c) are the principal axes at the center of mass and are body-fixed axes. $(I_{x_c x_c}, I_{y_c y_c}, I_{z_c z_c})$ are the principal moments of inertia and M_x, M_y, M_z are the moments of all forces acting on the mass about the center of mass C in the X-, Y- and Z-directions respectively.

For the general case the solution of such simultaneous differential equations is rather difficult.

The following examples are simple illustrations of the vibratory motions of rigid bodies.

Example 4

We shall study here the motion of a rigid body about a fixed axis in space. The body, Fig. 3-9a, is assumed symmetrical with respect to the x-y plane (i.e., has uniform thickness). The axis of rotation is therefore a principal axis because of this symmetry condition. Since no rotation is possible about the x or y axes, $\dot{\theta}_x = \dot{\theta}_y = \ddot{\theta}_x = \ddot{\theta}_y = 0$. The rigid body in this case has a single degree of freedom, and Euler's equations are reduced to the single equation:

$$\begin{aligned} I_{z_0 z_0}\ddot{\theta}_z &= M_z \\ &= -\text{mgh} \sin \theta_z \end{aligned}$$

where m is the mass of the body and h is the distance between the axis of rotation O and the center of mass C.

For small oscillations $\sin \theta_z = \theta_z$ and the equation of motion for the body becomes

$$I_{z_0 z_0}\ddot{\theta}_z = -mgh\,\theta_z$$

Example 5

The torsional pendulum, Fig. 3-9b, is also a case of rotation about a fixed axis in space which is a principal axis because of symmetry. The restoring moment in this case is due to the torsional resistance of the

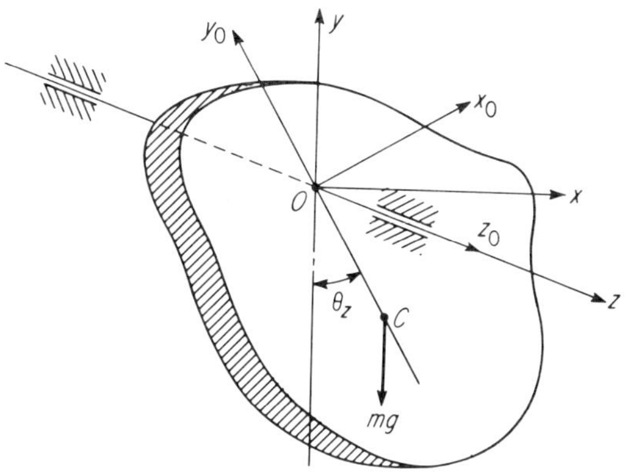

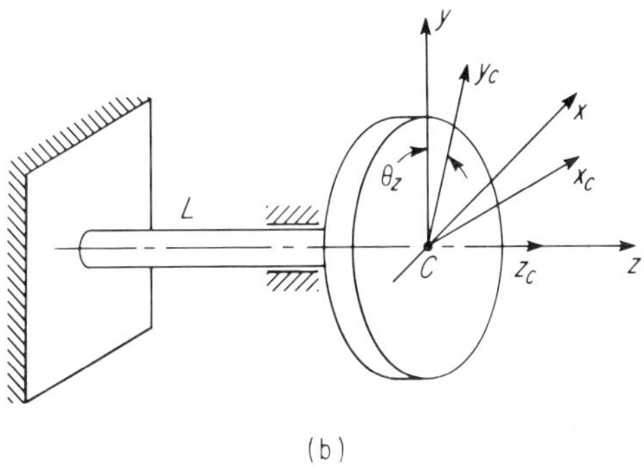

FIG. 3-9

supporting shaft. It is evident that $\dot{\theta}_x = \dot{\theta}_y = \ddot{\theta}_x = \ddot{\theta}_y = 0$ and the equation describing this single-degree-of-freedom system is

$$I_{z_c z_c}\ddot{\theta}_z = -\frac{GJ}{L}\theta_z$$

where G = modulus of shear for the shaft
J = polar moment of inertia for its cross section

Example 6

A cube having a mass m and side length a is suspended by six springs as shown in Fig. 3-10. In the equilibrium position, the lines of action of the springs as well as the sides of the cube are in the direction of the reference axes X, Y, Z.

Taking x_c, y_c, z_c as the body-fixed principal axes at the mass center, the equations representing the displacement of C are therefore, due to a

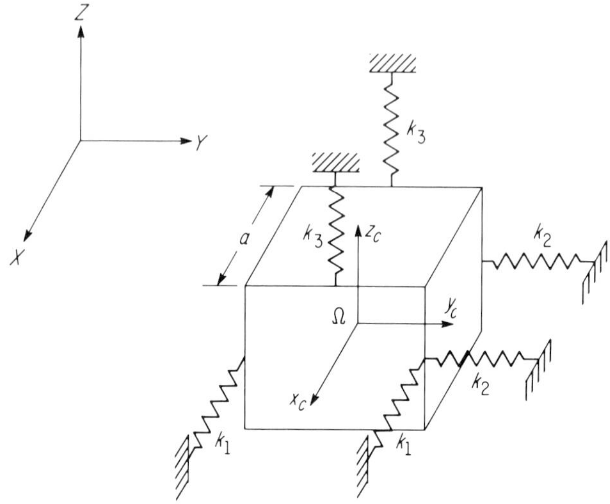

FIG. 3-10

small disturbance,

$$m\ddot{x} = -2k_1 x$$
$$m\ddot{y} = -2k_2 y$$
$$m\ddot{z} = -2k_3 z$$

Since $I_{x_c x_c} = I_{y_c y_c} = I_{z_c z_c}$, Euler's equations will in this case be

$$I_{x_c x_c}\ddot{\theta}_x = 0$$

$$I_{y_c y_c}\ddot{\theta}_y = -k_3 \frac{a^2}{2}\theta_y$$

$$I_{z_c z_c}\ddot{\theta}_z = -k_1 \frac{a^2}{2}\theta_z - k_2 \frac{a^2}{2}\theta_z$$

The Continuum

According to atomic theory, all systems are microscopically discrete. For macroscopic processes, however, we may regard solids, liquids, and gases as continuous systems or continuum. A finite continuum has infinite degrees of freedom, since it is composed of an infinite number of molecules, atoms, or electrons.

When an atom or molecule in a continuum is disturbed, its motion will affect neighboring atoms. This disturbance will bring into action forces that tend to restore the disturbed atoms, and vibrations of these particles will result. Although each atom or particle will vibrate in the neighborhood of its equilibrium positions, the disturbance will travel or propagate through the continuum. An example of this is the water surface wave discussed in Chapter 2.

The general equations of motion for a three-dimensional continuum are rather elaborate and can be found in texts on elasticity and fluid mechanics. We shall limit ourselves here to simple cases that are easy to visualize and analyze, and can therefore provide an introduction to the vibration of continuous systems.

The Solid Continuum

As examples of solid continuum we shall discuss briefly the vibration of strings, beams, membranes and plates.

The String

Considering an element δx (Fig. 3-11) of a string subjected to a small disturbance, the equation of motion in the Y direction can be written as

$$P \sin(\phi + \delta\phi) - P \sin \phi = m\delta x \frac{\partial^2 y}{\partial t^2}$$

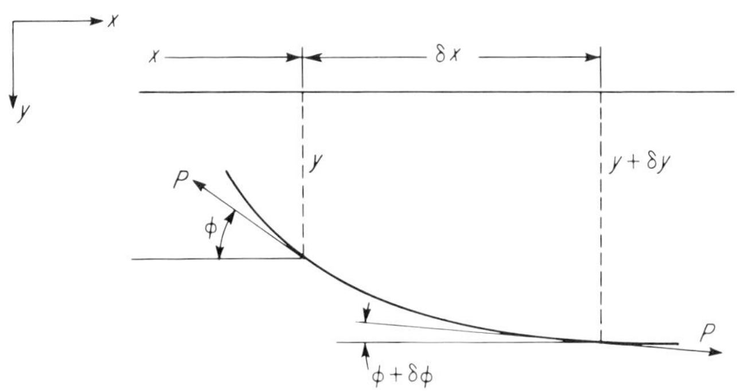

FIG. 3-11

where P = tension in string

y = deflection of string at a distance x along its length

ϕ = slope = $\dfrac{\partial y}{\partial x}$

m = mass per unit length

For small displacements, $\sin(\phi + \delta\phi) \cong \phi + \delta\phi$,

$$P \frac{\partial \phi}{\partial x} = m \frac{\partial^2 y}{\partial t^2}$$

since

$$\phi = \frac{\partial y}{\partial x}$$

The equation of motion of the string takes the form

$$\frac{\partial^2 y}{\partial x^2} = \frac{m}{P}\frac{\partial^2 y}{\partial t^2} \tag{3-5}$$

The Beam

While a string is capable only of lateral motion, the beam is capable of longitudinal, transverse, and torsional motions.

Considering the longitudinal displacement of a section at distance x to be u, the displacement of the section $(x + \delta x)$ will be $(u + \delta u)$, as shown in Fig. 3-12.

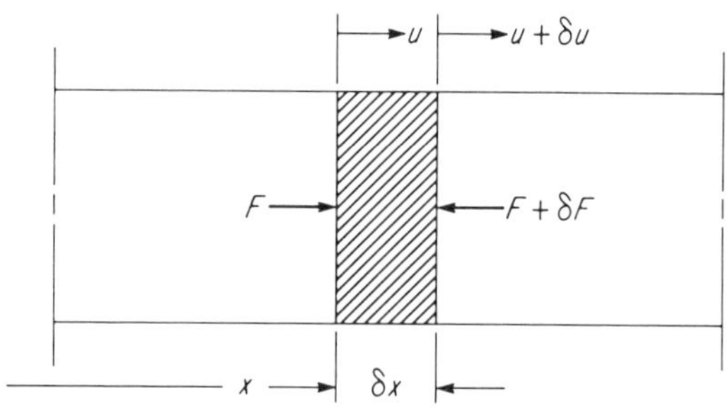

FIG. 3-12

Now the equation of motion of the element δx will be

$$\delta F = \rho A\,\delta x\frac{\partial^2 u}{\partial t^2} \tag{3-6}$$

where ρ = mass per unit volume

F = total force on the cross-sectional area A of the beam. From Hooke's law

$$F = EA\frac{\partial u}{\partial x}$$

where E is the modulus of elasticity.

Substituting in Eq. 3-6 we obtain

$$\frac{\partial^2 u}{\partial t^2} = \frac{E}{\rho}\frac{\partial^2 u}{\partial x^2} \tag{3-7}$$

The equation which describes the lateral or transverse motion of the beam can be derived from the bending theory of beams as

$$EI\frac{\partial^4 y}{\partial x^4} = m\frac{\partial^2 y}{\partial t^2} \tag{3-8}$$

where y = deflection of the beam at x
m = mass per unit length
E = modulus of elasticity
I = area moment of inertia about the bending axis

The equation for torsional motion, Fig. 3-13, can be derived from the relation between torque T and twist θ of an element δx. This equation

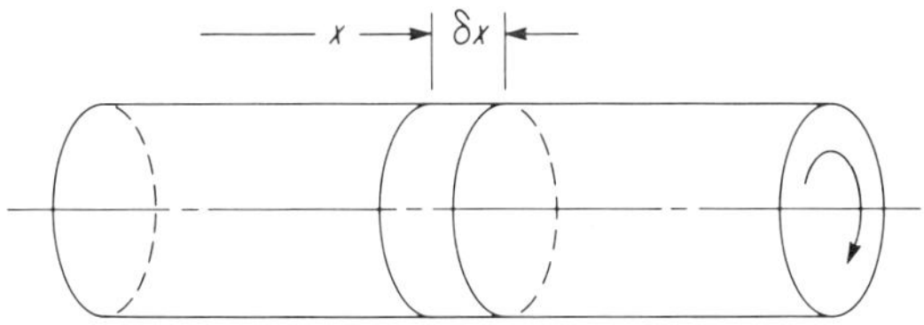

FIG. 3-13

is found from elementary strength of materials to be

$$T = GJ\frac{\partial \theta}{\partial x}$$

And by applying Euler's equation for rotation about the axis of twist we obtain

$$\frac{\partial T}{\partial x}\delta x = \frac{\partial\left(GJ\dfrac{\partial \theta}{\partial x}\right)}{\partial x}\delta x = \rho J \delta x \frac{\partial^2 \theta}{\partial t^2}$$

which simplifies to

$$\frac{\partial^2 \theta}{\partial t^2} = \frac{G}{\rho}\frac{\partial^2 \theta}{\partial x^2} \tag{3-9}$$

where G = shear modulus
J = polar moment of inertia of the cross section
ρ = mass per unit volume of the material
θ = angle of twist
T = twisting moment

Notice that Eqs. 3-5, 3-7, and 3-9 have the same form and are known as *wave equations*. The solution for these equations is discussed in Chapter 11. The constants P/m, E/ρ, G/ρ in these equations represent the square of the velocity of wave propagation for each case respectively.

The Membrane

The vibration of a membrane is one of great interest since many dynamic systems include membranes. Let us consider (Fig. 3-14) the

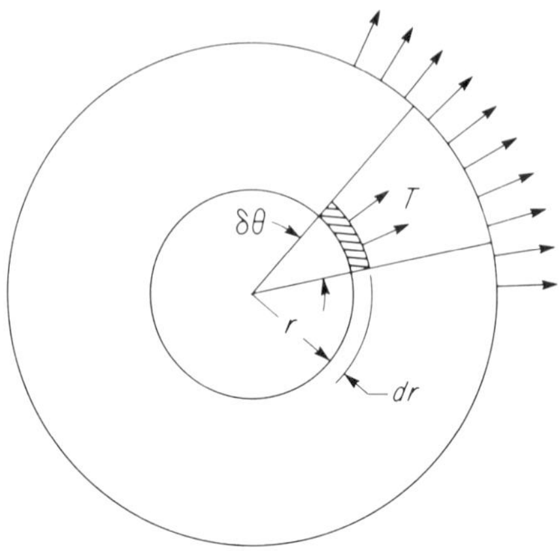

FIG. 3-14

case of the circular membrane for illustration. If T is the uniform tension in the membrane per unit length, the equation describing the symmetrical vibration due to a small disturbance can be developed by considering a section of the membrane as a string of variable cross section. For the sector shown in Fig. 3-14 the lateral component F of the tension is equal to

$$F = Tr\,\delta\theta \sin\phi \tag{3-10}$$

where $\phi = \dfrac{dz}{dr}$
$= $ slope of membrane at radius r
$= \sin\phi$ for small values of ϕ

and $z = $ deflection of membrane at same point

Newton's second law for the sector under consideration gives

$$\delta F = \frac{\partial F}{\partial r}\,\delta r$$

$$= -\rho r\,\delta\theta\,\delta r\,\frac{\partial^2 z}{\partial t^2} \tag{3-11}$$

where $\rho = $ mass per unit area of membrane

$\dfrac{\partial^2 z}{\partial t^2} = $ acceleration in lateral direction

By substituting the value of F from Eq. 3-10 in Eq. 3-11 we obtain

$$\frac{\partial}{\partial r}\left(Tr\,\delta\theta\,\frac{\partial z}{\partial r}\right)\delta r = -\rho r\delta\theta\,\delta r\,\frac{\partial^2 z}{\partial t^2}$$

and by dividing both sides by $(T\delta\theta\delta r)$, this equation becomes

$$\frac{\partial}{\partial r}\left(r\frac{\partial z}{\partial r}\right) = -\frac{\rho}{T}r\frac{\partial^2 z}{\partial t^2} \tag{3-12}$$

The Plate

While a membrane is only capable of resisting tension, a plate, Fig. 3-15, can support tension, bending, and shear. The following is the

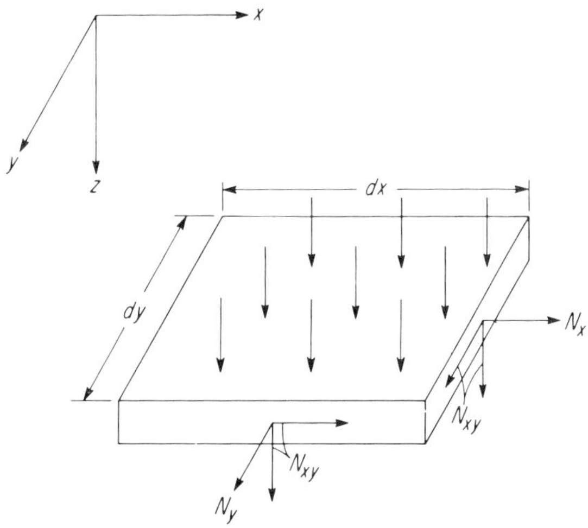

FIG. 3-15

general equation of pure bending of plates as given in texts on the theory of plates and shells (see Ref. 5 for example):

$$D\left(\frac{\partial^4 z}{\partial x^4} + 2\frac{\partial^4 z}{\partial x^2 \partial y^2} + \frac{\partial^4 z}{\partial y^4}\right) = q \tag{3-13}$$

where z = deflection
D = flexural rigidity of plate $\dfrac{Eh^3}{12(1 - \nu^2)}$
E = modulus of elasticity
h = thickness of plate
ν = Poisson ratio
q = load per unit area on the plate

When the plate is in a state of motion, the loading on the plate will be due to its inertia. In this case

$$q = m\frac{\partial^2 z}{\partial t^2}$$

where m is mass per unit area of the plate. The equation of plate vibration is therefore

$$D\left(\frac{\partial^4 z}{\partial x^4} + 2\frac{\partial^4 z}{\partial x^2 \partial y^2} + \frac{\partial^4 z}{\partial y^4}\right) = m\frac{\partial^2 z}{\partial t^2} \tag{3-14}$$

The solution of this equation involves elaborate computations and is dealt with later in Chapter 11.

The Fluid Continuum

The ability of a fluid to vibrate and propagate a disturbance depends on the existence of a relationship between small changes of pressure and small changes in volume. Since a change in pressure will induce a change in density, this relationship is expressed in terms of a *bulk modulus*, which is defined as the ratio between the change of pressure δP and the condensation s. The latter represents a measure of the change in density given by the ratio of this change to the equilibrium density. Analytically,

$$\begin{aligned} K &= \text{bulk modulus} \\ &= \frac{\delta P}{s} \\ \delta P &= P - P_0 \\ s &= \text{condensation} \\ &= \frac{\rho - \rho_0}{\rho_0} = \frac{\delta\rho}{\rho_0} \end{aligned}$$

According to the definition of bulk modulus,

$$K = \frac{\delta P}{\delta\rho}\rho_0$$

and in the limit

$$K = \frac{\partial P}{\partial\rho}\rho_0 \tag{3-15}$$

where P = pressure at any instant
P_0 = equilibrium pressure
ρ = density at any instant (mass/unit volume)
ρ_0 = equilibrium density

For a gas continuum we know that

$$P = RT\rho$$

where R = gas constant
T = absolute temperature

For an isothermal process T = const., therefore the bulk modulus is

$$K = \frac{\delta P}{s} = \frac{\partial P}{\partial \rho} \rho_0 = RT\rho_0 = P_0$$

If the process is adiabatic,

$$P = G\rho^{\gamma} \tag{3-16}$$

where G = const.

$\gamma = \dfrac{C_p}{C_v}$ = specific heat ratio

C_p, C_v = specific heat values under constant pressure and volume respectively

Therefore, for an adiabatic process the bulk modulus can be evaluated from Eqs. 3-15 and 3-16 as

$$\begin{aligned} K &= \frac{\partial P}{\partial \rho} \rho_0 \\ &= \gamma G \rho_0^{\gamma} \\ &= \gamma P_0 \end{aligned} \tag{3-17}$$

The adiabatic process is more realistic in describing the propagation of disturbance in a fluid.

The general three-dimensional equations of motion for the fluid continuum will not be considered. The simple one-dimensional problem of motion of a column of fluid in a rigid-wall tube, Fig. 3-16, will be discussed.

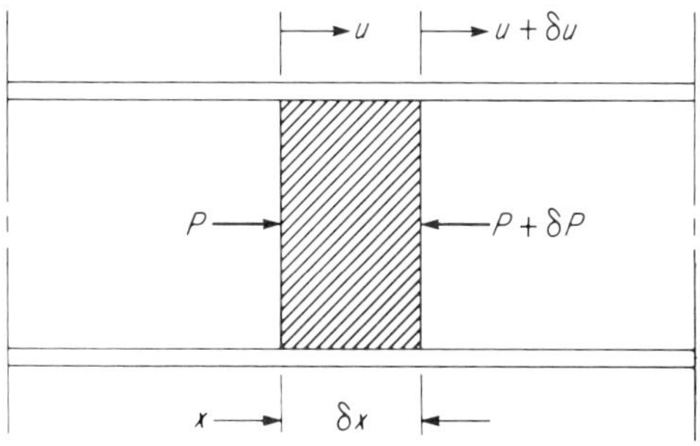

FIG. 3-16

The equation representing the condition under consideration is obtained from Newton's second law:

$$A\rho_0 \delta x \frac{\partial^2 u}{\partial t^2} = \delta F = -A\left[\left(\frac{\partial P}{\partial x}\right)\delta x\right] \tag{3-18}$$

where A = cross-sectional area of tube
P = pressure at distance x
u = particle displacement at distance x
ρ_0 = equilibrium density
δF = change in pressure force in distance δx.

The conservation of mass requires that the mass of the fluid element with length δx should be the same in the equilibrium state as when it is subjected to a deformation δu. This is expressed as

$$\rho_0 \delta x = \rho_0 (1 + s)(\delta x + \delta u)$$

or

$$\rho_0 \delta x = \rho_0(\delta x + \delta u + s\delta x + s\delta u)$$

For small displacements we can neglect $s\delta u$, and in the limit the condition of conservation of mass becomes

$$s = -\frac{\partial u}{\partial x} \tag{3-19}$$

From Eqs. 3-15 and 3-19 we get

$$\delta P = sK = -K\frac{\partial u}{\partial x}$$

and by substitution in Eq. 3-18 we have

$$A\rho_0 \delta x \frac{\partial^2 u}{\partial t^2} = +AK\frac{\partial^2 u}{\partial x^2}\delta x$$

from which

$$\frac{\partial^2 u}{\partial t^2} = \frac{K}{\rho_0}\frac{\partial^2 u}{\partial x^2} \tag{3-20}$$

This equation is also a wave equation where K/ρ_0 represents the square of the velocity of the wave propagation in this case. The containing pipe may be assumed rigid with sufficient accuracy when studying vibration or wave motion in gases. The elasticity of the container would, however, be of considerable value in certain problems as in the case of liquids. Neglecting friction, the following is the fundamental equation of motion for the fluid in an elastic pipe, Fig. 3-17, as obtained by applying Newton's law:

$$\frac{\pi D^2}{4}[w(H + \delta H) - wH] = -\frac{\pi D^2}{4}\frac{w}{g}\delta x\frac{\partial V}{\partial t} \tag{3-21}$$

where D = diameter of the pipe
H = total head at distance x along axis of pipe
w = weight per unit volume of liquid
V = velocity of liquid at distance x

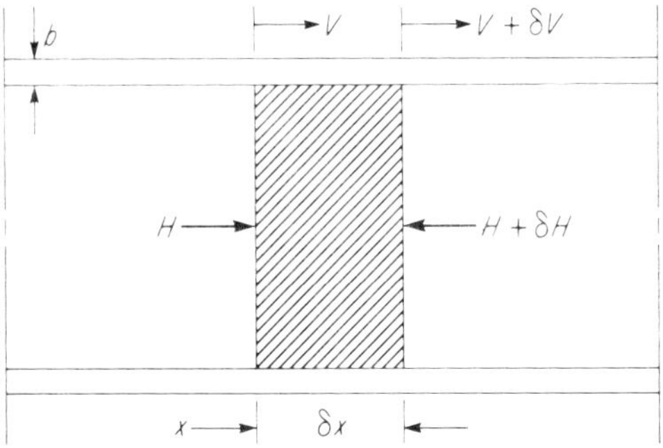

FIG. 3-17

and in the limit, Eq. 3-21 becomes

$$-\frac{\partial H}{\partial x} = \frac{1}{g}\frac{\partial V}{\partial t} \tag{3-22}$$

Knowing that the conservation of mass requires that any change of the volume of the liquid resulting from the extinction of the velocity (δV in the length δx and time δt) fills the space which is available by compressing the fluid and stretching the pipe, we therefore establish that change in volume δQ_v due to change in velocity $= (\pi/4)D^2\ \delta V\ \delta t$, and change in volume δQ_ρ due to the compression of the fluid $= (\pi/4)D^2 \delta x\ w\delta\ H/K$ where K = bulk modulus

$$= \frac{\delta P}{\delta \rho}\rho = \frac{\delta P}{\delta Q}Q$$

Q = volume of element

$$= \frac{\pi D^2}{4}\delta x$$

δP = change in pressure

$= w\delta H$

Also the change in volume δQ_e due to expansion of the tube $= \pi D \delta x e$, where e = change in radius under pressure

$$= \frac{w\delta H D}{2bE} \times \frac{D}{2}$$

Substituting for e in the previous equation we get

$$\delta Q_e = \frac{\pi D^3 w \delta x \delta H}{4bE}$$

The conservation of mass gives

$$-\delta Q_v = \delta Q_p + \delta Q_e$$

from which

$$\frac{\pi D^2}{4}\delta V \delta t = -\left(\frac{\pi D^3 w\, \partial x\, \partial H}{4\, bE} + \frac{\pi D^2 w\, \partial H\, \partial x}{4K}\right)$$

which simplifies to

$$-\frac{\partial V}{\partial x} = w\left(\frac{1}{K} + \frac{D}{bE}\right)\frac{\partial H}{\partial t} \tag{3-23}$$

Equations 3-22 and 3-23 give

$$\frac{\partial^2 V}{\partial x^2} = \frac{w}{g}\left(\frac{1}{K} + \frac{D}{bE}\right)\frac{\partial^2 V}{\partial t^2}$$

or

$$\frac{(K_e)}{\left(\frac{w}{g}\right)}\frac{\partial^2 H}{\partial x^2} = \frac{\partial^2 H}{\partial t^2} \tag{3-24}$$

where

$$\frac{1}{K_e} = \frac{1}{K} + \frac{D}{bE}$$

and K_e is considered as an equivalent bulk modulus which takes into account the elasticity of the wall. Equation 3-24 is the general equation used for studying the wave motion in pipes known as the *water-hammer phenomenon.* Notice that the velocity of wave propagation in this case is given by $\sqrt{[K_e/(w/g)]}$.

3-4. INFLUENCE OF FRICTION ON VIBRATION

It has been mentioned previously that frictional forces have a damping effect on the system vibrations. Frictional forces are resistive forces, but unlike restoring forces they are dissipative. They have the characteristic of transforming the energy of motion into heat and losing it to the surrounding atmosphere. Regardless of the type of frictional forces, they always have a direction which opposes the velocity. It should also be generally expected that they have a retarding effect on the vibrating system and continually reduce the amplitude of the motion.

There are three important types of frictional forces:

1. *Solid or dry friction* (Fig. 3-18a). This is generally known as Coulomb friction and exists in many mechanical systems. There are two common varieties of this type of friction. In the first, the dissipative force remains constant in magnitude and does not change with the amplitude or

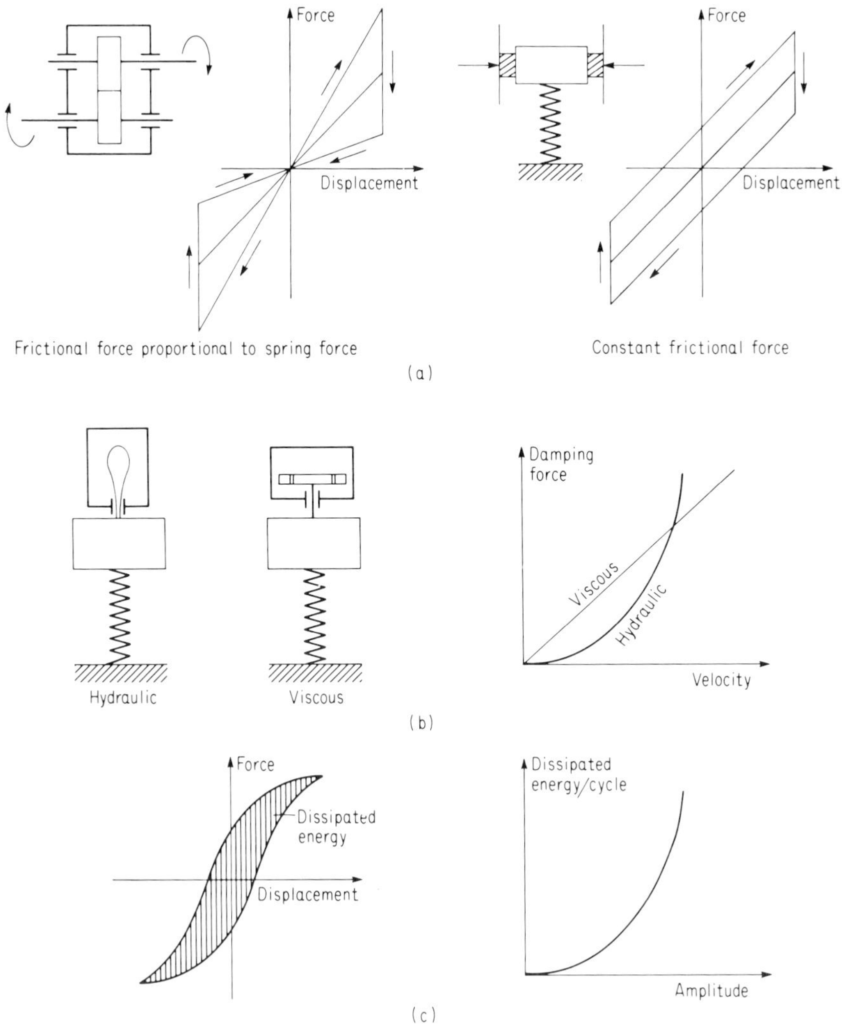

FIG. 3-18. (a) Coulomb damping. (b) Fluid damping. (c) Hysteresis damping.

frequency of oscillation. In the second, the frictional force is proportional to the spring force, as in the case of a geared system.

2. *Fluid friction* (Fig. 3-18b). This type of friction is a function of the velocity. When it is directly proportional to the velocity it is known as *viscous damping*. Otherwise it is proportional to higher powers of the speed. The viscous type frictional force is the simplest type to deal with mathematically.

3. *Hysteresis damping* (Fig. 3-18c). This is also known as *internal friction* and develops in a vibrating continuum as sliding occurs on the atomic level. In a mechanical system hysteresis is a deviation from Hooke's law and would not exist if the vibrating material were purely elastic.

If we now consider a single-degree-of-freedom system in a state of free vibration with friction, the equation of motion will be

$$m\ddot{x} = -f_1(x) - \operatorname{sgn}\dot{x}) \mid F_2 \mid$$

where $f_1(x)$ is a restoring force which is a function of the displacement, and F_2 is the frictional force whose direction always opposes the velocity.

Since the velocity changes direction twice in a vibration cycle, the duplicity in sign in the differential equation should be expected to cause complications in the mathematical solution. Only in the case of viscous forces will the sign change be automatically taken care of in the mathematical formulation.

3-5. THE DISTURBANCE

It was mentioned previously that a system satisfying all requirements of vibratory motion will remain at its position of stable equilibrium until it is subjected to a disturbance of some form or another. The magnitude, frequency, and mode of vibration will depend on this excitation.

Sources of disturbances in a mechanical system are numerous but can be generally placed in one of the following classifications.

1. Fluctuation of power input or output in a mechanical system, sometimes called *force excitation*. The motor-driven reciprocating compressor shown in Fig. 3-19a illustrates this type.

2. Motion of platform or foundation on which the system rests, denoted *foundation excitation*. (Fig. 3-19b)

3. Existence of unbalanced rotating parts, called *unbalanced excitation*. (Fig. 3-19c)

4. Fluctuation of system characteristics during operation as in the case of gearing systems where errors in manufacturing and change of mesh stiffness are a source of excitation. (Fig. 3-19d)

Another classification of the disturbances can be made as follows.

1. Transient
    - (a) Due to a sudden or instantaneous change in equilibrium conditions, equivalent to a set of initial conditions resulting from a sudden change of position or velocity.
    - (b) A nonperiodic disturbance that occurs during a finite time, after which it ceases to exist. When the system is subjected to this type of disturbance, it will continue to vibrate after the

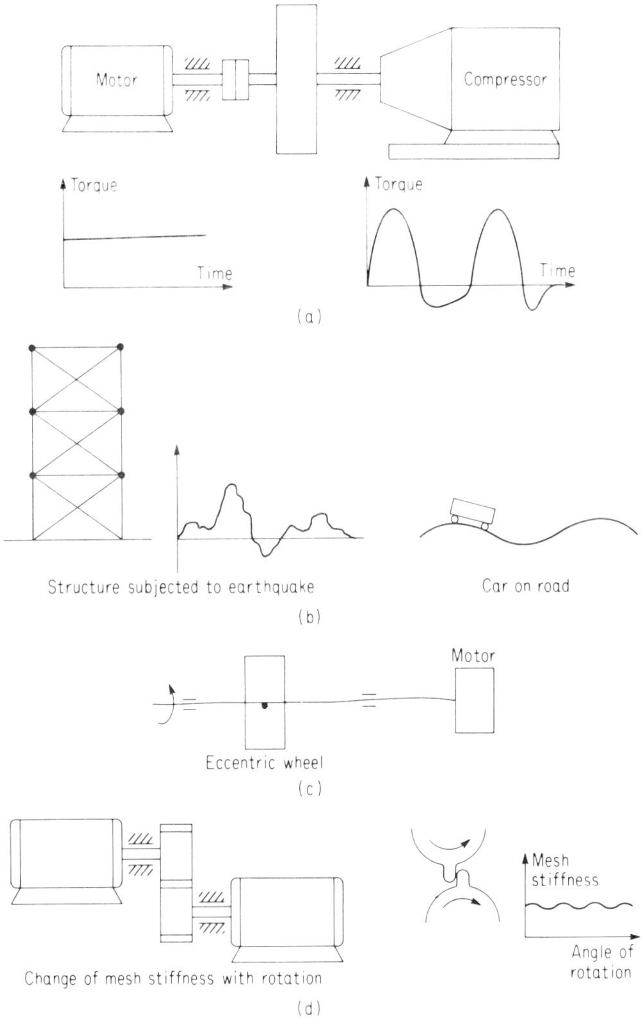

FIG. 3-19

disturbance has ceased and will be in a state of free vibration. With the existence of friction this free vibration will decay and eventually disappear.

2. Periodic
   Due to a disturbance which repeats itself in intervals of time called the *period of excitation.* When a system is subjected to this type, it will oscillate in a periodic fashion and the frequency of its oscillation will be dependent on the frequency of the excitation. This is usually known as *forced vibration.*

3. Random
This is an excitation characterized by a continued disturbance which, unlike the previous types, is unpredictable in an exact or precise manner. In this type, as would be expected, the motion of the system will be in a state of random vibration.

3-6. MATHEMATICAL MODELS FOR VIBRATORY SYSTEMS

An important first step in the analysis of vibratory systems is the transfer of the physical system into a mathematical model. This model may differ from one analyst to another, since it is usually based on simplifying assumptions which to a great extent are the result of the analyst's experience. The common rule in these models is that they should be simple enough to lend themselves to mathematical analysis, yet realistic enough to be representative of the behavior of the actual systems. The following examples illustrate the process of expressing physical situations into mathematical equations.

Example 7

Write the equations of motion of a car moving on tracks when it hits a buffer consisting of a spring and a dashpot as shown in Fig. 3-20a. In this case we may assume that the stand supporting the spring and dashpot is rigid. We may also assume that the spring has a constant stiffness k and that the dashpot provides essentially viscous damping. Neglecting air resistance as well as the spring hysteresis and the friction at the wheels, the system can be identified as shown in Fig. 3-20b. The illustration represents a mass m moving with a velocity v coming in contact with a spring with stiffness k and a viscous dashpot with a damping coefficient c. If we measure the displacement and time from the point of impact, the equations of motion will be

$$m\ddot{x} + kx + c\dot{x} = 0 \qquad \text{for } x > 0$$

$$m\ddot{x} = 0 \qquad \text{for } x < 0$$

and the initial conditions ($t = 0$) are

$$x = 0 \quad \text{and} \quad \dot{x} = v$$

If we choose in the same example to neglect the mass of the stand but consider its deflection as well as the friction on the car wheels, the system will be represented as in Fig. 3-20c. The wheel friction is assumed to be of the Coulomb type with a constant magnitude F. The spring representing the deflection of the stand is considered to have a constant stiffness k_s.

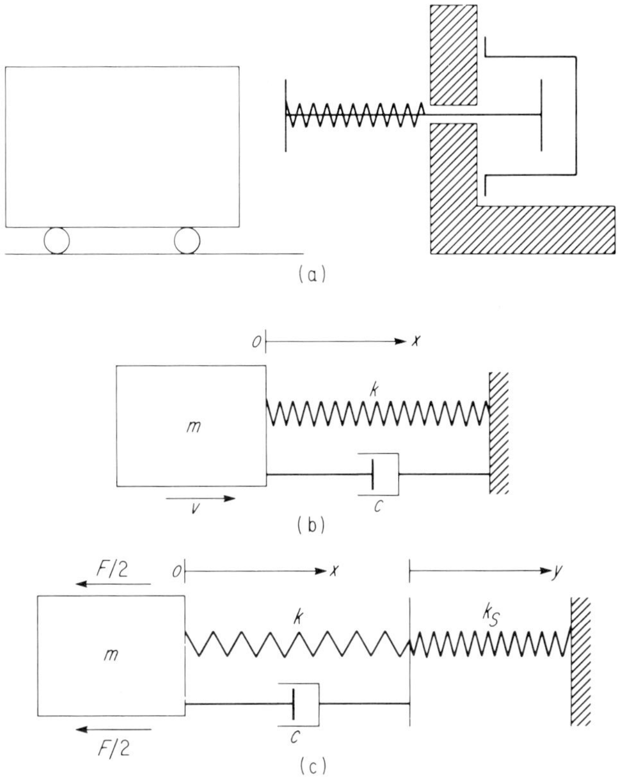

FIG. 3-20

The equations of motion in this case are therefore, for $x > 0$,

$$m\ddot{x} = -k(x - y) - c(\dot{x} - \dot{y}) - (\operatorname{sgn} \dot{x})F$$
$$k_s y = +k(x - y) + c(\dot{x} - \dot{y})$$

and for $x < 0$,

$$m\ddot{x} = -F$$

The initial conditions are

$$\text{at} \quad t = 0, \quad x = 0, \quad \dot{x} = v, \quad y = 0, \quad \dot{y} = 0$$

Example 8

A motor with unbalanced rotor is mounted on elastic pads as shown in Fig. 3-21a. Write the equations of motion.

Solution: This system is equivalent to an elastically supported rigid body with six degrees of freedom. If we assume that the mass of the rotor is m and that the center of mass c is at a distance e from the axis of rotation, the exciting force due to the unbalance will act at the center of

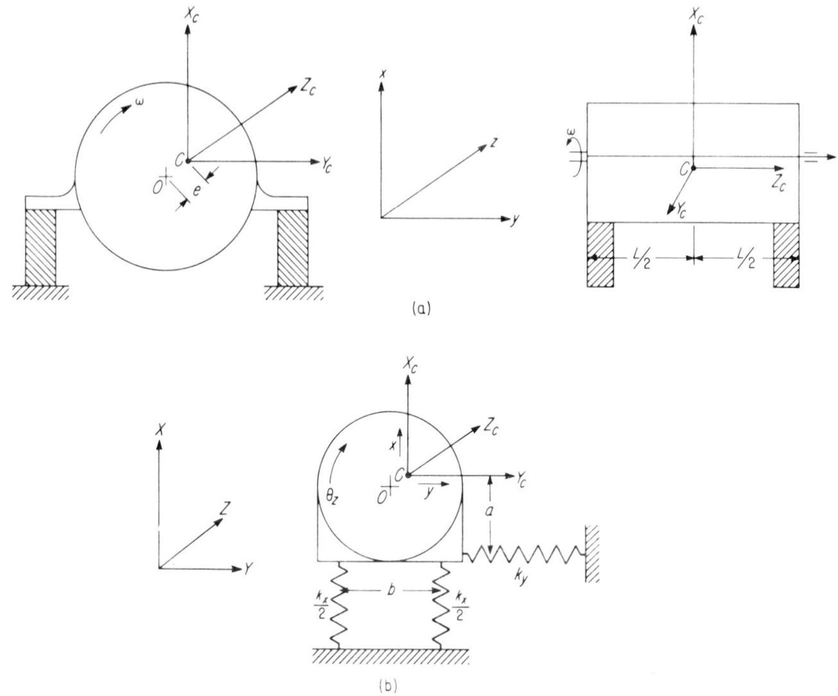

FIG. 3-21

mass and will have the components

$$m\omega^2 e \sin \omega t \text{ in the } X\text{-direction}$$

and

$$m\omega^2 e \cos \omega t \text{ in the } Y\text{-direction}$$

where ω is the angular velocity of the rotor and θ_z defines the angular position of $0c$ with respect to the Y-direction. Since the excitation has no component in the Z-direction there will be no translation of the center of mass in that direction. Also, due to the symmetry about the X_c-Y_c plane, there will be no rotation of the rigid body about the X- or Y-axes. The system will therefore be as shown in Fig. 3-21b and will have three degrees of freedom represented by the displacements x, y of the center of mass c and the angular displacement θ_z about the Z-axis. The resistance of the elastic pads can be represented by springs with elastic constants k_x, k_y in the X- and Y-directions respectively. By application of Eqs. 3-4 we obtain the equations of motion as

$$m\ddot{x} + k_x x = m\omega^2 e \sin \omega t$$

$$m\ddot{y} + k_y y = m\omega^2 e \cos \omega t$$

$$I_{z_c z_c}\ddot{\theta}_z = -k_y(y + a\theta_z)a - k_x\left(\frac{b}{2}\theta_z\right)\frac{b}{2}$$

where $I_{z_c z_c}$ is the moment of inertia about the axis z_c and a is the normal distance between the center of mass and the line of action of the equivalent spring k_y.

Example 9

The analysis of a reciprocating-engine drive is a common design problem. We shall consider here as an example the case of an electric generator driven by a four-cylinder engine, Fig. 3-22a. In this case the

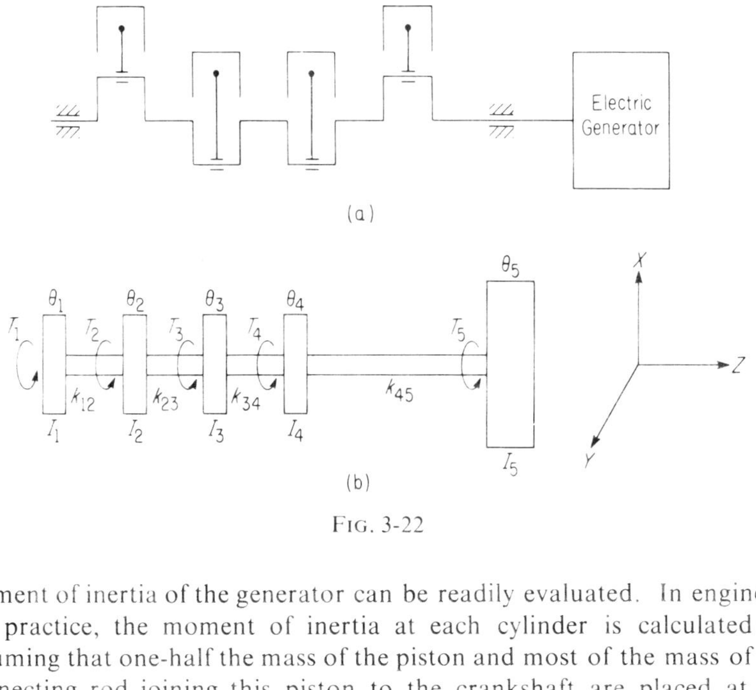

FIG. 3-22

moment of inertia of the generator can be readily evaluated. In engineering practice, the moment of inertia at each cylinder is calculated by assuming that one-half the mass of the piston and most of the mass of the connecting rod joining this piston to the crankshaft are placed at the junction of the connecting rod and crank. An equivalent torsional spring can also be evaluated between the different masses by computing the total twist between them resulting from a unit torque. Procedures for detailed calculation of equivalent inertias and stiffness are given in Ref. 6.

Neglecting damping, the system can now be diagrammatically represented, as shown in Fig. 3-22b. The mathematical model in this case is that of elastically connected rigid bodies which are free to rotate only about the Z-axis. The equations of motion are therefore

$$I_1\ddot{\theta}_1 = T_1 - k_{12}(\theta_1 - \theta_2)$$

$$I_2\ddot{\theta}_2 = T_2 - k_{12}(\theta_2 - \theta_1) - k_{23}(\theta_2 - \theta_3)$$
$$I_3\ddot{\theta}_3 = T_3 - k_{23}(\theta_3 - \theta_2) - k_{34}(\theta_3 - \theta_4)$$
$$I_4\ddot{\theta}_4 = T_4 - k_{34}(\theta_4 - \theta_3) - k_{45}(\theta_4 - \theta_5)$$
$$I_5\ddot{\theta}_5 = T_5 - k_{45}(\theta_5 - \theta_4)$$

where I_1, I_2, I_3, I_4, I_5 = moments of inertia

$k_{12}, k_{23}, k_{34}, k_{45}$ = stiffness values of the connecting torsional springs

$\theta_1, \theta_2, \theta_3, \theta_4, \theta_5$ = angular displacements about the Z-axis for the five masses

T_1, T_2, T_3, T_4, T_5 = external torques applied on the different masses

Example 10

In this example a mathematical model is required for a motor driving two rotary compressors by means of a gear unit as illustrated in Fig. 3-23a. The gear ratios are 3:1 and 2:1 as shown. The moments of

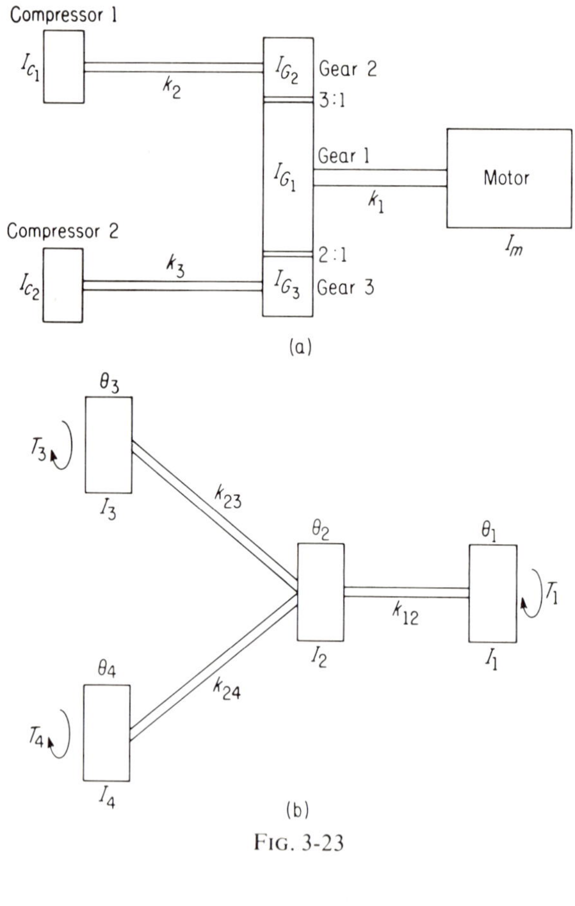

FIG. 3-23

inertia of the motor, the three gear wheels and the two compressors respectively are $I_m, I_{G_1}, I_{G_2}, I_{G_3}, I_{c_1}$, and I_{c_2}. The stiffness of the shafts are k_1, k_2, and k_3 as shown in the figure.

Neglecting the flexibility of the gear mesh as well as damping, the system can be diagrammatically represented as shown in Fig. 3-23b. Because of the different speeds of a geared system it is advisable to choose one shaft and refer all the angular displacements, inertias, and stiffnesses to that shaft. In the problem under consideration, it is easier to choose the motor shaft as reference. Now if the amplitude of oscillation of a disk on one of the geared shafts is θ, its equivalent θ_e referred to the motor shaft will be θ/n where n is the gear ratio (or the ratio of the speed of this shaft and the motor shaft). Similarly, if the torque in the geared shaft is T, the equivalent torque at the reference shaft T_e is T/n.

Because the kinetic energy as well as the potential energy should be the same in the shaft and its equivalent, then

$$\text{K.E.} = \frac{1}{2} I\dot{\theta}^2 = \frac{1}{2} I_e\dot{\theta}_e^2$$

from which

$$\frac{I_e}{I} = \frac{\dot{\theta}^2}{\dot{\theta}_e^2} = \frac{\theta^2}{\theta_e^2} = n^2$$

Also,

$$\text{P.E.} = \frac{1}{2} k\theta^2 = \frac{1}{2} k_e\theta_e^2$$

from which

$$\frac{k_e}{k} = \frac{\theta^2}{\theta_e^2} = n^2$$

The relationships expressed by these equations can be stated as: *The equivalent inertia as well as stiffness of a geared shaft with respect to a reference shaft is obtained by multiplying the actual inertias and stiffness by the square of the ratio between the speed of the actual shaft and that of the reference shaft.*

If we choose the motor shaft as reference in this problem, the different inertias and stiffnesses will be as shown in Fig. 3-23b, where

$$\begin{aligned}
I_1 &= I_m \\
I_2 &= I_{G_1} + (3)^2 I_{G_2} + (2)^2 I_{G_3} \\
k_{12} &= k_1 \\
k_{23} &= (3)^2 k_2 \\
k_{24} &= (2)^2 k_3 \\
I_3 &= (3)^2 I_{c_1} \\
I_4 &= (2)^2 I_{c_2}
\end{aligned}$$

Accordingly, the equations of motion for the undamped system are

$$\begin{aligned}
I_1\ddot{\theta}_1 &= T_1 - k_{12}(\theta_1 - \theta_2) \\
I_2\ddot{\theta}_2 &= -k_{12}(\theta_2 - \theta_1) - k_{23}(\theta_2 - \theta_3) - k_{24}(\theta_2 - \theta_4) \\
I_3\ddot{\theta}_3 &= T_3 - k_{23}(\theta_3 - \theta_2) \\
I_4\ddot{\theta}_4 &= T_4 - k_{24}(\theta_4 - \theta_2)
\end{aligned}$$

where $\theta_1, \theta_2, \theta_3$, and θ_4 are the different angular displacements referred to the motor shafts.

Example 11

The vibration of the knee system of a knee-type horizontal milling machine, Fig. 3-24a, can be analyzed by reducing the system to three elastically connected rigid bodies. The masses, Fig. 3-24b, are guided so that only displacements along the Z-axis and angular vibration about the

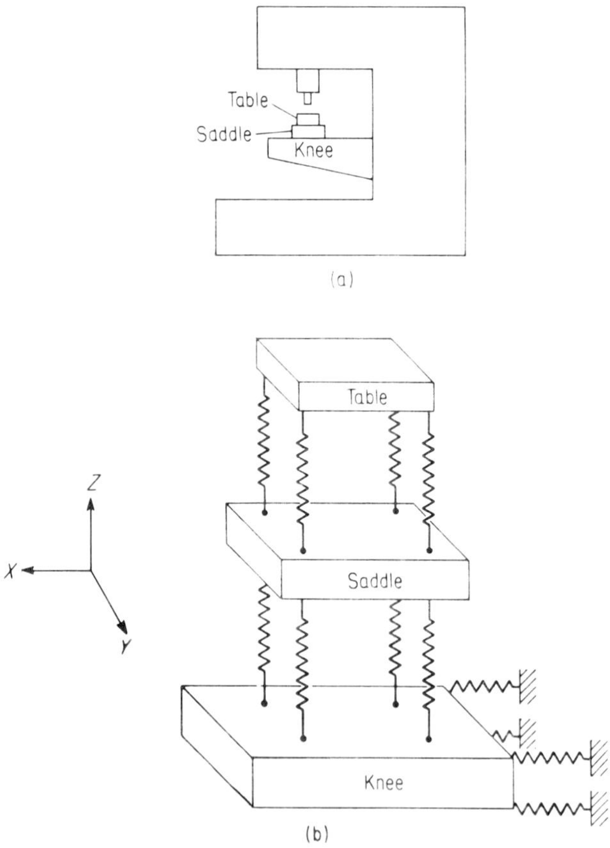

FIG. 3-24

Y- and X-axes are possible. Each mass will therefore have three degrees of freedom. If we assign the proper stiffness value to each spring, the equations of motion can be readily written by applying Eqs. 3-4 for each rigid body.

Example 12

This example deals with the dynamics of an aircraft in flight, Fig. 3-25. In order to analyze the motion of the craft as a whole, it can be idealized

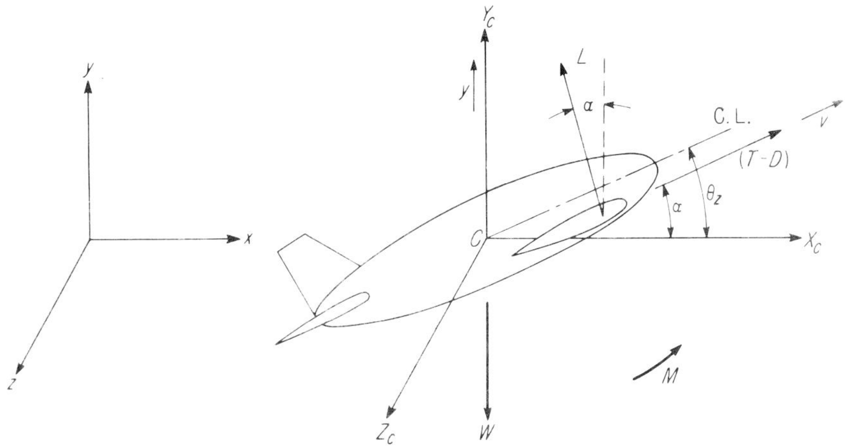

FIG. 3-25

as a rigid body with six degrees of freedom. The degrees of freedom can be reduced to three by assuming symmetry about the central plane X_c-Y_c. The coordinates of the motion in this case are the x, y displacement of the center of mass and the angular rotation θ_z about the Z-axis.

Using the notation

W = weight of plane
T = thrust
D = aerodynamic drag
L = aerodynamic lift
v = velocity of plane
α = angle between velocity vector and X-direction
M = moment of all forces about Z_c-axis
$I_{z_c z_c}$ = moment of inertia about Z_c-axis.

The equations of motion are therefore

$$\frac{W}{g}\ddot{x} = (T - D)\cos\alpha - L\sin\alpha$$

$$\frac{W}{g}\ddot{y} = L\cos\alpha + (T - D)\sin\alpha - W$$

$$I_{z_c z_c}\ddot{\theta}_z = M$$

Example 13

The vibration of a bridge due to a traveling load is studied here by idealizing the system as shown in Fig. 3-26. In this model, the mass of the

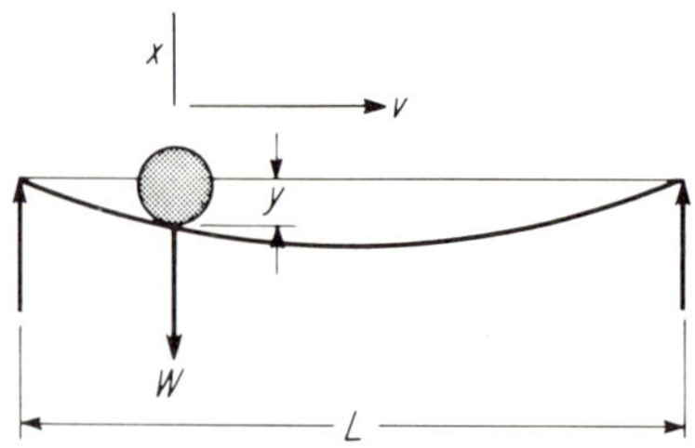

FIG. 3-26

traveling load is assumed large with respect to that of the bridge. The bridge is considered as a simple supported beam. The deflection y under the load is calculated from strength of materials as

$$y = \frac{Px^2(L - x)^2}{3LEI}$$

where L = length of beam

I = moment of inertia of cross section

E = modulus of elasticity of material

P = total load on beam

$$= W - \frac{W}{g}\frac{d^2y}{dt^2}$$

Since

$$\frac{dy}{dt} = \frac{dy}{dx}\frac{dx}{dt}$$

and

$$\frac{dx}{dt} = v = \text{velocity of the traveling load}$$

therefore

$$\frac{dy}{dt} = \frac{dy}{dx}v$$

and by similar differentiation we find, when v is constant, that

$$\frac{d^2y}{dt^2} = v^2\frac{d^2y}{dx^2}$$

By substitution in the deflection equation we can represent the vibration of the bridge in this case by the equation

$$y = W\left[1 - \frac{v^2}{g}\frac{d^2y}{dx^2}\;\frac{x^2(L-x)^2}{3LEI}\right]$$

3-7. ANALOGIES FOR DYNAMIC SYSTEMS

The Electrical Analogy

There is a definite analogy between the differential equations of motion of vibratory mechanical systems and those of electrical circuits. Mechanical analogies of electrical systems give the electrical engineer a useful visual aid for observing the behavior of electrical circuits. On the other hand, electrical analogies of mechanical systems provide a powerful tool of analysis for these systems, especially when they are considerably complex. Electrical circuits are more easily constructed than mechanical models, and the readings are more readily taken. Once the analogous electrical circuit is constructed, the behavior of the mechanical system can be readily studied by measuring the currents and voltages in the analog.

There are two common methods of simulation: (1) The classical mass-inductance analogy or the impedance method; (2) the mass-capacitance analogy or the mobility method. The basic components of a mechanical system are the inertial element or mass, the restoring element or spring, and the dissipating element or damper. Considering the damper to be of the viscous type, the simplest form of a mechanical system will be as shown in Fig. 3-27. When a disturbing force F is applied, it will be resisted by the inertia of the mass as well as the spring and damper forces. These resistive forces are given respectively as

$$\begin{aligned} F_m &= \text{inertial resistance} &&= m\ddot{x} \\ F_s &= \text{spring force} &&= kx \\ F_d &= \text{viscous damping force} &&= c\dot{x} \end{aligned}$$

where x is the displacement, m, k are the mass and spring constant respectively, and c is the proportionality constant of the viscous damper, generally known as the *damping coefficient*. The differential equation

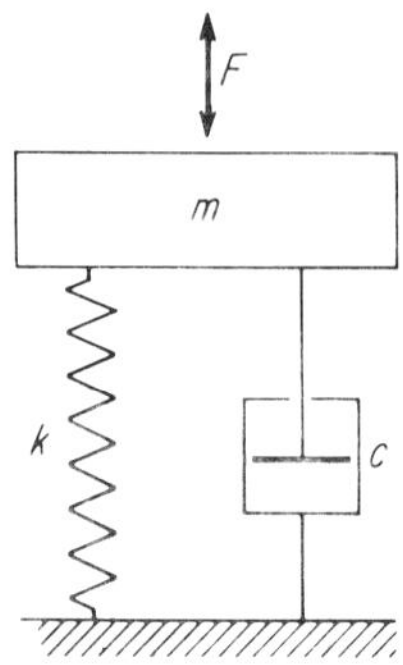

FIG. 3-27

describing the system will therefore be

$$F = m\ddot{x} + c\dot{x} + kx \tag{3-25}$$

which represents the application of Newton's second law to the system under consideration.

On the other hand, the basic electrical elements are the inductor L, the capacitor C and the resistor R. We know from elementary physics that these elements provide a resistance to the flow of electric current which is conveniently measured by the voltage drop E_L, E_C, E_R across these elements respectively. These are analytically given by

$$E_L = L\frac{di}{dt} \tag{3-26a}$$

$$E_C = \frac{1}{C}\int i\,dt \tag{3-26b}$$

$$E_R = iR \tag{3-26c}$$

where i is the current in each element.

If these elements are connected in series as shown in Fig. 3-28 and subjected to a voltage E, therefore by application of Kirchhoff's law we get

$$E - E_L - E_C - E_R = 0$$

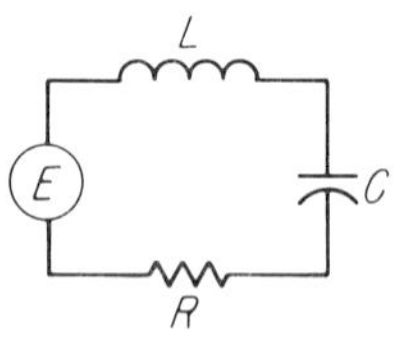

FIG. 3-28

and by substitution from Eqs. 3-26 we have

$$E = L\frac{di}{dt} + iR + \frac{1}{C}\int i\,dt \tag{3-27}$$

Since the electric charge q is the time integral of the current i,

$$q = \int i\,dt$$

or

$$i = \frac{dq}{dt}$$

Therefore Eq. 3-27 can be written as

$$E = L\frac{d^2q}{dt^2} + R\frac{dq}{dt} + \frac{1}{C}q \tag{3-28}$$

which is of the same form as Eq. 3-25 for the mechanical system. The analogy based on this equation requires that the applied voltage E corresponds to the applied force F, the electric charge q corresponds to the displacement x, and L, R, $1/C$ correspond to m, c, and k respectively. Other than this mathematical analogy, there is also a physical analogy based on the behavior of these elements. An inductor provides resistance to the flow of current in the same manner a mass resists motion. A resistor dissipates electric energy just as a viscous damper dissipates mechanical energy, and a capacitor stores energy in the same manner as a mechanical spring. This type of analogy is known as the *classical* or *impedance analogy*.

If these elements are connected in parallel as in Fig. 3-29, the voltage

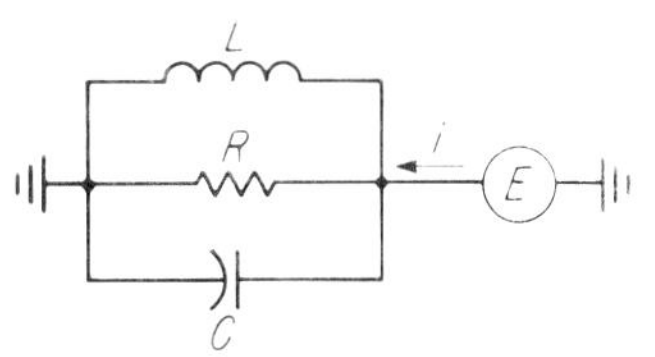

FIG. 3-29

drop across them will be the same but the current will be different. Assuming i_L, i_R, i_C to represent the current in the inductance, resistance, and capacitance respectively, Kirchhoff's law can then be expressed as

$$i = i_L + i_R + i_C$$

and from Eq. 3-26 we have

$$i = \frac{1}{L}\int E\,dt + \frac{E}{R} + C\frac{dE}{dt} \tag{3-29}$$

TABLE 3-1. ELECTRICAL ANALOGIES BY CLASSICAL AND MOBILITY METHODS

Rectilinear System	*Torsional System*	*Classical Analogy*	*Mobility Analogy*
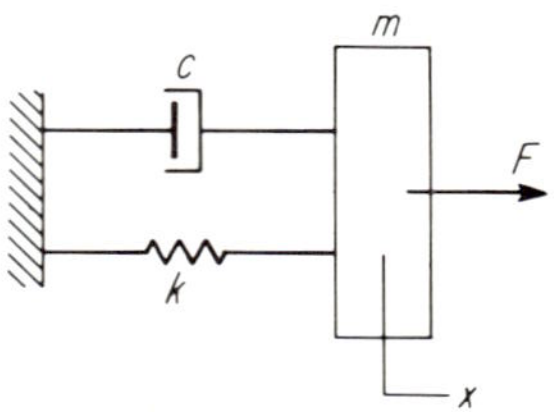	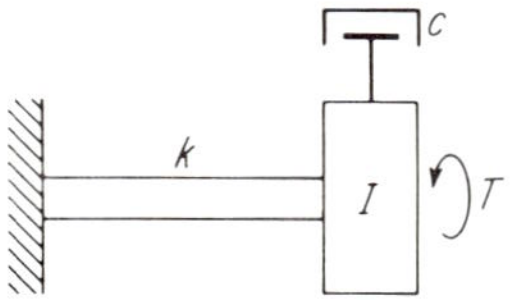	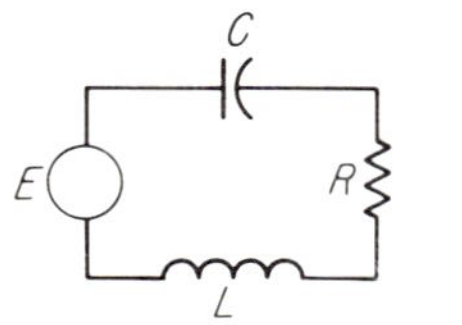	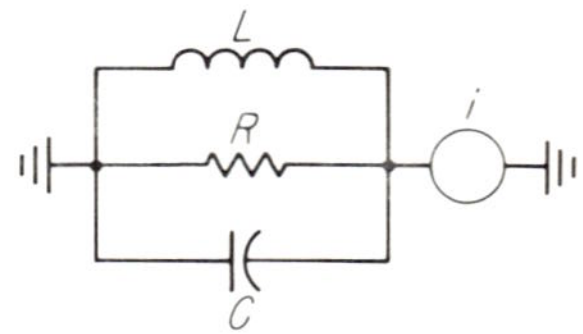
$m\ddot{x} + c\dot{x} + kx = F(t)$	$I\ddot{\theta} + c\dot{\theta} + k\theta = T(t)$	$L\frac{di}{dt} + Ri + \frac{1}{C}\int i\,dt = E(t)$	$C\frac{dE}{dt} + \frac{1}{R}E + \frac{1}{L}\int E\,dt = i(t)$
Force, F	Torque, T	Voltage, E	Current, i
Mass, m	Inertia, I	Inductance, L	Capacitance, C
Damping, c	Damping, c	Resistance, R	Conductance, $\frac{1}{R}$
Velocity, v	Angular velocity, ω	Current, i	Voltage, E
Compliance, $\frac{1}{k}$	Compliance, $\frac{1}{k}$	Capacitance, C	Inductance, L
Displacement, x	Angular displacement, θ	Charge, q	$\int E\,dt$

This equation is also analogous to Eq. 3-25. In this case the current will correspond to the force, the voltage to the velocity, and C, $1/R$, $1/L$, to the mass, the damping coefficient and the spring constant. This type of analogy is known as the *mobility analogy*. Table 3-1 shows the analogous quantities by both the impedance and mobility methods.

For complex systems, the classical analogy becomes rather difficult and the mobility method proves advantageous based as is on physical similarity of the systems rather than on the mathematical analogy of their differential equations. The mobility analogs have the same topology as

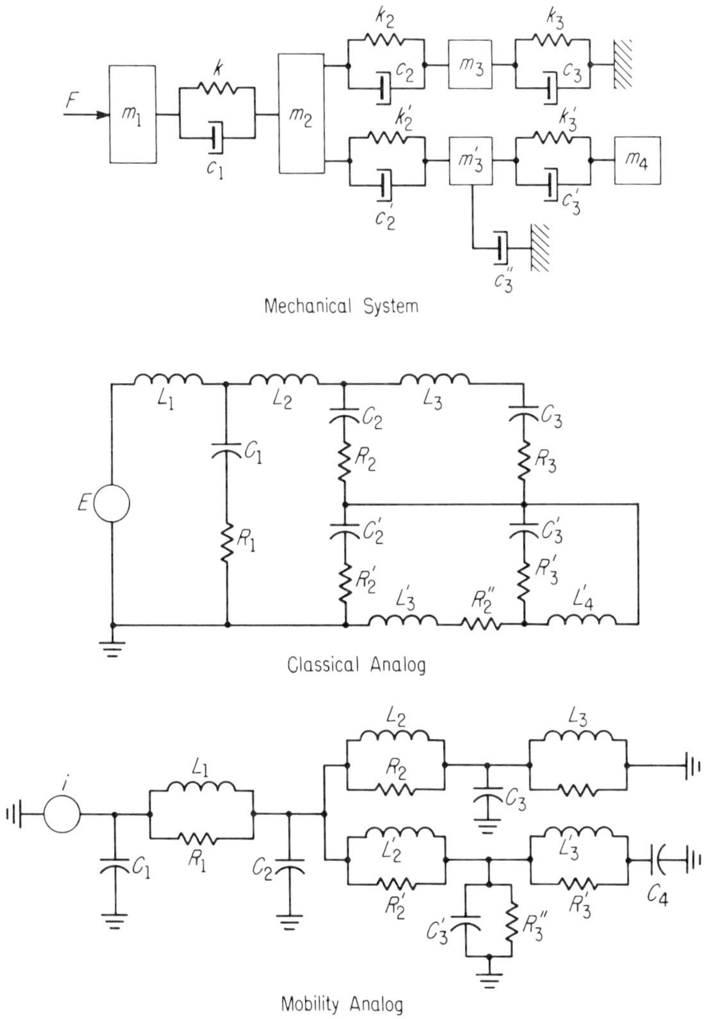

FIG. 3-30. Example of analogs.

the mechanical models, and the electrical circuits look identical to it. They are therefore simpler to construct, as can be seen in Figs. 3-30 and 3-31, which illustrate analogous circuits by both methods for multi-mass mechanical systems.

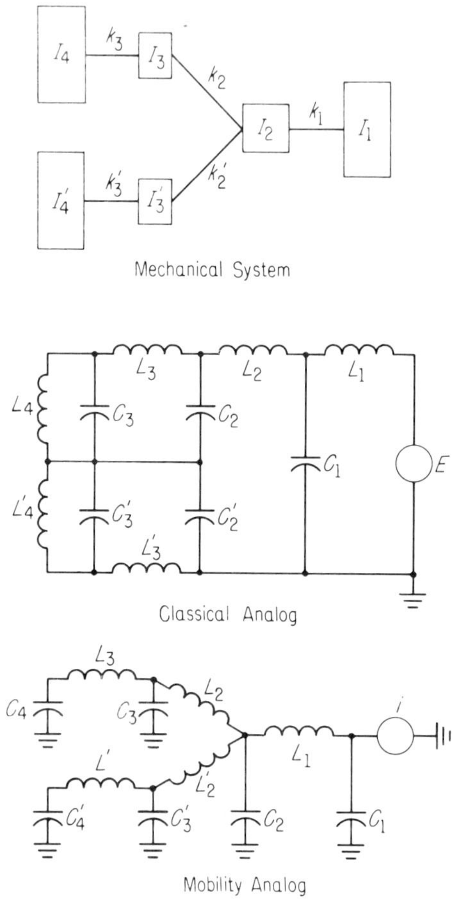

FIG. 3-31. Example of analogs.

Scale Factors

Whenever an analog is to be constructed, a decision must be made regarding the magnitudes of the electrical elements to be used for representing the mechanical system. The three fundamental quantities in a mechanical system are mass, displacement, and time. The scale factors chosen to represent these quantities are sufficient to obtain the scale factors for all other analogous quantities as they are dependent on these three factors. The following provides a tabulation of the scale factors for both methods.

A. Classical Analogy

The fundamental scale factors are

$$S_1 = \frac{L}{m} = \frac{\text{inductance}}{\text{mass}}$$

$$S_2 = \frac{q}{x} = \frac{\text{charge}}{\text{displacement}}$$

$$S_3 = \frac{t_e}{t_m} = \frac{\text{time (electrical circuit)}}{\text{time (mechanical circuit)}}$$

The other factors can therefore be given as

$$S_4 = \frac{i}{v} = \frac{\text{current}}{\text{velocity}} = \frac{S_2}{S_3}$$

$$S_5 = \frac{E}{F} = \frac{\text{voltage}}{\text{force}} = \frac{S_1 S_2}{(S_3)^2}$$

$$S_6 = \frac{R}{c} = \frac{\text{resistance}}{\text{damping coefficient}} = \frac{S_1}{S_3}$$

$$S_7 = \frac{C}{1/k} = \frac{\text{capacitance}}{\text{compliance}} = \frac{S_3^2}{S_1}$$

$$S_8 = \frac{\omega_e}{\omega_m} = \frac{\text{electrical frequency}}{\text{mechanical frequency}} = \frac{1}{S_3}$$

B. Mobility Method

The three fundamental factors are

$$S_1 = \frac{C}{m} = \frac{\text{capacitance}}{\text{mass}}$$

$$S_2 = \frac{\int E\,dt}{x} = \frac{\text{voltage integral}}{\text{displacement}}$$

$$S_3 = \frac{t_e}{t_m} = \frac{\text{time (electrical circuit)}}{\text{time (mechanical circuit)}}$$

The other factors can therefore be derived as

$$S_4 = \frac{i}{F} = \frac{\text{current}}{\text{force}} = \frac{S_1 S_2}{S_3^2}$$

$$S_5 = \frac{1/R}{c} = \frac{\text{conductance}}{\text{damping coefficient}} = \frac{S_1}{S_3}$$

$$S_6 = \frac{E}{v} = \frac{\text{voltage}}{\text{velocity}} = \frac{S_2}{S_3}$$

$$S_7 = \frac{L}{1/k} = \frac{\text{inductance}}{\text{compliance}} = \frac{S_3^2}{S_1}$$

$$S_8 = \frac{\omega_e}{\omega_m} = \frac{\text{electrical frequency}}{\text{mechanical frequency}} = \frac{1}{S_3}$$

The reader may want to prove these factors as an exercise.

The Acoustical Analogy

There is also a definite analogy between acoustical systems and mechanical or electrical systems. This analogy makes it possible to solve acoustical problems by means of their mechanical or electrical analogies.

To illustrate this similarity, we now study the Helmholtz resonator shown in Fig. 3-32. It consists of a container with a small opening. If a

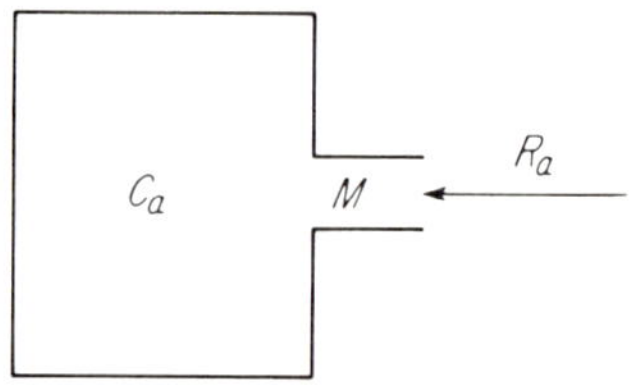

FIG. 3-32

certain volume of air Q is displaced in the opening and enters the container, it will cause the pressure in the container to increase by an amount P_1. The container volume will therefore provide the restoring element. The volume displacement Q corresponds to the spring displacement x in a spring-mass system and the pressure P_1 represents the restoring force. The ratio $C_a = \dfrac{Q}{P_1}$ is called the *acoustical capacitance* of the resonator, and is computed from the volume of the container and the gas properties.

The air in the opening provides the inertial element. If the cross-sectional area of the opening is A, the volume displacement due to a displacement x of the air in the opening can be written as $Q = Ax$. Thus $\dot{Q} = A\dot{x}$ is called the *volume current*, and $\ddot{Q} = A\ddot{x}$ is called the *volume acceleration.*

If a pressure P_2 is required to give a volume acceleration $\ddot{Q}$, the acoustical inertance M (which is analogous to the mass in a mechanical system) is given by

$$M = \frac{P_2}{\ddot{Q}}$$

and can be calculated from the dimensions of the opening and the density of the air.

Acoustical energy may be dissipated because of the internal friction of the air (or due to its viscosity). It can also be dissipated due to the radiation of sound into the external medium. The acoustical resistance R_a is therefore defined as

$$R_a = \frac{P_3}{\dot{Q}}$$

where P_3 is the pressure necessary to drive the volume current Q through the dissipative element.

Sound waves approaching the opening of the resonator provide the excitation in the acoustical system. If, for example, these sound waves apply an alternating pressure difference $P_0 \sin \omega t$, the equation representing the acoustical system can be written as

$$\begin{aligned} P_0 \sin \omega t &= P_2 + P_3 + P_1 \\ &= M\ddot{Q} + R_a\dot{Q} + \frac{1}{C_a}Q \end{aligned} \tag{3-30}$$

which is identical with the corresponding equation for mechanical or electrical systems. The analogous quantities in mechanical and acoustical systems are shown in Table 3-2.

TABLE 3-2

Mechanical		*Acoustical*	
Force	F	Pressure	P
Displacement	x	Volume displacement	Q
Velocity	$\dot{x}$	Volume current	$\dot{Q}$
Compliance	$1/k$	Capacitance	C_a
Mass	m	Inertance	M
Damping	c	Resistance	R_a

3-8. SUMMARY

In this chapter, mechanical systems were classified according to their dynamical categories and the mathematical representation of their equations of motion was discussed. The electrical and acoustical analogs for mechanical systems were also illustrated.

The reader is now prepared to study the behavior of these systems under different types of excitation. In the following chapters we shall investigate some of the techniques of solution, their limitations, and the results they render.

PROBLEMS

3-1. Write the equation of motion for the particle shown in the sketch.

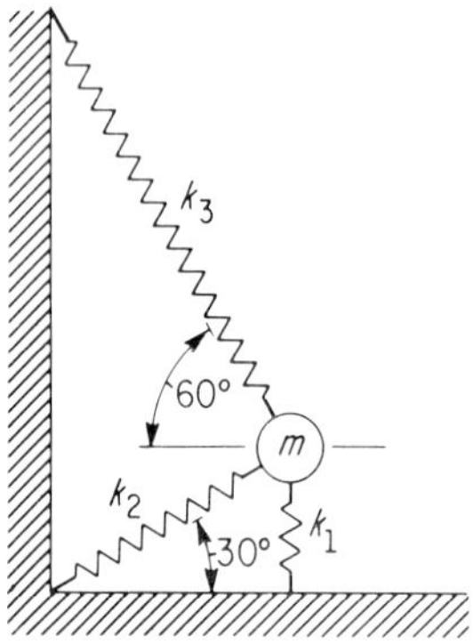

Prob. 3-1

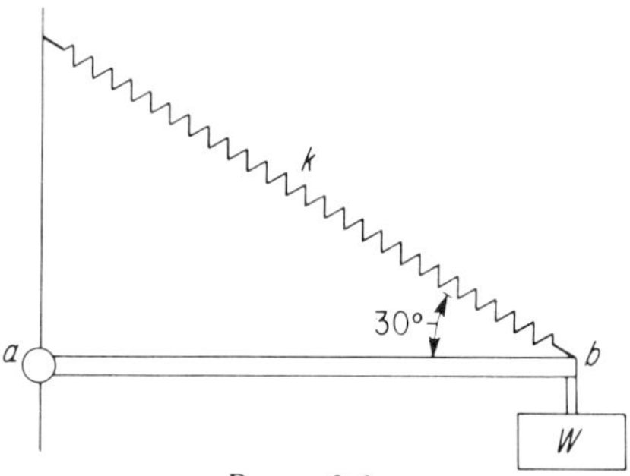

Prob. 3-2

3-2. A bracket ab is hinged at a and elastically supported by a spring k at b. If a weight W is rigidly connected to the bracket at b, write the equations of motion. Neglect the weight of the bracket.

3-3. A mass m is supported by the spring arrangement shown in the figure. Write the equations of motion.

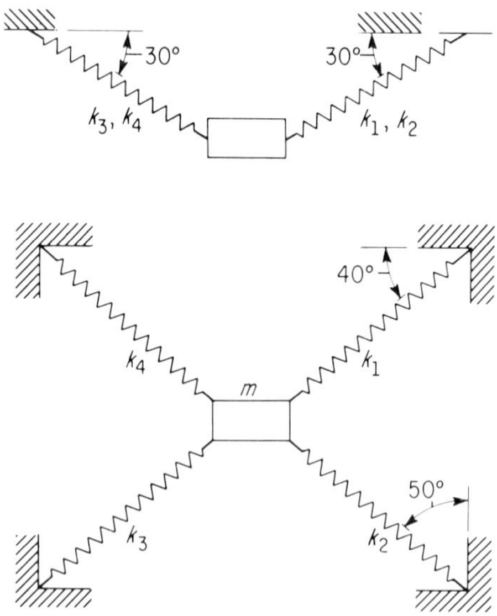

Prob. 3-3

3-4. Write the equations of motion for the two-mass system shown in the figure.

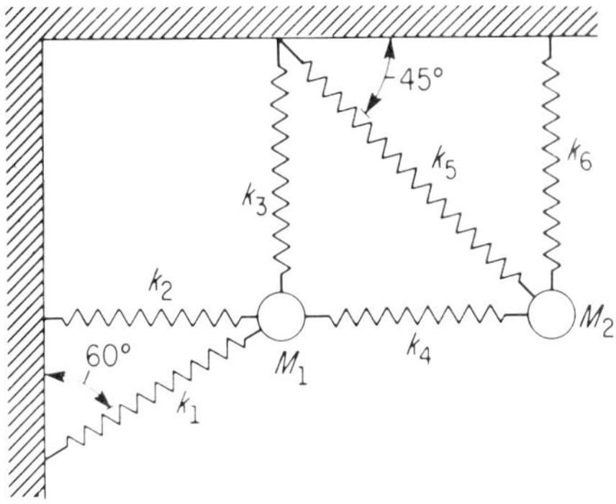

PROB. 3-4

3-5. A structure is idealized as shown in the sketch. Write the equations necessary to study its vibration.

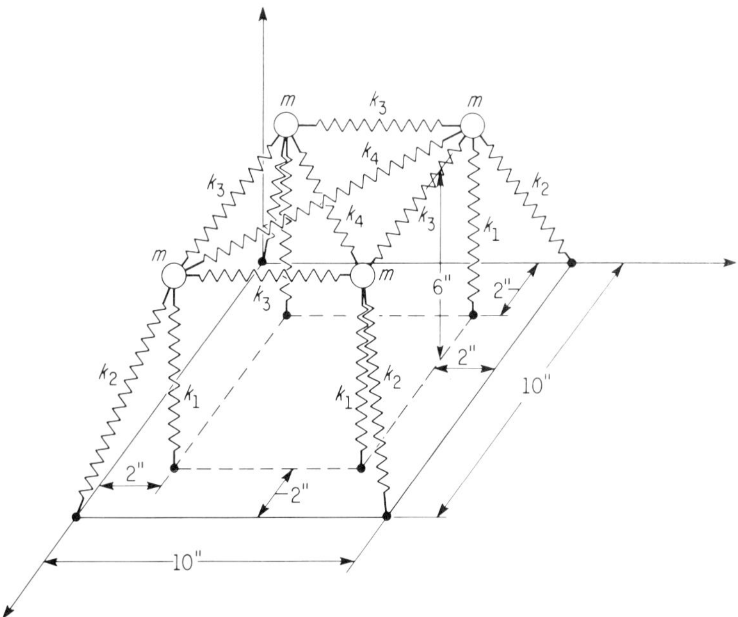

PROB. 3-5

3-6. Write the equations of motion for the system of elastically connected masses shown.

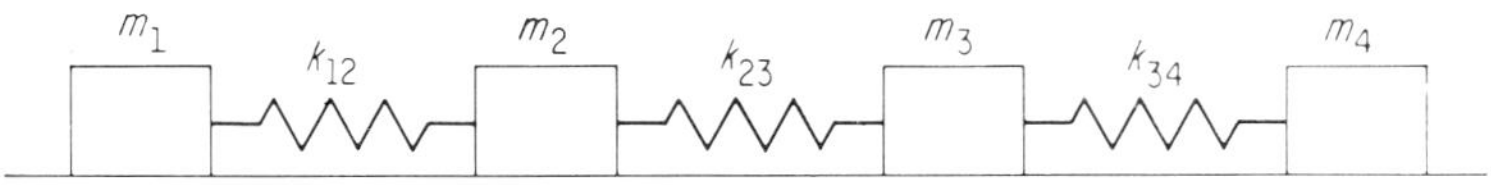

PROB. 3-6

3-7. Describe mathematically the motion of the double pendulum shown when subjected to a small disturbance.

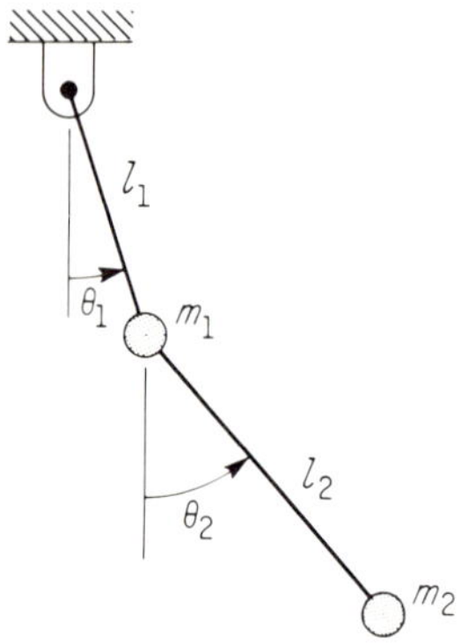

PROB. 3-7

3-8. A mechanical system is idealized as shown. Write the equations of motion for the system.

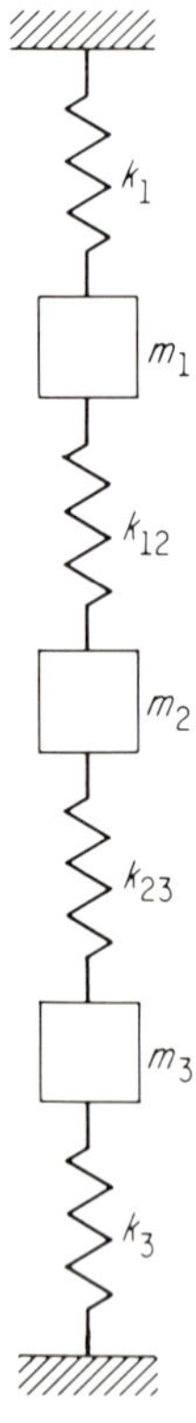

PROB. 3-8

3-9. If the moment of inertia of the two end masses of a four-mass torsional system are considerably larger than the middle masses, the system can be represented as shown in the sketch. Write the equations of motion.

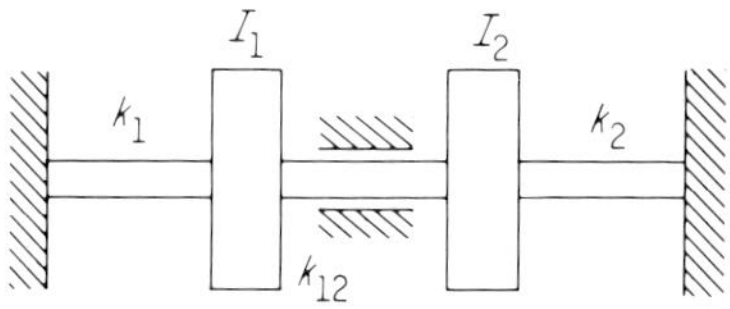

PROB. 3-9

3-10. The platform shown is supported by two springs k_1 and k_2. If the mass per unit volume of its material is e, write the equations necessary to study its vibration.

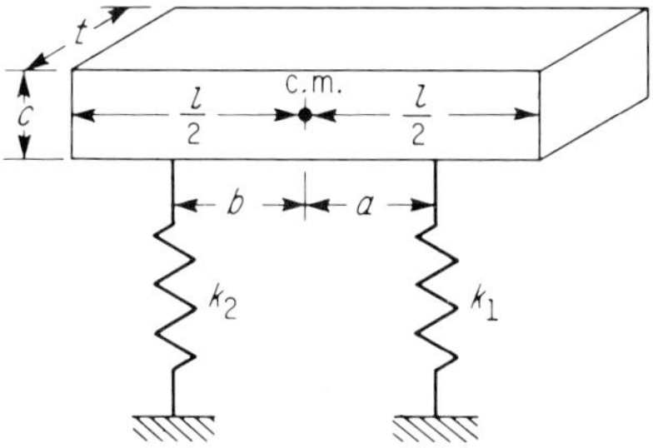

PROB. 3-10

3-11. Write the equation necessary to study the motion of the angle shown in the sketch, knowing that the weight of its material is w per unit volume.

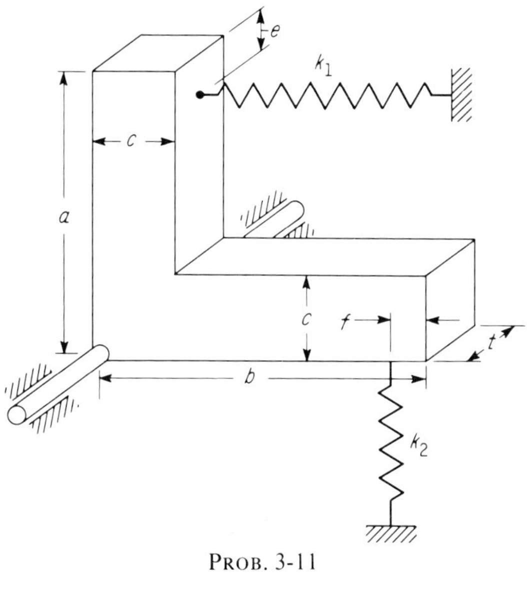

PROB. 3-11

3-12. The sketch shows the same angle as in Prob. 1-13 supported by a ball, a socket and three springs. Write its equations of motion.

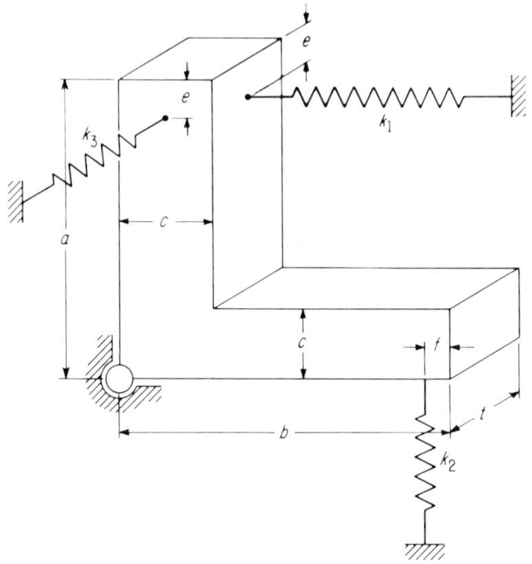

PROB. 3-12

3-13. A block having a mass m and a moment of inertia $I_{z_c z_c}$ about the z_c axis at the center of mass is supported by four springs as shown in the figure. Write the equation of motion under a small disturbance.

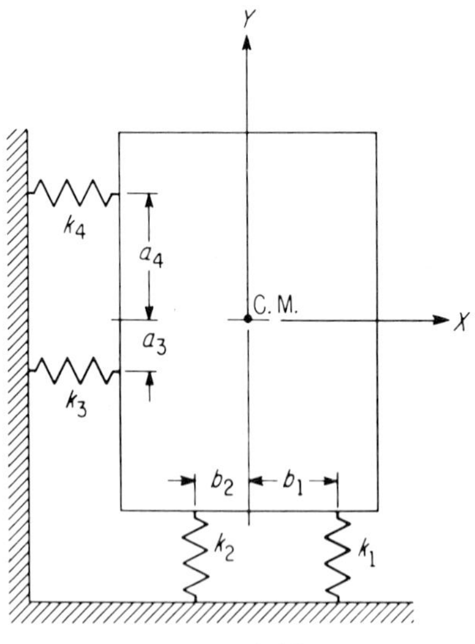

PROB. 3-13

3-14. A structure subjected to a sinusoidal force is diagrammatically represented in the figure. Write the mathematical model for the system.

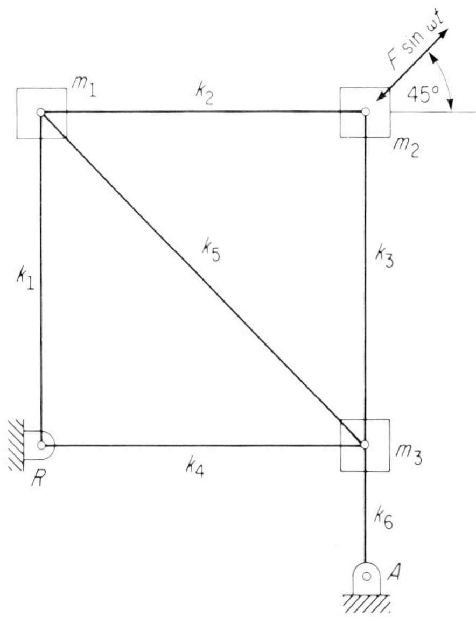

PROB. 3-14

3-15. A shake table consists of a mass m supported on four leaf springs as shown. The excitation is provided by two eccentric weight $m'/2$ rotating at N rpm in opposite directions as shown in the figure. Write the equations for the table vibration.

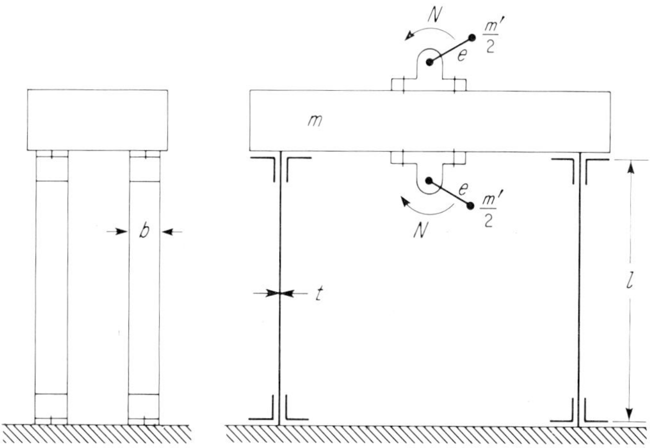

PROB. 3-15

3-16. A mass m is supported on three elastic members as shown. The axial stiffness of these members are k_1, k_2, k_3 respectively. If a force $F \sin \omega t$ acts on the mass in a line with directional cosines $l = m = 0.5$, write the equations of motion.

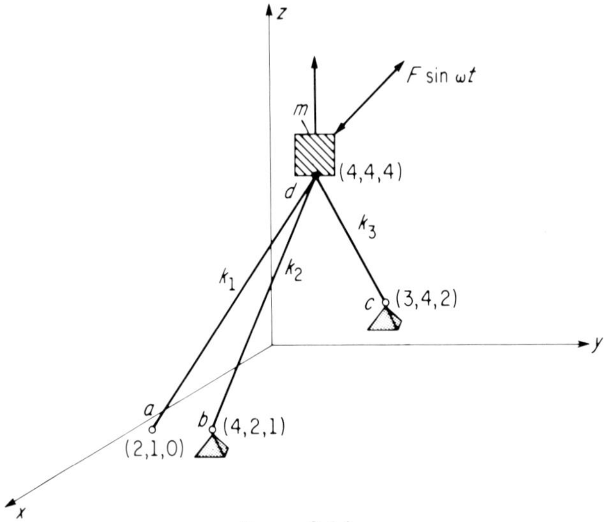

PROB. 3-16

3-17. A structure is represented by a mass m supported on six elastic members as shown. Write the mathematical model for the structure.

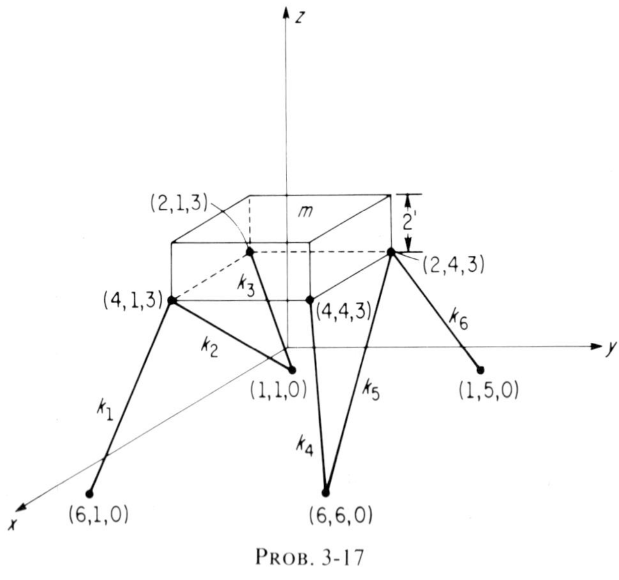

PROB. 3-17

3-18. The uniform beam shown in the figure weighs 200 lb and is supported by two strings *ST* and *PQ* and a universal joint support at *A*. (This is equivalent to a ball and socket with constraint against rotation about the *x*-axis.) If the elastic constant for both strings is 1000 lb/in., write the equations representing the response of the rod to an oscillating force 100 sin 5*t* in the *Z*-direction acting on the rod at *B*.

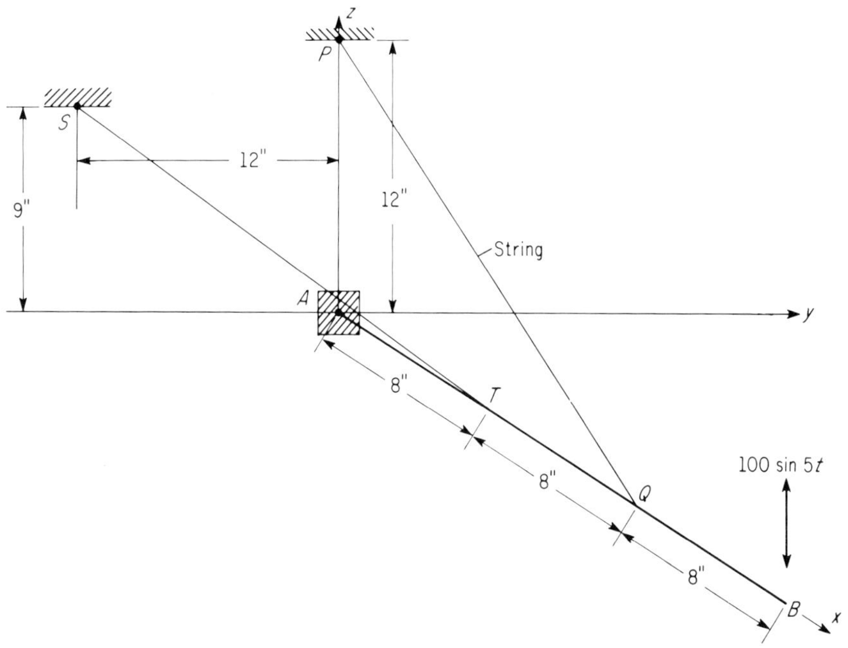

PROB. 3-18

3-19. Express mathematically the vertical motion of a motor car with shock absorbers on a rough road.

3-20. Write a mathematical model for a sailboat.

3-21. Analyze the motion of a motor boat. Consider the system to be in plane motion.

3-22. Calculate the velocity of sound wave propagation in air at 32°F at sea level. Given $C_p/C_v = 1.4$, atmospheric pressure = 14.7 psi and the air density as 0.081 lb/cu ft. Compare this with the velocity of sound in steel ($E = 30 \times 10^6$ density = 0.28 lb/cu in.).

3-23. Write the equations describing the motion of a single-span bridge under the effect of an earthquake.

3-24. A gear unit is equivalent to a transformer in the electrical analogy. What is the analogous circuit to the motor generator set shown in the figure by the impedance analogy?

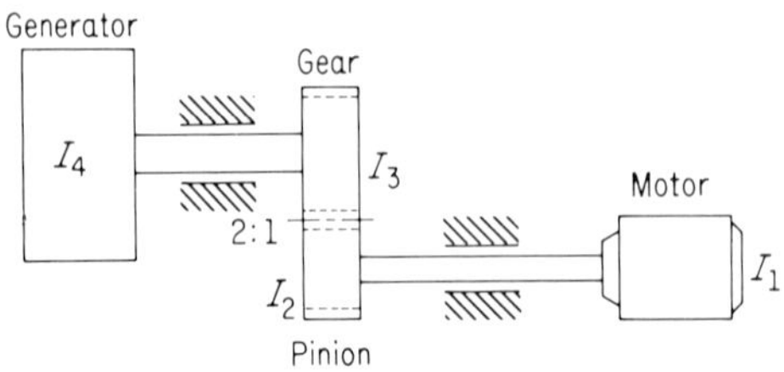

PROB. 3-24

3-25. Set up the equivalent electric circuits for the two-degree-of-freedom system shown in the sketch. Use both the voltage-force and current-force methods.

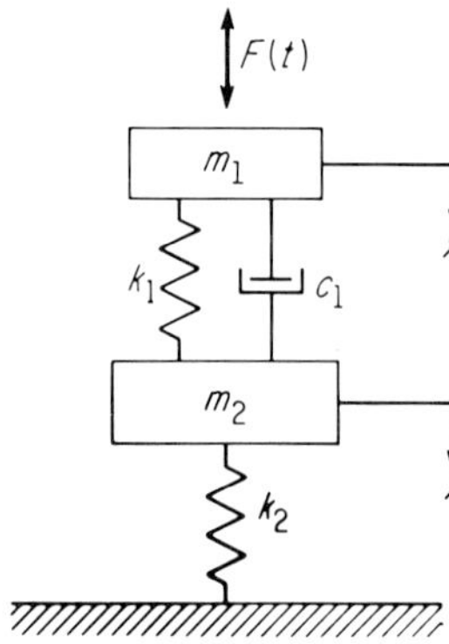

PROB. 3-25

3-26. Find the analogous electrical circuit for the two-mass system shown.

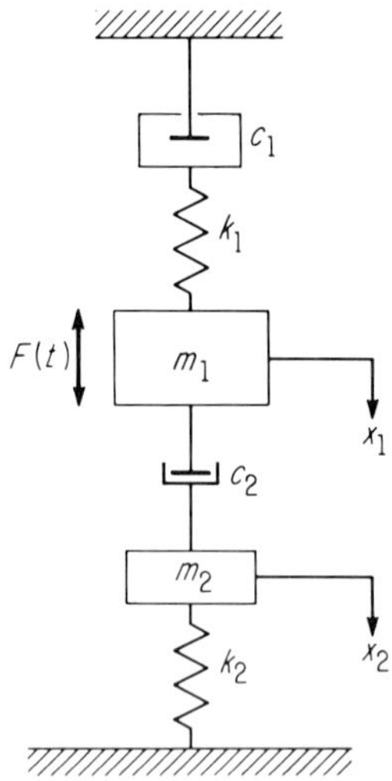

PROB. 3-26

3-27. Set up the analogous circuit for the two-mass torsional system shown.

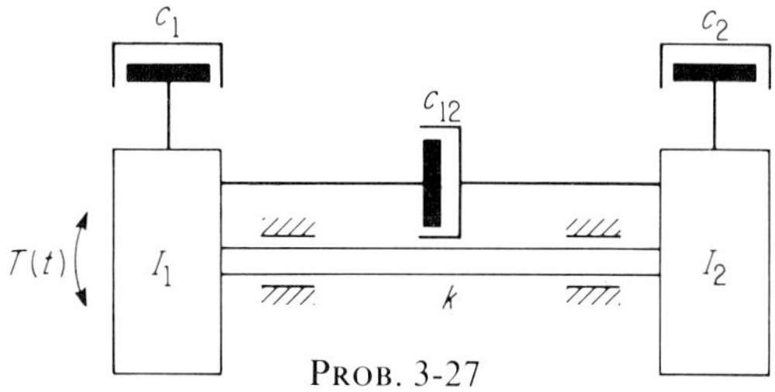

PROB. 3-27

3-28. Use the current-force analogy to set up the analogous circuit for the system shown in the figure.

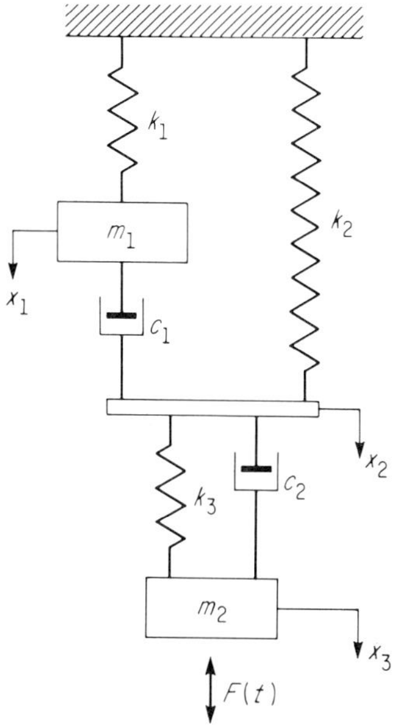

PROB. 3-28

3-29. A double-reduction motor drive is shown in the accompanying sketch. Find the analogous circuit. (A gear reduction is equivalent to an electrical transformer.)

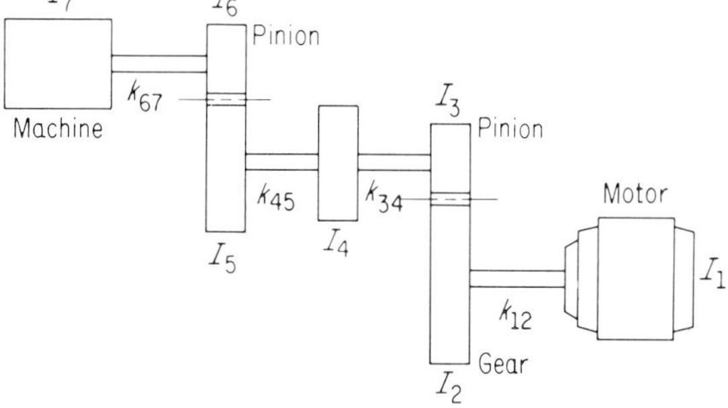

PROB. 3-29

3-30. Find the analogous circuit for the given system by both the classical and mobility analogy.

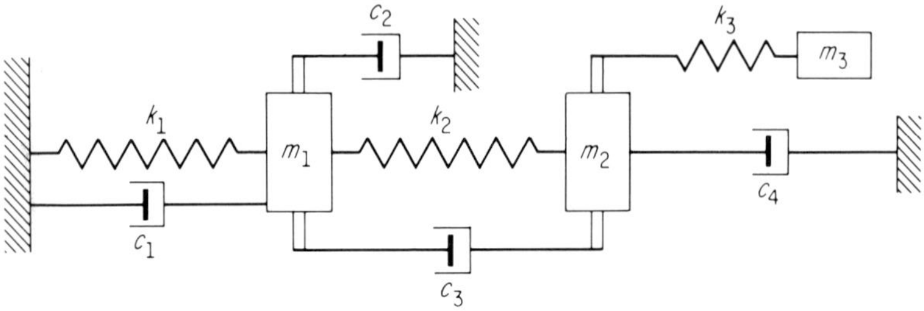

Prob. 3-30

3-31. Set up an electrical analog for the system shown. Each mass is supported by four leaf springs with thickness t and width b respectively.

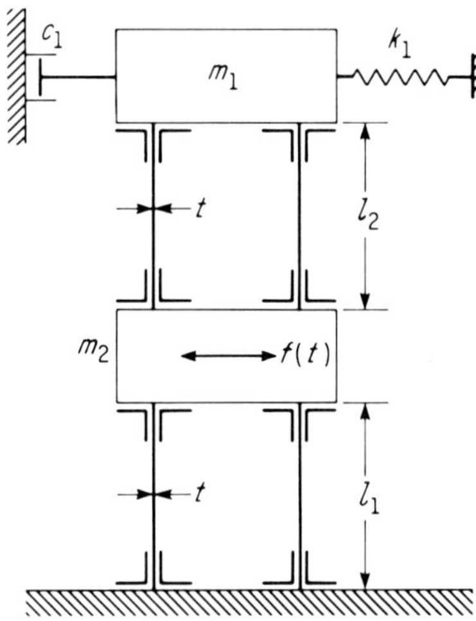

Prob. 3-31

3-32. Construct an analogous circuit to the system shown in Prob. 3-9.

REFERENCES

1. A. G. Webster, *The Dynamics of Particles and of Rigid, Elastic and Fluid Bodies.* New York: Dover, 1959.
2. I. H. Shames, *Engineering Mechanics.* Englewood Cliffs, N.J.: Prentice-Hall, 1960.
3. R. A. Waldron, *Waves and Oscillations.* Princeton, N.J.: Van Nostrand, 1964.
4. I. S. Sokolnikoff and F. S. Sokolnikoff, *Higher Mathematics for Engineers and Physicists.* New York: McGraw-Hill, 1941.

5. S. Timoshenko and S. Woinowsky-Krieger, *Theory of Plates and Shells.* New York: McGraw-Hill, 1959.
6. W. Ker Wilson, *Practical Solution of Torsional Vibration Problems.* New York: Wiley, 1956.
7. Theodore von Kármán and M. A. Biot, *Mathematical Methods in Engineering.* New York: McGraw-Hill, 1940.
8. S. Timoshenko, *Vibration Problems in Engineering.* 3d ed. Princeton, N.J.: Van Nostrand, 1955.
9. J. W. Strutt, Lord Rayleigh, *Theory of Sound.* 2d ed. New York: Dover, 1937.
10. R. Oldenburger. *Mathematical Engineering Analysis.* New York: Macmillan, 1950.
11. D. W. Ver Planck and B. R. Teare, *Engineering Analysis.* New York: Wiley, 1952.
12. J. J. Roark, *Formulas for Stress and Strain.* 3d ed. New York: McGraw-Hill, 1954.
13. S. Crandall, *Engineering Analysis.* New York: McGraw-Hill, 1954.
14. J. Reswick and C. Taft, *Introduction to Dynamic Systems.* Englewood Cliffs, N.J.: Prentice-Hall, 1967.
15. S. Timoshenko, *Strength of Materials*, Part II. 3d ed. Princeton, N.J.: Van Nostrand, 1956.

part II

The Analysis of Vibratory Systems with a Single Degree of Freedom

Mechanics is the paradise of mathematical science because here we come to the fruits of mathematics.

—Leonardo da Vinci (1452–1519)

chapter 4

Free Oscillations

4-1. UNDAMPED LINEAR SYSTEMS

Equation of Motion

The simplest case of a vibratory system is one whose motion is completely defined by a single coordinate. The examples shown in Fig. 4-1 represent such systems. The coordinates x, θ, α, y, z, and h respectively are all that is necessary to describe their motion.

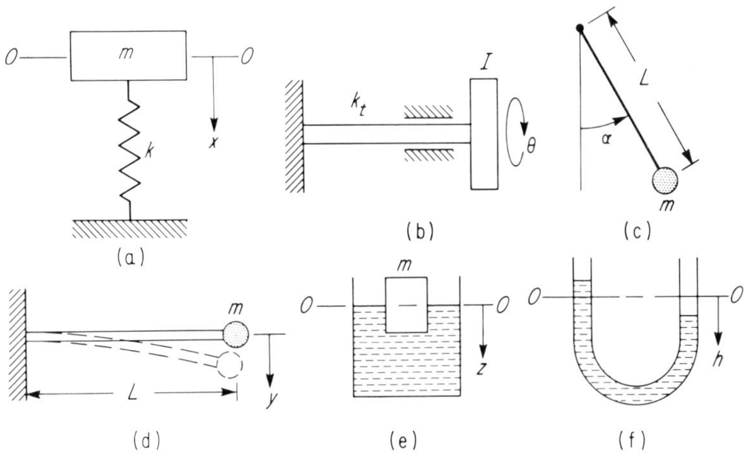

FIG. 4-1

All these examples satisfy the basic requirements of oscillating systems. They possess inertia and move under the influence of restoring forces which are dependent on the deviation from a stable equilibrium position. The restoring forces for small displacements can be considered as directly proportional to these displacements, and the systems are therefore called *linear*. By applying Newton's second law, the equations of motion for these systems are respectively given by

Case (a)

$$m\ddot{x} = -kx$$

$$\ddot{x} + (k/m)x = 0 \tag{4-1a}$$

where x = rectilinear displacement
m = oscillating mass
k = spring constant

Case (b)

$$I\ddot{\theta} = -k_t\theta$$

$$\dot{\theta} + (k_t/I)\theta = 0 \qquad (4\text{-}1\text{b})$$

where θ = angle of twist
I = polar moment of inertia of oscillating mass
k_t = GJ/L
= torsional spring constant
G = shear modulus for shaft material
J = polar moment of inertia of shaft cross section
L = length of shaft

Case (c)

$$mL\ddot{\alpha} = -mg(\sin\alpha)$$

where α = angular displacement
L = length of pendulum
m = suspended mass
g = gravitational acceleration

For small oscillations $\alpha = \sin\alpha$ and accordingly

$$mL\ddot{\alpha} = -mg\alpha$$

or

$$\ddot{\alpha} + \frac{g}{L}\alpha = 0 \qquad (4\text{-}1\text{c})$$

Case (d)

$$m\ddot{y} = -k_b y$$

$$\ddot{y} + (k_b/m)y = 0 \qquad (4\text{-}1\text{d})$$

where m = supported mass
y = mass displacement
k_b = bending stiffness of cantilever
= $3EI/L^3$

E, I, and L are the modulus of elasticity, moment of inertia of the cross section, and the length respectively. The weight of the beam is neglected in this case.

Case (e)

$$m\ddot{z} = -A\rho g z$$

$$\ddot{z} + (A\rho g/m)z = 0 \qquad (4\text{-}1\text{e})$$

where z = float displacement
m = mass of float
A = area of float
ρ = mass of fluid per unit volume
g = gravitational acceleration

Case (f)

$$\rho L a \ddot{h} = -\rho g 2hA$$

$$\ddot{h} + (2g/L)h = 0 \tag{4-1f}$$

where h = displacement of the free surface
ρ = mass per unit volume
A = cross-sectional area of the tube
L = total length of water column

These equations (4-1a through 4-1f) are of the form

$$\ddot{q} + \omega_n^2 q = 0 \tag{4-2}$$

where q is the coordinate of the motion and ω_n^2 is a constant describing the physical characteristics of the system. Equation 4-2 is a linear homogeneous differential equation whose solution can be obtained by assuming $q = e^{at}$. Accordingly, $\ddot{q} = a^2 e^{at}$ and Eq. 4-2 will be

$$(a^2 + \omega_n^2)q = 0$$

from which either $q = 0$, i.e., no deviation from equilibrium or

$$a = \pm i\omega_n$$

The general solution will therefore be

$$q = c_1 e^{i\omega_n t} + c_2 e^{-i\omega_n t}$$

where c_1, c_2 are integration constants. This can be rewritten as

$$q = A \sin(\omega_n t + \phi) \tag{4-3}$$

which represents a simple harmonic motion or a sinusoidal function of time Fig. 4-2. The motion has an amplitude A, a period $2\pi/\omega_n$, and a

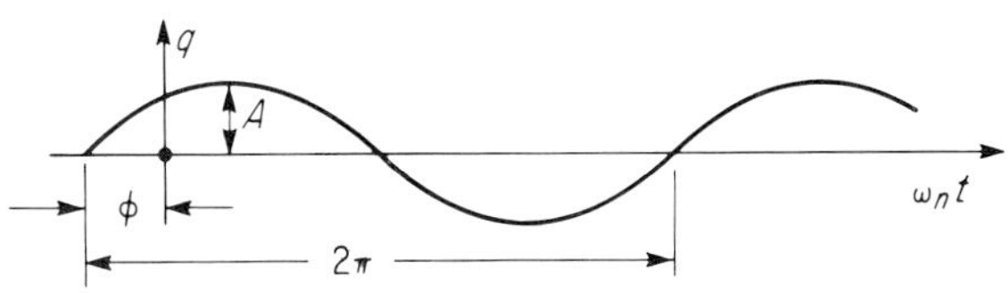

FIG. 4-2

phase angle ϕ. The constant ω_n is the frequency of vibration. Since ω_n is a characteristic of the physical model and is independent of the disturbance which sets the system in motion, it is referred to as the *natural frequency* of the system.

It should be noted here that the amplitude A and the phase angle ϕ are integration constants whose values are determined from the initial conditions of the motion.

The velocity $\dot{q}$ is obtained by differentiating Eq. 4-3 with respect to time:

$$\dot{q} = A\omega_n \cos(\omega_n t + \phi) \tag{4-4}$$

Consequently, if the initial conditions are given in terms of an initial position and velocity, Eqs. 4-3 and 4-4 can be used to determine A and ϕ and the motion will therefore be fully prescribed. For example, if at $t = 0$,

$$q = q_0 \qquad \text{and} \qquad \dot{q} = \dot{q}_0$$

Therefore

$$q_0 = A \sin \phi$$
$$\dot{q}_0 = A\omega_n \cos \phi$$

from which

$$A = q_0^2 + \frac{\dot{q}_0}{\omega_n}^{2} \tag{4-5}$$

and

$$\phi = \arctan \frac{q_0}{(\dot{q}_0/\omega_n)} \tag{4-6}$$

Example 1

A wooden block of mass M is suspended by a cord length L, to form a pendulum, Fig. 4-3. (a) If the pendulum is released from a position

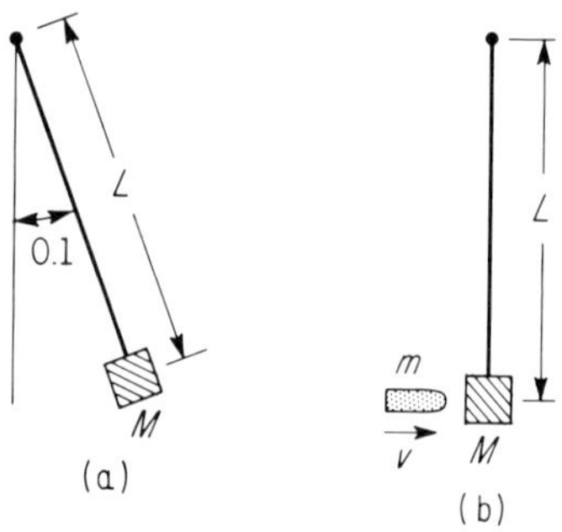

FIG. 4-3

making an angle 0.1 radian with the vertical, write the equation of motion. (b) If a bullet mass m is used to start the vibration, what bullet velocity v is required to give the same amplitude as in case (a)? Assume full plastic impact.

The equation of motion is

$$\ddot{\alpha} + (g/L)\alpha = 0$$

and the solution is

$$\alpha = A \sin(\sqrt{g/L}\,t + \phi)$$

(a) The initial conditions in this case are

$$\text{at } t = 0, \qquad \alpha = 0.1 \text{ radian}, \qquad \dot{\alpha} = 0$$

therefore

$$0.1 = A \sin \phi$$

and

$$0 = (\sqrt{g/L})\, A \cos \phi$$

from which

$$\cos \phi = 0, \phi = \pi/2, A = 0.1$$

and the displacement time function becomes

$$\begin{aligned} \alpha &= 0.1 \sin (\sqrt{g/L}\, t + \pi/2) \\ &= 0.1 \cos \sqrt{g/L}\, t \end{aligned}$$

(b) For full plastic impact the bullet and block will have the same velocity after impact. Taking this as v_f, the conservation of momentum requires that

$$mv + M \times 0 = (m + M)v_f$$

from which

$$v_f = \left(\frac{m}{M + m}\right)v$$

Now the initial conditions for the pendulum oscillation are

$$\text{at } t = 0, \qquad \alpha = 0, \qquad \dot{\alpha} = v_f/L$$

from which

$$0 = A \sin \phi \qquad \text{or} \qquad \phi = 0$$

$$\dot{\alpha} = v_f/L = \left(\frac{m}{M + m}\right)\frac{v}{L} = (\sqrt{g/L})\, A \cos \phi = (\sqrt{g/L})\, A$$

Therefore

$$v = AL\sqrt{g/L}\left(\frac{M + m}{m}\right)$$

and

$$\text{for } A = 0.1 \qquad v = 0.1L\sqrt{g/L}\left(\frac{M + m}{m}\right)$$

4-2. THE PHASE-PLANE METHOD

An interesting conclusion can be reached by studying Eqs. 4-3 and 4-4. Knowing that $\sin^2 (\omega_n t + \phi) + \cos^2 (\omega_n t + \phi) = 1$, the two equations can be reduced to the single equation

$$q^2 + (\dot{q}/\omega_n)^2 = A^2$$

which represents a circle in the q versus $(\dot{q}/\omega_n)$ plane. This type of plot is known as the *phase plane* and is a very powerful and useful tool of analysis for vibration problems. It could be seen in Fig. 4-4 that the horizontal projection of the motion on a time scale will give the velocity

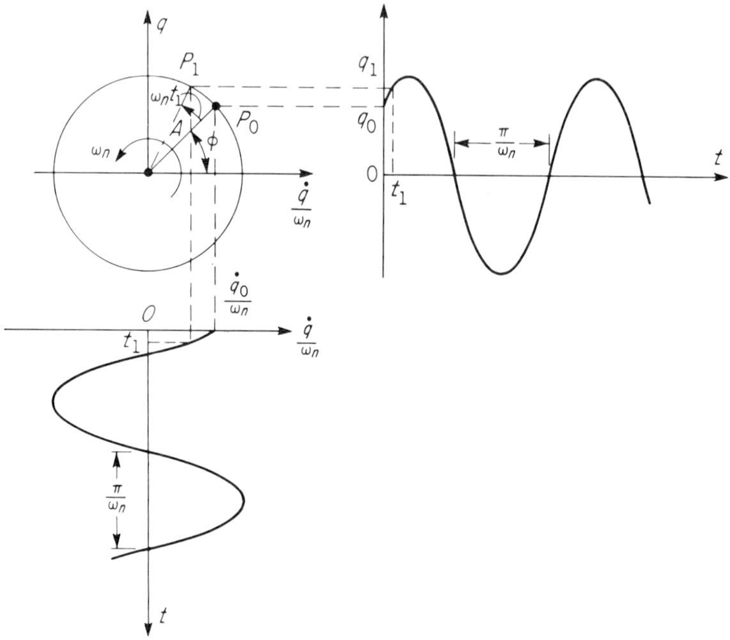

FIG. 4-4

curve plotted to a scale (ω_n). The starting point P_0 on the circle is defined by the initial conditions, namely: at $t = 0$, $q = q_0$, and $\dot{q}/\omega_n = \dot{q}_0/\omega_n$. From Eqs. 4-5 and 4-6, the amplitude of vibration is given by

$$A = \sqrt{q_0^2 + (\dot{q}_0/\omega_n)^2}$$

and the phase angle ϕ is defined as

$$\tan \phi = q_0/(\dot{q}_0/\omega_n)$$

The point P representing the motion in the phase-plane plot will then move on a circle with an angular velocity ω_n in a counterclockwise direction. The center of this circle represents the equilibrium condition and its radius is equal to A. When $t = t_1$, P will take the position P_1 having a radial angle $\omega_n t_1$ from P_0. The corresponding displacement and velocity will be q_1, $\dot{q}_1/\omega_n$ as shown in Fig. 4-4.*

*This technique is explained in detail in L. S. Jacobsen and R. S. Ayre, *Engineering Vibrations* (McGraw-Hill, New York, 1958).

The representation of a free vibration by the phase-plane method is summarized as follows:

1. Calculate the natural frequency ω_n of the system.

2. Plot two perpendicular axes representing q and $(\dot{q}/\omega_n)$ respectively. The scale of both axes will have the same dimensions as the coordinate q of the vibration. These two axes represent the phase plane.

3. The origin O of the phase plane represents the equilibrium condition of the system. This is evident, since the equilibrium condition is described by $q = 0$ and $\dot{q} = 0$.

4. Knowing the initial conditions of the motion in terms of the initial displacement and velocity, a point P_0 can be plotted in the phase plane to represent these conditions.

5. The free vibration of the system is then represented by a circle. The center of the circle is the equilibrium point (or the origin of the phase plane). The radius of the circle (which is also the amplitude of the vibration) is the distance OP_0 joining the origin to the initial point P_0.

6. The conditions of the motion at any time t_1 are represented by the coordinates of the point P_1 on the circle at a radial angle $\omega_n t_1$ from P_0 in the counterclockwise direction.

7. The projection of the phase-plane plot on a time scale in the direction of $q, (\dot{q}/\omega_n)$ gives the displacement and velocity functions respectively.

8. All information concerning the time history of the motion can be obtained from the phase-plane plot.

Example 2

A 386-lb weight is suddenly placed on a spring having a stiffness of 100 lb/in., (Fig. 4-5). Use the phase-plane method to plot the displacement-time and the velocity-time curves.

The mass on the spring form a vibratory system with a natural

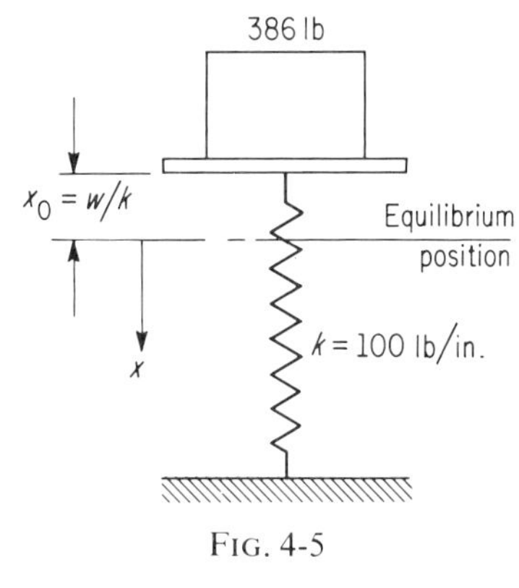

FIG. 4-5

frequency

$$\omega_n = \sqrt{\frac{100\,g}{386}} = 10 \text{ rad/sec}$$

Placing the weight suddenly would be equivalent to starting the motion at a distance

$$x_0 = \frac{386}{100} = 3.86 \text{ in.}$$

from the equilibrium position. Knowing the natural frequency and the initial conditions (at $t = 0$, $x = -x_0$ and $\dot{x} = 0$), the circle representing the motion can be plotted in the phase plane as shown in Fig. 4-6. The

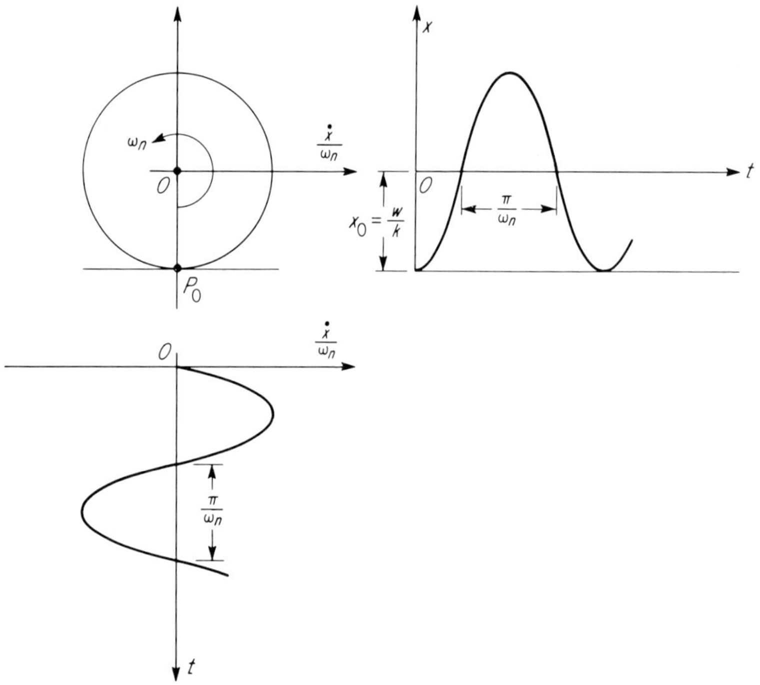

FIG. 4-6

displacement and velocity are then projected on a time scale to give the required curves.

Example 3

Solve Example 1 by utilizing the phase-plane method.

In this example, $\omega_n = \sqrt{g/L}$, and the phase plane represents α versus $\dot{\alpha}/\omega_n$.

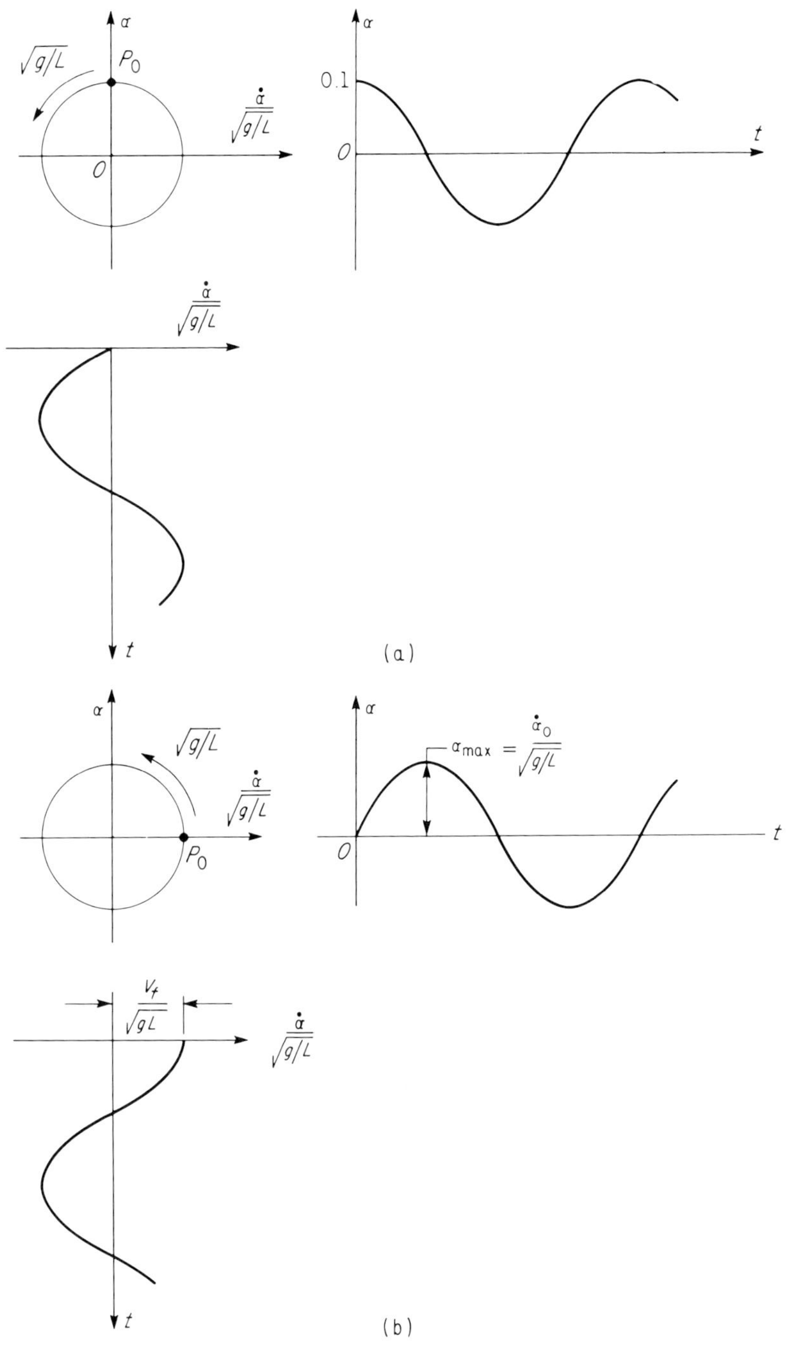
α
$\sqrt{g/L}$
P_0
$\frac{\dot{\alpha}}{\sqrt{g/L}}$
O
0.1
0
t
(a)
$\alpha_{max} = \frac{\dot{\alpha}_0}{\sqrt{g/L}}$
$\frac{V_f}{\sqrt{gL}}$
(b)

FIG. 4-7

(a) The initial conditions in this case are

$$\text{at } t = 0, \qquad \alpha = 0.1, \quad \dot{\alpha} = 0$$

The starting point is represented by point P_0 in phase plane (Fig. 4-7a). From the projection of the motion it can be seen that

$$\alpha = 0.1 \cos \omega_n t$$

and

$$\dot{\alpha} = -0.1\, \omega_n \sin \omega_n t$$

(b) In this case the initial conditions of the motion are

$$\text{at } t = 0 \qquad \alpha = 0 \qquad \dot{\alpha} = \frac{v_f}{L}$$

where v_f is the velocity of the pendulum after impact. The phase-plane plot as well as the starting point P_0 will be as shown in Fig. 4-7b. It can be seen from the figure that

$$\alpha_{\max} = \frac{v_f}{L\omega_n} = \frac{v_f}{\sqrt{gL}} = 0.1$$

It has been shown in Example 1 that

$$v_f = \frac{m}{M + m} v$$

Therefore

$$v = 0.1(m + M)\sqrt{gL}$$

which is the expected result.

4-3. UNDAMPED NONLINEAR SYSTEMS

It has been previously indicated that single-degree-of-freedom systems can be described by linear differential equations when the restoring force is proportional to the displacement from the equilibrium position, and when the frictional force is proportional to the velocity. Although this provides a good approximation for most systems under small oscillations, it is not always true for large oscillations. Many practical systems have restoring forces which cease to be proportional to the displacement when the displacements are large. Coulomb, hysteresis, and hydraulic damping (square function of the velocity) are also common in mechanical systems. The analysis of systems with nonlinear characteristics falls in the domain of solution of nonlinear differential equations. There are very few nonlinear equations with known exact solutions. For the general case, however, one must resort to approximate solutions such as the classical iteration and perturbation techniques. Averaging methods such as the Ritz (or Ritz-Galerkin) method and the equivalent linearization method of Kryloff and Bogoliuboff are frequently utilized in the literature on nonlinear vibrations.

Numerical solutions are extensively used especially with the availability of digital computers. Graphical solutions by the phase-plane method provide a powerful tool of analysis. The analog computer is also utilized effectively in many nonlinear problems.

It should be noted here that in nonlinear systems, unlike linear systems, the existence of a mathematical solution does not automatically prove the existence of a physical vibration. A necessary condition for the existence of a vibration is the stability of the motion.

In the following section we shall apply some of the methods of analysis of nonlinear problems to undamped systems with nonlinear restoring forces.

Exact Solutions*

The general equation of motion for undamped systems with nonlinear restoring characteristics can be written as

$$m\ddot{x} + kf(x) = 0$$

$$\ddot{x} + p^2 f(x) = 0$$

where $kf(x)$ is a nonlinear restoring function and p^2 is a constant of the system. Since

$$\begin{aligned} \frac{d\dot{x}^2}{dx} &= 2\dot{x}\,\frac{d\dot{x}}{dx} \\ &= 2\dot{x}\,\frac{d\dot{x}}{dt}\,\frac{dt}{dx} \\ &= 2\ddot{x} \end{aligned}$$

a solution for this equation in an integral form can be given as

$$\ddot{x} = \frac{1}{2}\,\frac{d\dot{x}^2}{dx} = -p^2 f(x)$$

$$\dot{x}^2 = -2p^2 \int_X^x f(\zeta)\,d\zeta = 2p^2 \int_x^X f(\zeta)\,d\zeta$$

*Can be omitted in a first reading without loss of continuity.

where X is the value of displacement when $\dot{x} = 0$ or the maximum displacement. Accordingly,

$$\dot{x} = \sqrt{2}p \sqrt{\int_x^X f(\zeta)\,d\zeta} = \frac{dx}{dt}$$

and

$$\int_{t_0}^{t} dt = \frac{1}{p\sqrt{2}} \int_{x_0}^{x} \frac{dx}{\sqrt{\left(\int_x^X f(\zeta)\,d\zeta\right)}} \tag{4-7}$$

where $x = x_0$ at $t = t_0$ or x_0 is the initial position.

Example 4

We shall consider for illustration the case when $f(x) = x^n$ and the initial conditions are such that $x = 0$ at $t = t_0$. Equation 4-7 is then

$$t - t_0 = \frac{1}{\sqrt{2}p} \int_0^x \frac{dx}{\sqrt{\left(\int_x^X f(\zeta)\,d\zeta\right)}} = \frac{1}{\sqrt{2}p} \int_0^x \frac{dx}{\sqrt{\left(\int_x^X \zeta^n\,d\zeta\right)}}$$

$$= \frac{1}{p}\sqrt{\frac{n+1}{2}} \int_0^x \frac{dx}{\sqrt{X^{n+1} - x^{n+1}}}$$

$$t - t_0 = \frac{1}{p}\sqrt{\frac{n+1}{2}} \frac{1}{\sqrt{(X^{n+1})}} X \int_0^{x/X} \frac{du}{\sqrt{1 - u^{n+1}}}$$

$$= \frac{1}{p}\sqrt{\frac{n+1}{2}} \frac{1}{\sqrt{(X^{n-1})}} \int_0^{x/X} \frac{du}{\sqrt{(1 - u^{n+1})}}$$

where $u = \dfrac{x}{X}$, $dx = X\,du$. This solution also applies to binomials such as $f(x) = x^n + \mu x^m$ where $m > n \geq 0$.

Example 5

Another example to be considered here is the general vibration of the pendulum, Fig. 4-8. The equation of motion is

$$\ddot{x} + (g/L)\sin x = 0$$

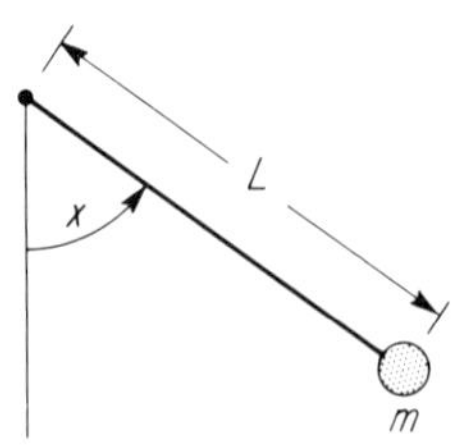

FIG. 4-8

The boundary conditions are taken as $x = 0$ at $t = t_0$, and $x = X$ when $\dot{x} = 0$. In Eq. 4-7 $f(\zeta)$ will be $\sin\zeta$ and p will be $\sqrt{g/L}$. Therefore,

$$t - t_0 = \frac{1}{\sqrt{2}\,p}\int_0^x \frac{dx}{\sqrt{\left(\int_x^X \sin\zeta\,d\zeta\right)}}$$

$$= \frac{1}{\sqrt{2}\,p}\int_0^x \frac{dx}{\sqrt{[-(\cos\zeta)]_x^X}}$$

Knowing that ($\cos\zeta = \cos^2\zeta/2 - \sin^2\zeta/2 = 1 - 2\sin^2\zeta/2$), we obtain

$$t - t_0 = \frac{1}{\sqrt{2}\,p}\int_0^x \frac{dx}{\sqrt{2\left(\sin^2\dfrac{X}{2} - \sin^2\dfrac{x}{2}\right)}}$$

Introducing the notation

$$u\sin\phi = \sin\frac{x}{2}$$

where $u = \sin X/2 = \text{const.}$

and by differentiation we get

$$dx = \frac{2u\cos\phi d\phi}{\cos\dfrac{x}{2}}$$

$$= \frac{2u\cos\phi d\phi}{\sqrt{(1 - u^2\sin^2\phi)}}$$

Substituting these values for dx, $\sin x/2$ and $\sin X/2$ in the $(t - t_0)$ equation and simplifying we obtain

$$t - t_0 = \frac{1}{p}\int_0^{\phi_x} \frac{d\phi}{\sqrt{(1 - u^2\sin^2\phi)}} \tag{4-8}$$

where $\phi_x = \arcsin(1/u\sin x/2)$.

The integral $\displaystyle\int \frac{d\phi}{\sqrt{(1 - u^2\sin^2\phi)}}$ is an elliptic integral of the first kind. It can be obtained from tables of functions (Ref. 44) or evaluated as follows:

$$(1 - u^2\sin^2\phi)^{-1/2} = 1 + \frac{1}{2}u^2\sin^2\phi + \frac{3}{8}u^4\sin^4\phi + \frac{5}{16}u^6\sin^6\phi + \cdots$$

In order to evaluate the time necessary for the pendulum to perform 1/4 cycle we shall consider the boundary conditions

$$\text{at } t = 0 \qquad x = 0$$

therefore

$$t_0 = 0$$

and

$$\text{at } \dot{x} = 0 \qquad x = \alpha = X$$

therefore

$$\begin{aligned}\phi_x &= \arcsin\left(\frac{1}{u}\sin\frac{x}{2}\right)\\ &= \arcsin\left(\frac{1}{u}u\right)\\ &= \frac{\pi}{2}\end{aligned}$$

Accordingly, Eq. 4-8 gives

$$\begin{aligned}\tau = \frac{T}{4} &= \frac{1}{p}\int_0^{\pi/2} d\phi\left(1 + \frac{1}{2}u^2\sin^2\phi + \cdots\right)\\ &= \frac{1}{p}\left[\frac{\pi}{2} + \frac{1}{2}u^2\frac{\pi}{4} + \frac{3}{8}u^4\frac{3}{8}\frac{\pi}{2} + \frac{5}{16}u^6\frac{5}{16}\frac{\pi}{2} + \cdots\right]\\ &= \frac{\pi}{2p}\left[1 + \frac{1}{4}u^2 + \frac{9}{64}u^4 + \frac{25}{256}u^6 + \cdots\right]\end{aligned}$$

Note that

$$\int_0^{\pi/2} \sin^n\phi\, d\phi = \frac{n-1}{n}\int_0^{\pi/2} \sin^{n-1}\phi\, d\phi$$

Substituting $\tau_0 = 2\pi/p =$ the natural period of the simple pendulum with small oscillations we get

$$\frac{\tau}{\tau_0} = \left[1 + \frac{1}{4}u^2 + \frac{9}{64}u^4 + \frac{25}{256}u^6 + \cdots\right]$$

where $u = \sin X/2 = \sin \alpha/2$. This equation shows the change of the period of the pendulum with the amplitude which is characteristic of nonlinear vibrations. The plot of this result is given in Fig. 4-9.

Classical Approximate Solutions for Nonlinear Systems*

Taking as an example a system with the restoring force $(kx - \lambda x^2)$ where k and λ are constants, the equation of motion in a free vibration is

$$m\ddot{x} + (kx - \lambda x^2) = 0 \tag{4-9}$$

which is obviously nonlinear. The degree of nonlinearity depends on the value of λ. We shall try a solution for this problem by the classical iteration and perturbation methods to illustrate their application.

*Can be omitted in a first reading without loss of continuity.

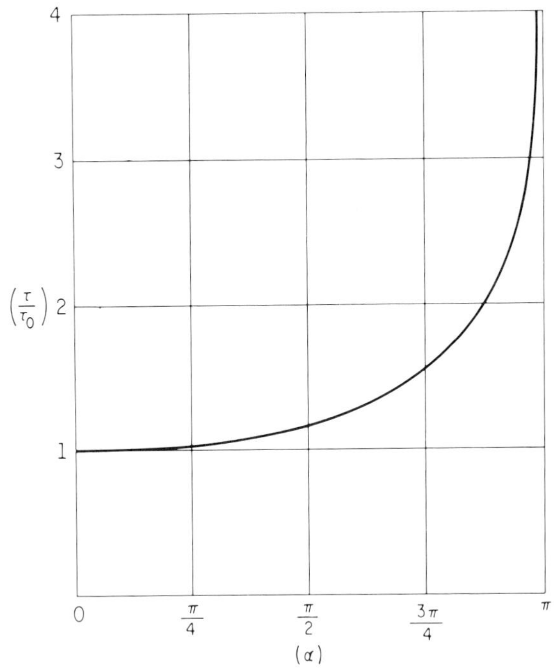

FIG. 4-9. Change of natural period with amplitude of vibration.

Iteration or Successive Approximations

As a first approximation, consider λx^2 to be small. Accordingly, a first approximation x_1 for x will be harmonic, since Eq. 4-9 will be linear. Therefore

$$x_1 = A \cos(\omega_n t + \phi)$$

where $\omega_n = (\sqrt{k/m})$ and A, ϕ are integration constants. Now by substitution in the original equation we can obtain a second approximation x_2 as follows:

$$m\ddot{x}_2 = -kx_2 + \lambda A^2[\cos^2(\omega_n t + \phi)]$$

$$= -kx + \frac{1}{2}\lambda A^2[1 + \cos^2(\omega_n t + \phi)] \tag{4-10}$$

In seeking a solution for this equation, we must bear in mind that the solution must be periodic and it also must contain a term with the frequency $2\omega_n$. Therefore we shall take as a second approximation

$$x_2 = A\cos(\omega_n t + \phi) + B\cos 2(\omega_n t + \phi) + C$$

By substitution in Eq. 4-10 we find

$$C = \lambda A^2/2k$$

$$B = \lambda A^2/6k$$

and the solution becomes

$$x_2 = A\cos(\omega_n t + \phi) - \frac{\lambda A^2}{6k}\cos 2(\omega_n t + \phi) + \frac{\lambda A^2}{2k}$$

A and ϕ are determined by initial conditions. This motion is no longer simple harmonic and is called *anharmonic*.

Methods of Perturbation

In this method, the desired quantity is expressed in powers of a small quantity and the solution is obtained in a step-by-step manner. If we expand the solution in a series involving the parameter λ, we obtain

$$x = f_1 + \lambda f_2 + \lambda^2 f_3 + \cdots$$

Then regarding λ as being small, we only retain the first two terms

$$x = f_1 + \lambda f_2$$

Thus λ serves as an indicator of perturbation terms, f_1 is the solution to the first approximation, λf_2 is the term added in the second approximation, and so on. By substituting in Eq. 4-9 we obtain

$$m\ddot{f_1} + \lambda m\ddot{f_2} = -kf_1 - \lambda kf_2 + \lambda(f_1 + \lambda f_2)^2 = -kf_1 - \lambda kf_2 + \lambda f_1^2$$

where we neglected the term in λ^2 and λ^3 as being still of smaller order of magnitude.

Since λ is arbitrary, the coefficient of each power of λ on the left- and right-hand sides must be equal—that is,

$$m\ddot{f_1} = -kf_1$$

and

$$m\ddot{f_2} = -kf_2 + f_1^2$$

The first equation gives

$$f_1 = A\cos(\omega_n t + \phi)$$

Substituting this result in the second equation we get

$$m\ddot{f_2} = -kf_2 + A^2\cos^2(\omega_n t + \phi)$$

$$= -kf_2 + \frac{1}{2}A^2[\cos 2(\omega_n t + \phi) + 1]$$

Taking $f_2 = B' \cos 2(\omega_n t + \phi) + C'$, we obtain

$$-4kB' \cos 2(\omega_n t + \phi) = -kB' \cos 2(\omega_n t + \phi) - kC' + \frac{1}{2}A^2 + \frac{1}{2}A^2 \cos 2(\omega_n t + \phi)$$

from which

$$C' = A^2/2k$$
$$B' = -A^2/6k$$

and the solution becomes

$$x = A \cos(\omega_n t + \phi) - \frac{\lambda A^2}{6k} \cos 2(\omega_n t + \phi) + \frac{\lambda A^2}{2k}$$

It is evident that this solution is identical to the one obtained by iteration. The perturbation method, however, is more elegant, and using λ as a label for small quantities makes it easier to distinguish different orders of small quantities—particularly in further approximations.

4-4. THE PHASE-PLANE-δ METHOD

The phase-plane method previously used to study the free vibrations of a linear system will be applied here to study the free vibration of nonlinear systems.

Let us assume the general case of the free vibration of a mass supported on a nonlinear spring.

$$m\ddot{x} + kf(x) = 0$$

where $kf(x)$ is the given nonlinear restoring function. By adding and subtracting a linear function kx this equation can be rewritten as

$$\ddot{x} + \frac{k}{m}[x + f(x) - x] = 0$$

or

$$\ddot{x} + p^2[x + \delta_x] = 0 \tag{4-11}$$

where $\delta_x = [f(x) - x]$ and is a known function of x.

Equation 4-11 is of the same form as Eq. 4-2 where p^2 takes the place of ω_n^2 and x has to be continually modified by an amount δ_x which is a function of x in this case. Therefore, at any instant of time the nonlinear equation can be represented in the phase plane as a free vibration of a linear system with a continually changing datum.

The phase-plane-δ technique as applied to a single-degree-of-freedom system with nonlinear spring can be summarized as follows:

1. Write the equation of motion in the form

$$\ddot{x} + p^2[x + \delta_x] = 0$$

where p^2 is a constant.

2. Express δ_x as a function of x.
3. Plot the phase-plane axes to represent x, $(\dot{x}/p)$ Fig. 4-10.

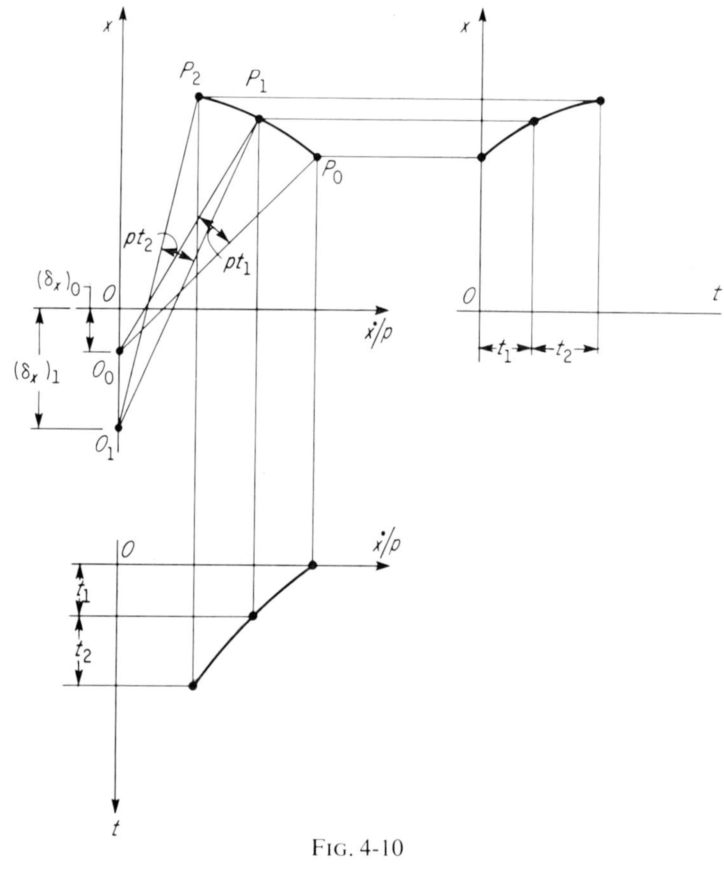

FIG. 4-10

4. Knowing the initial conditions of the motion in terms of the initial displacement x_0 and velocity $\dot{x}_0$, a point P_0 can be plotted in the phase plane to represent the starting conditions.

5. The value of $(\delta_x)_0$ is also calculated for the initial condition. Since in the case under consideration δ_x is a function of x only, therefore $(\delta_x)_0$ is obtained by substituting x_0 in the expression for δ_x.

6. The center O_0 of the instantaneous linear vibration at the initial

phase of the motion is therefore located on the x-axis at a distance $(\delta_x)_0$ from the origin. When δ is positive, the distance is taken on the negative x-axis and vice versa. The initial value of δ shown in Fig. 4-10 is assumed to be positive.

7. The instantaneous free vibration at the initial phase of the motion can be represented by a small arc of a circle with center O_0 and radius O_0P_0 in the counterclockwise direction.

8. A new condition P_1 is now reached after a small increment of time t_1 where the displacement is x_1 and the velocity is $\dot{x}_1$.

9. The value of $(\delta_x)_1$ corresponding to this new condition can be calculated, and a new center of oscillation O_1 is determined.

10. The new condition P_2 corresponding to an additional increment of time t_2 can be obtained from the arc of a circle with center O_1, having a radius O_1P_1 and subtending an angle pt_2 in the counterclockwise direction.

11. This procedure is then repeated until the point P_0 is reached again. Now a full cycle is attained. The total of the time increments required to accomplish this is the period of the oscillation.

12. It should be noted here that the accuracy of this procedure depends on the magnitude of the angular displacement pt. Better accuracy can be attained also by iteration. This means that after P_1 is determined, a value of δ is calculated which corresponds to a point midway between P_0 and P_1. Using this new value of δ, a better approximation for the new position P_1 after a small increment of time t_1 can be obtained.

13. It can be seen that this technique transforms a nonlinear vibration into small increments of linear vibrations with continually changing centers of oscillation.

14. The phase-plane-δ procedure is also suitable for programming on high-speed digital computers. This is explained in Part III of the text.

Example 6

In order to illustrate this procedure we shall again consider the general case of the simple pendulum. The equation of motion is

$$\ddot{\alpha} + (g/L)\sin\alpha = 0$$

which is clearly nonlinear. It can be rewritten in the δ form as

$$\ddot{\alpha} + g/L\,[\alpha + (\sin\alpha - \alpha)] = 0$$

Therefore

$$p = \sqrt{g/L} \qquad \text{and} \qquad \delta_x = (\sin\alpha - \alpha)$$

Taking the initial conditions as

$$\text{at } t = 0, \quad \alpha = \frac{3\pi}{4} \qquad \text{and} \qquad \dot{\alpha} = 0$$

the graphical solution can be pursued in small increments of time, modifying the datum for x (or the center of the arc) at each increment by the corresponding amount δ_x. This is shown in Fig. 4-11 and the graph-

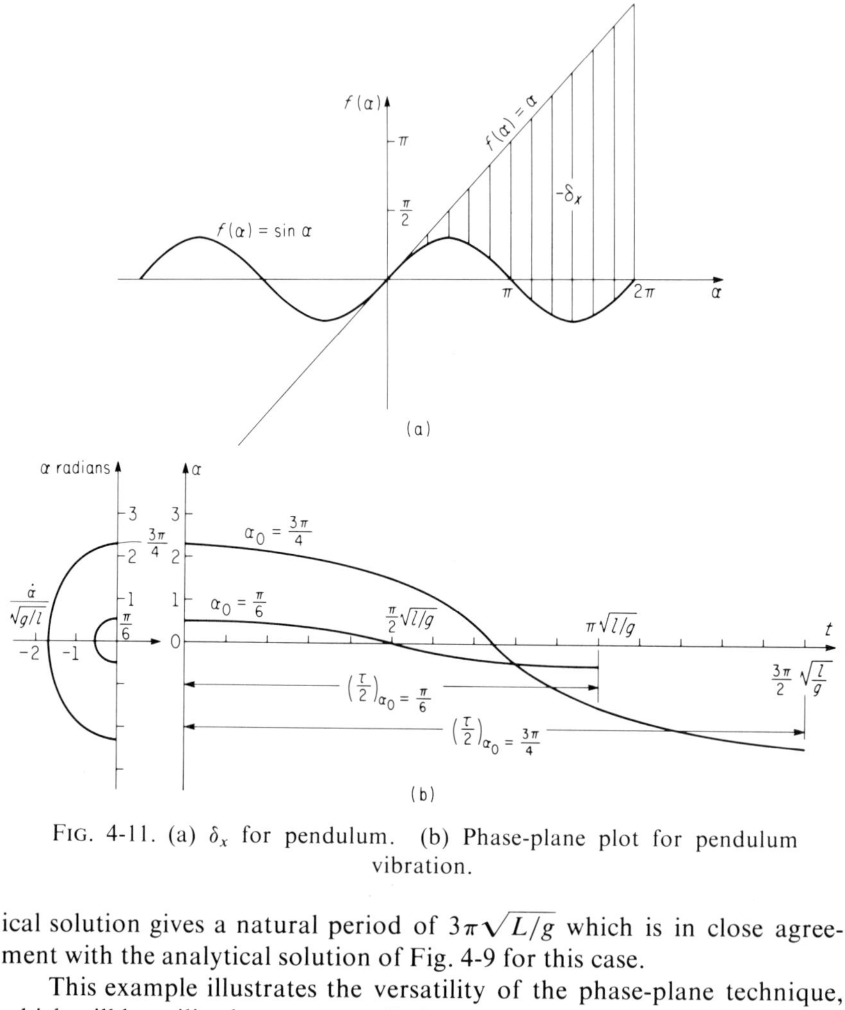

FIG. 4-11. (a) δ_x for pendulum. (b) Phase-plane plot for pendulum vibration.

ical solution gives a natural period of $3\pi\sqrt{L/g}$ which is in close agreement with the analytical solution of Fig. 4-9 for this case.

This example illustrates the versatility of the phase-plane technique, which will be utilized more generally later in the text.

Example 7

Study the time history of the motion of a mass vibrating between two equal springs with backlash d (Fig. 4-12a). Take the following initial conditions, at $t = 0$, $x = 0$ and $\dot{x} = \dot{x}_0$. Systems of this type are known as *piecewise-linear systems*, due to the nature of the restoring function. The equation of motion is $m\ddot{x} + f(x) = 0$ which can be rewritten in the

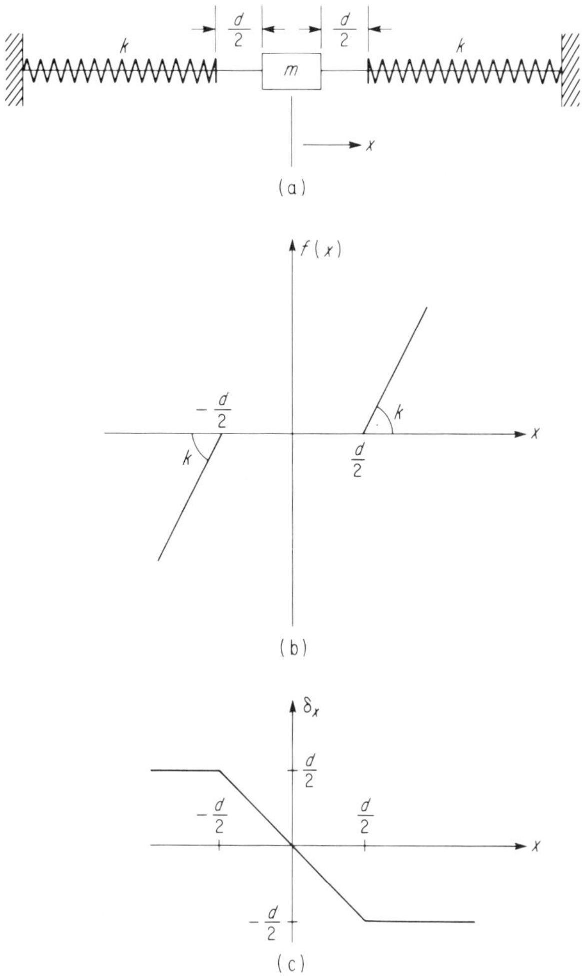

FIG. 4-12

δ form as

$$\ddot{x} + \frac{k}{m}\left[x + \frac{f(x)}{k} - x\right] = 0$$

$$x + p^2[x + \delta_x] = 0$$

where $p = \sqrt{k/m}$. The δ_x function is plotted in Fig. 4-12c and the equation can now be analyzed by the phase-plane method as shown in Fig. 4-13. Point b represents the initial contact between the mass and the spring. From b to c, δ is constant and is equal to $(-d/2)$.

It is obvious that the mass will travel through the backlash with a constant velocity $\dot{x}_0$. This can be found to be true by continuing the plot

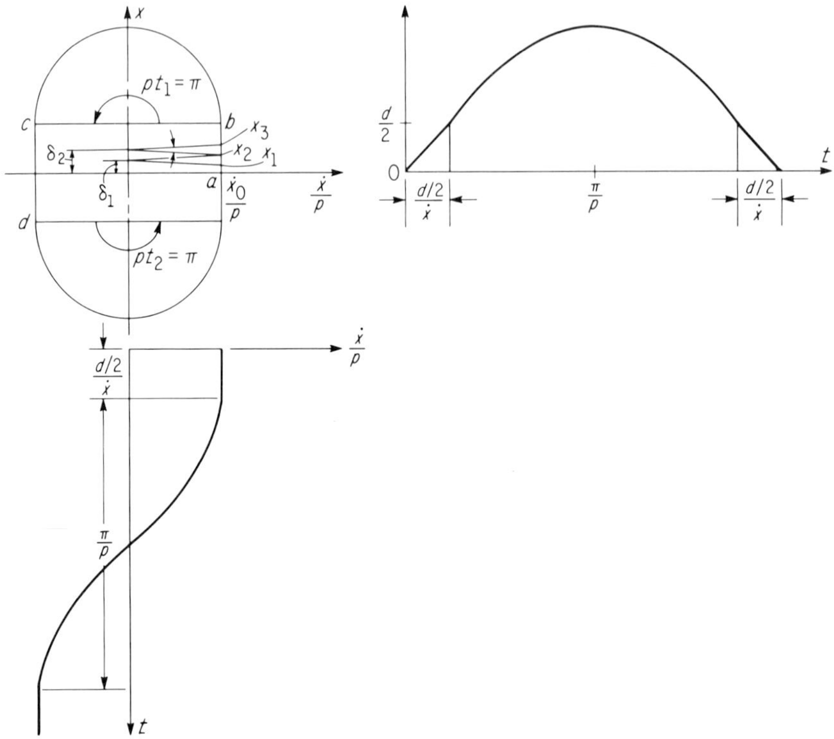

FIG. 4-13. Phase-plane solution for system with backlash.

from a to b according to the δ function. It can be seen from Fig. 4-13 that for any increment of time Δt, the corresponding displacement $\Delta x = p\Delta t\, \dot{x}_0/p = \dot{x}_0 \Delta t$, which indicates a constant velocity motion.

The displacement-time plot is therefore obtained by projection from the phase diagram. The period of oscillation is found to be

$$\text{Period} = 2\left(\frac{d}{\dot{x}_0} + \frac{\pi}{p}\right) = \frac{2\pi}{p} + 2\frac{d}{\dot{x}_0}$$

and in this case, the period is dependent on the initial condition of the motion.

Numerical Solution

A solution for the nonlinear vibration can also be obtained in a numerical form. This is particularly useful in view of the availability of the digital computers. There are several methods and techniques for the numerical solution of differential equations of which the step-by-step methods of Adams and Runge-Kutta are among the best-known examples (Refs. 2 and 48). The latter is discussed in Chapter 9.

We shall illustrate the concept of numerical solutions by the following introductory example. Take the equation of the free vibration of a system with nonlinear restoring force,

$$m\ddot{x} + f(x) = 0$$

and assume that the initial conditions are such that at $t = 0$,

$$x = x_0 \quad \text{and} \quad \dot{x} = \dot{x}_0$$

By substitution in the equation of motion, the initial acceleration is found as

$$\ddot{x}_0 = -\frac{1}{m}f(x_0)$$

Assuming this to be constant as a first approximation over a short period of time Δt, the velocity at the end of this period will be

$$\dot{x}_1 = \ddot{x}_0\,\Delta t + \dot{x}_0$$

and

$$x_1 = x_0 + \frac{\dot{x}_0 + \dot{x}_1}{2}\Delta t$$

where $\frac{\dot{x}_0 + \dot{x}_1}{2}$ is the average velocity in the time interval Δt. A better approximation can be obtained by substituting the value of x_1 in the equation of motion to obtain the acceleration at $t = \Delta t$

$$\ddot{x}_1 = -\frac{1}{m}f(x_1)$$

Accordingly, a better value for the velocity $\dot{x}_1$ and displacement x_1 can be found as

$$\dot{x}_1 = \dot{x}_0 + \frac{\ddot{x}_0 + \ddot{x}_1}{2}\Delta t$$

and

$$x_1 = x_0 + \frac{\dot{x}_0 + \dot{x}_1}{2}\Delta t$$

This can be repeated until sufficient accuracy is reached between the substituted x_1 and the resulting value.

Next, we can proceed in the same way to evaluate x_2 after another interval of time Δt or at $t = 2\Delta t$, and so on.

In this method of numerical solution sufficient accuracy can usually be obtained, except for cases of strong nonlinearities, by repeating the calculations twice for every increment of time.

It should also be noted that this method which has been applied to the free undamped vibration, is general and can be applied as well when the systems are damped and subjected to a general forcing function. Although more sophisticated methods are utilized in numerical solution of differential equations, the one described here serves as an illustration for numerical techniques.

Example 8

A simple pendulum with length 61.8 in. is released from a position 0.1 radian from the equilibrium position. Utilize the numerical integration method described above to find the time history of the motion.

The equation of motion in this case is

$$\ddot{\alpha} + \frac{386}{61.8}\alpha = 0$$

or

$$\ddot{\alpha} + 6.25\alpha = 0$$

and the natural frequency is 2.5 rad/sec. The initial conditions are

$$\text{at} \quad t = 0, \quad \alpha = 0.1, \qquad \dot{\alpha} = 0$$

Taking $\Delta t = 0.1$ sec and following the procedure shown above, the second approximation values for x, $\dot{x}$ are tabulated below for values of t up to 0.6 sec.

t, sec	*α, rad*	*α̇, rad/sec*	*α̈, rad/sec²*
0	0.1	0	−0.6250
0.1	0.09690	−0.06153	−0.60563
0.2	0.08788	−0.11825	−0.54925
0.3	0.07344	−0.16950	−0.4590
0.4	0.05449	−0.20945	−0.3400
0.5	0.03220	−0.23643	−0.20125
0.6	0.00794	−0.24885	−0.049625

It is left to the reader to check this solution with the analytical solution given in Example 1, namely:

$$\alpha = 0.1 \cos 2.5t \qquad \dot{\alpha} = -0.25 \sin 2.5t$$

and

$$\ddot{\alpha} = -0.625 \cos 2.5t$$

4-5. FREE VIBRATION UNDER THE INFLUENCE OF FRICTION

We have discussed in the previous section the free vibration of systems with linear as well as nonlinear restoring functions. It was found that once a vibration is started it will continue indefinitely in the absence of frictional forces.

In this section we shall study the effect of these frictional forces on the free vibration. Starting with a system having a linear spring under the influence of a frictional force which is proportional to the velocity, Fig. 4-14, the equation of motion will be linear:

$$m\ddot{x} + c\dot{x} + kx = 0 \tag{4-12}$$

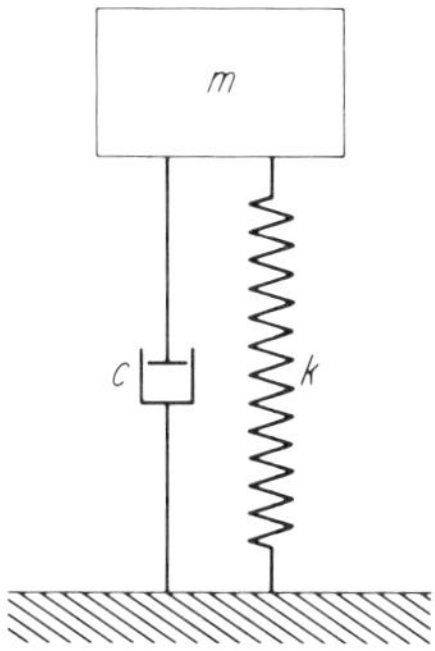

FIG. 4-14

The damping in this case is viscous and c is known as the *damping coefficient*. Assuming a solution to Eq. 4-12 of the form $x = Ac^{at}$ therefore

$$(ma^2 + ca + k)x = 0$$

from which either $x = 0$ or $(ma^2 + ca + k) = 0$. The last condition requires that

$$\begin{aligned} a &= -c/2m \pm \sqrt{(c/2m)^2 - k/m} \\ &= \sqrt{k/m}\left[-\frac{c}{2\sqrt{km}} \pm \sqrt{\left(\frac{c}{2\sqrt{km}}\right)^2 - 1}\right] \end{aligned}$$

Using the notation

$$\sqrt{k/m} = \omega_n, \; 2\sqrt{km} = c_c, \quad \text{and} \quad c/c_c = \zeta$$

therefore

$$a = \omega_n[-\zeta \pm \sqrt{\zeta^2 - 1}]$$

There are four different possibilities for the solution, depending on the value of ζ.

A. $\zeta > 1$

The solution will be

$$x = e^{-\zeta\omega_n t}[A_1 e^{+\omega_n\sqrt{\zeta^2-1}\,t} + A_2 e^{-\omega_n\sqrt{\zeta^2-1}\,t}] \tag{4-13}$$

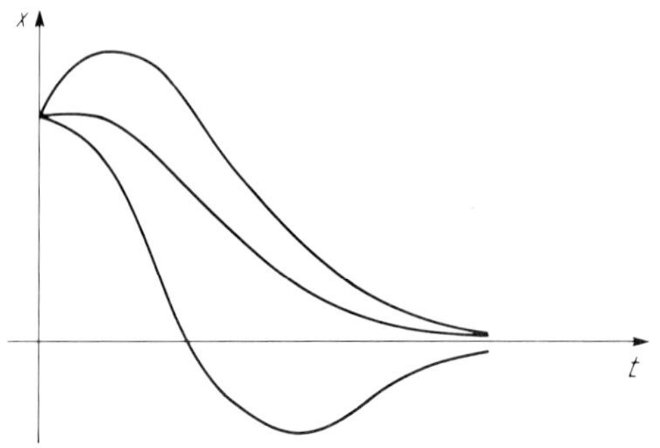

FIG. 4-15

which represents an overdamped motion, Fig. 4-15. A_1 and A_2 are integration constants whose values depend on the initial conditions. It should be noted here that under certain initial conditions, one and only one crossing of $x = 0$ axis is possible.

B. $\zeta = 1$

The solution is therefore

$$x = (A_1 + A_2 t)e^{-\omega_n t} \tag{4-14}$$

which represents a critically damped motion, Fig. 4-16. A_1 and A_2 are also integration constants. A critically damped case represents the transition between vibratory and nonvibratory motions.

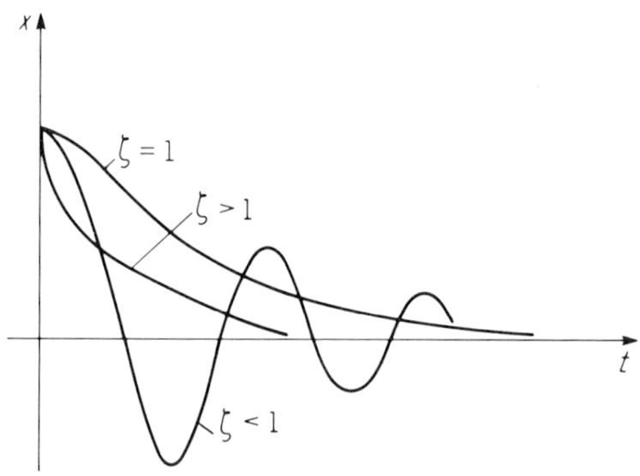

FIG. 4-16

C. $\zeta < 1$

The solution in this case is

$$\begin{aligned} x &= e^{-\omega_n t \zeta}[A_1 e^{i(\omega_n \sqrt{1-\zeta^2})t} + A_2 e^{-i(\omega_n \sqrt{1-\zeta^2})t}] \\ &= A e^{-\omega_n t \zeta} \sin (\omega_n \sqrt{1 - \zeta^2}\, t + \phi) \end{aligned} \tag{4-15}$$

where A and ϕ are integration constants. This describes a vibration with a frequency of

$$\omega_{nd} = \omega_n \sqrt{1 - \zeta^2} \tag{4-16}$$

and a continually decreasing amplitude, Figs. 4-17 and 4-18.

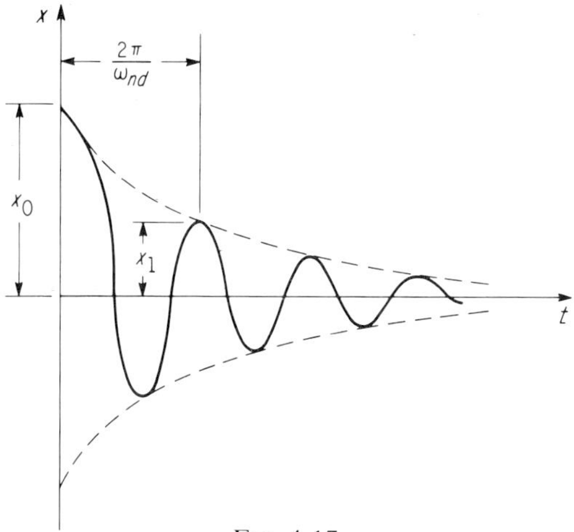

FIG. 4-17

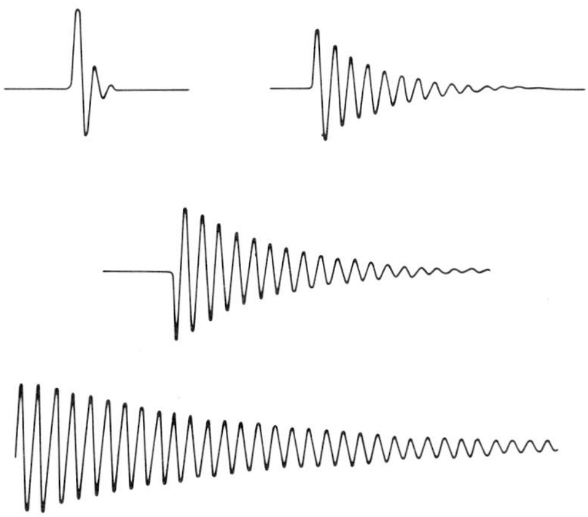

FIG. 4-18. Examples of vibration decay with different damping coefficients.

The decrease in amplitude after one cycle can be calculated by substituting

$$t = \frac{2\pi}{\omega_n \sqrt{1 - \zeta^2}} = \text{period of oscillation}$$

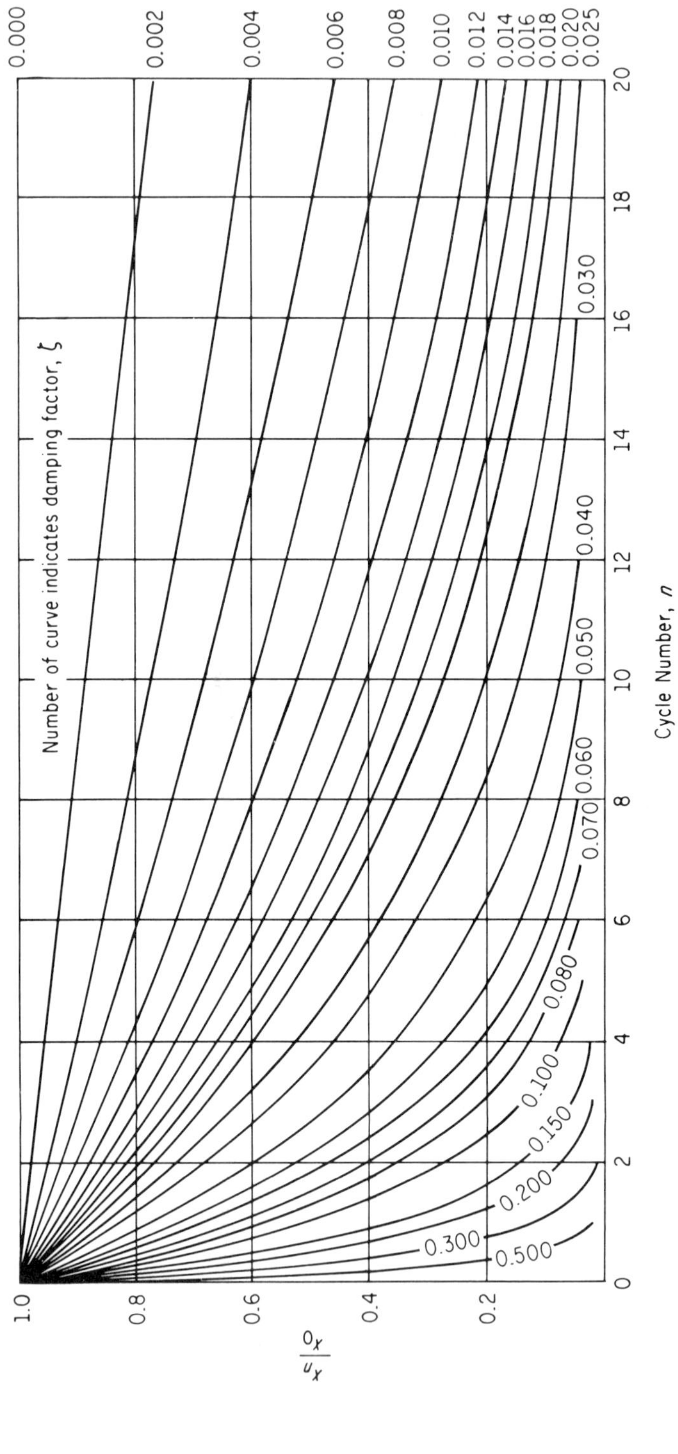

FIG. 4-19. Theoretical decay curves for viscous damping.

and we obtain

$$\frac{x_0}{x_1} = e^{\frac{2\pi\zeta}{\sqrt{1-\zeta^2}}}$$

and

$$\log\frac{x_0}{x_1} = \delta = \text{logarithmic decrement} = \frac{2\pi\zeta}{\sqrt{1-\zeta^2}} \qquad (4\text{-}17)$$

The amplitude x_n after n cycles is similarly obtained from

$$\frac{x_n}{x_0} = e^{-\frac{2\pi\zeta}{\sqrt{1-\zeta^2}}n} \qquad (4\text{-}18)$$

which gives the equation for the decay curve. The relationship is plotted in Fig. 4-19 for different damping factors. This is a useful graph for determining the damping coefficient from the decay curves of a transient vibration. Figure 4-20 shows the plot of the damped natural frequency, ω_{nd}, as calculated by Eq. 4-16.

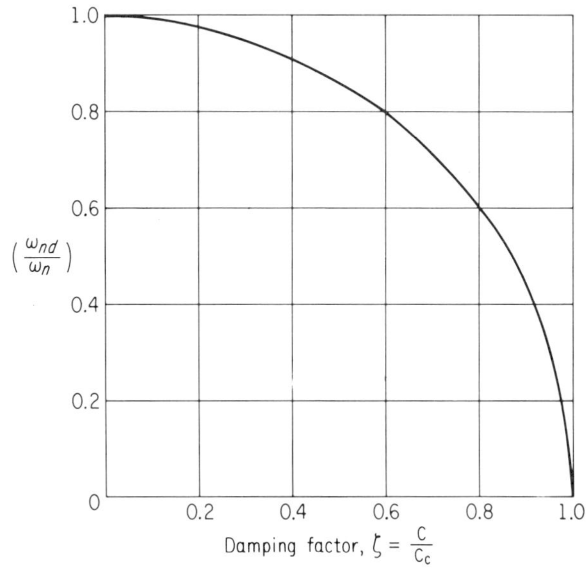

FIG. 4-20

D. $\zeta = 0$

Needless to say, this condition will result in the displacement-time relation of the free undamped vibration, Eq. 4-3.

The Phase-Plane Method for Solution of Damped Free Vibration

The viscous type damping which has been discussed is the only type of frictional resistance which allows the determination of the motion by simple analytical methods. A general solution for any damped free vibration can be obtained by the phase-plane technique.

Let us consider the general equation

$$m\ddot{x} + f_1(\dot{x}) + f_2(x) = 0 \tag{4-19}$$

where $f_1(\dot{x})$ = a general damping force
$f_2(x)$ = a general restoring force

Equation 4-19 can be rewritten in the δ form as

$$\ddot{x} + (1/m)[x + f_1(\dot{x}) + (f_2(x) - x)] = 0$$

or

$$\ddot{x} + (1/m)[x + \delta_{\dot{x}} + \delta_x] = 0$$

which is again a free vibration with a continually modified x. The modification can generally have two components: δ_x as a function of the displacement, and $\delta_{\dot{x}}$ as a function of the velocity. We shall try several examples to illustrate the use of this method in damped vibration.

Example 9

A mass 4 lb-in.$^{-1}$ sec^2 is supported by a linear spring $k = 100$ lb/in. Its motion is damped by a viscous dashpot having a damping coefficient c, Fig. 4-21a. If the initial conditions of the motion are such as at $t = 0$, $x = x_0$, $\dot{x} = 0$, find the displacement-time curve.

The differential equation is

$$4\ddot{x} + 100x + c\dot{x} = 0$$

which is rewritten in the δ form as

$$\ddot{x} + 5^2[x + c\dot{x}/100] = 0$$

from which

$$\delta_{\dot{x}} = (c/100)x, \quad \delta_x = 0, \quad \text{and} \quad p = 5 \text{ rad/sec}$$

$\delta_{\dot{x}}$ can be represented by a straight line as shown in Fig. 4-21b. This straight line can also be represented in the phase plane by the line *ab* for simplifying the graphical construction. The plot of the phase trajectory is carried on as shown in Fig. 4-21c from which the x-t curve is projected.

Example 10

The frictional force in the previous system is provided by means of a Coulomb damper, Fig. 4-22a. The normal force is taken equal to N and the coefficient of friction is μ. Find the displacement-time function.

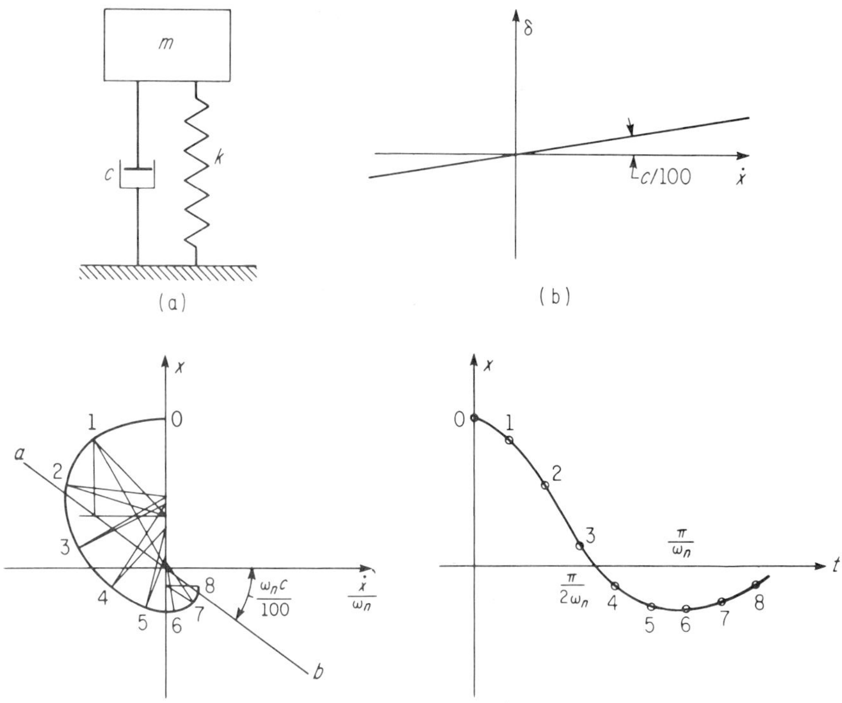

(c)

FIG. 4-21

The equation of motion in this case is

$$m\ddot{x} + 100\,x + (\operatorname{sgn}\dot{x})\mu N = 0$$

where $\operatorname{sgn}\dot{x} = (\dot{x}/\,|\,\dot{x}\,|\,)$.

Therefore

$$\ddot{x} + 5^2\left(x + \frac{(\operatorname{sgn}\dot{x})}{100}\mu N\right) = 0$$

where $p = 5$

$$\delta = \frac{(\operatorname{sgn}\dot{x})}{100}\mu N$$

and the phase-plane plot is as shown in Fig. 4-22b.

It can be easily seen that the envelope of the displacement-time curve consists of straight lines. This is an important characteristic of a Coulomb-damped motion. The decay in amplitude is $2\mu N/k$, every half-cycle or $4\mu N/k$ per cycle.

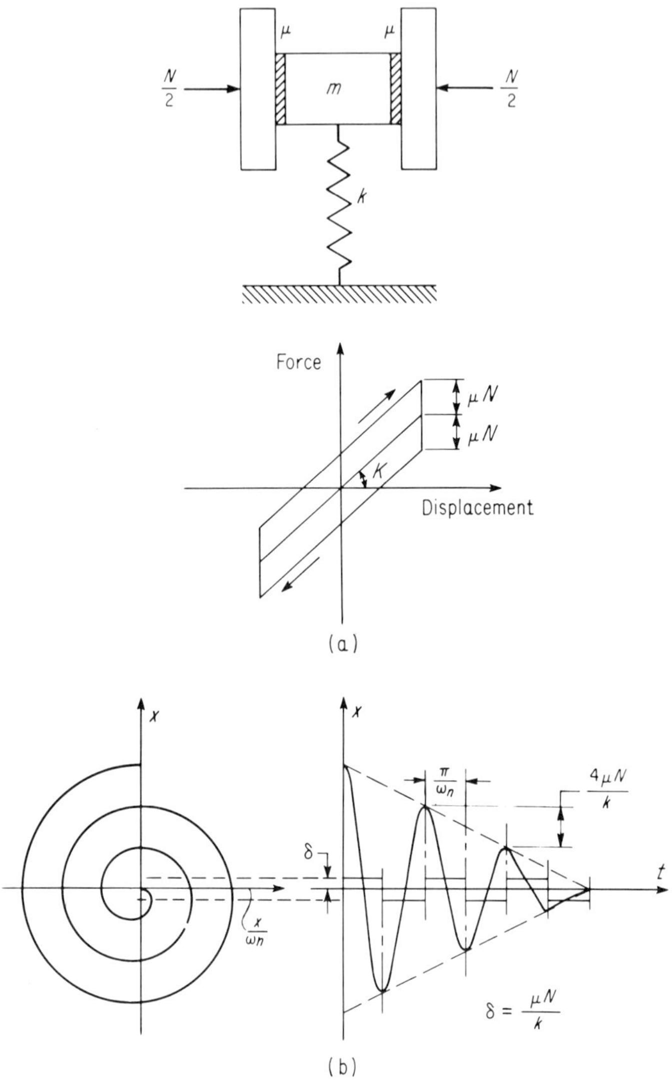

FIG. 4-22. Vibration decay with Coulomb damping.

Example 11

In this example the frictional resistance to the motion for the previous mass-spring system will be assumed proportional to the spring force. This type occurs frequently in mechanical systems, especially in certain types of couplings, geared systems, and other power-transmitting elements. The shape of the static hysteresis loop in this case is as shown in Fig. 4-23a. The equation of motion can be written as follows.

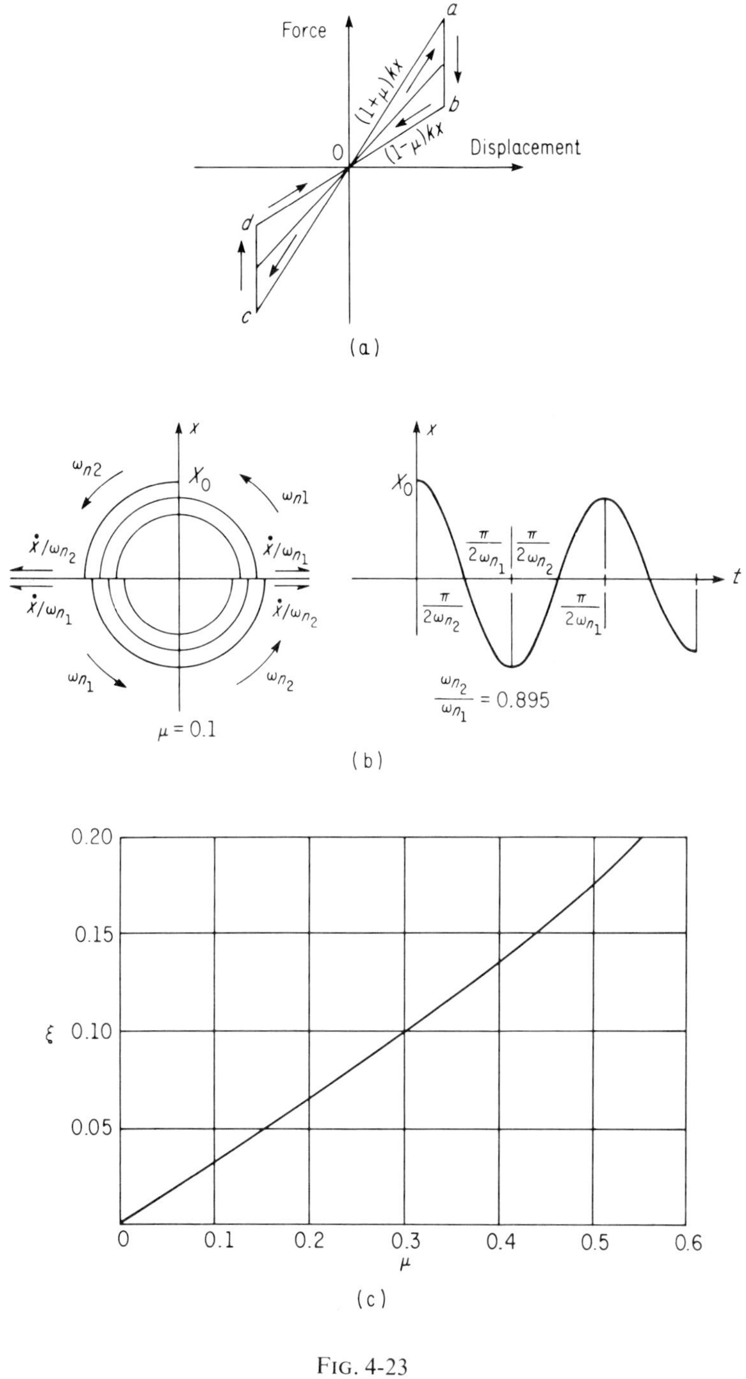
Force
Displacement
0
a
b
c
d
$(1+\mu)kx$
$(1-\mu)kx$
(a)
x
X_0
ω_{n2}
ω_{n1}
$\dot{x}/\omega_{n_2}$
$\dot{x}/\omega_{n_1}$
$\dot{x}/\omega_{n_1}$
$\dot{x}/\omega_{n_2}$
ω_{n_1}
ω_{n_2}
$\mu = 0.1$
t
$\frac{\pi}{2\omega_{n_1}}$
$\frac{\pi}{2\omega_{n_2}}$
$\frac{\pi}{2\omega_{n_2}}$
$\frac{\pi}{2\omega_{n_1}}$
$\frac{\omega_{n_2}}{\omega_{n_1}} = 0.895$
(b)
ξ
0.20
0.15
0.10
0.05
0
0.1
0.2
0.3
0.4
0.5
0.6
μ
(c)

FIG. 4-23

From $0 \to a$

$$x > 0 \quad \text{and} \quad \dot{x} > 0$$

$$m\ddot{x} + k(1 + \mu)x = 0$$

or

$$\ddot{x} + \omega_{n_1}x = 0$$

From $b \to 0$

$$x > 0, \quad \dot{x} < 0$$

$$m\ddot{x} + k(1 - \mu)x = 0$$

or

$$\ddot{x} + \omega_{n_2}x = 0$$

From $0 \to c$

$$x < 0, \quad \dot{x} < 0$$

$$m\ddot{x} + k(1 + \mu)x = 0$$

or

$$\ddot{x} + \omega_{n_1}x = 0$$

From $d \to 0$

$$x < 0, \quad \dot{x} > 0$$

$$m\ddot{x} + k(1 - \mu)x = 0$$

or

$$\ddot{x} + \omega_{n_2}x = 0$$

where

$$\omega_{n_1} = \sqrt{\frac{k}{m}(1 + \mu)}$$

$$\omega_{n_2} = \sqrt{\frac{k}{m}(1 - \mu)}$$

Assuming that at $t = 0$, $x = x_0$, $\ddot{x} = 0$, the phase plane can be plotted as shown in Fig. 4-23b. In this case the diagram is composed of quarter circles in the four quadrants. It can be seen from the equations of motion that the natural frequency (and consequently the scale of the velocity axis) changes in each quadrant as illustrated. The displacement history is also plotted. The period for a complete cycle can be calculated as

$$\tau_d = \frac{\pi}{\omega_{n1}} + \frac{\pi}{\omega_{n2}}$$

and the damped frequency becomes

$$\omega_{nd} = \frac{2\pi}{\tau_d} = \frac{2(\omega_{n1}\omega_{n2})}{\omega_{n1} + \omega_{n2}} = 2\sqrt{\frac{k}{m}}\left[\frac{\sqrt{1 - \mu^2}}{\sqrt{1 - \mu} + \sqrt{1 + \mu}}\right]$$

The amplitude decay for any arbitrary number of cycles n can be readily calculated from the phase-plane plot as follows:

$$\frac{x_n}{x_0} = \left(\frac{\omega_{n2}^2}{\omega_{n1}^2}\right)^n$$

$$= \left[\frac{1 - \mu}{1 + \mu}\right]^n$$

The amplitude decay under condition of viscous damping is found from Eq. 4-18 to be

$$\frac{x_n}{x_0} = e^{-\frac{2\pi\zeta}{\sqrt{1-\zeta^2}} n}$$

Therefore, for equal decay in both types of damping,

$$\frac{1 - \mu}{1 + \mu} = e^{-\left(\frac{2\pi\zeta}{\sqrt{1-\zeta^2}}\right)}$$

or

$$\mu = \frac{e^{\left(\frac{2\pi\zeta}{\sqrt{1-\zeta^2}}\right)} - 1}{e^{\left(\frac{2\pi\zeta}{\sqrt{1-\zeta^2}}\right)} + 1}$$

giving an expression for the equivalent viscous damping ratio ζ for a particular coefficient of friction μ for the case where the friction force is proportional to the spring force. Figure 4-23c gives a plot of this relationship.

Example 12

Damping of vibratory motion by the energy dissipated in an impact can be illustrated (Fig. 4-24a) by the simple example of a pendulum released from an angle θ_0 to repeatedly impact a stop placed at the equilibrium position of the pendulum. Assuming the coefficient of restitution to be e_r, the pendulum oscillation can be readily represented in the phase plane as shown in Fig. 4-24b. The amplitude decay after n impacts is therefore evaluated from

$$\frac{\theta_n}{\theta_0} = e_r^n$$

and the equivalent viscous damping factor ζ for a particular coefficient of restitution e_r can be evaluated from

$$e_r = e^{-\left(\frac{2\pi\zeta}{\sqrt{1-\zeta^2}}\right)}$$

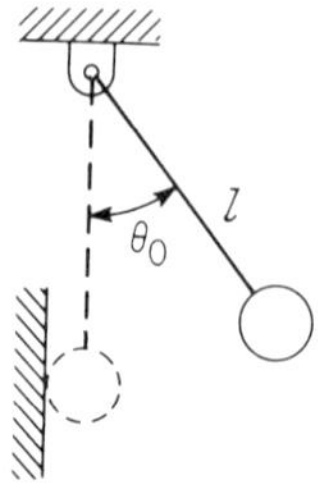

(a)

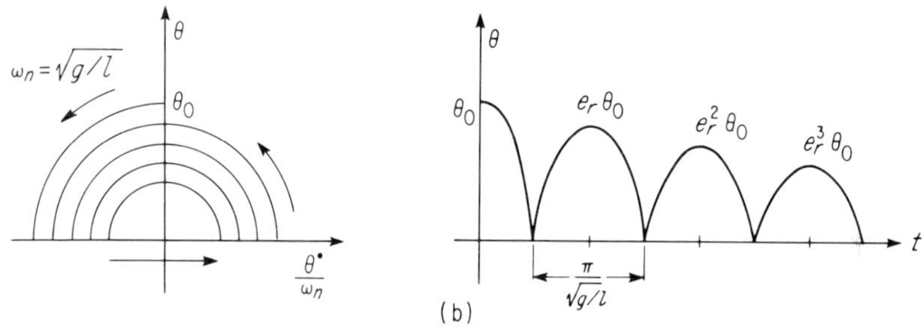

(b)

FIG. 4-24

4-6. SUMMARY

We have discussed in this chapter the free vibration of single-degree-of-freedom systems with linear as well as nonlinear characteristics. The effect of damping forces on the natural frequency and the amplitude decay of the free vibration was also discussed. Some of the analytical, graphical, and numerical techniques used for the solution of the second-order differential equation of motion were illustrated in an introductory manner in order to familiarize the reader with these methods. In the following chapter we shall deal with the response of these systems to different types of disturbances which are described by their time functions.

PROBLEMS

4-1. A system can be represented by a mass m supported by two springs with stiffness factors k_1, k_2 as shown. Write an expression for the natural frequency of the system.

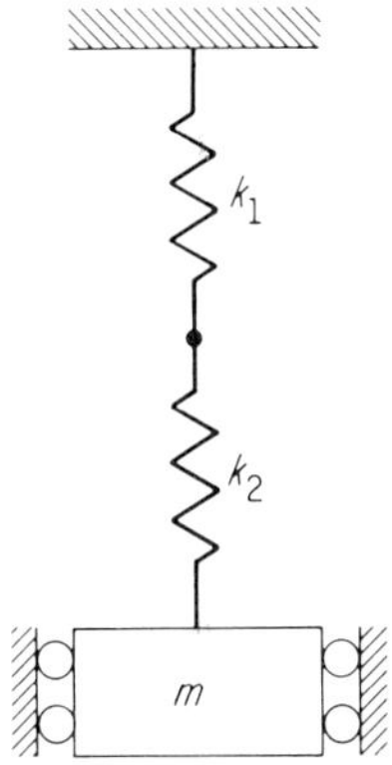

PROB. 4-1

4-2. A mass m is suspended by two springs with stiffness k_1 and k_2 respectively. Write an expression for the period of its natural oscillation.

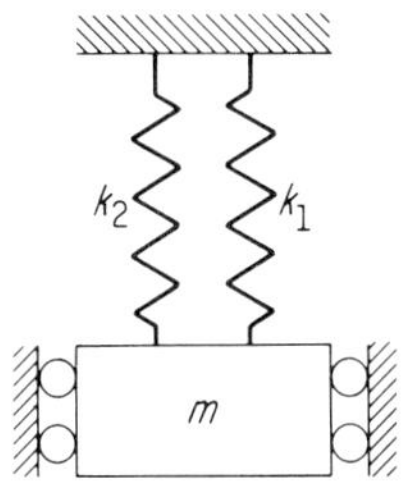

PROB. 4-2

4-3. Calculate the natural frequency for the system shown.

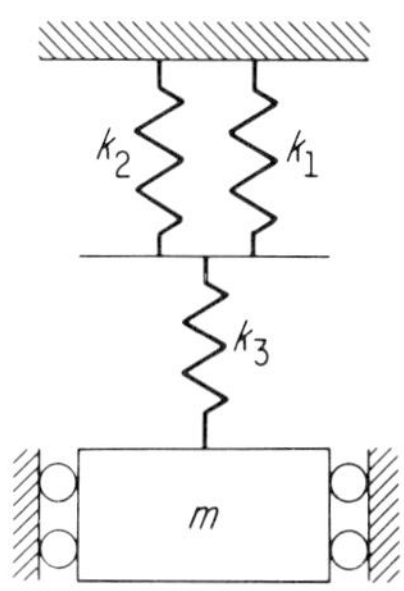

PROB. 4-3

4-4. A spring is attached horizontally to a simple pendulum at a distance a from the supporting point. Calculate the spring constant k necessary to double the frequency of its oscillation. Assume small displacements.

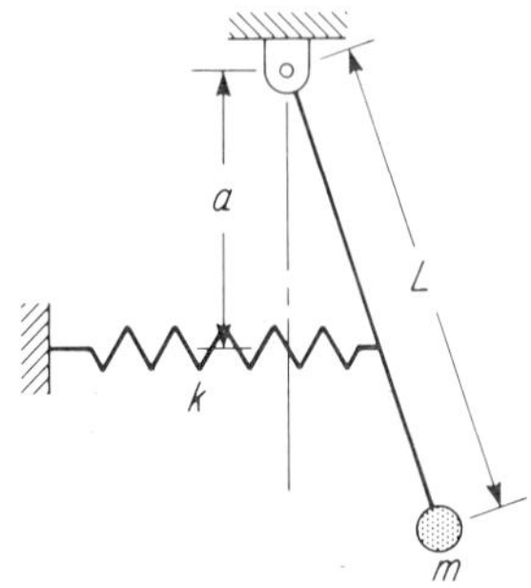

PROB. 4-4

4-5. A mass m lb in.$^{-1}$ sec^2 is supported on a functionless pulley with radius R and moment of inertia I (lb in. sec.^{2}) by means of an inextensible string ab. The string is attached at b to a spring with stiffness k lb/in. If the weight was suddenly released from the position where there was neither slack nor tension in the string ab, find the maximum spring force by utilizing the phase plane.

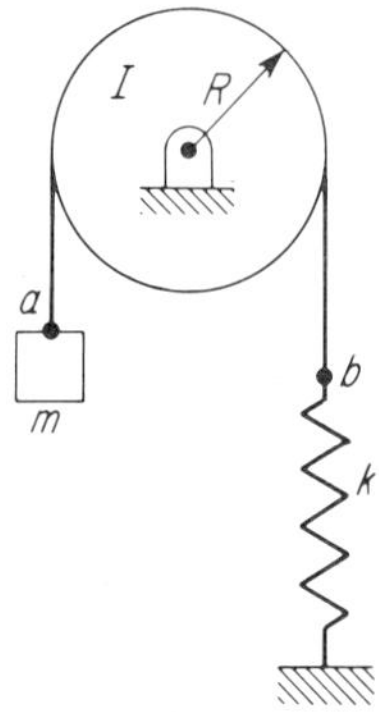

PROB. 4-5

4-6. A 3.86-lb particle is supported by two equal strings oa and ob as shown. The equilibrium position is 100 lb on both sides. The particle is pulled 0.1 in. and then released. Represent the motion in the phase plane, assuming that the tension remains constant during the oscillation.

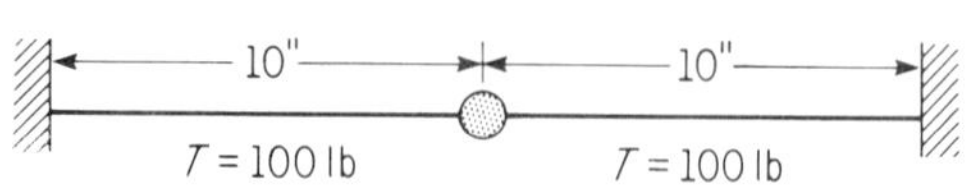

PROB. 4-6

4-7. A beam 20 in. long and 1 × 1/8 in. cross section is simple supported at both ends and carries a 100-lb weight at midspan. If the mass is subjected to an initial velocity of 10 in./sec, calculate the maximum stress in the beam due to this velocity. Compare this value to the static stress due to the weight.

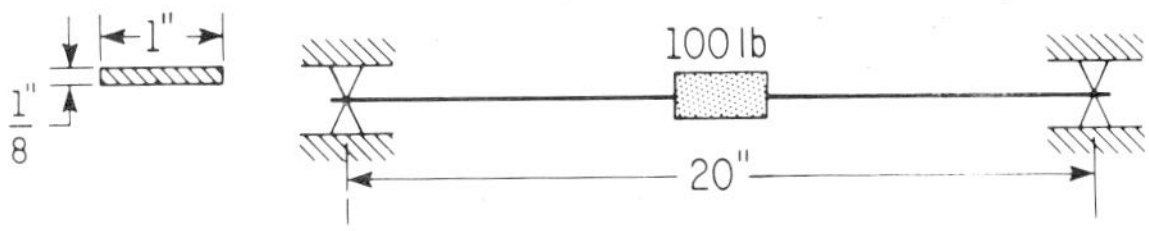

PROB. 4-7

4-8. Utilize the phase-plane-δ method to evaluate the period of oscillation for the given system when the mass is given an initial velocity of 10 in/sec to the right. What is the maximum force in each spring?

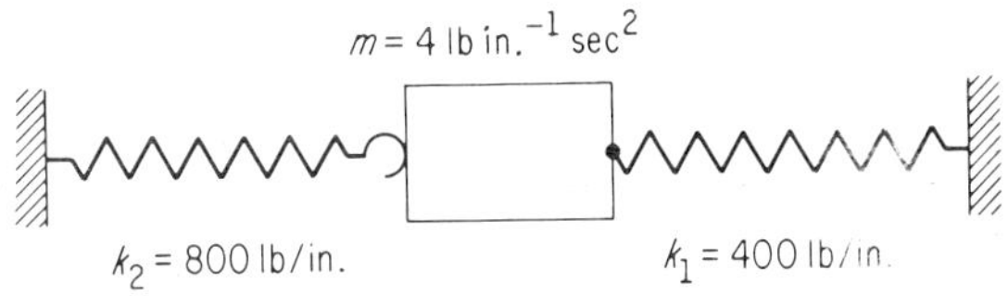

PROB. 4-8

4-9. The pendulum shown is released from rest when $\theta = 30°$. Calculate the maximum spring compression and the period of the oscillation.

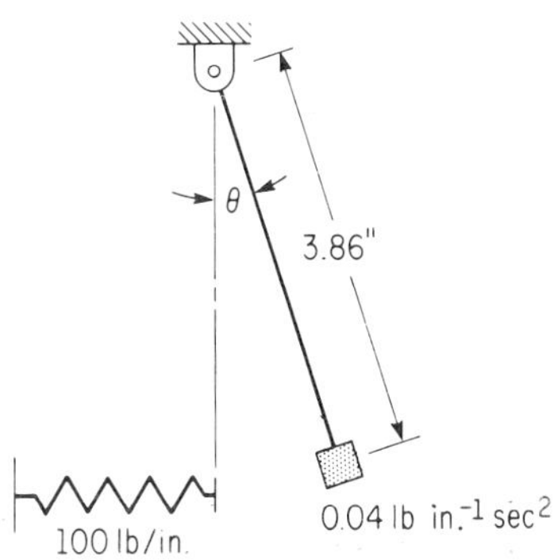

PROB. 4-9

4-10. The mass in the sketch is displaced a distance 1.5 in. to the left and then released. Evaluate the period of oscillation by utilizing the phase-plane-δ method.

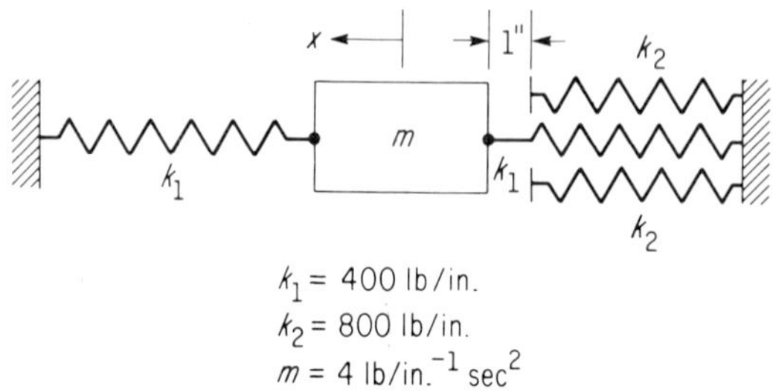

PROB. 4-10

4-11. A cantilever beam carries a weight of 3.86 lb and has a maximum static deflection of 0.01 in./lb. In order to limit the maximum stress in the cantilever a stop is provided on both sides of the mass at a distance 0.1 in. from the equilibrium position. If the mass is given an initial velocity of 20 in./sec, calculate the period of oscillation.

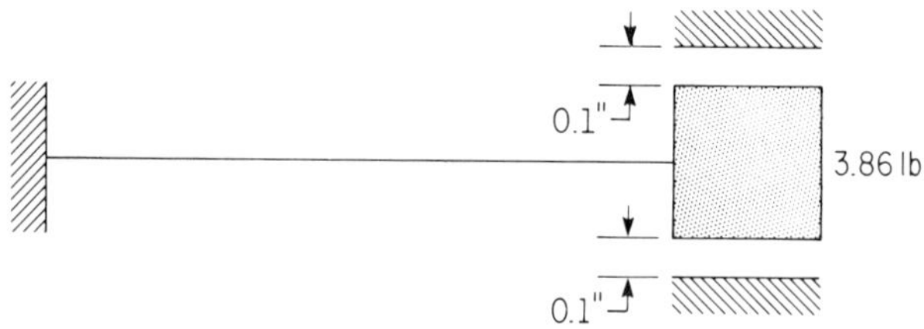

PROB. 4-11

4-12. A steel rod 25 in. long with $1/2 \times 1/2$ in. cross section carries a 50-lb weight at one end. The rod is supported by a frictionless hinge and a spring with stiffness 400 lb/in. as shown. Calculate the damping coefficient of a viscous dashpot necessary for critical damping of the system oscillation.

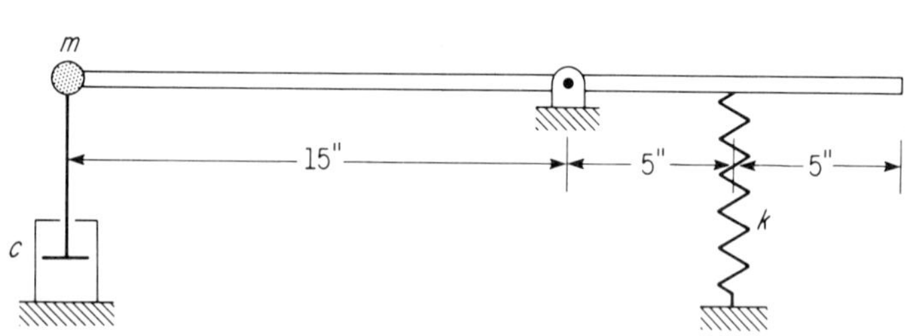

PROB. 4-12

4-13. A weightless rod carries a mass m at its midspan. The rod is maintained at a horizontal position by a rod stiffness k. It is required to design a viscous dashpot to provide an amplitude decay of 40 per cent after two cycles of natural vibrations caused by any disturbance. Calculate the damping coefficient c for the dashpot. What is the maximum Coulomb frictional torque necessary to satisfy this requirement for a maximum amplitude of d in.

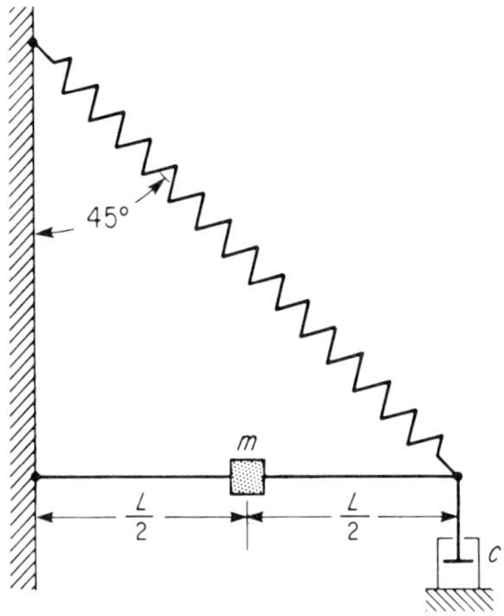

PROB. 4-13

4-14. A 165-lb mass is supported by four leaf springs with a combined stiffness of 54 lb/in. Calculate the Coulomb frictional force which can be applied on both sides of the mass to cause a rate of decay of 0.1 in. per cycle in a natural vibration. A 5-lb sphere supported as a ballistic pendulum is released from a 30° position. The angle of rebound after striking the mass is 25°. How many cycles will the system complete before it comes to rest due to the calculated Coulomb friction?

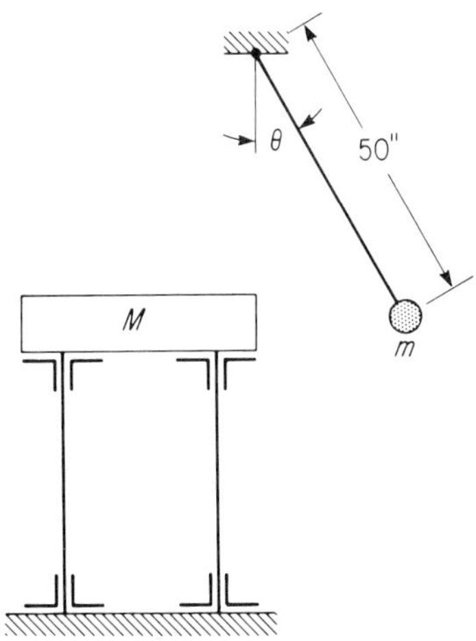

PROB. 4-14

4-15. The sphere in the given system slides without rolling under the influence of a spring on a surface which is partly smooth and partly rough as shown. If the mass is pulled 1 in. to the left and then released, describe its motion in the phase plane.

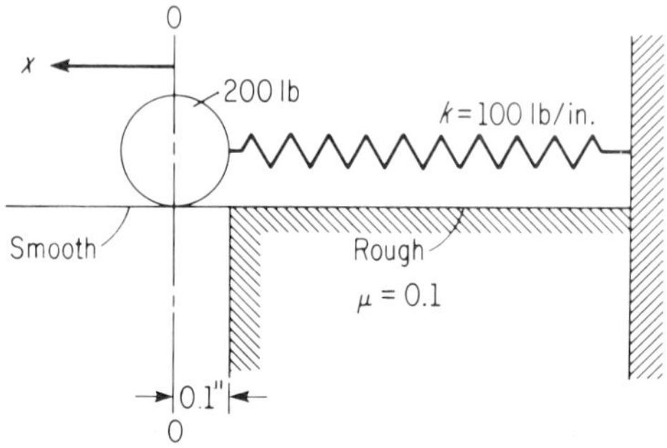

PROB. 4-15

4-16. The mass in the given system is subjected to an impact causing an initial velocity of 5 in./sec. It is required to utilize a Coulomb friction damper to stop the mass before it changes the direction of motion due to this impact. Use the phase-plane method to find the minimum frictional force to accomplish this.

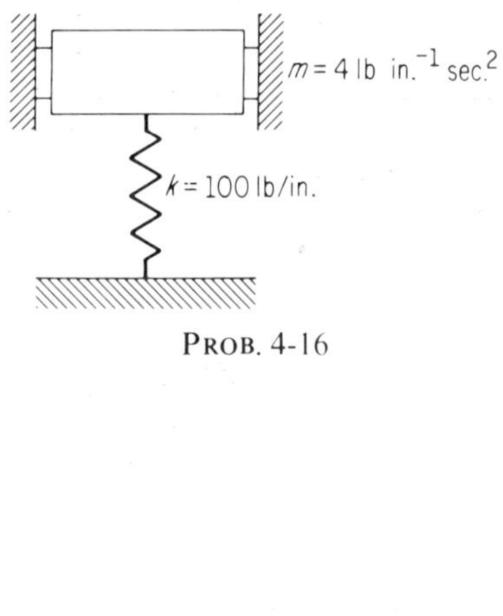

PROB. 4-16

chapter 5

System Response to General Disturbances

5-1. INTRODUCTION

We have discussed in the previous chapter the behavior of a disturbed single-degree-of-freedom oscillating system after the disturbance ceased to exist. The effect of the disturbance in this case is to give the system the initial values of displacement and velocity causing the subsequent free vibration. In the following chapters we shall study the response of the system to the external disturbing function itself. Chapter 5 deals with the system behavior under the influence of a general force which is specified by its time function. In Chapter 6 the special case of a periodic disturbance is discussed. The latter is of considerable practical interest because of its frequent occurrence in rotating machinery.

5-2. METHODS OF ANALYSIS

In analyzing the motion of a single-degree-of-freedom vibratory system under the influence of a general forcing function, three of the methods of solution for the differential equation of motion will be discussed. The first is the general form of the phase-plane-δ method, which is applicable to linear as well as nonlinear systems. The second is the Laplace transform technique, which is restricted to the analysis of linear systems. The third is electronic integration by utilizing the analog computer.

The Phase-Plane-δ Method

This method has been applied to the free vibration of single-degree-of-freedom systems in Chapter 4. It will be used here in its general form to include the effect of general forcing functions.

The general equation describing the motion of the system under consideration is

$$m\ddot{x} + f_1(\dot{x}) + f_2(x) = f_3(t) \tag{5-1}$$

where m = mass of vibrating element
x = displacement of mass
$f_1(\dot{x})$ = damping force represented by a general function of velocity
$f_2(x)$ = restoring force represented by a general function of displacement
$f_3(t)$ = general forcing function

This equation is rewritten in the δ form by adding and subtracting x and arranging as follows:

$$\ddot{x} + \frac{1}{m}[x + f_1(\dot{x}) + f_2(x) - x - f_3(t)] = 0$$

By using the notation

$$\begin{aligned} \delta_{\dot{x}} &= f_1(\dot{x}) &&= \text{general function of velocity} \\ \delta_x &= f_2(x) - x &&= \text{general function of displacement} \\ \delta_t &= -f_3(t) &&= \text{general function of time} \end{aligned}$$

and

$$\delta = \delta_{\dot{x}} + \delta_x + \delta_t$$

Therefore the previous equation takes the form

$$\ddot{x} + \frac{1}{m}[x + \delta] = 0$$

This equation is readily treatable in the phase-plane as explained in Chapter 4. The displacement x in this case would be continually modified by the amount $\delta = (\delta_{\dot{x}} + \delta_x + \delta_t)$ where $\delta_{\dot{x}}$, δ_x, and δ_t are obtained from the values of x, $\dot{x}$, and time at every step in the phase-plane plot.

It should be noted here that this is the general form of the phase-plane procedure which is basically a graphical technique. It is an excellent tool of analysis and is applicable to the general case of a single-degree-of-freedom system subjected to a general forcing function. It is also applicable to systems with multiple degrees of freedom, as will be explained later in Part III of the text. The only limitations imposed on this technique are the considerable labor expended in complex problems and the graphical inaccuracies that may result when the chosen increments are not sufficiently small. The phase-plane procedure can also be programmed for solution by high-speed digital computers to make it feasible for complex systems.

Example 1

The system shown in Fig. 5-1a is subjected to a platform disturbance. As illustrated in the figure, the platform is suddenly raised a distance d_0 and then suddenly lowered to its original position after a time t_0. Find the displacement-time curve for the mass.

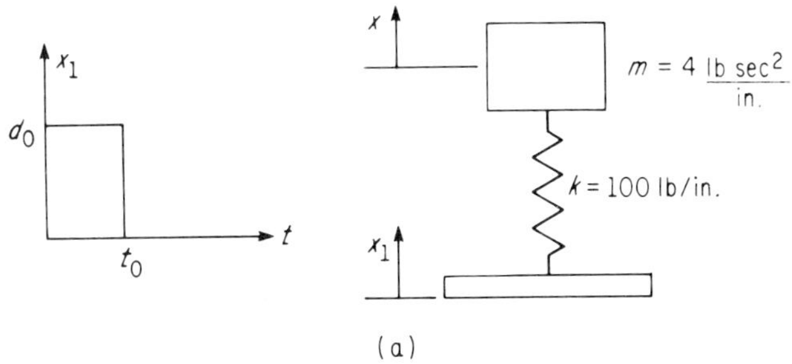

FIG. 5-1a

The differential equation of motion in this case is

$$4\ddot{x} + 100(x - x_1) = 0$$

which can be rewritten in the δ form as

$$\ddot{x} + p^2(x + \delta) = 0$$

where $p = 5$ rad/sec
$\delta = -d_0$ for $0 < t < t_0$
$= 0$ for $t > t_0$

The initial conditions are

$$\text{at} \quad t = 0, \quad x = 0, \quad \dot{x} = 0$$

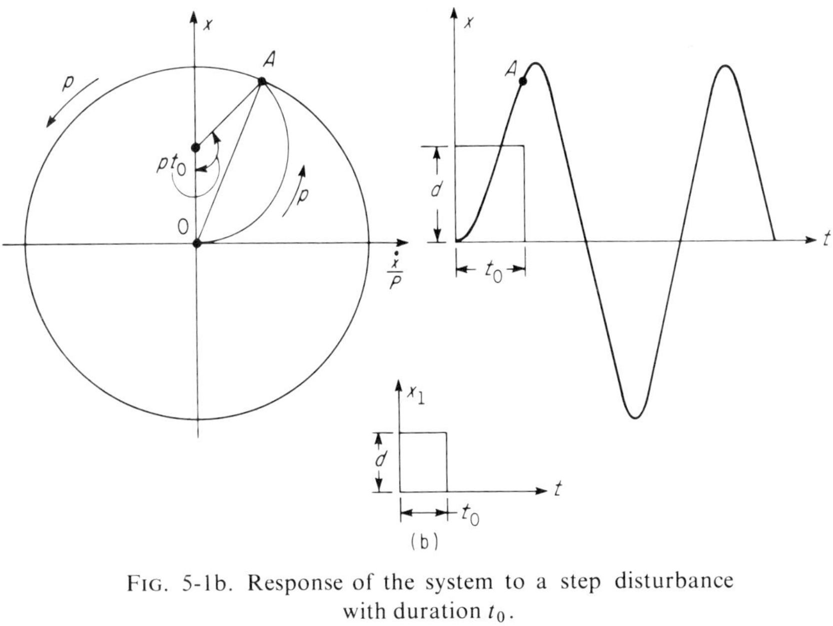

FIG. 5-1b. Response of the system to a step disturbance with duration t_0.

The origin of the phase plane will accordingly represent the starting point in this case.

For the interval between $t = 0$ and $t = t_0$, δ will have a constant value equal to $-d_0$. The center of the circular arc representing the vibration during this interval will fall at a distance d_0 above the origin of the phase-plane figure (5-2b). Since the starting point is at the origin, the radius of the circle will be equal to d_0. The angular displacement during the interval t_0 is pt_0 in the counterclockwise direction. The terminal conditions for this interval are represented by point A in the phase-plane plot.

For $t > t_0$, δ will be zero and the center of oscillation will be at the origin. The motion in this case will be represented by a circle with its center at the origin O and radius equal to OA. This motion will be a free vibration with a frequency $p = 5$ rad/sec and initial conditions represented by point A.

Example 2

Plot the displacement-time and velocity-time curves for the system of Fig. 5-2a when subjected to the pulse given in Fig. 5-2b.

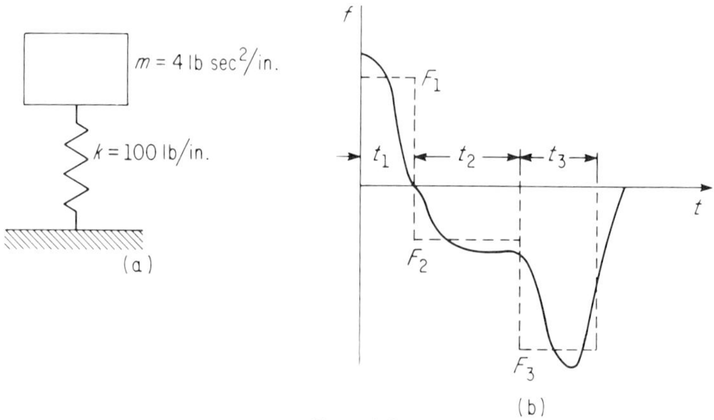

FIG. 5-2

The forcing function of Fig. 5-2b can be approximated to three steps, as follows:

$$F_1 = 500 \text{ lb} \qquad 0 < t < 0.2 \text{ sec}$$
$$F_2 = -200 \text{ lb} \qquad 0.2 < t < 0.5 \text{ sec}$$
$$F_3 = -700 \text{ lb} \qquad 0.5 < t < 0.7 \text{ sec}$$

The equation of motion in this case will be

$$4\ddot{x} + 100x = f(t)$$

which can be arranged in the δ form as

$$\ddot{x} + p^2[x + \delta_t] = 0$$

where

$$\delta_t = -\frac{f(t)}{100}$$

$$\delta_{\dot{x}} = \delta_x = 0$$

$$p = 5 \text{ rad/sec}$$

The initial conditions in this case are

$$\text{at } t = 0, \quad x = 0, \quad \dot{x} = 0$$

The phase-plane plot is then performed in three steps. The respective values of δ_t for these steps are -5 in., $+2$ in., and $+7$ in. The corresponding angular displacements are $pt_1 = 1$ rad, $pt_2 = 1.5$ rad, and $pt_3 = 1$ rad. Point P in the phase-plane plot will therefore represent the condition when the forcing function ceases to exist and accordingly will define the initial conditions for the subsequent free vibration. This is represented in the phase plane by a circle with its center at the origin O having a radius OP. Figure 5-3 shows the phase-plane plot together with the displacement-time and velocity-time curves for this example.

Example 3

A single-degree-of-freedom system has the following characteristics:

$$\text{Mass} = 4 \text{ lb/in./sec}^2$$

$$\text{Restoring force} = 100x^3 \text{ lb}$$

$$\text{Damping force} = 5\dot{x}^2 \text{ lb}$$

where x and $\dot{x}$ are the displacement and velocity respectively.

When at rest, the mass was subjected for a period of 5 sec to a force represented by

$$f(t) = 50t - t^3 \text{ lb}$$

where t = time in seconds. Study the history of the motion during the period of force application.

The equation of motion in this case is

$$4x + (\text{sgn } \dot{x})\, 5\dot{x}^2 + 100x^3 = 50t - t^3$$

from which

$$\ddot{x} + 25[x + (\text{sgn } \dot{x})(0.05(x)^2) + (x^3 - x) + (0.01t^3 - 0.5t)] = 0$$

which is written in the δ form as

$$\ddot{x} + p^2[x + \delta] = 0$$

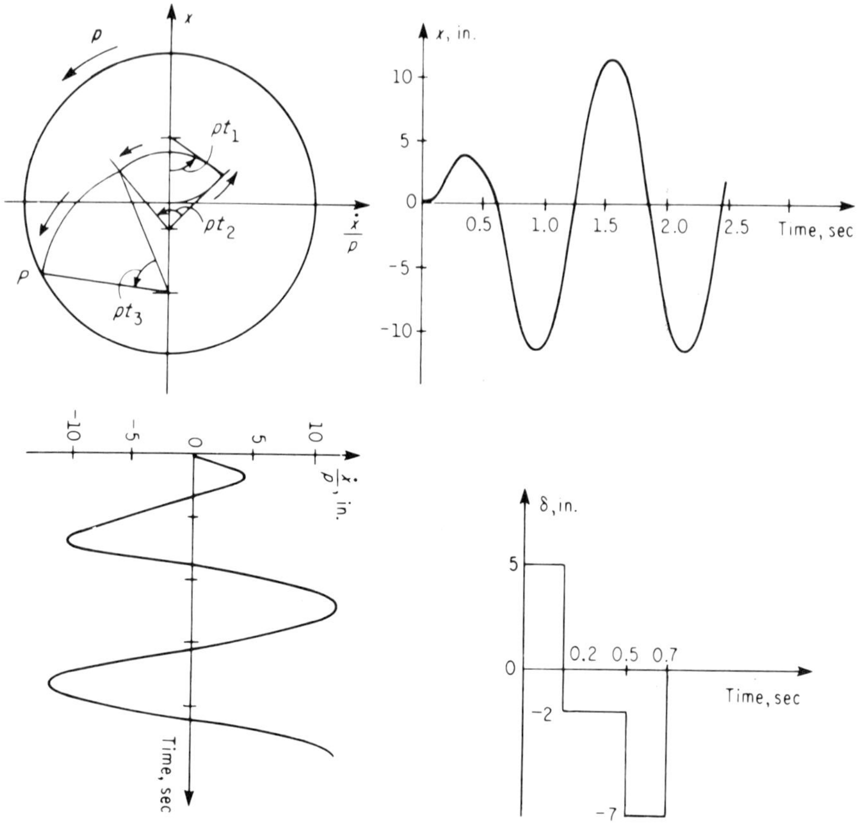

FIG. 5-3

where $p = 5$ rad/sec

$$\delta = \delta_{\dot{x}} + \delta_x + \delta_t$$
$$\delta_{\dot{x}} = 0.05(\dot{x})^2 \text{ in.}$$
$$\delta_x = x^3 - x \text{ in.}$$
$$\delta_t = 0.01t^3 - 0.5t \text{ in.}$$

The initial conditions are

$$\text{at} \quad t = 0, \quad x = 0, \quad \dot{x} = 0$$

from which

$$\delta_0 = \text{initial value of } \delta = 0$$

and the starting position in the phase plane will be at the origin.

Taking for example an incremental angular displacement $p(\Delta t) = 0.2$ rad, the corresponding increment of time will be $\Delta t = 0.2/5 = 0.04$ sec. Assuming δ as constant during the first increment of time, both the center and the starting point will be the same and there will be no motion. The conditions at the end of this increment will be

$$\text{at} \quad t = 0.04 \text{ sec}, \quad x = 0, \quad \dot{x} = 0$$

from which

$$\delta_{\dot{x}} = 0$$
$$\delta_x = 0$$
$$\delta_t = 0.01\,(0.04)^3 - 0.5\,(0.04) = -0.02 \text{ in. approx.}$$

Now the phase-plane plot can be started. The center O_1 for this increment will be at a distance 0.02 in. above the origin and the starting point will be at the origin. An arc can be plotted for an angular displacement 0.2 radian in the counterclockwise direction. At the end of this increment a point is reached with a displacement x_1 and velocity $\dot{x}_1$ at $t = 2\Delta t = 0.08$ sec. At this step, δ can be calculated by substituting these values in the expressions for δ_x, $\delta_{\dot{x}}$, and δ_t respectively. Now an additional increment can be plotted, and so on. This is continued for the 125 increments to give the required 5-sec history of the motion.

It should be noted here that the value of δ calculated at the beginning of each increment is assumed constant throughout the increment. Better accuracy is obtained by using an average value of δ calculated from the conditions at the beginning and the end of each increment. This however, requires two plots for each step. It will be left to the reader to do the actual plotting in this problem as an exercise.

It can be seen from the previous examples that the phase-plane-δ method can be utilized effectively in the analysis of systems with linear or nonlinear characteristics under the influence of a general disturbance. In many instances the system characteristics as well as the force may be given in terms of numerical data or a recorded curve. In such cases the system analysis lends itself more readily to graphical techniques than the phase-plane method. The accuracy of the solution depends on the number of increments used to fit the given functions.

Laplace Transform

There are some analytical techniques known as the methods of operational calculus which provide powerful tools for solution of differential equations. The Laplace transform which is introduced in this section is the basis of the operational calculus. This method is, however, limited to the analysis of linear systems.

The Laplace transform is a technique by which linear differential equations can be transformed into algebraic equations which are easy to manipulate. The transformation automatically takes into account the initial conditions of the problem. The solution to the differential equation is then obtained by an inverse transformation. The reader will be introduced here to the general concept of the Laplace transform and how to apply it to the solution of problems with the help of transformation tables. No attempt will be made to provide a rigorous discussion of the method. The reader who is interested in understanding this technique in depth should refer to one of the many texts on the subject. (Refs. 17–20).

The Laplace transform of a function $f(x)$ is defined as

$$Lf(x) = \int_0^\infty f(x)e^{-pt}\,dt = F(p)$$

By performing the prescribed integration, a function of a real variable is transformed into a function of an arbitrary complex variable $p = a + ib$. As will be shown, the transformation of an equation containing derivatives results in an algebraic equation in the independent complex variable which may be solved by utilizing algebraic techniques.

As an illustrative example, the Laplace transform of e^{kt} is

$$L(e^{kt}) = \int_0^\infty e^{kt}\,e^{-pt}\,dt$$

$$= \int_0^\infty e^{-(p-k)t}\,dt$$

For $a < k$ the integral diverges and for $a > k$ the integral converges. Therefore, for $a > k$,

$$L(e^{kt}) = \left[\frac{-e^{-(p-k)t}}{p-k}\right]_0^\infty$$

$$= \frac{1}{p-k}$$

Also, the transformation of $f(t) = \sin\omega t$ is obtained from

$$F(p) = L(\sin\omega t) = \int_0^\infty \sin\omega t\,e^{-pt}\,dt$$

The above integration can be evaluated by integration by parts. Let $u = e^{-pt}$ and $dv = \sin\omega t\,dt$. Then

$$du = -pe^{-pt}$$

and

$$v = -\frac{\cos\omega t}{\omega}$$

Therefore

$$\int_0^\infty \sin\omega t e^{-pt}\,dt = \frac{1}{\omega} - \frac{p}{\omega}\int_0^\infty \cos\omega t e^{-pt}\,dt$$

Similarly, $\left[\int_0^\infty \cos \omega t e^{-pt}\, dt\right]$ can be integrated by parts. If we let $u = e^{-pt}$ and $dv = \cos \omega t\, dt$, then $du = -pe^{-pt}\, dt$ and $v = \dfrac{\sin \omega t}{\omega}$, then

$$\int_0^\infty \cos \omega t e^{-pt}\, dt = \frac{p}{\omega}\int_0^\infty \sin \omega t e^{-pt}\, dt$$

The previous equations give

$$F(p) = L(\sin \omega t) = \frac{\omega}{p^2 + \omega^2}$$

It can also be seen from the previous equations that the transformation for $\cos \omega t$ is

$$L(\cos \omega t) = \int_0^\infty \cos \omega t e^{-pt}\, dt = \frac{p}{\omega}\int_0^\infty \sin \omega t e^{-pt}\, dt = \frac{p}{p^2 + \omega^2}$$

The transformations for some important functions are given in the appendix. Most of the transformations necessary for engineering purposes are available in tables of functions (Refs. 19, 20).

We can now proceed to transform as an example the differential equation of motion describing the response of an undamped linear system to a step-loading function F_0:

$$m\ddot{x} + kx = F_0$$

Using the transformation tables we find that

$$L(\ddot{x}) = p^2X(p) - \dot{X}_0 - pX_0$$

where $X(p) = Lx(t)$ and X_0, $\dot{X}_0$ are the initial conditions of the displacement and velocity respectively. Also,

$$L(F_0) = \frac{F_0}{p}$$

and

$$L(x) = X(p)$$

Accordingly, the transformation of the equation will be

$$mp^2X(p) - mpX_0 - m\dot{X}_0 + kX(p) = \frac{F_0}{p}$$

If the initial displacement X_0 and velocity $\dot{X}_0$ are zero, this equation reduces to

$$X(p) = F_0/mp(p^2 + \omega_n^2), \qquad \omega_n = \sqrt{k/m}$$

Now that we have expressed the transformation of the required function explicitly, we are ready for the inverse transformation. The transform tables show that the inverse transformation of $1/p$ is $\int_0^t dt$ and that of $\dfrac{\omega_n}{p^2 + \omega_n^2}$ is $\sin \omega_n t$. Therefore the required function is given by

$$x_{(t)} = \frac{F_0}{m\omega_n}\int_0^t \sin \omega_n t dt = \frac{F_0}{m\omega_n}\left(\frac{-\cos \omega_n t}{\omega_n}\right)_0^t$$

$$= \frac{F_0}{k}(1 - \cos \omega_n t)$$

When $F_0 = 1$, the force function is known as the *unit step function* and is denoted $u(t)$.

The response of the system to the unit step function $u(t)$ will therefore be

$$H(t) = \frac{1}{k}[1 - \cos \omega_n t] \tag{5-2}$$

If we consider the step function as composed of small impulses $(\Delta\tau)u'(t)$, where $u'(t)$ is the unit impulse (Fig. 5-4), $H_{(t)}$ could be thought of as the

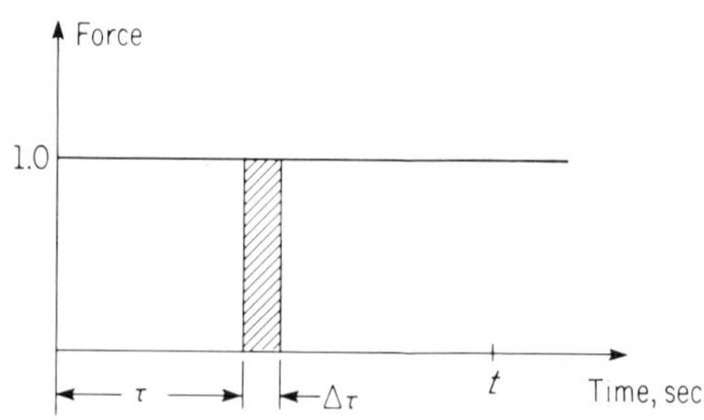

FIG. 5-4

integrated effect of all these small impulses between 0 and t or

$$H(t) = \int_0^t h(t - \tau)d\tau$$

where $h(t - \tau)$ is the response of the system to the unit pulse. Therefore

$$h(t - \tau) = \frac{dH(t)}{dt}$$

and if the time is measured from the instant when the pulse is applied,

$$h(t) = \frac{dH(t)}{dt} = \frac{\sin \omega_n t}{\sqrt{km}}$$

This conclusion could be arrived at directly by assuming that the unit pulse represents a large force with very short duration such that its time integral is unity:

$$u'(t) = \text{unit impulse} = \int_0^{\Delta t} f(t)\, dt = 1$$

Knowing that the impulse equals the change in momentum, the effect of the unit pulse is equivalent to an initial velocity $\dot{X}_0 = \frac{u'(t)}{m}$. The equation of motion of the system is

$$m\ddot{x} + kx = 0$$

and the initial conditions are

$$\text{at} \quad t = 0, \quad \dot{X}_0 = \frac{u'(t)}{m}, \quad X_0 = 0$$

The transformation of the differential equation is

$$mp^2X(p) - m\dot{X}_0 + kX(p) = 0$$

$$X(p) = \frac{\dot{X}_0}{p^2 + \omega_n^2} = \frac{\dot{X}_0}{\omega_n} \frac{\omega_n}{p^2 + \omega_n^2}$$

and the inverse transform will therefore be

$$x(t) = \frac{\dot{X}_0}{\omega_n} \sin \omega_n t$$

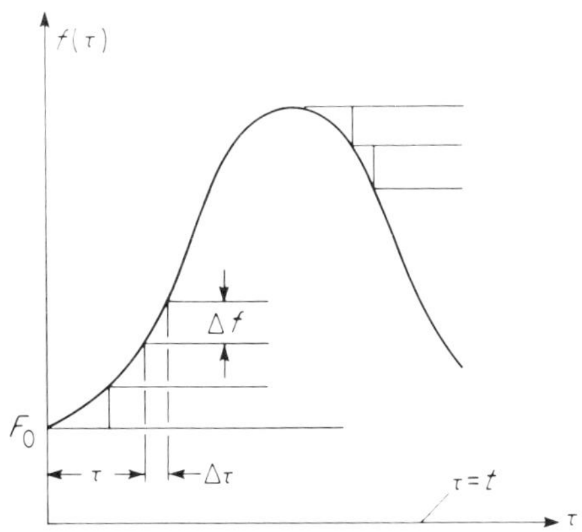

FIG. 5-5

and by substituting for the value of $\dot{X}_0$ we have

$$h(t) = [1/\sqrt{km}] \sin \omega_n t \tag{5-3}$$

Equations 5-2 and 5-3 are of considerable interest since any general function can be reduced to either a series of step functions or a series of impulses as shown in Figs. 5-5 and 5-6 respectively.

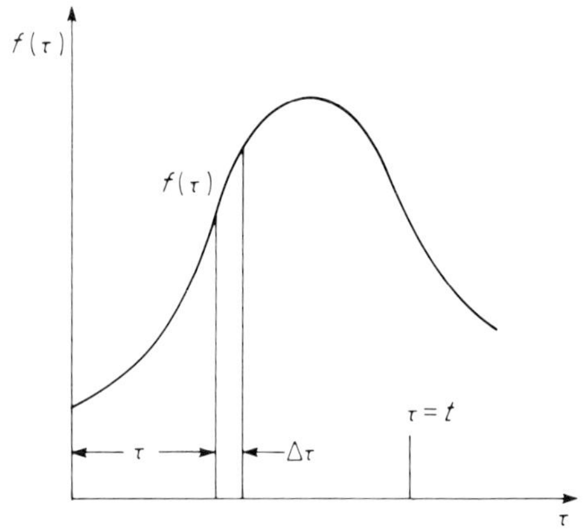

FIG. 5-6

Duhamel's Integral

In linear systems, where the principle of superposition* applies, the response of the system to any forcing function can be obtained by integrating the effects of the step functions or impulses. This integration is usually referred to as *Duhamel's integral* or *convolution integral.*

Using the step function approach, the forcing function $f(t)$ can be divided into an initial step function F_0 and incremental step functions:

$$\Delta f(\tau) = \frac{\Delta f}{\Delta \tau} \Delta \tau \quad \text{at} \quad t = \tau$$

Therefore by superposition the response at $\tau = t$ is

$$x(t) \cong F_0 H(t) + \sum_{\tau = \Delta\tau}^{t} \frac{\Delta f(\tau)}{\Delta \tau} \Delta \tau H(t - \tau)$$

*The principle of superposition can be simply stated as: *The sum of the effects of the force components is equal to the effect of the total force.* This rule applies only to linear systems.

where $H(t - \tau)$ is the response of the unit step measured from τ, and in the limit when $\Delta\tau \to 0$,

$$x(t) = F_0 H(t) + \int_0^t f'(\tau) H(t - \tau)\, d\tau \tag{5-4}$$

Notice that when the slope of the curve is negative the effect of the steps is automatically subtracted. This is illustrated in Fig. 5-5. After maximum point of the force curve is reached, the incremental steps are subtracted.

The solution can also be obtained by dividing the forcing function into incremental impulses, and the total response is obtained by superposition as

$$x(t) = \int_0^t f(\tau) h(t - \tau) d\tau \tag{5-5}$$

Example 4

Find the response of a mass m initially at rest supported on a spring with stiffness k when subjected to a force $f(t) = ct$, where c is a constant, Fig. 5-7.

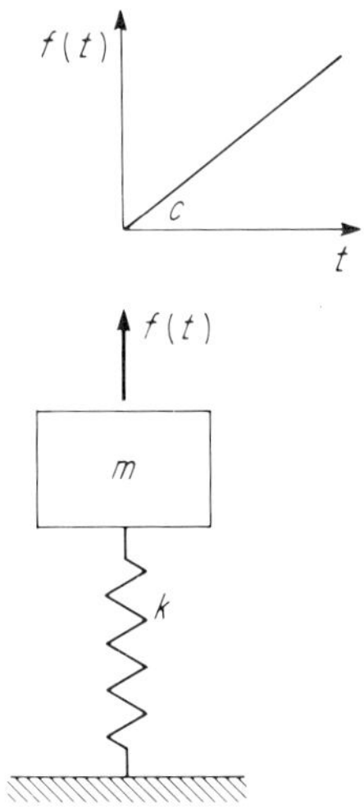

FIG. 5-7

The equation of motion is

$$m\ddot{x} + kx = ct$$

(a) Using the Laplace transform,

$$x(p) = \frac{c}{mp^2(p^2 + \omega_n^2)}$$

and the inverse transform gives

$$x(t) = \frac{c}{m\omega_n^3}(\omega_n t - \sin \omega_n t)$$
$$= (c/k)(t - \sin \omega_n t/\omega_n)$$

(b) Using Duhamel's integral with incremental steps (Eq. 5-2). Since $F_0 = 0, f'(t) = c$, and $H(t - \tau) = 1/k[1 - \cos \omega_n(t - \tau)]$, therefore, by substituting in Eq. 5-2 we obtain

$$x(t) = \frac{1}{k}\int_0^t c[1 - \cos \omega_n(t - \tau)]d\tau$$
$$= \frac{1}{k}\int_0^t c(1 - \cos \omega_n \tau)d\tau$$
$$= \frac{c}{k}\left[t - \frac{\sin \omega_n t}{\omega_n}\right]$$

(c) From Duhamel's integral with incremental impulses, Eq. 5-3, and knowing that

$$f(\tau) = c\tau, \quad h(t - \tau) = \frac{\sin \omega_n(t - \tau)}{m\omega_n}$$

we get

$$x(t) = 1/m\omega_n \int_0^t c\tau \sin \omega_n(t - \tau)d\tau$$

Introducing the notation

$$t - \phi = \tau, \quad t - \tau = \phi, \quad d\tau = -d\phi$$

we obtain

$$x(t) = c/m\omega_n \int_0^t (t - \phi) \sin \omega_n \phi \, d\phi$$
$$= c/m\omega_n \left[-t \frac{\cos \omega_n}{\omega_n}\right]_0^t - \int_0^t \phi \sin \omega_n \phi \, d\phi$$
$$= c/m\omega_n \left[\frac{t}{\omega_n} - \frac{t}{\omega_n}\cos \omega_n t\right] - \frac{\sin \omega_n t}{\omega_n^2} + \frac{t}{\omega_n}\cos \omega_n t$$
$$= c/k\left[t - \frac{\sin \omega_n t}{\omega_n}\right]$$

which is identical to the previous result.

Example 5

Find the response of the system with mass m and spring constant k to the forcing function, Fig. 5-8, given by

$$f(t) = ct \quad \text{for} \quad 0 < t < t_0$$
$$= ct_0 \quad \text{for} \quad t \geq t_0$$

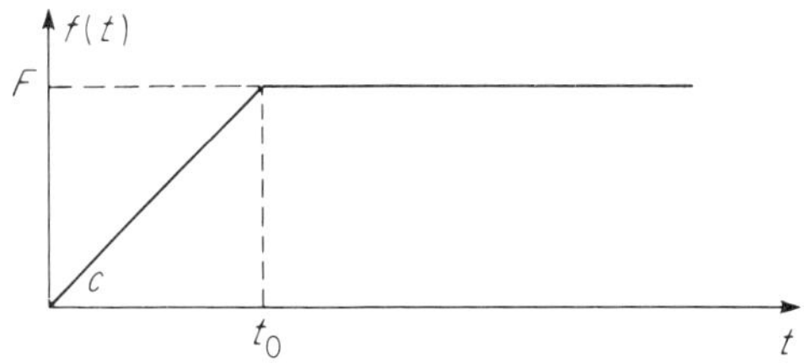

FIG. 5-8

1. **FOR $t < t_0$**

   The result is the same as in the previous case (Example 4).

2. **FOR $t > t_0$**

   (a) Using incremental steps and knowing that

$$F_0 = 0, \quad f'(t) = c, \quad 0 < t < t_0$$
$$f'(t) = 0, \quad t \geq t_0$$

therefore Eq. 3-5 becomes

$$x(t) = 0 + \int_0^t f'(\tau)H(t - \tau)d\tau$$

Knowing that $f'(\tau) = c$ and substituting for

$$H(t - \tau) = \frac{1}{k}[1 - \cos \omega_n(t - \tau)]$$

we get

$$x(t) = \frac{1}{k}\int_0^{t_0} c[1 - \cos \omega_n(t - \tau)]d\tau + 1/k \int_{t_0}^t 0$$

$$= \frac{c}{k}\left\{[t_0] + \left[\frac{\sin \omega_n(t - \tau)}{\omega_n}\right]_0^{t_0}\right\}$$

$$= \frac{c}{k}\left[t_0 + \frac{\sin \omega_n(t - t_0)}{\omega_n} - \frac{\sin \omega_n t}{\omega_n}\right]$$

and by trigonometric manipulation we obtain

$$x(t) = \frac{c}{k}\left[t_0 - \frac{1}{\omega_n} 2 \sin \frac{\omega_n t_0}{2} \cos \omega_n\left(t - \frac{t_0}{2}\right)\right]$$

$$= \frac{F}{k}\left[1 - \frac{\sin \frac{\omega_n t_0}{2}}{\frac{\omega_n t_0}{2}} \cos \omega_n \left(t - \frac{t_0}{2}\right)\right] \tag{5-6}$$

For $t > t_0$.

(b) Using incremental impulses,

$$x(t) = \int_0^t f(\tau)h(t - \tau)d\tau$$

Knowing that $f(\tau) = c\tau$ for $\tau = 0 \to t_0$, $f(\tau) = ct_0$, for $\tau = t_0 \to t$, and $h(t - \tau) = \frac{1}{m\omega_n} \sin \omega_n(t - \tau)$, we have

$$x(t) = 1/m\omega_n\left[\int_0^{t_0} c\tau \sin \omega_n(t - \tau)d\tau + \int_{t_0}^t ct_0 \sin \omega_n(t - \tau)d\tau\right]$$

Substituting $\phi = t - \tau$, $d\tau = -d\phi$, we obtain

$$\left[\frac{m\omega_n}{c}\right]x(t) = -\int_t^{t-t_0} (t - \tau) \sin \omega_n \phi\, d\phi + t_0 \int_{t_0}^t \sin \omega_n(t - \tau)d\tau$$

$$= -t\left[\frac{\cos \omega_n \phi}{\omega_n}\right]_{t-t_0}^t + \left[\frac{1}{\omega_n 2} \sin \omega_n \phi - \frac{1}{\omega_n} \phi \cos \omega_n \phi\right]_t^{t-t_0}$$

$$+ t_0\left[\frac{\cos \omega_n(t - \tau)}{\omega_n}\right]_{t_0}^t$$

$$= -t \frac{\cos \omega_n t}{\omega_n} + t \frac{\cos \omega_n(t - t_0)}{\omega_n} + \frac{1}{\omega_n 2} \sin \omega_n(t - t_0)$$

$$- \frac{1}{\omega_n 2} \sin \omega_n t - \frac{1}{\omega_n} (t - t_0) \cos \omega_n(t - t_0)$$

$$+ \frac{1}{\omega_n} t \cos \omega_n t + \frac{t_0}{\omega_n} - \frac{t_0}{\omega_n} \cos \omega_n(t - t_0)$$

By arranging we obtain

$$\left[\frac{m\omega_n}{c}\right] x(t) = \frac{t_0}{\omega_n} - \frac{1}{\omega_n 2} [\sin \omega_n(t - t_0) - \sin \omega_n t]$$

$$= \frac{t_0}{\omega_n}\left[1 - \frac{\sin \frac{\omega_n t_0}{2}}{\frac{\omega_n t_0}{2}} \cos \omega_n\left(t - \frac{t_0}{2}\right)\right]$$

and by substituting $F = ct_0$ and $\omega_n = \sqrt{k/m}$, we get

$$x(t) = \frac{F}{k}\left[1 - \frac{\sin\dfrac{\omega_n t_0}{2}}{\dfrac{\omega_n t_0}{2}} \cos \omega_n \left(t - \frac{t_0}{2}\right)\right]$$

for $t > t_0$, which is identical with Eq. 5-6.

Equation 5-6 is plotted in Fig. 5-9 for different values of the ratio of t_0 to the natural period of the system $2\pi/\omega_n$.

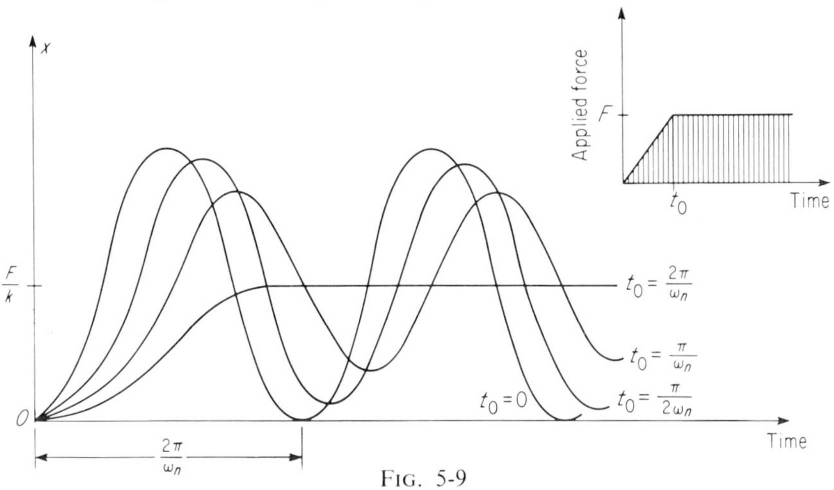

FIG. 5-9

The plot of the maximum values of $x(t)/(F/k)$ versus $t_0/(2\pi/\omega_n)$, which is known as the *response spectrum*, is shown in Fig. 5-10.

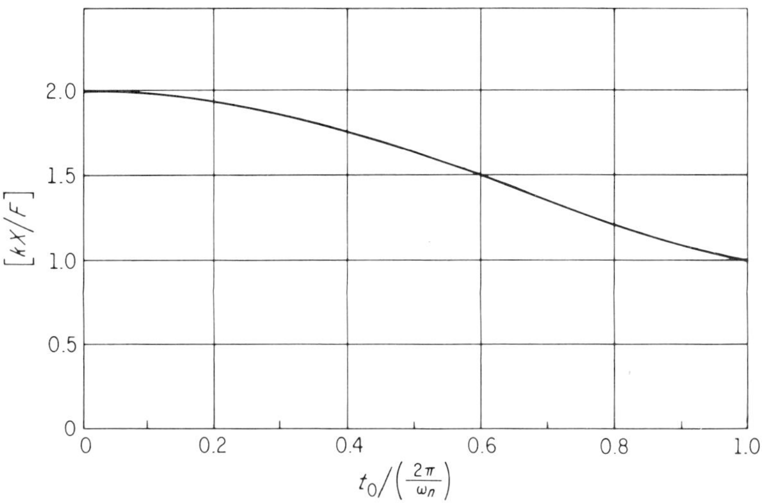

FIG. 5-10

This problem is of considerable practical interest since it gives the maximum dynamic displacements resulting from different rates of load application. A special section of this chapter will be devoted to presenting some response spectra for common types of disturbances.

5-3. THE ANALOG COMPUTER

Computers of the analog type provide a useful tool for solution of differential equations of motion for dynamic systems. They are equally useful for analysis of simple and complex systems of linear or nonlinear characteristics subjected to a given disturbance. It is not intended here to study in depth the computer or its operation. The objective is to give a general description of the computer and show how to utilize it for the solution of vibration problems.

The computer consists of a few basic components which perform mathematical operations such as addition, subtraction, multiplication, division, and integration. The basic elements in a computer are the amplifiers, potentiometers, multipliers, and function generators. These components can be connected to solve a variety of system equations.

On an analog computer the physical variables are represented by voltages. Arbitrary scale factors relate these voltages to the variables of the problem. The computer elements are designed to operate within an output range of ± 100 volts, and the scale factors are selected such that the variables fall within this range. In order to solve the equations derived from the dynamics of the system the computer elements are interconnected by means of a patch panel. Every component has its input and output terminals on this panel. The panel is removable to allow the problems to be patched external to the computer.

Components of Analog Computers

The main components of an electronic analog computer are listed and described as follows.

1. High-Gain Amplifiers

These are the basic elements in a computer and are represented in Fig. 5-11a. Notice that one of the terminals of the input and output voltages are grounded. Accordingly, one lead is all that is necessary for

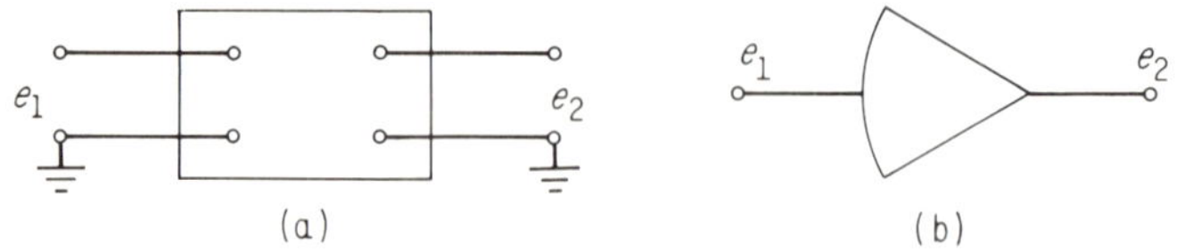

FIG. 5-11

a circuit diagram. High-gain amplifiers are considered as constant-gain amplifiers with a specified range of frequencies. An amplifier with a gain factor A is shown schematically in Fig. 5-11b. Its operation is described by the equation

$$e_2 = -Ae_1$$

2. Multipliers

A combination of two resistors and an amplifier (Fig. 5-12a) act as a constant multiplier. This is schematically represented in Fig. 5-12b and

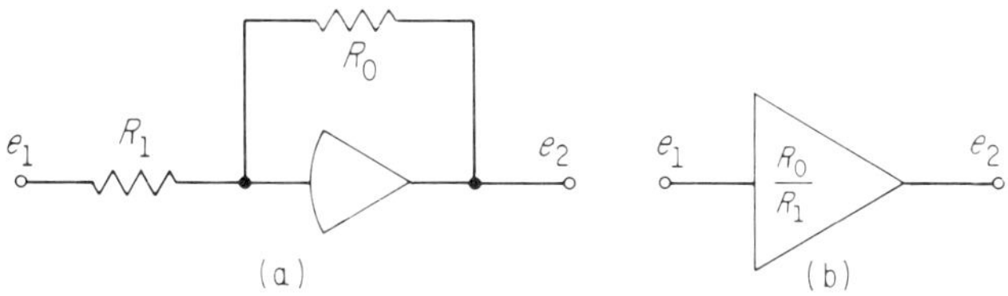

FIG. 5-12

its operation is given by

$$e_2 = -\frac{R_0}{R_1} e_1$$

Summation operations can also be performed by connecting the different input voltages through additional resistors. The diagram in this case will be as in Fig. 5-13a, which is schematically represented in Fig. 5-13b.

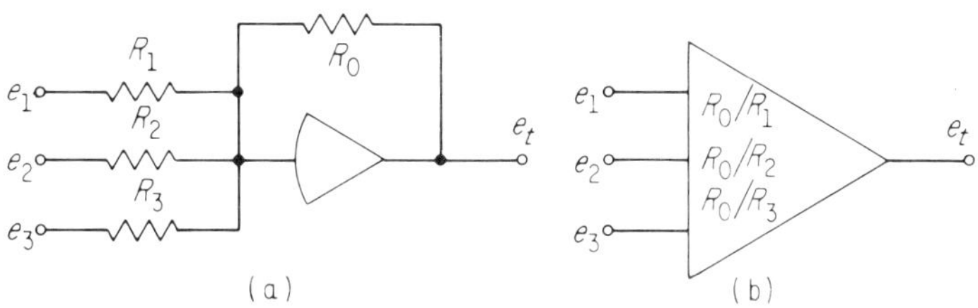

FIG. 5-13

It can be seen that

$$e_t = -\frac{R_0}{R_1} e_1 - \frac{R_0}{R_2} e_2 - \frac{R_0}{R_3} e_3, \ldots, \text{etc.}$$

3. Integrators

An integrator is composed of an amplifier, a resistor, and a low-loss capacitor, as shown in Fig. 5-14a. Its schematic representation is given in Fig. 5-14b.

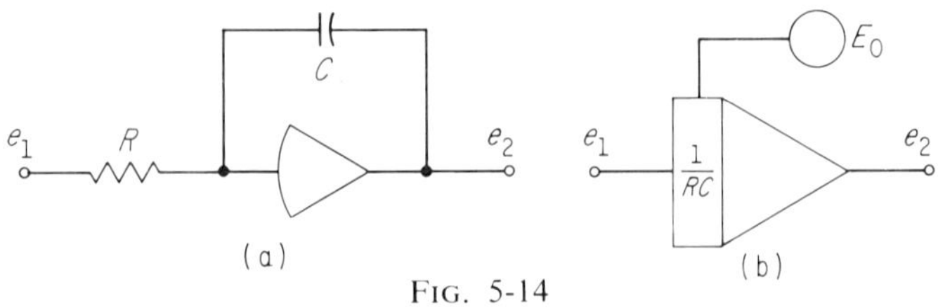

FIG. 5-14

In this case

$$e_2 = -\frac{1}{RC}\int_0^t e_1(\tau)\,d\tau + E_0$$

where E_0 is the initial value of e_2 which is introduced by an initial charge in the capacitor.

4. Potentiometers

A potentiometer is used for multiplication by a factor smaller than 1. This is shown in Fig. 5-15a and schematically represented in Fig. 5-15b.

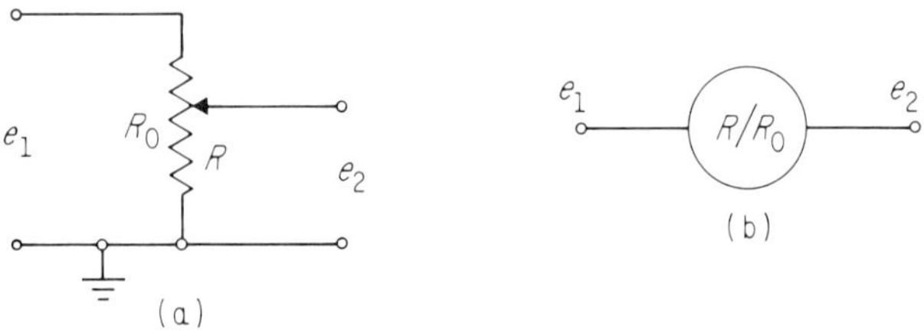

FIG. 5-15

It is evident that

$$e_2 = \frac{R}{R_0}\,e_1$$

5. Function Generators

Several devices and circuits are used to generate a specific function, depending on the complexity of the function. The most common types are those employing diode networks. The function generation operation is schematically represented in Fig. 5-16.

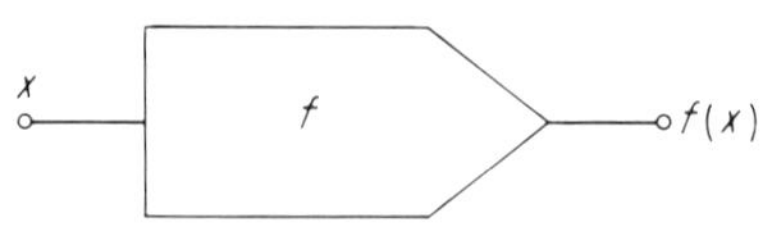

FIG. 5-16

6. Function Multipliers

Devices known as *function multipliers* are used when multiplication of two functions such as e_1 and e_2 is required. A common type of function multiplier is the *servomultiplier*. It has been shown previously that a variable voltage can be multiplied by a constant coefficient by using a potentiometer ("pot"). To multiply one varying quantity by another we can use a "pot" whose slider is automatically positioned by an electric motor to follow a second variable. If we have two varying voltages e_1 and e_2 and wish to produce a voltage proportional to $e_1 e_2$ we use one potentiometer whose input is e_2 and whose sliding setting is proportional to e_1. The schematic representation of a function multiplier is shown in Fig. 5-17. In this case

$$e_3 = -k e_1 e_2$$

where k is a constant.

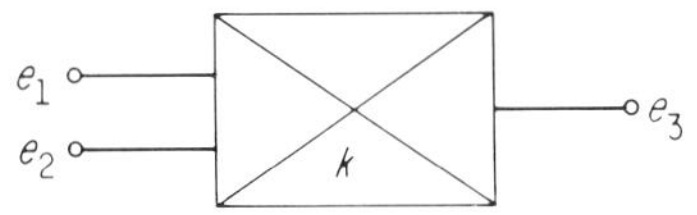

FIG. 5-17

Application of Analog Computers in System Analysis

In all dynamic systems the motion starts with a set of given initial conditions and is governed by one or more differential equations. The purpose of the analysis is to obtain the time history of the motion for different parameters and to come to conclusions which are useful in the design of the system or the control of its motion.

Let us take for example the simple case of a damped linear system whose equilibrium is disturbed by a given force $f(t)$. The initial conditions are such as at $t = 0$, $x = 0$, and $\dot{x} = 0$. The motion is governed by the equation

$$m\ddot{x} = -c\dot{x} - kx + f(t)$$

or

$$\ddot{x} = -(c/m)\dot{x} - (k/m)x + f(t)/m \tag{5-7}$$

Since the analog computer contains integrators, inverters for change of sign, potentiometers, function generators, and multipliers, this equation can be easily simulated on the computer as shown by the diagram in Fig. 5-18. The computer will be set in motion as soon as the forcing function $f(t)$ is switched on. The readout can be tapped to give displacement, velocity, or acceleration of the motion.

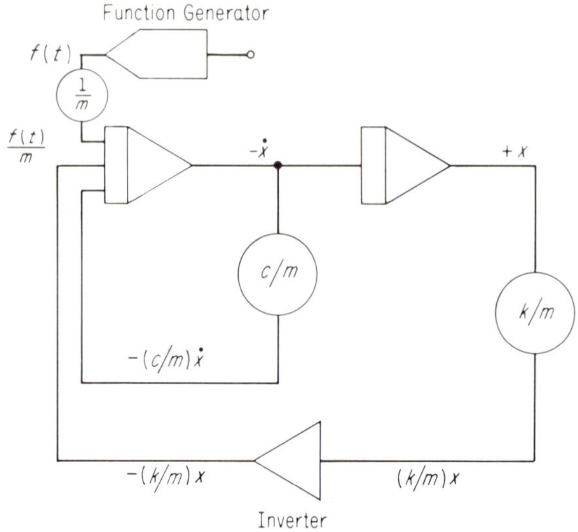

FIG. 5-18

If the initial conditions are nonzero, they should be introduced at the corresponding location on the integrators. (See Fig. 5-14b.)

For systems with nonlinear restoring and damping functions, diode function generators should be used instead of the potentiometers to approximate the given functions.

Scaling

Scaling of voltage and time is an important aspect of analog computations. The amplitude-voltage scale should be chosen such as the output is readable or recordable without overloading the components of the computer or the recording equipment. Time-scaling is also important, since some of the computing and recording elements have limited range of frequency response. The time-scale factor may be greater than 1, smaller than 1, or equal to 1, and the scale is called *advanced time*, *slow time*, and *true time* respectively.

5-4. SPECIAL CASES OF NONPERIODIC EXCITATION

We have presented in the previous sections some of the methods applicable to the analysis of systems which are subjected to general excitations. We shall now consider two cases of special interest to the engineer. The first case is the response of systems to shock loading in the form of step functions with different fronts and pulses with simple shapes. The results in this case will be represented as "response spectra," a useful concept in describing system performance as a dimensionless function of the significant characteristics of the system as well as the excitation.

The second case is self-excited vibration. In this case a system with particular characteristics when subjected to a finite disturbance is capable of building up its motion indefinitely or attaining very large amplitudes and sustaining a vibration at these amplitudes.

Response Spectra

In many instances of forced vibration, the excitation may be easily described by a simple time function. The examples shown in Figs. 5-19a through 5-19g represent such forcing functions and can be generally

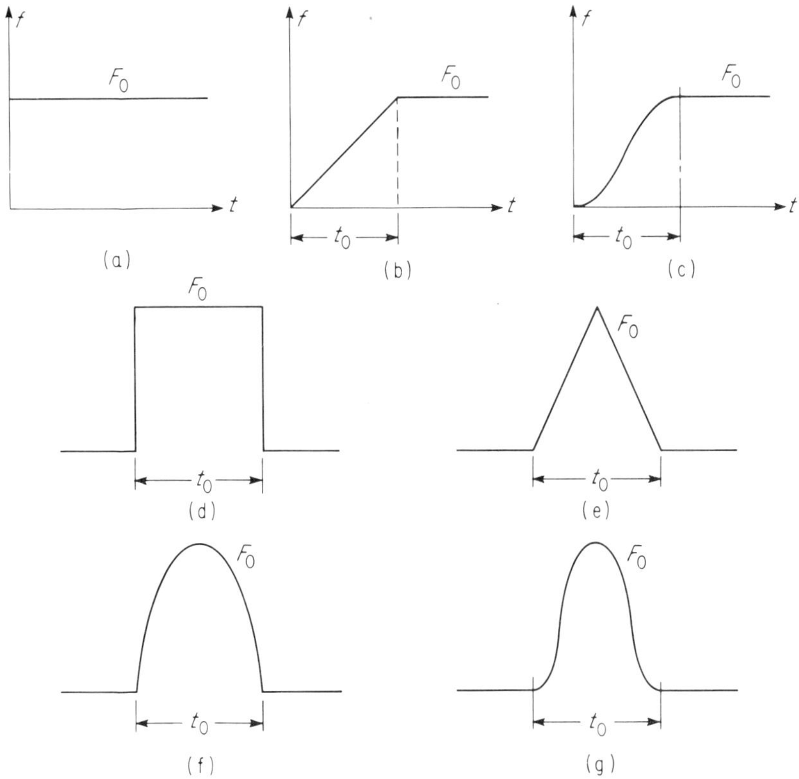

FIG. 5-19

classified as shock loads. The figures correspond respectively to a step function, a ramp function, a step with versed sine front, a rectangular pulse, an equilateral triangular pulse, a sinusoidal pulse, and a versed sine pulse. Within any of these categories, the forcing function is defined by its peak value F_0 and its characteristic time interval t_0.

When a system is subjected to such excitations, the designer is usually interested in evaluating the maximum system response. The response of interest may be in the form of displacement, velocity, or acceleration. Any of the previously discussed techniques can be utilized for evaluation of the required response for known values of F_0 and t_0.

It is useful, however, in situations where forcing functions of the same shape with different characteristic values F_0, t_0 are applied to the same type of system; to present the maximum response data in a plot known as the *response spectrum*. This type of plot shows the maximum response—as evaluated by any of the previous techniques—in a dimensionless manner versus the characteristic time interval t_0, also in a nondimensional manner. The nondimensional response is usually expressed as the ratio of the peak response to the static response of the system to the maximum force. The nondimensional time scale is usually represented by the ratio of the characteristic time interval t_0 of the forcing function to a characteristic period of the system. For linear systems the characteristic period is conveniently taken as the natural period of oscillation.

This type of plot can be constructed for linear as well as nonlinear systems and for transient as well as steady-state excitations. Once available, it gives the designer a very useful tool for judging the performance of existing systems or for choosing adequate parameters of a new design. In general, a response spectrum summarizes all the necessary information concerning the response of a certain type of system to a certain type of excitation.

The graphs shown in Fig. 5-20 are the response spectra for a single-degree-of-freedom system subjected to the loads given in Fig. 5-19. The system in this case is linear with a mass m and a spring constant k and with negligible damping. The nondimensional response is represented by $\frac{X_{max}}{F_0/k}$ which is the ratio of the maximum dynamic response X_{max} to the maximum static deflection F_0/k. This ratio can be considered as a dynamic magnifier. The nondimensional time ratio is given by t_0/τ where t_0 is the characteristic time interval of the excitation and $\tau = 2\pi/\sqrt{k/m}$ is the natural period for the system.

As an illustrative example of the usefulness of the response spectra, consider the response of a linear system with a natural period τ to a rectangular pulse with duration t_0. It can be easily seen in Fig. 5-20 that the maximum dynamic displacement is twice the static displacement for pulse durations in excess of half the natural period of the system. This ratio will never exceed 2 for this type of excitation, and will vary between 0 and 2 when t_0/τ is smaller than 1/2. Within this range the dynamic magnifier will decrease as t_0/τ decreases and will approach zero as t_0/τ

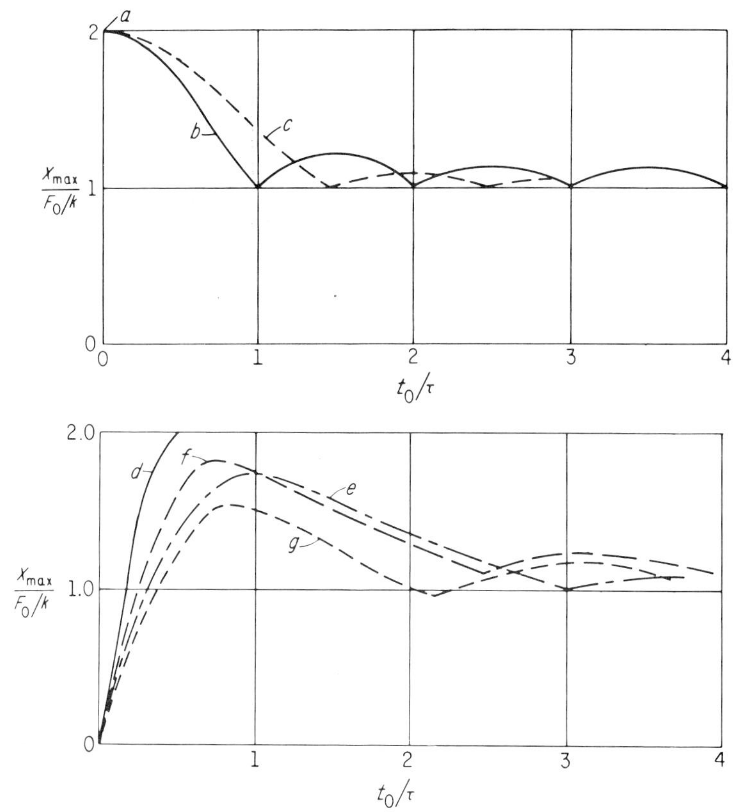

FIG. 5-20

approaches zero. In other words, when the pulse duration is very small compared to the natural period of the system, the dynamic displacement will be much smaller than the displacement due to the maximum force of the pulse if applied statically.

Self-Excited Vibration—Limit Cycles and Stability Analysis

There are numerous cases in physical systems where vibrations are developed and sustained by a nonalternating source of energy. Such cases are generally known as *self-excited vibrations*. They are described as such since the vibration of the system itself causes the nonalternating source to provide the necessary energy for sustaining the motion. The frequency of the vibration is therefore equal to (or close to) the natural frequency of the system.

Some of the common examples of self-excited (or self-sustained) vibrations are the chatter vibration in machine tools, the vibration of the violin strings due to the motion of the bow, the vibration of mechanical systems subjected to kinetic friction, the vibration of smokestacks and suspension bridges, wing flutter in aircraft, and feedback in electrical circuits.

One type of wing flutter, for example, will occur when the wing configuration in vibration is as shown in Fig. 5-21. In this case, bending of

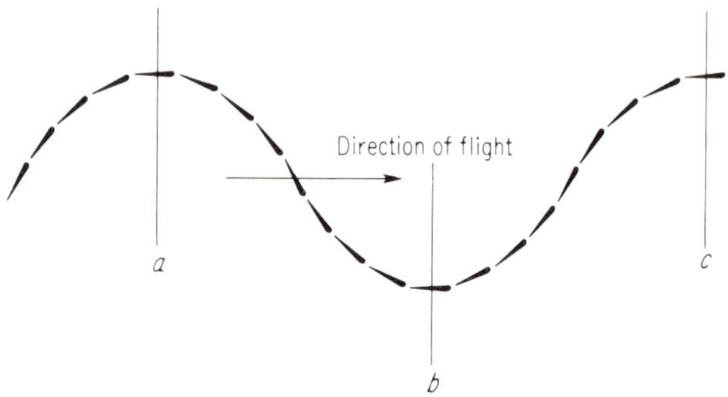

FIG. 5-21

the wing is accompanied by torsion in such a way that the angle of attack is negative between a and b and positive between b and c as shown in the figure. It can be easily seen that the aerodynamic forces resulting from this change in configuration will be in the same direction as the wing motion, causing a continuous buildup in the amplitude of the vibration. In other words, if the wing undergoes a small oscillation for one reason or another, this vibration will build itself up due to the very nature of the vibration and is therefore called *self-excited.*

The most widely investigated case of self-excited vibration is the so-called *relaxation oscillation.* It occurs in physical systems obeying Van der Pol's or related equations. This equation has the form

$$m\ddot{x} + c(x^2 - 1)\dot{x} + kx = 0 \tag{5-8}$$

which is nonlinear because of the damping function. The self-excitation aspect of this type of system also lies in the damping term. If x is small, the damping is negative and the vibration builds up. When the amplitude is sufficiently large, the damping will be positive. It is also possible to obtain an undamped oscillation in this damped system due to the particular behavior of the damping term. This latter type is known as *relaxation oscillation.*

We shall now consider the well-known case of a single-mass system vibrating in a self-excited manner under the influence of kinetic friction, Fig. 5-22. Assuming that μ is the coefficient of kinetic friction and N is the normal force between the mass m and the frictional wheel, the uni-

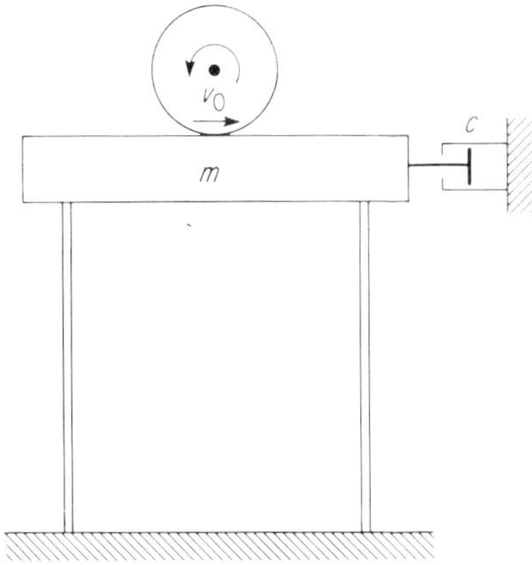

FIG. 5-22. Frictional drive for self-excited vibration.

directional frictional force acting on the mass will therefore be equal to μN. It is well known that the coefficient of kinetic friction is not a constant value but diminishes slightly as the velocity of relative sliding increases (see Fig. 5-23). If due to some slight disturbance the mass starts

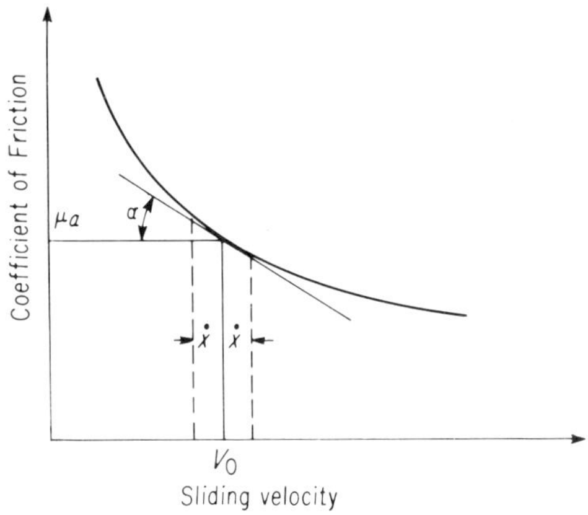

FIG. 5-23. Coefficient of kinetic friction.

to vibrate, the frictional force μN will not remain constant but will be larger when the mass moves in the direction of the tangential velocity V_0 of the wheel than when it moves opposite to it.

Assuming that the velocity of the oscillation $\dot{x}$ is much smaller than the tangential velocity of the frictional wheel, the frictional force μN

(which is a function of the relative velocity ($V_0 - \dot{x}$) will therefore always be in the direction of V_0. Over a complete cycle of vibration, the frictional force will therefore produce net positive work on the mass and the amplitude of its vibration will build up.

In order to study this vibration, the equation of motion can be written as

$$m\ddot{x} + c\dot{x} + kx = \mu N$$
$$= [\mu_0 + \alpha\dot{x}]N$$

where c = coefficient of viscous damping
μ_0 = coefficient of kinetic friction at V_0 relative velocity
α = slope of the friction curve at V_0 and can be considered constant for a small $\dot{x}$

This equation can be rearranged as

$$m\ddot{x} + [c - \alpha N]\dot{x} + kx = \mu_0 N$$

The term $[c - \alpha N]$ which represents a net damping coefficient will determine whether the vibration will be stopped or built up in a self-excited manner. If $\alpha N < c$ the resultant damping term will be positive and the vibration will decay signifying the stability of the system. On the other hand, if $\alpha N > c$, a negative damping term will exist and the vibration will build up as shown in Fig. 5-24a. The system in this case is unstable.

It is quite clear from the previous example that a quantitative knowledge of the frictional force and damping functions is essential for any analysis of this type of self-excited vibration. Any variation in either function due to increase in amplitude or velocity of the vibration can have considerable effect on the vibration.

Study of Self-Excited Vibrations by the Phase-Plane-δ Method

Since the frictional and damping forces are usually complex functions which require experimentally obtained information of a nonlinear character, the phase-plane-δ method of analysis is particularly well suited for studies of self-excited vibration.

Assume that in the previous equation the resultant function $[c - \alpha N]$ was not a linear function of the velocity but rather a complicated function $f(\dot{x})$ obtained experimentally. The equation of motion is now

$$m\ddot{x} + f(\dot{x}) + kx = \mu_a N$$

This equation can be written in the δ form as

$$\ddot{x} + \omega_n^2\left[x + \frac{1}{k}(f(\dot{x}) - \mu_a N)\right] = 0$$

$$\ddot{x} + \omega_n^2[x + \delta] = 0$$

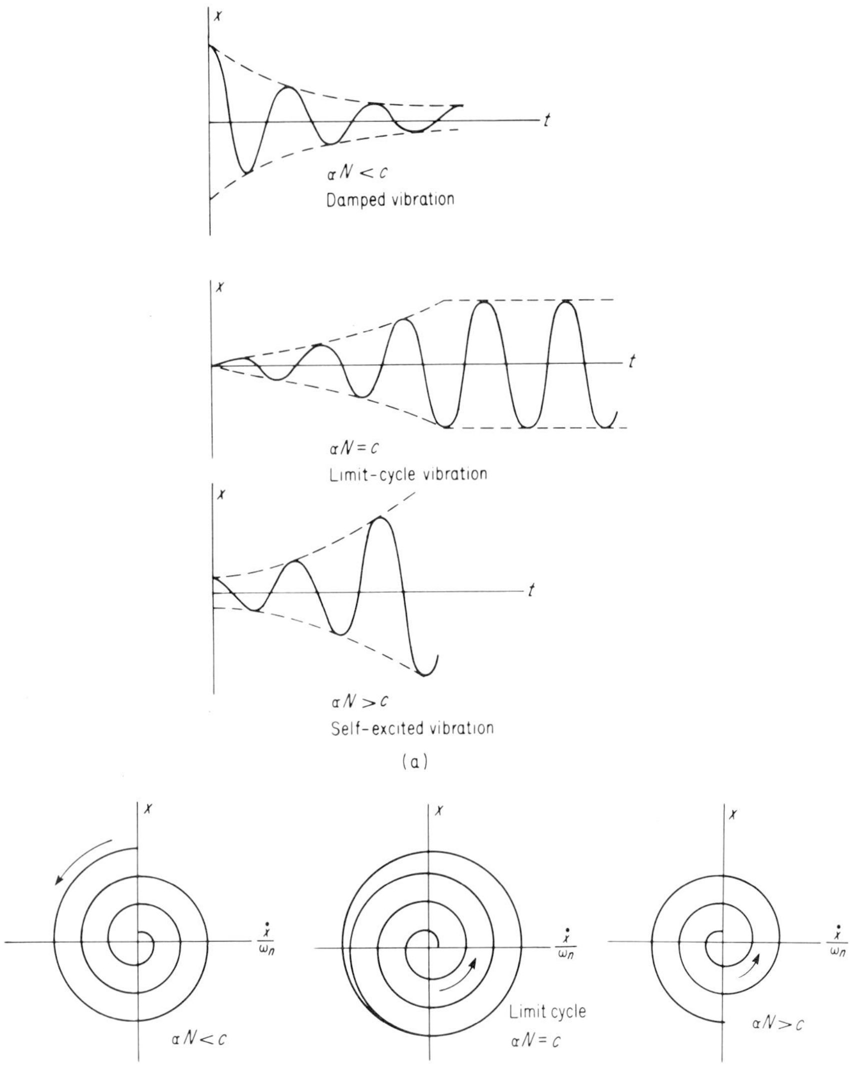

FIG. 5-24

where $\delta = \delta_{\dot{x}} + \delta_0$

$$\delta_{\dot{x}} = \frac{1}{k} f(\dot{x})$$

$$\delta_0 = \frac{-1}{k} \mu_a N = \text{const.}$$

and the problem accordingly will transform to a free vibration with a continuously changing datum. The datum variation δ which in this hypothetical case is a function of the velocity $\dot{x}$ can be obtained at any instant of the motion in the phase plane from the δ-curve. By successively plotting small segments of the locus and changing the datum according to the new position, the motion can be graphically represented in the phase plane. This is shown in Fig. 5-24b, which also shows that the motion is stable when the vibration decays or a limit cycle is reached. The motion at the limit cycle will be similar to a free undamped vibration.

This case of self-excited vibration can also be explained by an energy approach. It is well known that the energy dissipated per cycle due to the positive part of the damping term is a squared function of the displacement. If the energy due to the negative part (or energy input from the nonalternating source) is higher at low amplitudes of vibration than the dissipated energy and lower at high amplitudes (see Fig. 5-25), the intersection of the two curves gives the limit cycle or the amplitude x_L at which the vibration is free and undamped due to the equality of the damping and the energy input.

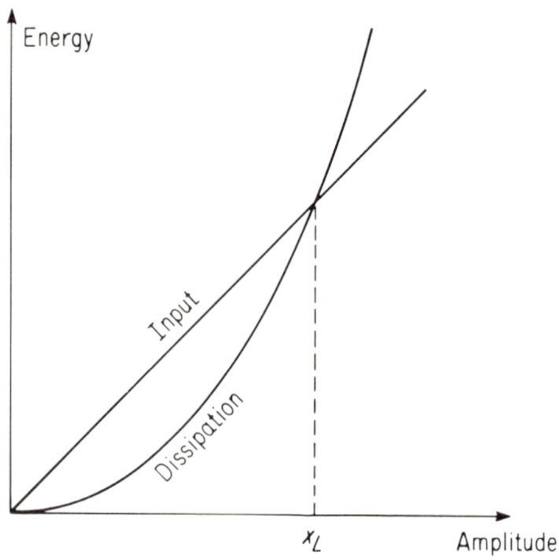

FIG. 5-25

Phase-plane Method for Stability Analysis of Nonlinear Systems

The phase-plane method can be effectively utilized to determine stability criteria for a particular class of nonlinear systems. Considering, for example, that as the class of systems described by the equation

$$\ddot{x} + A\dot{x}^2 + Bx = 0$$

are subjected to an initial displacement. The effect of the parameters A and B on stability of the motion can be easily investigated in the phase plane.

The system equation can be rewritten as

$$\ddot{x} + Bx + \frac{A}{B}\dot{x}^2 = 0$$

$$\ddot{x} + \omega_n^2\left[x + A\left(\frac{\dot{x}}{\omega_n}\right)^2\right] = 0$$

where $B = \omega_n^2$ represents a reference natural frequency.

Since the phase-plane diagrams are plotted for x versus $\dfrac{\dot{x}}{\omega_n}$, it can be therefore seen that A is the only parameter which influences the shape of the curves.

Assuming the initial conditions to be

$$\text{at} \quad t = 0, \quad x = x_0, \quad \dot{x} = 0$$

a family of phase-plane plots for different values of the dimensionless parameter Ax_0 can be obtained as shown in Fig. 5-26. It can be seen that for $Ax_0 < 1/2$ the system is stable with limit cycles having peak

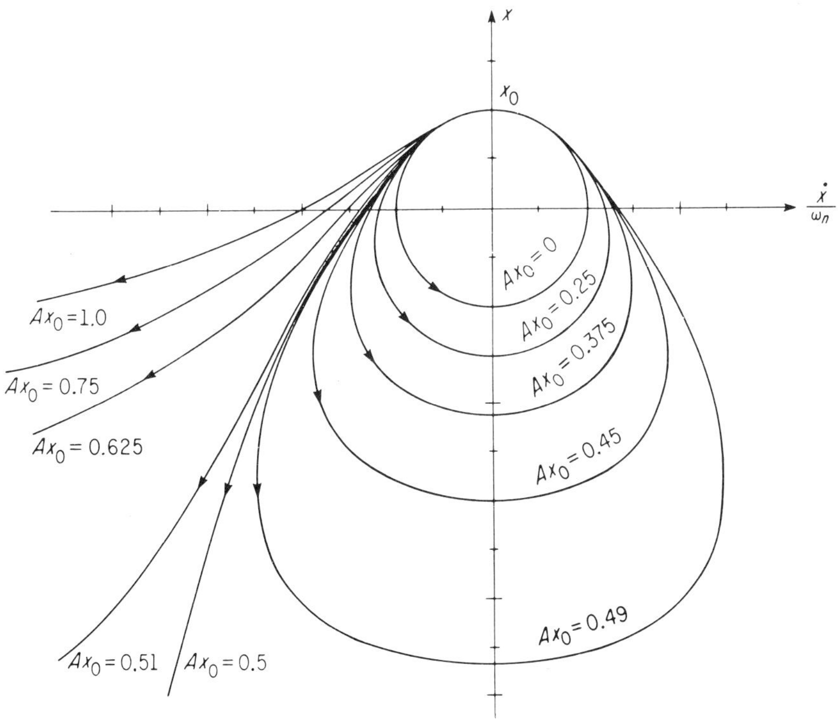

FIG. 5-26

values of x in the negative direction which progressively increase as Ax_0 approaches 1/2. For $Ax_0 > 1/2$ the system is unstable. This example shows that for the system under consideration (which is subjected to an initial displacement only) the value of the parameter A as well as the initial displacement x_0 determine the stability of the motion and the shape of the limit cycle.

It can be similarly illustrated that if the previous system is subjected only to an initial velocity v_0 (i.e., at $t = 0$, $x = 0$, $\dot{x} = v_0$) the stability criterion in this case becomes

$$\left(\frac{Av_0}{\omega_n}\right)^2 < \frac{1}{2}$$

Consequently if the system is set into motion by a general initial disturbance (at $t = 0$, $x = x_0$, and $\dot{x} = v_0$), the stability criterion can be derived as

$$\left(\frac{Av_0}{\omega_n}\right)^2 + Ax_0 < \frac{1}{2}$$

which is clearly a function of the system parameter A as well as the initial conditions of the motion.

5-5. SUMMARY

The response of single-degree-of-freedom systems to general excitations was discussed in this chapter. The general phase-plane-δ method as well as the Laplace transform and analog-computer techniques were utilized in the analysis. The shock-response spectra, self-excited vibrations, and system stability were presented as special cases of system response to nonperiodic disturbances.

PROBLEMS

5-1. A structure can be represented as a single-degree-of-freedom system with a mass of 25 lb in.$^{-1}$ sec^2 and a natural frequency of 200 rad/sec. The

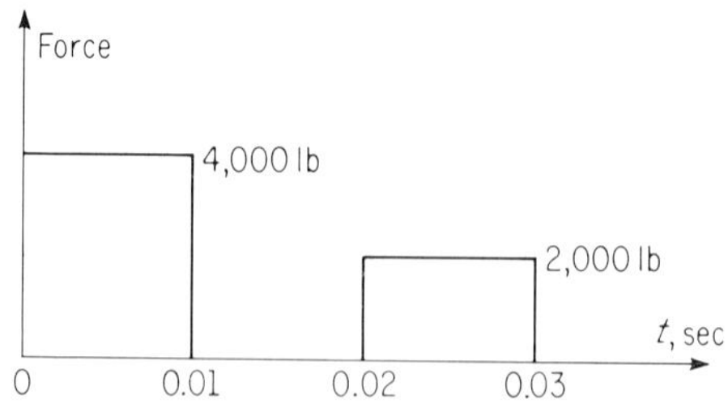

PROB. 5-1

structure is subjected to a pulse as shown. Use the phase-plane method to determine the maximum deflection due to this pulse.

5-2. A rectilinear single-degree-of-freedom system has a mass of 4 lb/in.$^{-1}$ sec^2 and a restoring force of 1,600 lb/in. If the system is disturbed from its equilibrium position by a triangular pulse as shown, find the ratio of the maximum spring force to the peak pulse force. Use the phase-plane-δ method.

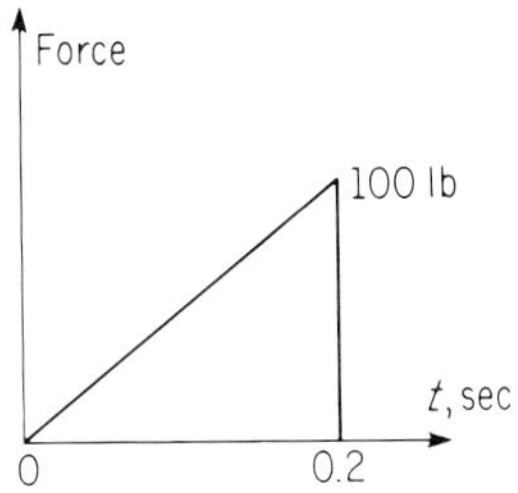

PROB. 5-2

5-3. Solve Prob. 5-2 when the pulse shape is as shown.

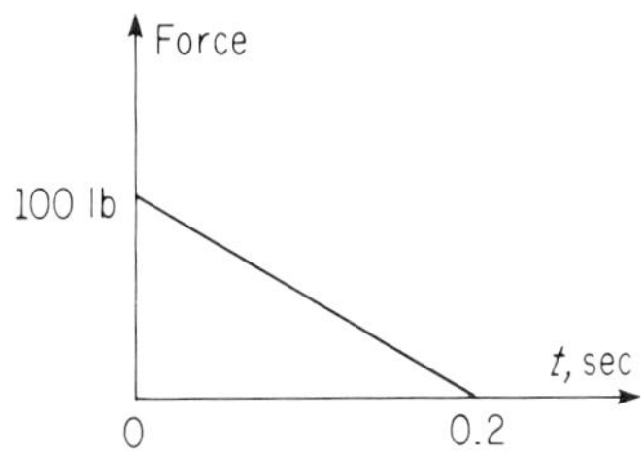

PROB. 5-3

5-4. Solve Prob. 5-2 when the pulse shape is as shown.

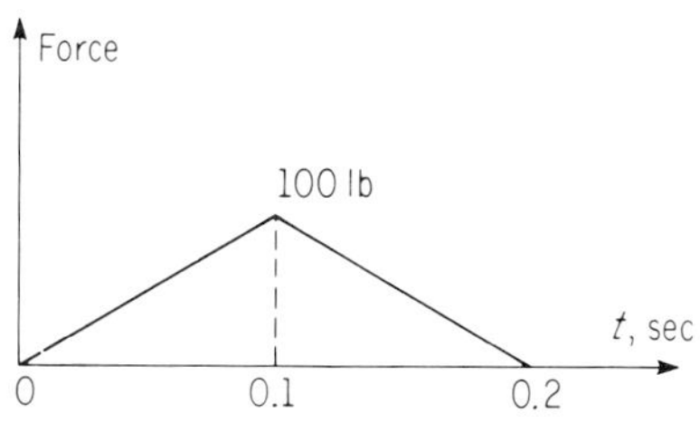

PROB. 5-4

5-5. Calculate the maximum spring force if the pulse in Probs. 5-2, 5-3, 5-4, is treated as producing initial velocity of the mass without regard to its shape.

5-6. A rectilinear single-degree-of-freedom system has the following characteristics:

$$\text{Mass} = 40 \text{ lb in.}^{-1} \text{ sec}^2$$
$$\text{Restoring force} = 25{,}000 \text{ lb/in.}$$
$$\text{Damping force} = 500\, \dot{x}^2 \text{ lb/in./sec}$$

where $\dot{x}$ represents the velocity of the motion. Calculate the peak displacement under the square pulse shown in the figure.

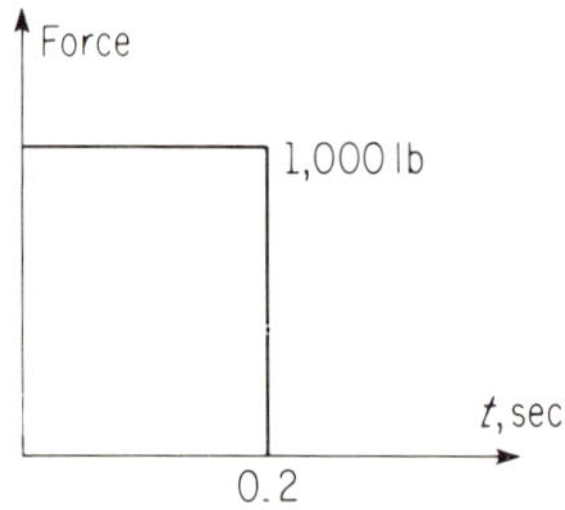

PROB. 5-6

5-7 through 5-10. Use the Laplace transform technique to solve Probs. 5-2 through 5-5.

5-11 through 5-13. Solve Probs. 5-2 through 5-4 by utilizing Duhamel's integrals.

5-14. A test setup for torsional flexible couplings is constructed in the form of a torsional pendulum. The flywheel used has a moment of inertia of 36 lb in. sec.^{2} The combined torsional stiffness of the shafts and coupling is 2×10^6 lb in./rad. A torque with the characteristics shown is applied at the flywheel. Calculate the maximum torque in the coupling.

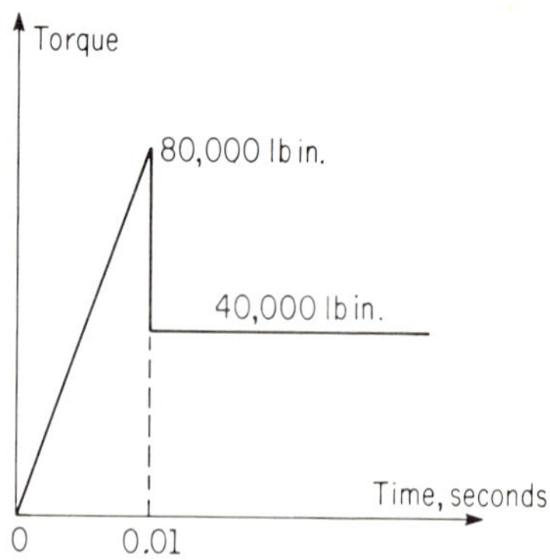

PROB. 5-14

5-15. A sinusoidal pulse with a 100-microsec duration and 10,000 lb peak force is applied to a 10-lb weight supported by a steel cantilever beam as shown. Calculate the maximum stress in the cantilever beam.

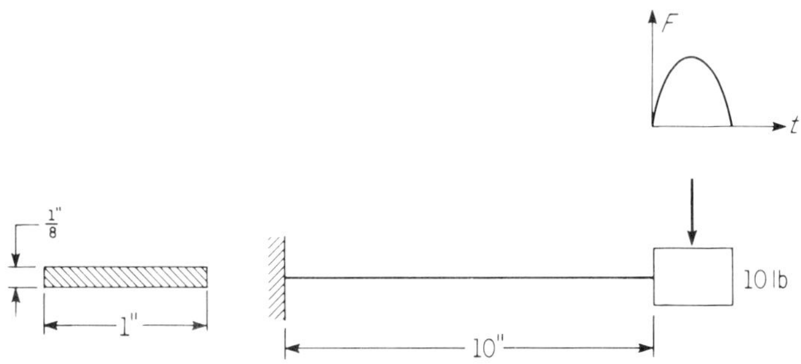

PROB. 5-15

5-16. If the allowable stress for the cantilever of the previous problem is 30,000 lb/in.^{2}, what is minimum thickness of the cantilever to withstand the applied pulse?

5-17. Plot the response spectra for the maximum force in the nonlinear system shown in Fig. 4-11 when subjected to a square pulse with a maximum force F_0 and duration t_0.

5-18. Construct a computer circuit to give

$$e = -5(e_1 - 2e_2 + 3e_3 - 4e_4)$$

5-19. If the input voltage e_1 in the adjacent sketch is sinusoidal with amplitude E_1, evaluate e_3/e_1 and e_2/e_1.

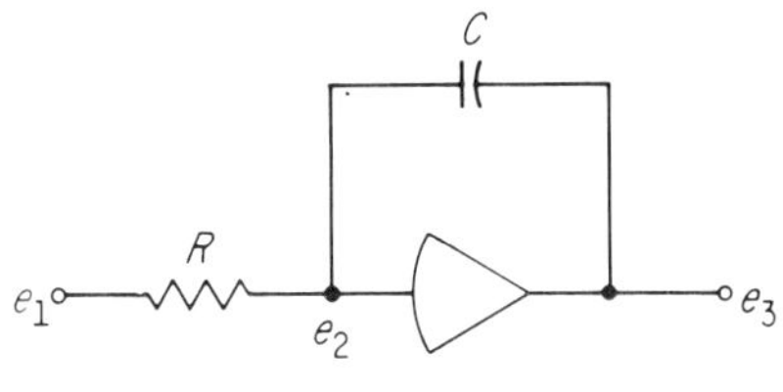

PROB. 5-19

5-20. The circuit shown has the following values:

$$\begin{aligned} e_1 &= 10 \sin \omega t \text{ volts} \\ e_2 &= 5.0 \cos \omega t \text{ volts} \\ e_3 &= 7.5 \sin (\omega t - \pi/4) \text{ volts} \\ e_4 &= 12.0 \sin (\omega t + \pi/4) \text{ volts} \\ R_1 &= R_2 = 400 \times 10^3 \text{ ohms} \end{aligned}$$

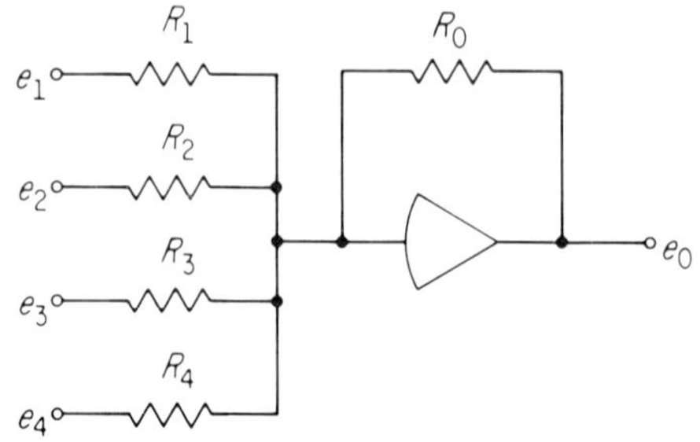

PROB. 5-20

$$R_3 = R_4 = 200 \times 10^3 \text{ ohms}$$
$$R_0 = 10^6 \text{ ohms}$$

Calculate e_0.

5-21. Plot the phase-plane diagram representing the vibration of the given system when subjected to a pulse with magnitude 2000 lb and a duration of 0.157 sec. Assume coefficient of restitution ($e = 0.8$) between the mass m and the stop A.

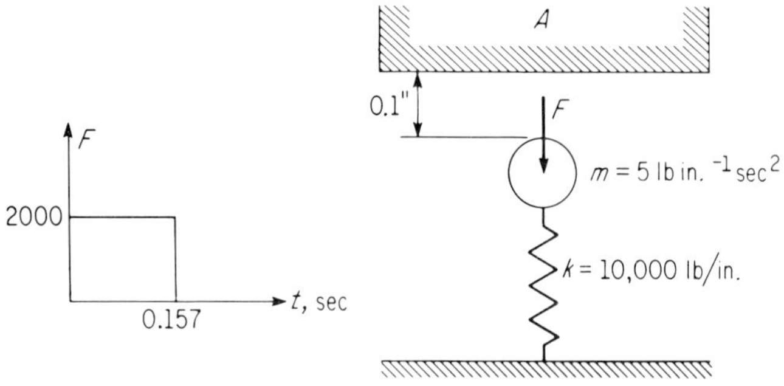

PROB. 5-21

chapter 6

Periodic Disturbances

6-1. INTRODUCTION

A special class of disturbances repeat themselves indefinitely in an exact manner over a specific increment of time. They are therefore known as *periodic*; and the constant increment of time is their period. It has been indicated in Chapter 2 that all periodic functions can be analyzed into harmonic components by the well-known method of Fourier analysis.

If $f(t)$ is a periodic function with period τ, as shown in Fig. 6-1, the

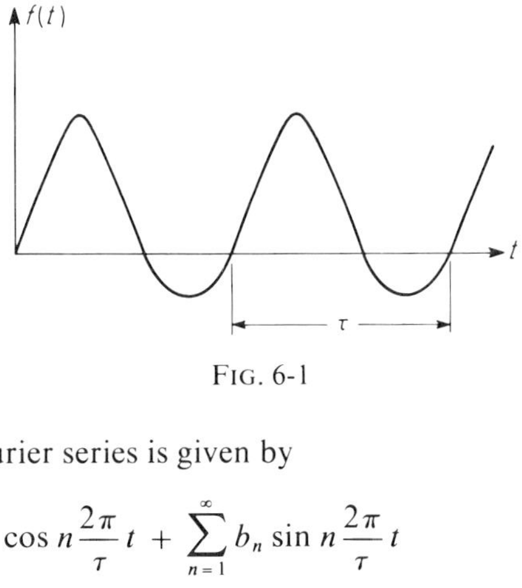

FIG. 6-1

corresponding Fourier series is given by

$$f(t) = a_0 + \sum_{n=1}^{\infty} a_n \cos n\frac{2\pi}{\tau}t + \sum_{n=1}^{\infty} b_n \sin n\frac{2\pi}{\tau}t$$

$$= a_0 + \sum_{n=1}^{\infty} a_n \cos n\omega t + \sum_{n=1}^{\infty} b_n \sin n\omega t = a_0 + \sum_{n=1}^{\infty} C_n \sin(n\omega t + \phi_n)$$

where $\omega = \dfrac{2\pi}{\tau}$

$$a_0 = \frac{1}{\tau}\int_0^{\tau} f(t)\,dt$$

$$a_n = \frac{1}{n\pi}\int_0^{2\pi} f(t)\cos n\omega t\,d(n\omega t)$$

$$b_n = \frac{1}{n\pi}\int_0^{2\pi} f(t) \sin n\omega t \, d(n\omega t)$$

$$C_n = \sqrt{a_n^2 + b_n^2}$$

$$\phi_n = \arctan\frac{a_n}{b_n}$$

A considerable number of periodic disturbances in practice are of harmonic nature. Alternating current, sound waves, unbalance disturbances in mechanical drives, and the like, are common occurrences. In linear systems the principle of superposition applies, and the response to a periodic function can be obtained by adding the responses due to its harmonic components. It is therefore of basic interest to study the behavior of the vibratory system under harmonic disturbances. These disturbances will be classified as sinusoidal force functions, foundation excitations, and unbalanced disturbances respectively.

The analysis of nonlinear systems in this chapter will be limited to harmonic disturbances only since the principle of superposition does not apply to these systems.

6-2. UNDAMPED LINEAR SYSTEMS SUBJECTED TO HARMONIC EXCITATION

Sinusoidal Forcing Function

If we consider that a mass m resting on a spring with stiffness k is disturbed from its equilibrium position by a sinusoidal force with amplitude F and frequency ω, the equation of motion is

$$m\ddot{x} + kx = F \sin \omega t \tag{6-1}$$

This is a linear differential equation whose solution has two parts: a general solution of the homogeneous equation ($m\ddot{x} + kx = 0$) and a particular integral to satisfy the forcing function.

The general solution x_1 was obtained previously (Eq. 4-3) as

$$x_1 = X_1 \sin (\omega_n t + \phi_1)$$

or

$$x_1 = A \cos \omega_n t + B \sin \omega_n t$$

where ω_n is the natural frequency of the system and A, B are integration constants.

A particular solution x_2 can be taken as

$$x_2 = X \sin \omega t$$

which represents a harmonic motion with the same frequency as the excitation. Substituting the particular solution in the differential equation, we get

$$-m\omega^2 X \sin \omega t + kX \sin \omega t = F \sin \omega t \tag{6-2}$$

from which

$$X = \frac{F}{k - m\omega^2} = \frac{F/k}{1 - (\omega/\omega_n)^2}$$

The solution for Eq. 6-1 will therefore be

$$\begin{aligned} x &= x_1 + x_2 \\ &= (A \cos \omega_n t + B \sin \omega_n t) + \frac{F/k}{1 - (\omega/\omega_n)^2} \sin \omega t \end{aligned} \tag{6-3}$$

The first term of the solution is known as the *transient part* of the displacement function and is dependent on the initial disturbance. The second term is the *steady-state* solution and is due to the periodicity of the excitation.

The constants of integration A, B in Eq. 6-3 are determined from the initial conditions of the motion.

If at $t = 0, \quad x = \dot{x} = 0$

then

$$0 = A$$

and

$$0 = \omega_n B \cos (0) + \frac{\omega F/k}{1 - (\omega/\omega_n)^2}$$

from which

$$B = \frac{-\omega}{\omega_n} \frac{F}{k} \frac{1}{1 - (\omega/\omega_n)^2}$$

By substituting in Eq. 6-3 we have

$$\begin{aligned} x(t) &= -\frac{F}{k} \frac{\omega/\omega_n}{1 - (\omega/\omega_n)^2} \sin \omega_n t + \frac{F/k}{1 - (\omega/\omega_n)^2} \sin \omega t \\ &= \frac{F}{m\omega_n(\omega_n^2 - \omega^2)} [\omega_n \sin \omega t - \omega \sin \omega_n t] \end{aligned} \tag{6-4}$$

and by trigonometric manipulation

$$\begin{aligned} x(t) &= \frac{F}{m\omega_n(\omega_n^2 - \omega^2)} \left[(\omega_n - \omega) \sin \left(\frac{\omega_n + \omega}{2} t \right) \cos \left(\frac{\omega - \omega_n}{2} t \right) \right. \\ &\quad \left. + (\omega_n + \omega) \sin \left(\frac{\omega - \omega_n}{2} t \right) \cos \left(\frac{\omega + \omega_n}{2} t \right) \right] \\ &= \frac{F}{m\omega_n} \left[\left(\frac{1}{\omega_n + \omega} \right) \sin \left(\frac{\omega_n + \omega}{2} t \right) \cos \left(\frac{\omega_n - \omega}{2} t \right) \right. \\ &\quad \left. + \frac{1}{\omega_n - \omega} \sin \left(\frac{\omega - \omega_n}{2} t \right) \cos \left(\frac{\omega + \omega_n}{2} t \right) \right] \end{aligned}$$

In the case when ω approaches ω_n or $\omega = (\omega_n - \Delta\omega)$ where $\Delta\omega/\omega_n$ is a small value, the previous equation will be

$$x(t) = \frac{F}{m\omega_n}\left[\frac{1}{\omega_n + \omega} \sin \frac{\omega_n + \omega}{2} t \cos \frac{\Delta\omega}{2} t + \frac{1}{\Delta\omega} \sin \frac{\Delta\omega}{2} t \cos \frac{\omega + \omega_n}{2} t\right]$$

Since $[1/\Delta\omega]$ is large compared to $1/(\omega_n + \omega)$, the first term can be neglected. Therefore

$$x(t) = \frac{F}{m\omega_n \Delta\omega} \sin \frac{\Delta\omega}{2} t \cos \left(\frac{\omega + \omega_n}{2}\right) t \qquad (6\text{-}5)$$

which represents a harmonic motion with frequency $(\omega + \omega_n)/2$ and having a sinusoidally changing amplitude. This is known as the *beat phenomenon*, Fig. 6-2. The beat frequency is given by $\Delta\omega/2$.

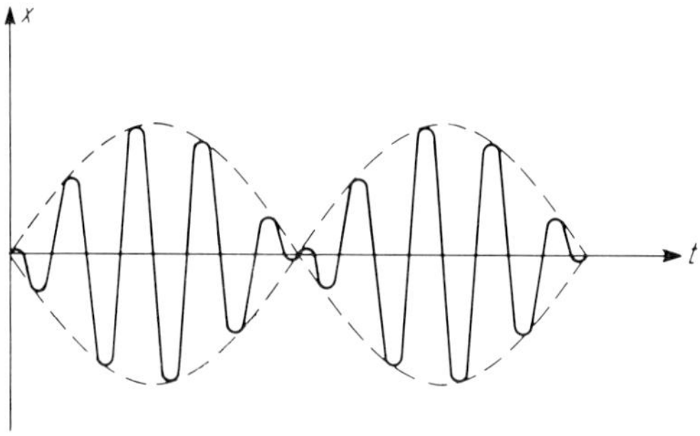

FIG. 6-2. Vibration beats.

In the limit when $\omega \to \omega_n$, $\Delta\omega \to 0$ and $\sin \frac{\Delta\omega}{2} t = \frac{\Delta\omega}{2} t$, therefore Eq. 6-5 becomes

$$x(t) = \frac{F}{2m\omega_n} t \cos \omega_n t \qquad (6\text{-}6)$$

which represents a vibration with indefinitely increasing amplitude (Fig. 6-3) and is the equation of motion when the exciting frequency coincides with the natural frequency of the system. This is known as *resonance* where finite disturbances at the natural frequency can produce infinitely large displacements.

Example

The Laplace transform method can be effectively used to obtain a solution for Eq. 6-1.

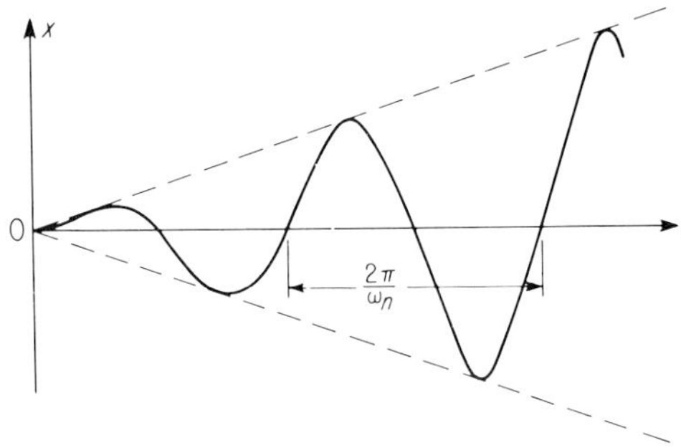

FIG. 6-3 Increase of amplitude at resonance.

The transformation of the equation as obtained by use of the tables is

$$mp^2X(p) - mpX_0 - m\dot{X}_0 + kX(p) = \frac{F\omega}{p^2 + \omega^2}$$

where X_0 and $\dot{X}_0$ are the initial values of x and $\dot{x}$. The transform for x is

$$\begin{aligned} X(p) &= \frac{F}{(p^2 + \omega^2)(mp^2 + k)} + \frac{mpX_0}{(mp^2 + k)} + \frac{m\dot{X}_0}{(mp^2 + k)} \\ &= \frac{F\omega}{m(p^2 + \omega^2)(p^2 + \omega_n^2)} + \frac{pX_0}{(p^2 + \omega_n^2)} + \frac{\dot{X}_0\omega_n}{\omega_n(p^2 + \omega_n^2)} \end{aligned}$$

and the inverse transform gives

$$\begin{aligned} x(t) &= \frac{F_0}{m(\omega_n^2 - \omega^2)}\left[\sin \omega t - \frac{\omega}{\omega_n}\sin \omega_n t\right] + X_0 \cos \omega_n t \\ &\quad + \frac{\dot{X}_0}{\omega_n}\sin \omega_n t \end{aligned}$$

for $X_0 = \dot{X}_0 = 0$

$$x(t) = \frac{F_0}{m\omega_n(\omega_n^2 - \omega^2)}[\omega_n \sin \omega t - \omega \sin \omega_n t]$$

which is identical to Eq. 6-4.

Steady-State Vibrations and Resonance

Vibrations due to periodic disturbances will eventually reach a steady-state condition where all parts of the system are assumed to partake of the motion at the frequency of the disturbance. It is the only frequency at which energy is being applied. The transient vibrations at the natural frequency will be assumed nonexistent, since the natural damping within the system will cause these transient vibrations to die out. The steady-state solution is therefore

$$x(t) = \frac{F/k}{1 - (\omega/\omega_n)^2} \sin \omega t \tag{6-7}$$

where F/k represents the displacement of the mass if the force was statically applied.

It is obvious from Eq. 6-7 that the dynamic amplitude and consequently the dynamic spring force are equal to their respective static values if the frequency of disturbance is very low compared to the natural frequency of the system. This is expected, since at these low frequencies the inertia forces would be negligible and the disturbing forces will be resisted mainly by the restoring forces. As the speed increases the inertia effects become more and more important and the dynamic amplitude attains higher values than would be expected if the disturbance was very slow. The ratio of the dynamic value at a certain frequency to the corresponding static value is conventionally known as the dynamic magnifier or the magnification factor. It can be seen from Eq. 6-7 that that

$$\text{Magnification factor} = \frac{1}{1 - (\omega/\omega_n)^2} \tag{6-8}$$

This magnification approaches infinity when the frequency of the excitation ω approaches the natural frequency. At this frequency the system will be in *resonance*. The infinite dynamic amplitude resulting from finite disturbances at resonance can be easily explained by the following pendulum analogy.

It is well known that a simple pendulum, once started, will oscillate at its natural frequency and will sustain its original amplitude indefinitely as long as no resistance to its motion exists. If a disturbing force is applied to it in the same direction and at that same frequency, it is expected that the amplitude of oscillation of the pendulum will start to build up unrestricted, reaching extremely high values irrespective of the magnitude of the disturbing force.

The dynamic magnifier will once again attain finite values as the frequency of disturbance increases beyond the natural frequency. The negative sign resulting from Eq. 6-8 merely indicates that the direction of oscillation will be out of phase. The dynamic amplitude once again becomes equal to the static amplitude when $(\omega/\omega_n) = 1.414$.

As the disturbing frequency increases the magnification continues to decrease reaching zero at infinitely high frequencies. It is obvious that at such high frequencies very small vibratory amplitudes will introduce sufficient inertia forces to resist the disturbing forces and the restoring forces will be of negligible value.

The variation of magnification with the frequency of external forces is plotted in Fig. 6-4 as derived from Eq. 6-8. The graph represents a response spectrum for a harmonic forcing function.

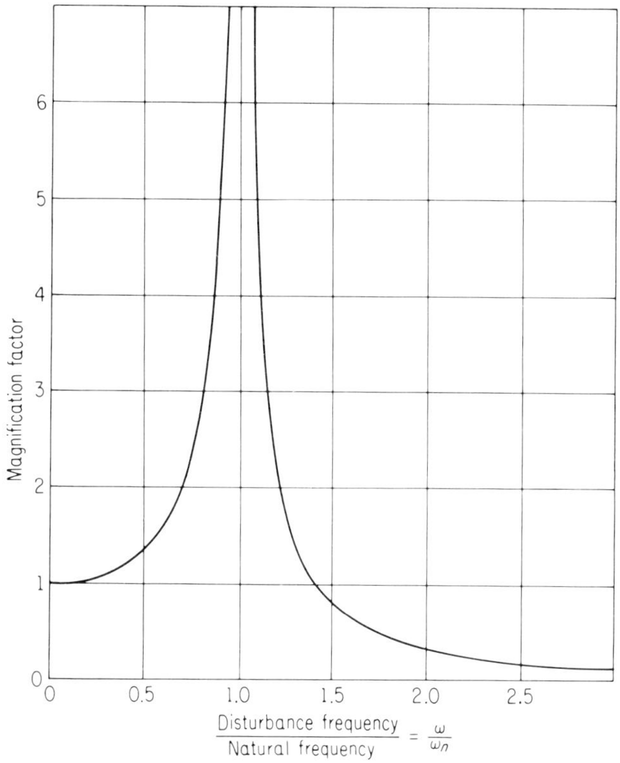

FIG. 6-4. Magnification factors.

Graphical Solution

A graphical solution of the steady-state vibration can be devised from Eq. 6-1. Substituting the solution $x(t) = X \sin \omega t$ in the equation and dividing both sides by $\sin \omega t$ we get

$$-m\omega^2 X + kX = F$$

or

$$F + m\omega^2 X = kX$$

Each side of this equation represents a straight line in the force-displacement plot, Fig. 6-5. The right-hand side is the spring-force curve, and the left-hand side is a line with the intercept F and slope $m\omega^2$. The independent variable in this problem is ω, and the intersection of the two lines for any ω gives the corresponding dynamic amplitude X. It can be

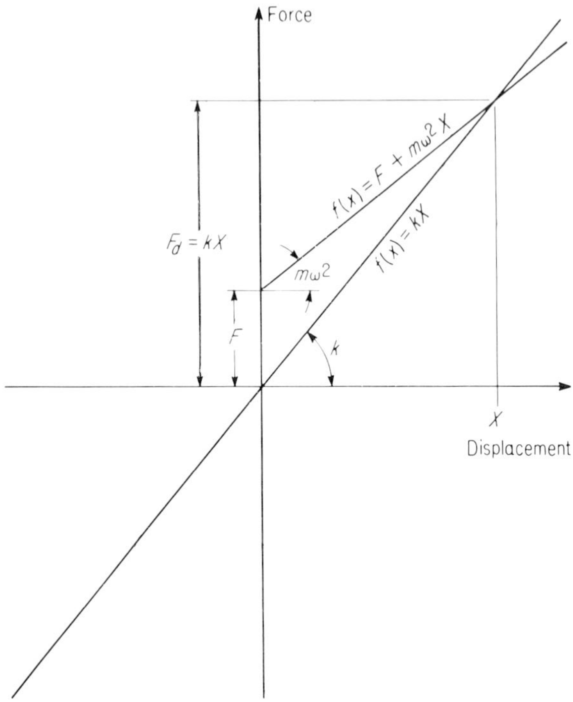

FIG. 6-5. Graphical solution.

easily seen in Fig. 6-6a that at $\omega = 0$, $X = X_{st} = F/k$. As ω increases X will increase until it becomes infinite when $\omega = \sqrt{k/m}$. For $\omega > \sqrt{k/m}$, X becomes finite again and will change sign indicating a 180° phase change. The amplitude X will continue to decrease as ω increases until it approaches 0 when ω becomes infinitely large. This graphical procedure can therefore be utilized to obtain the response spectrum plot of Fig. 6-4. It will also be extended later in the chapter to find approximate solutions for systems with nonlinear springs. Figure 6-6 shows the use of the graphical method to study the effect of the different parameters on the steady-state amplitude. The diagrams are self-explanatory.

Steady-State Vibrations by the Phase-Plane Method

The response of linear systems to periodic excitations can also be obtained by the phase-plane method. The solution in this case requires considerable labor. The technique however can be usefully applied in particular types of excitation such as in the case of a square-wave forcing function where the phase plane provides obvious advantages over other methods of solution. By the nature of the square wave, the phase-plane plot of the motion during the positive or negative phase of the excitation

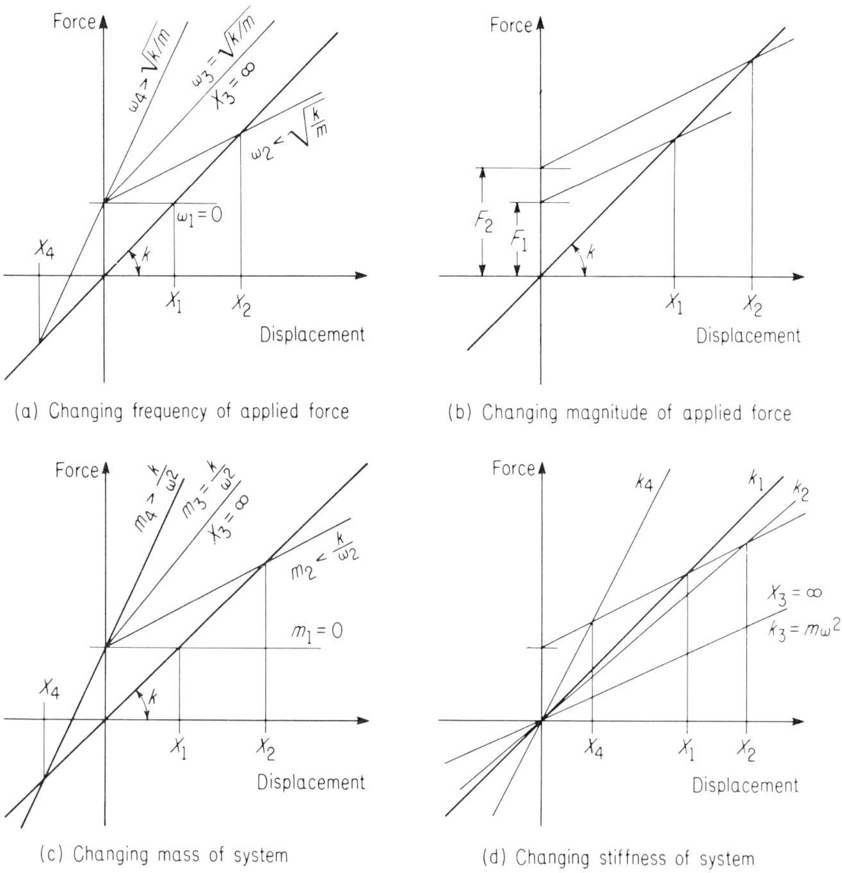

FIG. 6-6. Effect of changing system parameters on amplitude of vibration.

will be a circular arc with its center displaced an amount δ, from the velocity axis, given by

$$\delta = +\frac{F_1}{k}$$

$$= -\frac{F_2}{k}$$

where F_1 and F_2 are the force levels for the positive and negative phases respectively.

If the duration of each of the phases is given by $C_1\tau$ and $C_2\tau$ respectively, then the corresponding angles in the phase plane are

$$\theta_1 = \omega_n C_1 \tau = 2\pi C_1 \frac{\omega_n}{\omega}$$

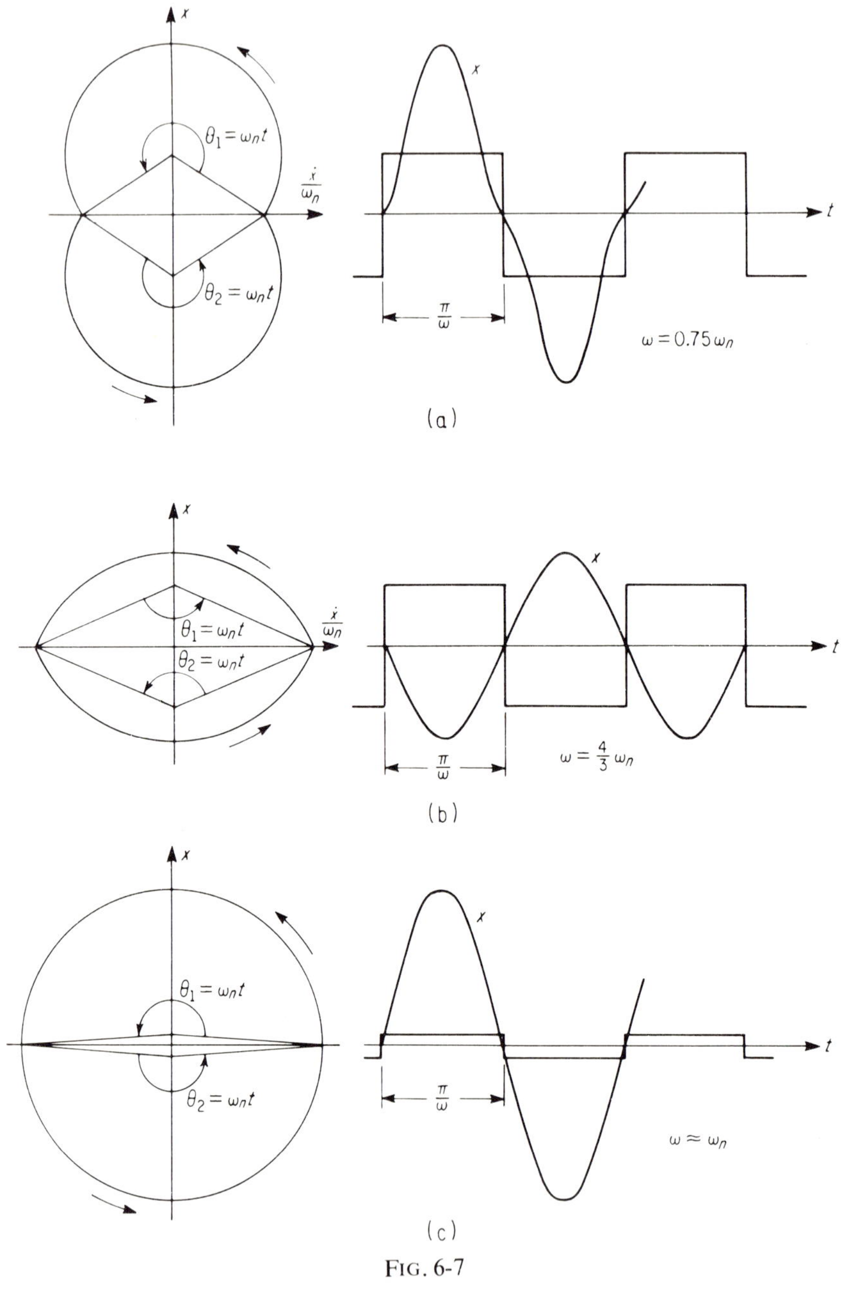
x
$\theta_1 = \omega_n t$
$\frac{\dot{x}}{\omega_n}$
$\theta_2 = \omega_n t$
t
$\frac{\pi}{\omega}$
$\omega = 0.75\omega_n$
(a)
$\omega = \frac{4}{3}\omega_n$
(b)
$\omega \approx \omega_n$
(c)

FIG. 6-7

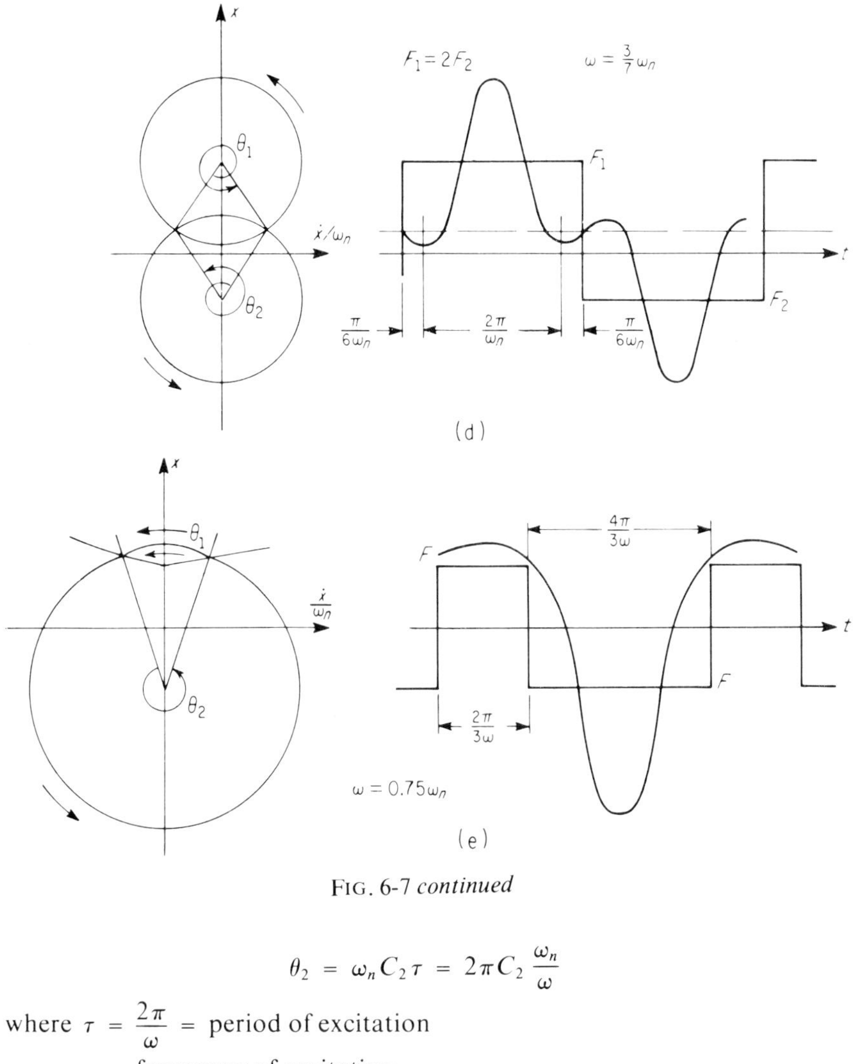

FIG. 6-7 *continued*

$$\theta_2 = \omega_n C_2 \tau = 2\pi C_2 \frac{\omega_n}{\omega}$$

where $\tau = \dfrac{2\pi}{\omega}$ = period of excitation

ω = frequency of excitation

C_1, C_2 = fraction of total period for the positive and negative phases respectively ($C_1 + C_2 = 1$)

Since the initial condition for each phase is the end condition for the other, the steady motion of the system can be completely described in the phase plane when δ_1, δ_2, θ_1, and θ_2 are known. It should be kept in mind that the phase-plane plot is symmetrical about the x-axis, and that the angles θ_1 and θ_2 are measured in the counterclockwise direction.

The following examples (shown in Fig. 6-7) illustrate the solution procedure.

Case (a)

Given

$$\omega = 0.75\omega_n, \quad C_1 = C_2 = 0.5, \quad F_1 = F_2 = F$$

The phase plot shows

$$\theta_1 = \theta_2 = \frac{4}{3}\pi \qquad \delta_1 = \delta_2 = \frac{F}{k}$$

Case (b)

Given

$$\omega = \frac{4}{3}\omega_n, \quad C_1 = C_2 = 0.5, \quad F_1 = F_2 = F$$

The phase plot is therefore constructed from

$$\theta_1 = \theta_2 = \frac{3\pi}{4}$$

$$\delta_1 = \delta_2 = \frac{F}{k}$$

Notice that the motion is out of phase with the excitation in this case.

Case (c)

Given

$$\omega \to \omega_n, \quad C_1 = C_2 = 0.5, \quad F_1 = F_2 = F$$

In this case

$$\theta_1 = \theta_2 \cong \pi$$

and

$$\delta_1 = \delta_2 = \frac{F}{k}$$

The amplitude x for this condition is much greater than $\frac{F}{k}$ and would approach infinity as $\theta \to \pi$.

Case (d)

Given

$$\omega = \frac{3}{7}\omega_n, \quad C_1 = C_2 = 0.5, \quad F_1 = 2F_2$$

Therefore

$$\theta_1 = \theta_2 = \frac{7}{3}\pi = 2\pi + \frac{\pi}{3}$$

$$\delta_1 = 2\delta_2$$

Case (e)

Given

$$\omega = 0.75\omega_n, \quad C_1 = \frac{1}{3}, \quad C_2 = \frac{2}{3}, \quad F_1 = F_2 = F$$

Accordingly

$$\theta_1 = \frac{8}{9}\pi$$

$$\theta_2 = \frac{16}{9}\pi$$

$$\delta_1 = \delta_2 = \frac{F}{k}$$

and the displacement in this case is as shown in the figure.

Sinusoidal Foundation Disturbance

In this case, Fig. 6-8, the disturbance is a harmonic movement of the foundation on which the single-degree-of-freedom system (with mass m

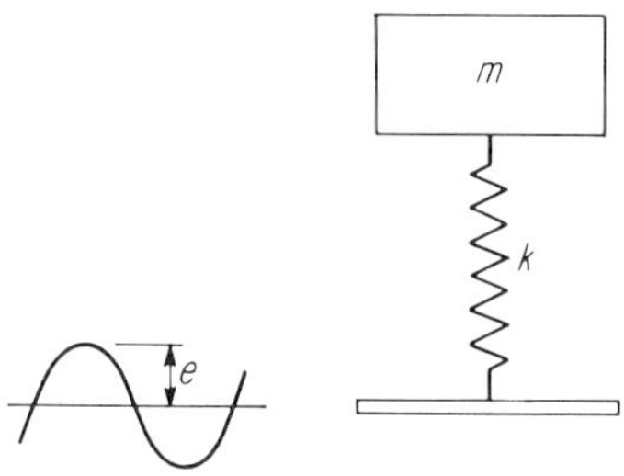

FIG. 6-8

and spring stiffness k) rests. If this disturbance is given by

$$x_f = e \sin \omega t$$

the equation of motion will be

$$m\ddot{x} + k(x - x_f) = 0$$

or

$$m\ddot{x} + kx = ke \sin \omega t \tag{6-9}$$

Assuming a steady-state solution $x = X \sin \omega t$ and substituting in the equation of motion, we obtain

$$X = \frac{ke}{k - m\omega^2}$$

and

$$x = \frac{ke \sin \omega t}{k - m\omega^2} = \frac{e}{1 - (\omega/\omega_n)^2} \sin \omega t$$

The dynamic spring deflection x_s is calculated from

$$x_s = x - x_f = \frac{e}{1 - (\omega/\omega_n)^2} \sin \omega t - e \sin \omega t$$

$$= \frac{e(\omega/\omega_n)^2}{1 - (\omega/\omega_n)^2} \sin \omega t$$

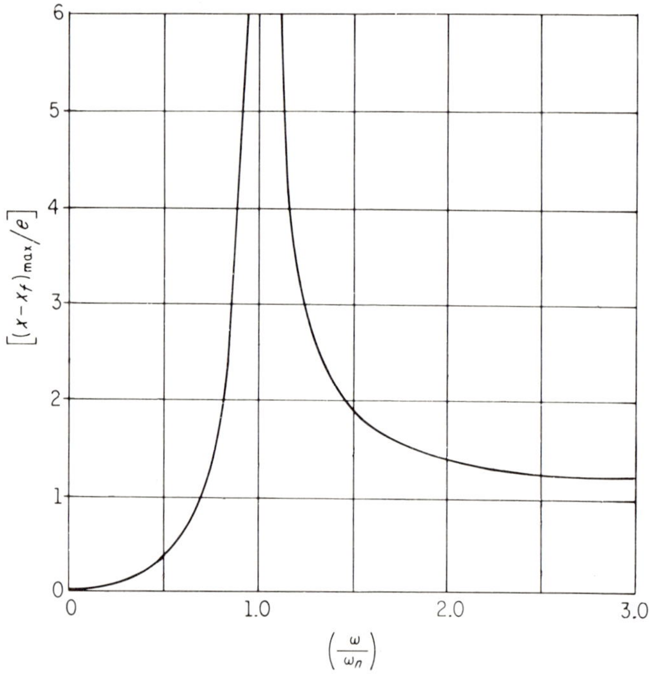

FIG. 6-9. Magnification factors for foundation excitation.

This equation is plotted in Fig. 6-9. A graphical solution for this equation can be obtained by rewriting Eq. 6-9 in the form

$$-m\omega^2 X + kX = ke$$

or

$$k(X - e) = m\omega^2 X$$

The corresponding straight lines in the force displacement plot are shown in Fig. 6-10. The amplitude of mass displacement X and the maximum

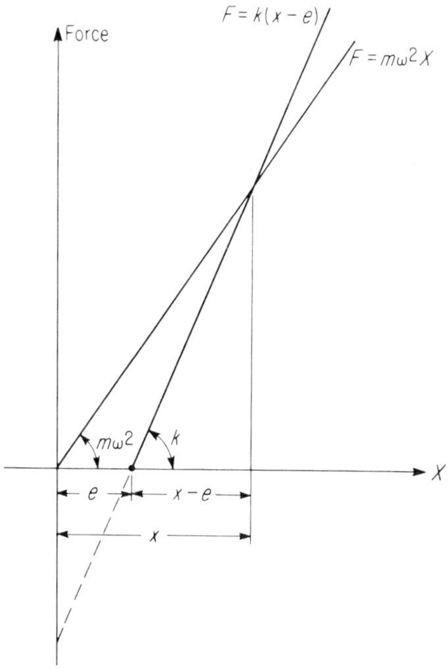

FIG. 6-10

spring deflection $(X - e)$ can be readily found for any frequency of excitation, as illustrated in the diagram.

Excitation Due to Unbalance

A common type of excitation in mechanical and structural systems is that resulting from unbalanced rotating parts. We shall consider the single-degree-of-freedom system shown in Fig. 6-11 which is subjected to the excitation due to an eccentric mass m with eccentricity e and rotating

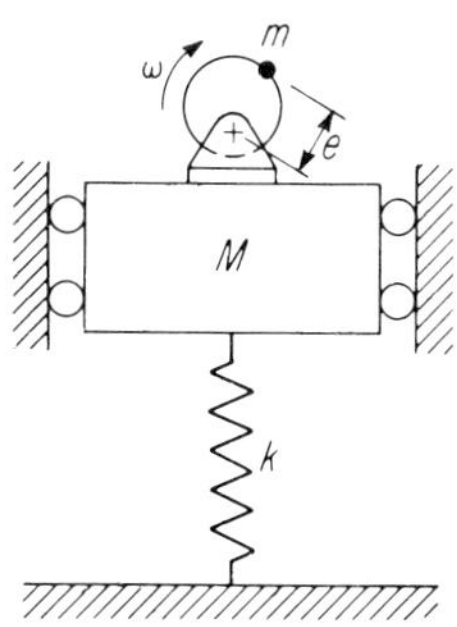

FIG. 6-11

at a speed at ω rad/sec. Assuming that the total mass supported by the spring is M and neglecting friction, the equation of motion will be

$$M\ddot{x} + kx = me\omega^2 \sin \omega t$$

Taking a steady-state solution $x = X \sin \omega t$, we find by substituting in the previous equation that

$$-M\omega^2 X + kX = me\omega^2 \tag{6-10}$$

The expression for the amplitude is therefore

$$X = \frac{\frac{m}{k}\,\omega^2 e}{1 - (\omega/\omega_n)^2} = \frac{\frac{m}{M}\,(\omega/\omega_n)^2 e}{1 - (\omega/\omega_n)^2}$$

from which

$$\frac{X}{e}\left(\frac{M}{m}\right) = \frac{(\omega/\omega_n)^2}{1 - (\omega/\omega_n)^2}$$

The value of $[(X/e)(M/m)]$ can, therefore, be calculated for any ω. The plot of the dynamic magnifier $[(\omega/\omega_n)^2/1 - (\omega/\omega_n)^2]$ is given in Fig. 6-12.

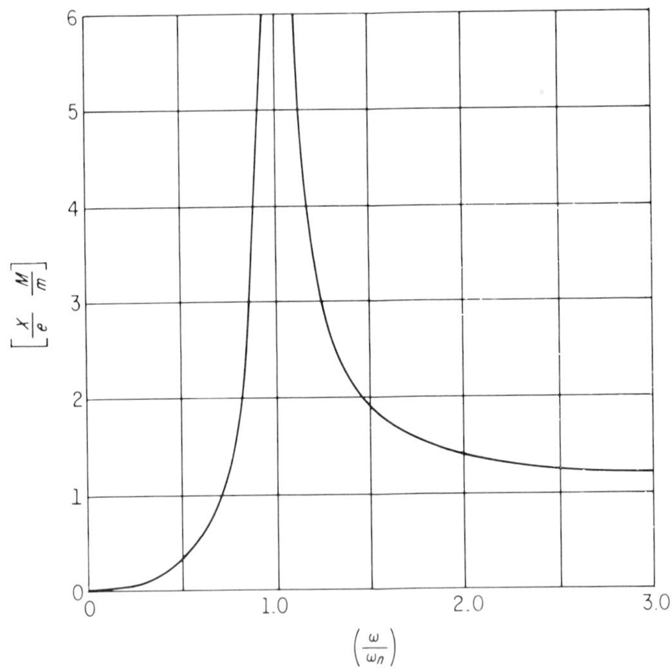

FIG. 6-12. Magnification factors for unbalance excitation.

A graphical solution can also be obtained for this case by rearranging Eq. 6-10 as

$$\begin{aligned} kX &= MX\omega^2 + me\omega^2 \\ &= M\omega^2(X + m/Me) \end{aligned}$$

The value of X for any ω is obtained as shown in Fig. 6-13 from the intersection of the two straight lines given by the right-hand side and left-hand side of the equation in a force-displacement plot.

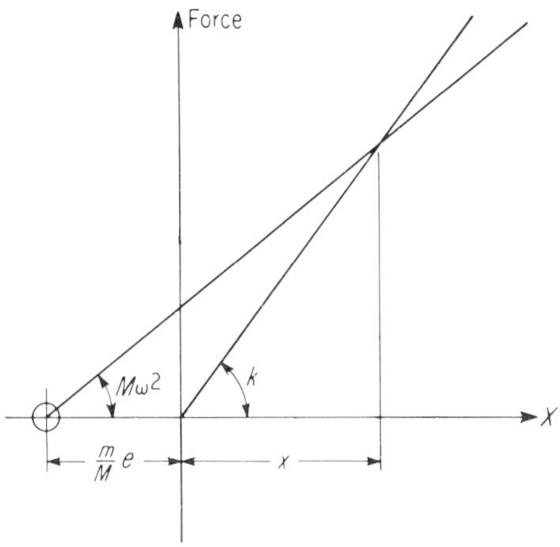

FIG. 6-13

6-3. VISCOUS-DAMPED LINEAR SYSTEMS

Sinusoidal Forcing Function

Considering the system shown in Fig. 6-14, the equation of motion in this case is

$$m\ddot{x} + c\dot{x} + kx = F \sin \omega t \tag{6-11}$$

where c is the viscous-damping coefficient. The solution will be composed of the general solution of the homogeneous equation and a particular solution satisfying the forcing function.

The general solution will be the same as in Eqs. 4-13, 4-14, and 4-15 for $\zeta > 1$, $\zeta < 1$, $\zeta = 1$ respectively. The particular solution may be assumed as

$$x = X \sin (\omega t - \phi)$$

which is a harmonic motion with an amplitude X having the same frequency as the exciting force and lagging behind it by the phase angle ϕ.

Substituting this in Eq. 6-11, we obtain

$$-\omega^2 X \sin (\omega t - \phi) + \frac{c}{m}\omega X \cos (\omega t - \phi) + \frac{k}{m} \sin (\omega t - \phi) = \frac{F}{m} \sin \omega t$$

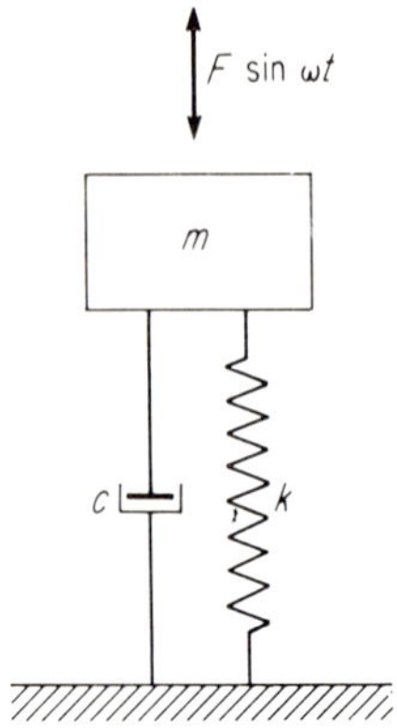

FIG. 6-14

Expanding the sine and cosine terms and arranging, we find

$$[X(k - m\omega^2)\cos\phi + Xc\omega\sin\phi - F]\sin\omega t + [X(k - m\omega^2)\sin\phi - Xc\omega\cos\phi]\cos\omega t = 0$$

The coefficients of sin ωt and cos ωt must vanish independently in order to satisfy the previous equation. This requires that

$$X = \frac{F}{(k - m\omega^2)\cos\phi + c\omega\sin\phi}$$

and

$$\tan\phi = \frac{c\omega}{(k - m\omega^2)}$$

which is represented graphically in Fig. 6-15. By substituting for cos ϕ and sin ϕ from Fig. 6-15 X is rewritten as

$$X = \frac{F}{\sqrt{[(k - m\omega^2)^2 + (c\omega)^2]}}$$

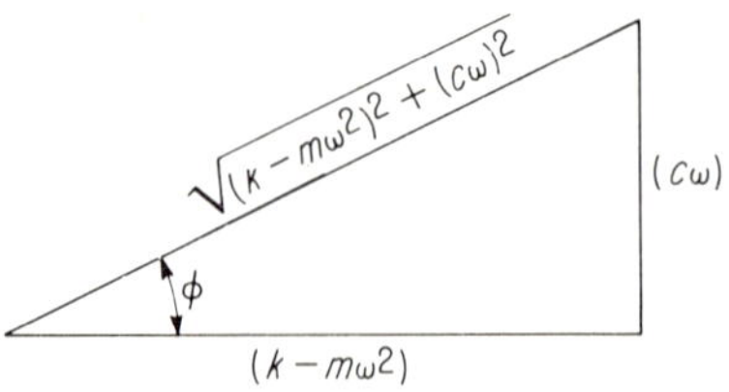

FIG. 6-15

Now using the notation $c_c = 2m\omega_n$, $\zeta = c/c_c$, we obtain

$$X = \frac{F/k}{\sqrt{\{[1 - (\omega/\omega_n)^2]^2 + (2\zeta\omega/\omega_n)^2\}}} \tag{6-12}$$

and

$$\tan\phi = \frac{2\zeta\omega/\omega_n}{1 - (\omega/\omega_n)^2} \tag{6-13}$$

Note that

$$c\omega/k = c\omega/m\omega_n^2 = [c/(\frac{1}{2}\,c_c)](\omega/\omega_n) = 2\zeta\omega/\omega_n$$

The complete solution for Eq. 6-11, including the transient part for $\zeta < 1$, will therefore be

$$x_{(t)} = Ae^{-\omega_n t\zeta}\sin(\omega_n\sqrt{1 - \zeta^2}\,t + \phi') + X\sin(\omega t - \phi) \tag{6-14}$$

where A and ϕ' are integration constants to be evaluated from initial conditions.

Method of Complex Algebra

The complex algebra method proves to be useful in problems involving phase relations as in the case under consideration.

It has been shown in Chapter 2 that a harmonic motion can be represented by the projection in a fixed direction of a vector rotating at a constant speed.

If we take this vector in the complex plane Fig. 6-16 as

$$x = Xe^{i(\omega t)} = X\cos\omega t + iX\sin\omega t$$

its projection on the imaginary axis will give a sinusoidal wave ($X\sin\omega t$) and the projection on the real axis will give ($X\cos\omega t$), as in Fig. 6-16.

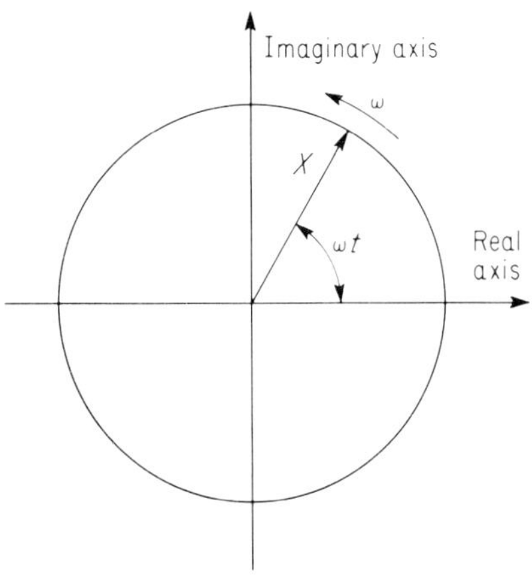

FIG. 6-16

Complex functions are easy to manipulate when phase relations exist and are therefore useful in analyzing damped systems. We shall demonstrate their use in obtaining the steady-state solution for the damped system described by Eq. 6-11.

Representing the forcing function by the imaginary part of $Fe^{i\omega t}$, the steady-state response of the system is expected to be harmonic at the same frequency as the excitation. Because of the damping effects it will be lagging by an angle ϕ behind the forcing function. Accordingly,

$$x = Xe^{i(\omega t - \phi)}$$

Substituting in Eq. 6-11, we obtain

$$(-m\omega^2 + i\omega c + k)Xe^{i(\omega t - \phi)} = Fe^{i\omega t}$$

or

$$(-m\omega^2 + i\omega c + k)Xe^{-i\phi} = F \tag{6-15}$$

Since $e^{-i\phi} = \cos\phi - i\sin\phi$, Eq. 6-15 becomes

$$[(k - m\omega^2) + ic\omega](\cos\phi - i\sin\phi)X = F$$

and by multiplying and rearranging we get

$$[(k - m\omega^2)\cos\phi + c\omega\sin\phi]X + i[c\omega\cos\phi - (k - m\omega^2)\sin\phi]\,X = F$$

By equating the real and imaginary terms on both sides of this equation we find that

$$X = \frac{F}{(k - m\omega^2)\cos\phi + c\omega\sin\phi}$$

$$\tan\phi = \frac{c\omega}{k - m\omega^2}$$

and ϕ can be represented as shown in Fig. 6-15.

Substituting for $\sin\phi$ and $\cos\phi$ in the expression for X gives

$$X = \frac{F}{\sqrt{(k - m\omega^2)^2 + (c\omega)^2}}$$

These expressions for the amplitude X and the phase angle ϕ can be rewritten in a nondimensional form which is identical with Eqs. 6-12 and 6-13.

Transfer Function

The main concern in the analysis of dynamic systems is to evaluate system response to given excitations. In other words, the purpose of the analysis is to determine the outputs corresponding to specific inputs. The expression which describes the relationship between outputs and inputs based on the characteristics of the system and the nature of the input is called the *transfer function*.

We shall illustrate the transfer function concept by assuming that the required output is the dynamic spring force F_d for the damped linear system (Fig. 6-14) due to a harmonic force input $F_i = Fe^{i\omega t}$. The dynamic spring force is calculated from Eq. 6-15 as

$$\begin{aligned} F_d &= kX \\ &= \frac{kFe^{i\omega t}}{(k - m\omega^2) + ic\omega} \\ &= \frac{1}{[1 - (\omega/\omega_n)^2] + i(2\zeta\omega/\omega_n)} F_i \\ &= T(i\omega)F_i \end{aligned} \tag{6-16}$$

where

$$T(i\omega) = \frac{1}{[1 - (\omega/\omega_n)^2]^2 + (2\zeta\omega/\omega_n)^2} e^{-i\phi} \tag{6-17}$$

It can be easily seen that $T_{(i\omega)}$ is the transfer function in this case and describes the frequency response of the system. It is a complex function of the frequency ω with magnitude

$$1/\sqrt{\left[1 - \left(\frac{\omega}{\omega_n}\right)^2\right]^2 + \left(2\zeta\frac{\omega}{\omega_n}\right)^2}$$

and

$$\text{phase lag } \phi = \tan^{-1} 2\zeta\frac{\omega}{\omega_n} \left[1 - \left(\frac{\omega}{\omega_n}\right)^2\right]$$

The magnitude of the transfer function is the gain or magnification factor, and the phase lag ϕ is the angle by which the output lags behind the input. When the transfer function is plotted in the complex plane for a specific value of ζ, the locus for all values of ω (from 0 to ∞) is known as the *Nyquist locus*, Fig. 6-17a.

If the locus is plotted for a specific value of ω/ω_n with ζ changing between 0 and ∞, it is called *Argand's locus*, Fig. 6-17b.

A very useful design diagram can be obtained by assembling a number of these plots as in Fig. 6-18. The response of any single-degree-of-freedom damped system to a harmonic excitation can be readily evaluated from this chart. Such a chart also enables the designer to choose the system parameters necessary to produce a specific response without performing many calculations.

Example

Find the magnification factor (or gain) and the phase lag—by the use of the Nyquist diagram—for the system with

$$m = 4 \text{ lb in.}^{-1}\text{ sec}^2, \quad k = 100 \text{ lb/in.}, \quad \text{and} \quad c = 5 \text{ lb in.}^{-1}\text{ sec}$$

when the exciting force is harmonic with a frequency of 3 rad/sec.

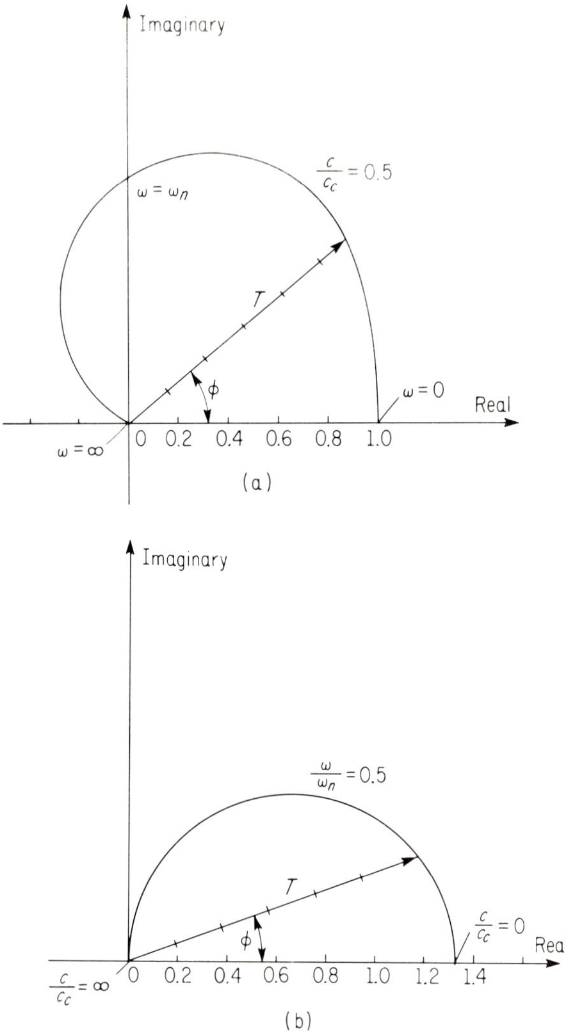

FIG. 6-17. (a) Nyquist locus, (b) Argand's locus.

Solution: The natural frequency for the system is

$$\omega_n = \sqrt{\frac{100}{4}} = 5 \text{ rad/sec}$$

Therefore

$$\frac{\omega}{\omega_n} = \frac{3}{5} = 0.6$$

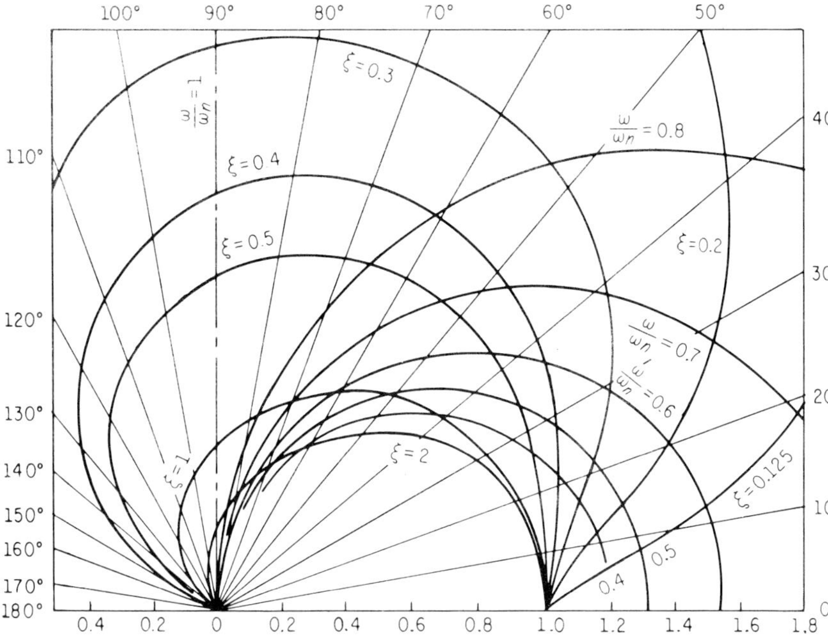

FIG. 6-18

The critical damping factor is calculated from

$$c_c = 2\sqrt{km} = 40 \text{ lb in.}^{-1} \text{ sec}$$

from which

$$\zeta = \frac{c}{c_c} = \frac{5}{40} = 0.125$$

Figure 6-18 shows that the magnification factor in this case is 1.52 and the phase lag is approximately 12.5°.

Laplace Transform

The Laplace transform method can also be used to analyze the motion in this case since it is described by a linear differential equation.

The transform for the equation

$$m\ddot{x} + c\dot{x} + kx = F \sin \omega t$$

is obtained by using the transformation tables as

$$mp^2X(p) - mpX_0 - m\dot{X}_0 + cpX(p) - c\dot{X}_0 + kX(p) = \frac{F\omega}{p^2 + \omega^2}$$

where X_0, $\dot{X}_0$ are the initial values of the displacement and velocity respectively. Now the transform for x is

$$X(p) = \frac{F\omega}{m(p^2 + \omega^2)(p^2 + (c/m)p + \omega_n^2)} + \frac{(p + c/m)X_0}{(p^2 + (c/m)p + \omega_n^2)} + \frac{\dot{X}_0}{(p^2 + (c/m)p + \omega_n^2)} \tag{6-18}$$

and the inverse transform gives the complete solution for the time history of the motion including the transient vibration. This part will be left to the reader as an exercise.

Graphical Solution

The steady-state solution for the damped vibration of the single-degree-of-freedom system under harmonic excitation can be graphically obtained by the use of a vector diagram as shown in Fig. 6-19. If we

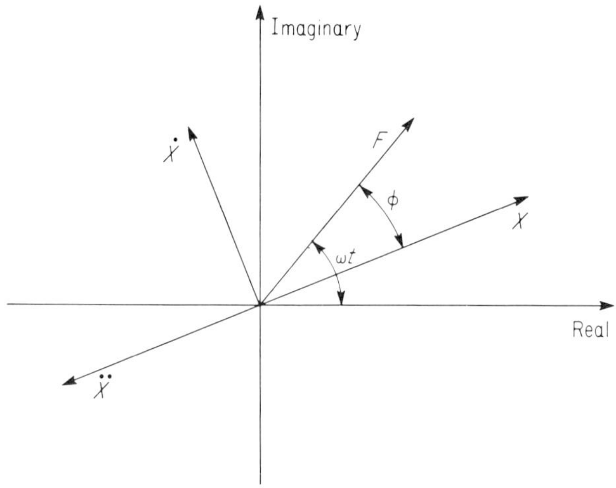

FIG. 6-19

represent the forcing function at any time t by a vector with magnitude F and argument ωt, the displacement will be a vector with magnitude X, lagging by an angle ϕ. It has been shown in Chapter 2 that the phase angle between the velocity and displacement vectors always is $(\pi/2)$ while the acceleration is 180° behind the displacement. The force diagram describing the motion will therefore be as in Fig. 6-20. The phase relationships are taken care of in the diagram by the orientation of the vectors representing the different forces. It is clear from the figure that at low speeds both the damping and inertia forces are small compared to the restoring force. The latter will practically balance the applied force.

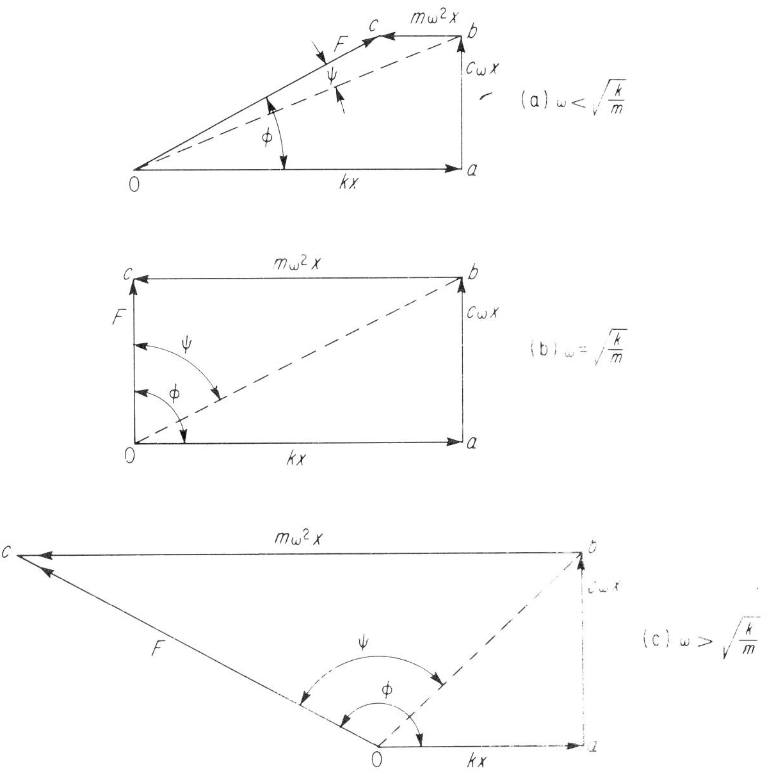

FIG. 6-20

As the speed increases, the damping and the inertia effects get larger. When $m\omega^2 = k$, or when $\omega = \sqrt{k/m}$, which is the resonance condition, the applied force will be balanced by the damping forces and the equilibrium amplitude X will be decided by the damping coefficient and the natural frequency of the system. Figure 6-20b shows that at resonance

$$F = c\omega_n X$$
$$= c\sqrt{k/m}\, X \tag{6-19}$$

or

$$X = \frac{F}{c\sqrt{k/m}}$$

Multiplying both sides by k, we obtain the dynamic spring force at resonance:

$$F_d = kX$$
$$= \frac{F\sqrt{km}}{c}$$

and the corresponding dynamic magnifier is

$$\frac{F_d}{F} = \frac{\sqrt{km}}{c}$$

This equation illustrates very clearly that in a resonance condition the magnitude of the dynamic force depends on the damping constant of the system as well as its mass and spring constant.

Using the notation for the critical damping coefficient,

$$c_c = 2\sqrt{km}$$

The magnification factor can therefore be written as

$$\begin{aligned} \text{M.F.} &= \frac{c_c}{2c} \\ &= \frac{1}{2c/c_c} = \frac{1}{2\zeta} \end{aligned} \tag{6-20}$$

where ζ is the "damping factor" for the system.

The dynamic forces (or displacements) and phase angles for a damped linear system are calculated at different frequencies of disturbance from Eq. 6-12 and plotted in Fig. 6-21a. The magnification factors at resonance coincide with Eq. 6-20. These values for different c/c_c ratios are also plotted in Fig. 6-21b.

The phase angles at resonance are always equal to $\pi/2$ as can be derived from Eq. 6-13. It should be noted here that the amplitudes at resonance for the damped system do not represent peak amplitudes. The latter can be obtained by minimizing the denominator in Eq. 6-12. This gives the condition for the frequency ω_p of excitation necessary to produce the peak amplitude as

$$\frac{\omega_p}{\omega_n} = \sqrt{1 - 2\zeta^2} \tag{6-21}$$

The peak amplitude X_p and the corresponding phase angle can be calculated by substituting the value of ω_p into Eqs. 6-12 and 6-13 respectively. This gives

$$X_p = \frac{F/k}{2\zeta\sqrt{1 - \zeta^2}} \tag{6-22}$$

and

$$\phi \quad (\text{for } X_p) = \frac{1}{\zeta}\sqrt{1 - 2\zeta^2} \tag{6-23}$$

The values for the peak amplitudes and the frequencies at which they occur are plotted in Fig. 6-22. The natural frequency of the damped system ω_{nd} as calculated from Eq. 4-16 is also plotted in Fig. 6-22b for comparison.

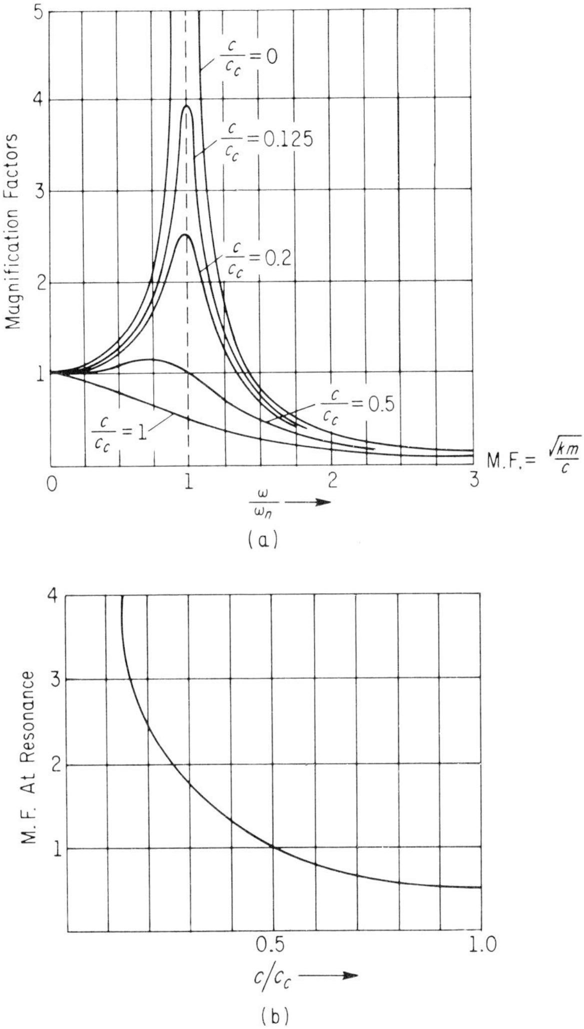

FIG. 6-21. Magnification factors for damped systems.

The velocity of the vibrating mass rather than its displacement is often used to represent the response spectrum of the system. Figure 6-23 shows such a graph in a log-log plot. Notice the symmetry of the response curves about the resonance line ($\omega = \omega_n$) in this case.

Energy Dissipation

The energy dissipated per cycle E_c due to viscous damping can be obtained by integrating the work done by the frictional force F_c over one

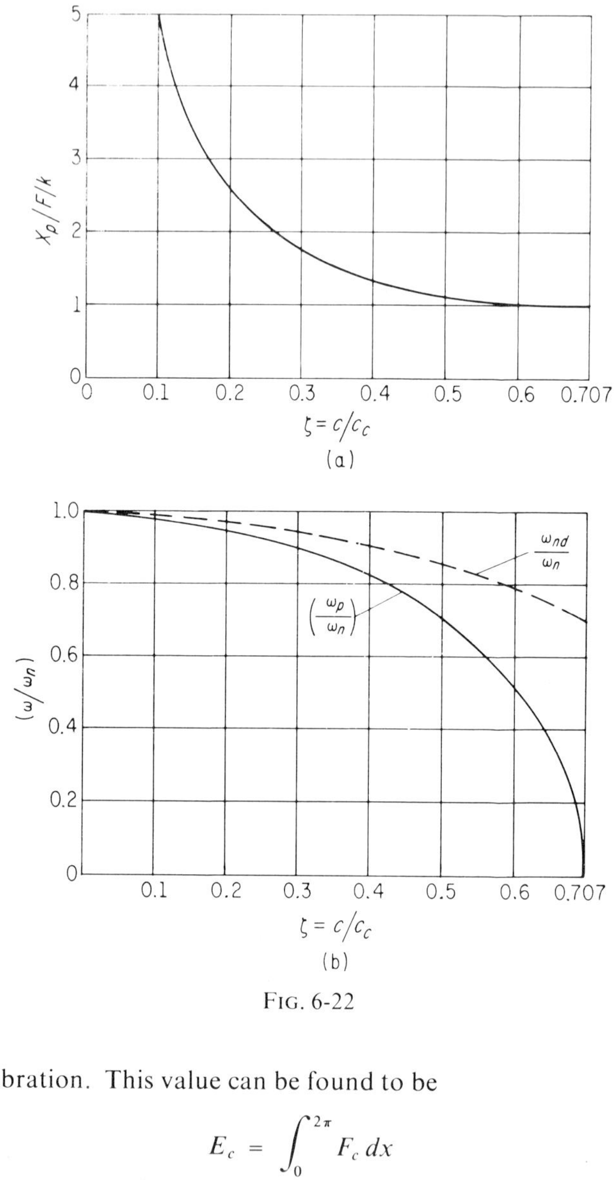

FIG. 6-22

cycle of vibration. This value can be found to be

$$E_c = \int_0^{2\pi} F_c \, dx$$

where $F_c = c\dot{x} = c\omega X \cos(\omega t - \phi)$

$dx = X \cos(\omega t - \phi)\, d\omega t$

c = viscous-damping coefficient

ω = frequency of vibration

X = amplitude of vibration

ϕ = phase angle

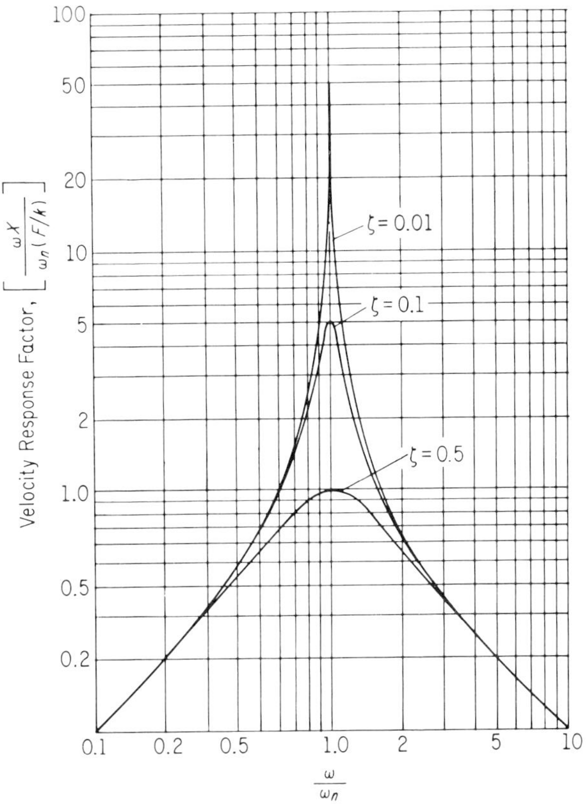

FIG. 6-23

Substituting in the previous equation we obtain

$$E_c = \int_0^{2\pi} c\omega X \cos(\omega t - \phi)\, X \cos(\omega t - \phi) d\omega t$$

$$= \pi c \omega X^2 \qquad (6\text{-}24)$$

The work done E_f by the disturbing force F is also found to be

$$E_f = \int_0^{2\pi} F \sin \omega t \, dx$$

$$= \int_0^{2\pi} F \sin \omega t \, \dot{x} \, dt$$

$$= \int_0^{2\pi} F \sin \omega t \, \omega X \cos(\omega t - \phi) \, dt$$

$$= \pi F X \sin \phi \qquad (6\text{-}25)$$

The equilibrium amplitude can be obtained from energy considerations as

$$E_f = E_c$$
$$\pi F X \sin \phi = \pi\, c\omega X^2$$

or

$$X = (F/c\omega) \sin \phi$$

At resonance, $\phi = \pi/2$ and the previous equation will be identical with Eq. 6-19.

Transmissibility

The magnitude of the force F_t transmitted to the foundation (Fig. 6-24) when a force $F \sin \omega t$ is applied to the mass of a damped system is

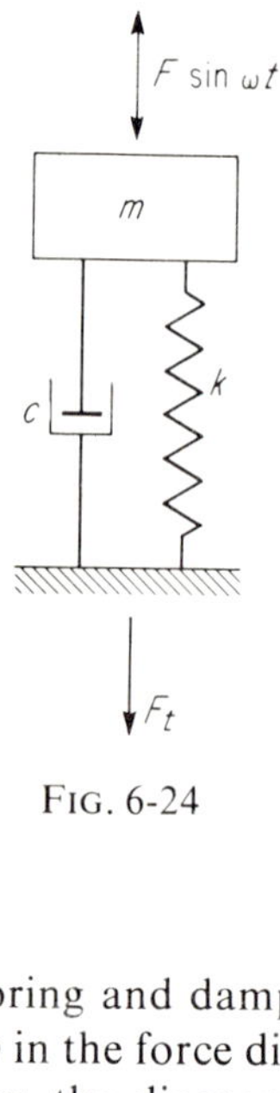

FIG. 6-24

the vector summation of the spring and damping forces. F_t can therefore be represented by the vector *ob* in the force diagrams, Fig. 6-20.

It can be readily seen from the diagram that the transmitted force has a magnitude given by

$$F_t = X\sqrt{k^2 + c^2\omega^2}$$

and a phase lag ψ given by

$$\psi = \phi - \tan^{-1} \frac{c\omega}{k}$$

$$= \phi - \tan^{-1} 2\zeta \frac{\omega}{\omega_n} \qquad (6\text{-}26)$$

The ratio F_t/F is known as the *transmissibility* and has the value

$$\frac{F_t}{F} = \frac{X\sqrt{k^2 + c^2\omega^2}}{F}$$

$$= \frac{\sqrt{1 + \left(2\zeta\dfrac{\omega}{\omega_n}\right)^2}}{\sqrt{\left[\left(1 - \dfrac{\omega}{\omega_n}\right)^2\right]^2 + \left(2\zeta\dfrac{\omega}{\omega_n}\right)^2}} \tag{6-27}$$

Figure 6-25 shows the plot of this equation for different damping factors ζ.

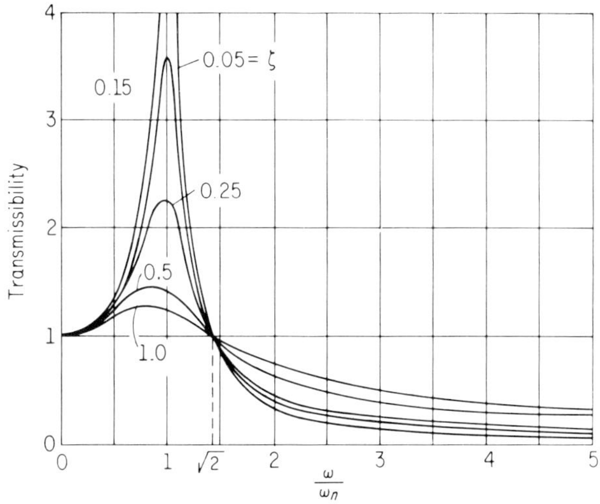

FIG. 6-25

Foundation Disturbances

In this case the system will be as shown in Fig. 6-26 and the equation describing the motion will be

$$m\ddot{x} + c(\dot{x} - \dot{x}_f) + k(x - x_f) = 0$$

where x = absolute displacement of mass
$(x - x_f)$ = displacement of mass relative to base (or spring deflection)

To find the dynamic deflection of the spring, the equation can be written as

$$m(\ddot{x} - \ddot{x}_f) + c(\dot{x} - \dot{x}_f) + k(x - x_f) = -m\ddot{x}_f$$

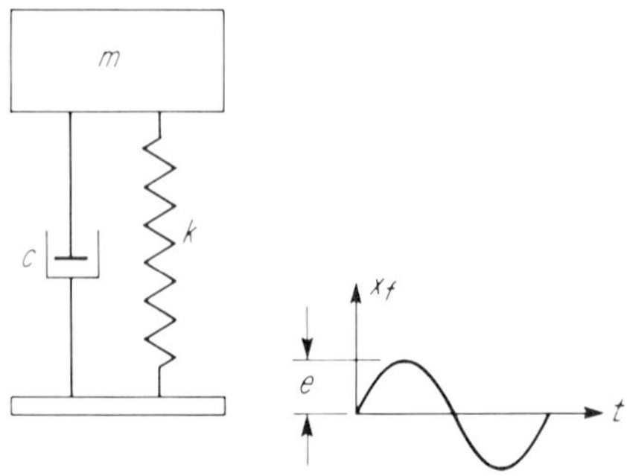

FIG. 6-26

Knowing that

$$x_f = e \sin \omega t$$

which represents a motion with amplitude e and frequency ω, we obtain

$$\ddot{x}_f = -\omega^2 e \sin \omega t$$

and by using the notation

$$x_s = x - x_f = X_s \sin (\omega t - \phi)$$

the equation of motion takes the form

$$m\ddot{x}_s + c\dot{x}_s + kx_s = m\omega^2 e \sin \omega t$$

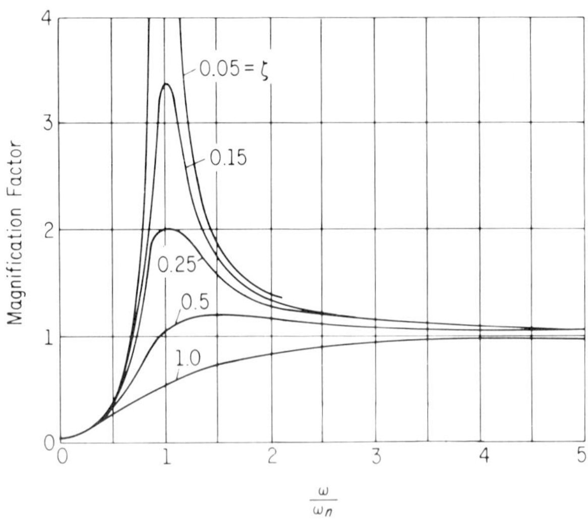

FIG. 6-27

which is of the same form as Eq. 6-11. The amplitude X_s and the phase angle ϕ for the steady-state solution will be

$$X_s = \frac{m\omega^2 e}{\sqrt{(k - m\omega^2)^2 + (c\omega)^2}} = \frac{e(\omega/\omega_n)^2}{\sqrt{[1 - (\omega/\omega_n)^2]^2 + (2\zeta\,\omega/\omega_n)^2}} \tag{6-28a}$$

and

$$\tan\phi = \frac{2\zeta(\omega/\omega_n)}{1 - (\omega/\omega_n)^2} \tag{6-28b}$$

The solution is plotted as a response spectrum in Fig. 6-27 for different values of the damping factor. The dynamic magnifier represents the ratio (X_s/e) in this case. The absolute displacement of the mass can also be found from

$$x = x_s + x_f = X_s \sin(\omega t - \phi) + e\sin\omega t$$

as represented graphically in Fig. 6-28.

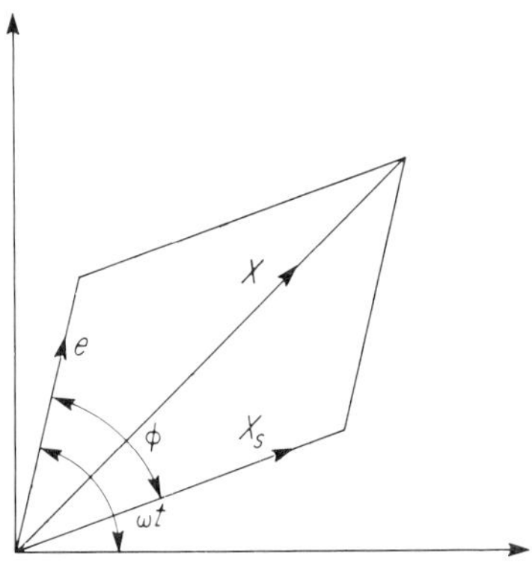

FIG. 6-28

Disturbance Due to Unbalance

A single-degree-of-freedom damped system under the influence of a rotating unbalance is shown in Fig. 6-29. M is the total supported mass, m is the eccentric mass, r its eccentricity, and ω the frequency of the rotation. The equation of motion will be

$$M\ddot{x} + c\dot{x} + kx = (m\omega^2 r)\sin\omega t$$

which is identical to the previous case.

The solution can therefore be given as

$$x = \text{mass displacement} = X\sin(\omega t - \phi)$$

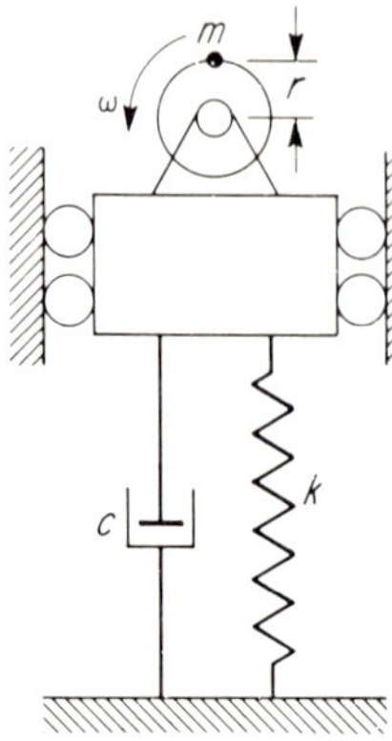

FIG. 6-29

where

$$X = \frac{m\omega^2 r}{\sqrt{(k - M\omega^2)^2 + (c\omega)^2}} = \frac{(m/M)r\,(\omega/\omega_n)^2}{\sqrt{(1 - (\omega/\omega_n)^2)^2 + (2\zeta\,\omega/\omega_n)^2}} \tag{6-29a}$$

and

$$\tan\phi = \frac{2\zeta\,\omega/\omega_n}{1 - (\omega/\omega_n)^2} \tag{6-29b}$$

The response spectrum will be the same as Fig. 6-27. The dynamic magnifier in this case represents the ratio $\left(\frac{X}{r} \times \frac{M}{m}\right)$.

Whirling of Shafts

A special case of the unbalance excited vibration is the whirling of a shaft due to an unbalanced rotor as shown in Fig. 6-30.

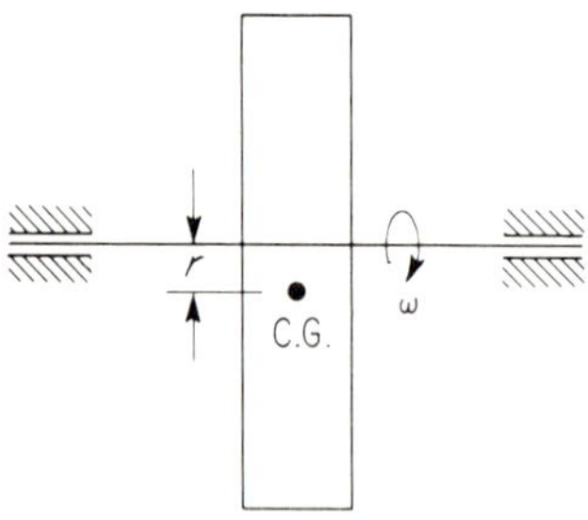

FIG. 6-30

The total mass is eccentric in this case at an eccentricity r. Accordingly, m is equal to M and k will be the restoring shaft force per unit deflection. For example, in the case of a simple supported shaft with the rotor at its midspan the stiffness is

$$k = \frac{48EI}{L^3}$$

The solution for the case of whirling is the same as in the previous case.

The dynamic deflection of the shaft is calculated from Eq. 6-29 as

$$x = X \sin (\omega t - \phi)$$

where

$$X = \frac{r(\omega/\omega_n)^2}{\sqrt{[1 - (\omega/\omega_n)^2]^2 + (2\zeta\omega/\omega_n)^2}} \tag{6-30a}$$

and

$$\phi = \tan^{-1} \frac{2\zeta\omega/\omega_n}{1 - (\omega/\omega_n)^2} \tag{6-30b}$$

It can be easily seen that large vibration can result from small eccentricity when the shaft rotates in the neighborhood of its natural frequency. The speed of rotation corresponding to the natural frequency is known as the *critical speed*. It is interesting to note from Eq. 6-30a that when (ω/ω_n) is very large, $X = -r$ and the rotor will rotate about its mass center.

6-4. FORCED VIBRATIONS OF NONLINEAR SYSTEMS

Exact Solutions

The only class of nonlinear systems with known exact solutions under harmonic excitations are those represented in a stepwise linear manner such as the system shown in Fig. 6-31. The motion in this case is repre-

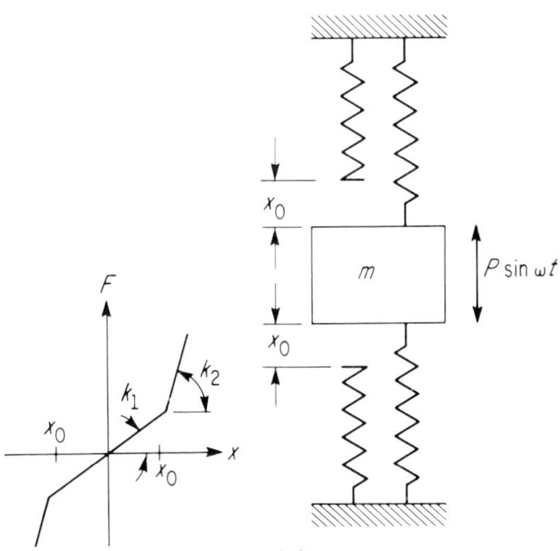

FIG. 6-31

sented by the two equations

$$m\ddot{x}_1 + k_1 x_1 = \pm P \sin \omega t$$

for $x_0 \geq x_1 \geq 0$, and

$$m\ddot{x}_2 + (k_1 - k_2)x_0 + k_2 x_2 = \pm P \sin \omega t$$

for $x_2 \geq x_0$.

The $\pm$ signs are for the in-phase and out-of-phase motions respectively. The boundary conditions are

$$\begin{aligned} \text{at} \quad t &= 0, & x_1 &= 0 \\ t &= t_0, & x_1 &= x_2 = x_0 \\ t &= t_0, & \dot{x}_1 &= \dot{x}_2 \\ t &= \pi/2\omega, & \dot{x}_2 &= 0 \end{aligned}$$

and the solution has the form

$$x_1 = \frac{\pm P/k_1}{1 - (\omega/\omega_1)^2} \sin \omega t + A_1 \cos \omega_1 t + B_1 \sin \omega_1 t$$

$$x_2 = \frac{\pm P/k_2}{1 - (\omega/\omega_2)^2} \sin \omega t + A_2 \cos \omega_2 t + B_2 \sin \omega_2 t + \left(1 - \frac{k_1}{k_2}\right) X_0 \tag{6-31}$$

where A_1, B_1, A_2, and B_2 are integration constants to be determined from the boundary condition.

Approximate Methods*

Duffing's Method

As in the case of free vibration, we can use iteration or perturbation procedures for the solution of the forced vibration of nonlinear systems.

We shall illustrate here as an example Duffing's method of solution for the equation

$$\ddot{x} + k^2 (x \pm \mu^2 x^3) = P \cos \omega t \tag{6-32}$$

describing an undamped system with nonlinear spring. The $\pm$ signs indicate increasing or decreasing stiffness respectively. Assuming as a first approximation

$$x_1 = A \cos \omega t \quad \text{(subscript indicates order of approximation)}$$

A second approximation is obtained from

$$x_2 = -(k^2 A \pm \frac{3}{4} k^2 \mu^2 A^2 - P) \cos \omega t \pm \frac{1}{4} k^2 \mu^2 A^3 \cos 3 \omega t$$

*Can be omitted in a first reading without loss of continuity.

$$\left(Note{:}\ \cos^3 \omega t = \frac{3}{4} \cos \omega t + \frac{\cos 3\,\omega t}{4}\right)$$

By integration of the previous equation, a second approximation solution is obtained as

$$x_2 = \frac{1}{\omega^2}\left[(k^2 A \pm \frac{3}{4} k^2 \mu^2 A^3 - P) \cos \omega t \pm \frac{1}{36} k^2 \mu^2 A^3 \cos 3\,\omega t\right]$$

Then by substitution in original equation, a third approximation is calculated from

$$x_3 = \cdots$$

This is an iteration procedure. In order for it to be convergent k^2, μ^2, A and P must be small. For the slightly nonlinear case the solution may be assumed as

$$x = A \cos \omega t$$

from which

$$A = 1/\omega^2(k^2 A \pm \frac{3}{4} k^2 \mu^2 A^3 - P)$$

or

$$\omega^2 = k^2(1 \pm \frac{3}{4} \mu^2 A^2) - P/A \tag{6-33}$$

which describes the response curve as a first approximation.

Ritz Averaging Method

This is a very useful and versatile method for the analysis of nonlinear problems. It was used by K. Klötter in 1951 (Refs. 37, 39) and has since been used extensively. This is an energy procedure which requires that the average value of the virtual work per cycle must vanish. To illustrate this technique we shall start with the general equation

$$\ddot{x} + \frac{1}{m} f_1(\dot{x}) + \frac{1}{m} f_2(x) - \frac{1}{m} F_3(t) = 0 \tag{6-34}$$

This is an equation of equilibrium in which the inertia force is balanced by the resisting force and the disturbing force. The work done by these forces in a virtual displacement δx must vanish, therefore

$$\left[\ddot{x} + \frac{1}{m} f_1(\dot{x}) + \frac{1}{m} f_2(x) - \frac{F_3(t)}{m}\right] \delta x = 0$$

Assume the approximate steady-state solution

$$x = a_1 \phi_1(t) + a_2 \phi_2(t) + a_3 \phi_3(t) + \cdots \tag{6-35}$$

from which

$$\delta x = \delta a \phi_n(t)$$

Since the solution is approximate, the virtual work cannot be expected to vanish at all times, but we can select the parameters $a_1, a_2, \ldots, a_n$ so as to make the average value of the virtual work per cycle vanish. This condition gives

$$\int_0^\tau \left[\ddot{x} + \frac{1}{m} f_1(\dot{x}) + \frac{1}{m} f_2(x) - \frac{F}{m}(t)\right] \phi_n(t)\, dt = 0 \qquad (6\text{-}36)$$

By substituting for x from Eq. 6-35 in this equation, as many algebraic equations as the number of terms in the assumed solution are obtained, and by solving them we find the proper values of the parameters $a_1, a_2 \ldots,$ etc. This procedure is illustrated by the following example.

Example

Find a solution for Duffing's equation

$$\ddot{x} + p^2(x + \beta x^3) = q \cos \omega t$$

using the Ritz method with one-term approximation. The solution will take the form

$$x = \pm a_1 \cos \omega t$$

Substituting in Eq. 6-36 we have

$$\int_0^\tau [\ddot{x} + p^2 (x + \beta x^3) - q \cos \omega t] \cos \omega t\, dt = 0$$

Knowing

$$\int_0^\tau \cos^2 \omega t\, dt = \pi/\omega$$

and

$$\int_0^\tau \cos^4 \omega t\, dt = 3/4\pi/\omega$$

the integration therefore yields

$$p^2 a_1 + \frac{3}{4} \beta p^2 a_1^3 = \pm q + a_1 \omega^2$$

or

$$a_1 = 1/\omega^2 [p^2 a_1 + \frac{3}{4} \beta p^2 a_1^2 \pm q]$$

which is identical with Eq. 6-33 for the response curve in this case.

Two-Term Solution for Nonlinear Systems Under a Steady Alternating Force by the Ritz Method

Considering the general nonlinear system described by Eq. 6-34 and taking $F_3(t) = F \sin \omega t$, the equation of motion becomes

$$\ddot{x} + \frac{1}{m} f_1(\dot{x}) = \frac{1}{m} f_2(x) - \frac{F}{m} \sin \omega t = 0$$

Assuming a solution with two-term approximation as

$$\begin{aligned} x &= X \sin(\omega t - \phi) \\ &= a_1 \sin \omega t + a_2 \cos \omega t \end{aligned} \tag{6-37}$$

and using the notation

$$E = \ddot{x} + \frac{1}{m} f_1(\dot{x}) + \frac{1}{m} f_2(x) - \frac{F}{m} \sin \omega t$$

it can be seen that E will be in the neighborhood of but not equal to zero at all times, since x in this expression is only an approximate solution for Eq. 6-34.

The Ritz conditions given by Eq. 6-36 require in this case that

$$\begin{aligned} \int_0^{2\pi} E \sin \omega t \, d\omega t &= 0 \\ \int_0^{2\pi} E \cos \omega t \, d\omega t &= 0 \end{aligned} \tag{6-38}$$

These two equations will furnish the following two general algebraic equations in X and ϕ

$$-(\omega^2) \cos \phi + \sin \phi \, G(X, \omega) + \cos \phi \, H(X) = \frac{F}{mX} \tag{6-39}$$

$$-(\omega^2) \sin \phi - \cos \phi \, G(X, \omega) + \sin \phi \, H(X) = 0 \tag{6-40}$$

where

$$G(X, \omega) = \frac{1}{mX\pi} \int_0^{2\pi} f_1(\omega X \cos \sigma) \cos \sigma \, d\sigma \tag{6-41}$$

$$H(X) = \frac{1}{mX\pi} \int_0^{2\pi} f_2(X \sin \sigma) \sin \sigma \, d\sigma \tag{6-42}$$

By squaring Eqs. 6-39 and 6-40 we obtain

$$[H(X) - (\omega^2)]^2 + G^2(X, \omega) = \left[\frac{F}{mX}\right]^2 \tag{6-43}$$

and ϕ can be evaluated from Eq. 6-40 as

$$\tan \phi = \frac{G(X, \omega)}{H(X) - (\omega^2)} \tag{6-44}$$

Equation 6-43 describes the response of the system, while Eq. 6-44 gives the phase lag.

For the undamped vibration $G(X, \omega) = 0$ and the response of the system is obtained by substitution in Eq. 6-43:

$$H(X) - \omega^2 = \frac{F}{mX} \tag{6-45}$$

Also, by taking $F = 0$, the equation for the free undamped vibration becomes

$$H(X) = \omega^2 \tag{6-46}$$

The curve described by this equation is known as the *backbone curve.* Since in a free vibration the frequency is the natural frequency, Eq. 6-46 describes the variation in the natural frequency with the amplitude of vibration. Some response curves for stepwise linear systems calculated by the Ritz method are shown in Fig. 6-32. The backbone curves are shown in broken lines. Notice for the undamped case of Fig. 6-32 that the response curve for a particular excitation has two branches as in the linear case. One branch is in phase with the excitation while the other is out of phase with it. When the system is damped, the response curves are continuous and therefore intersect the backbone curve. The points of intersection of the backbone curve and the response curves are known as the *resonant points.*

The amplitude of vibration and phase angle at these resonance points can be calculated by substituting Eq. 6-46 in Eqs. 6-43 and 6-44. The result is

$$X = \frac{F}{mG(X, \omega)} \quad \text{and} \quad \phi = \frac{\pi}{2}$$

In systems where the functions $f_1(x)$ and $f_2(x)$ are not easily integrable in an exact manner, $H(X)$ and $G(\omega, X)$ can be calculated by numerical integration as shown in the following.

For a given value of X, assume an incremental angle $\Delta\sigma = \dfrac{2\pi}{n}$ where n is a convenient number of steps. In order to evaluate $H(X)$, tabulate σ, $\sin\sigma$, $X\sin\sigma$, $f_2(X\sin\sigma)$ and $f_2(X\sin\sigma)\sin\sigma$ as shown in the accompanying table.

σ	$\sin\sigma$	$X\sin\sigma$	$f_2(X\sin\sigma)$	$f_2(X\sin\sigma) \times \sin\sigma$
0	0	0	$f_2(0)$	$f_0 = 0$
$\Delta\sigma$	$\sin\Delta\sigma$	$X\sin\Delta\sigma$	$f_2(X\sin\Delta\sigma)$	f_1
$2\Delta\sigma$	...	...	...	...
...	...	...	...	...
...	...	...	...	...
$(n-1)\Delta\sigma$	...	...	...	f_{n-1}
$2\pi = n\Delta\sigma$	0	...	...	$f_n = 0$

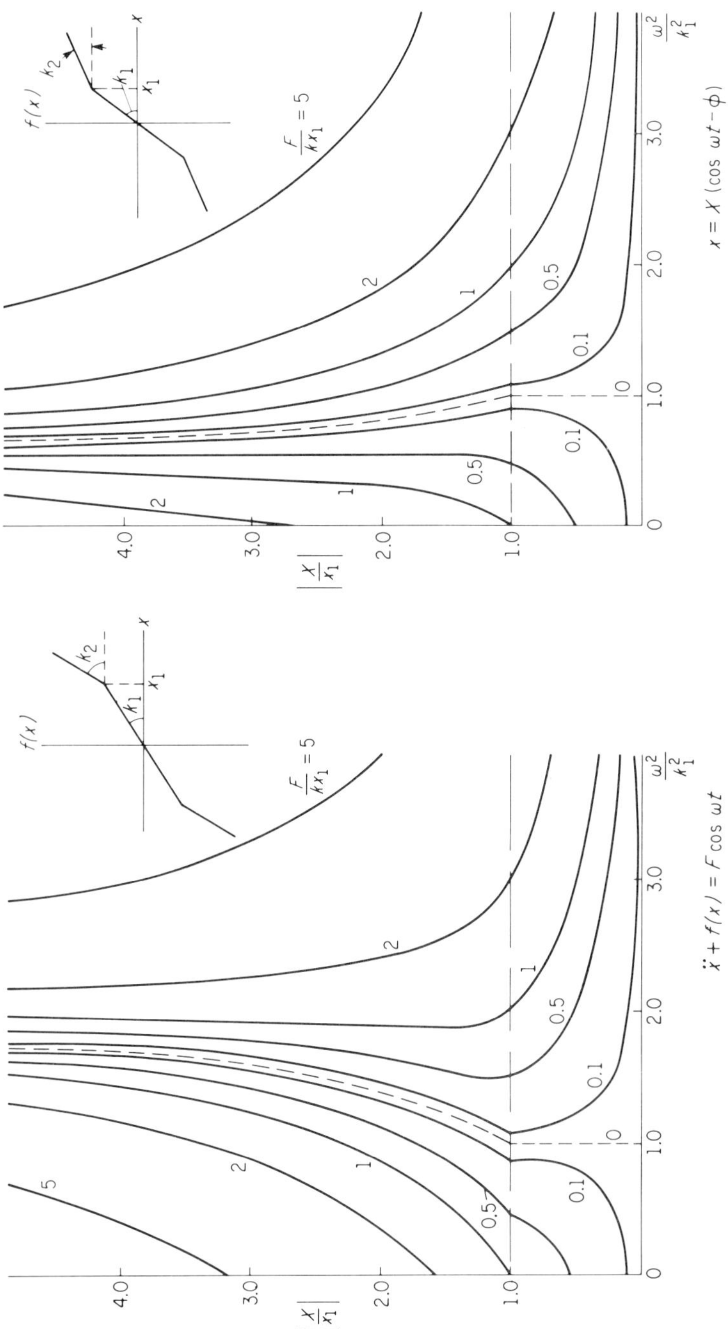

FIG. 6-32. Stepwise linear systems (Ref. 37).

By means of the trapezoidal rule of integration, $H(X)$ is therefore calculated as

$$H(X) = \frac{\Delta\sigma}{mX\pi} \sum \frac{f_0 + f_n}{2} + f_1 + \cdots + f_{n-1}$$

$$= \frac{2}{nmX} \sum \frac{f_0 + f_n}{2} + f_1 + \cdots + f_{n-1}$$

The value of $G(X, \omega)$ can also be obtained in a similar manner.

Example

Use the Ritz method to calculate the frequency response of a linear damped system due to harmonic force excitation.

Solution: In this case

$$E = \ddot{x} + \frac{c}{m}\dot{x} + \frac{k}{m}x - \frac{F}{m}\sin\omega t$$

where

$$f_1(\dot{x}) = c\dot{x} \quad \text{and} \quad f_2(x) = kx$$

From Eqs. 6-41 and 6-42 we have

$$G(X, \omega) = \frac{1}{mX\pi} \int_0^{2\pi} \omega c x \cos^2\sigma \, d\sigma$$

$$= \frac{\omega c}{m}$$

$$H(X) = \frac{1}{mX\pi} \int_0^{2\pi} kx \sin^2\sigma \, d\sigma$$

$$= \frac{k}{m}$$

The response equation 6-43 therefore becomes

$$\left(\frac{k}{m} - \omega^2\right)^2 + \left(\frac{\omega c}{m}\right)^2 = \left[\frac{F}{mX}\right]^2 \tag{6-47}$$

and the phase angle is calculated from Eq. 6-44 as

$$\tan\phi = \frac{\omega c/m}{k/m - \omega^2}$$

which can be written as

$$X = \frac{F}{\sqrt{(k - m\omega^2)^2 + (c\omega)^2}} = \frac{F/k}{\sqrt{(1 - (\omega/\omega_n)^2)^2 + (2\zeta\omega/\omega_n)^2}}$$

and

$$\tan\phi = \frac{2\zeta\omega/\omega_n}{1 - (\omega/\omega_n)^2}$$

This is the exact solution for this problem as would be expected since the assumed solution for the Ritz method $x = X \sin(\omega t - \phi)$ is the exact one.

The undamped free-vibration or backbone curve will be represented by the following equation:

$$\omega^2 = H(X) = \frac{k}{m}$$

and the amplitude of vibration at the resonant point can be evaluated from the intersection of the backbone curve and the response curve. This is done by substituting the previous equation in Eq. 6-47 which gives

$$X = \frac{F}{\omega c} = \frac{F/k}{2\zeta}$$

which is the same as obtained by the linear theory.

Example

Find the equation for the backbone curve for the vibration of a mass m under the effect of a restoring function $f(x) = kx^3$. Calculate the natural period when the amplitude is 1 in.

Solution: From Eq. 6-42 we obtain

$$\begin{aligned} H(X) &= \frac{1}{mX\pi} \int_0^{2\pi} kx^3 \sin^4 \sigma \, d\sigma \\ &= \frac{kx^2}{m\pi} \left(\frac{3}{4} \pi \right) \\ &= \frac{3}{4} \frac{k}{m} X^2 \end{aligned}$$

Equation 6-46 for the backbone curve is therefore

$$\omega^2 = \frac{3}{4} \frac{k}{m} X^2$$

The natural period of oscillation $= \tau = 2\pi/\omega$

$$\text{for} \quad X = 1, \quad \omega^2 = \frac{3}{4} \frac{k}{m} \quad \text{and} \quad \tau = 4\pi/\sqrt{3k/m}$$

Graphical Solution

An approximate solution can be obtained for the undamped non-linear system subjected to harmonic excitation by extending the graphical method used for the linear problem in Fig. 6-5.

The solution for the nonlinear system will be assumed harmonic ($x = X \sin \omega t$) as an approximation and the amplitude of vibration for any force amplitude F and frequency ω is therefore found from the inter-

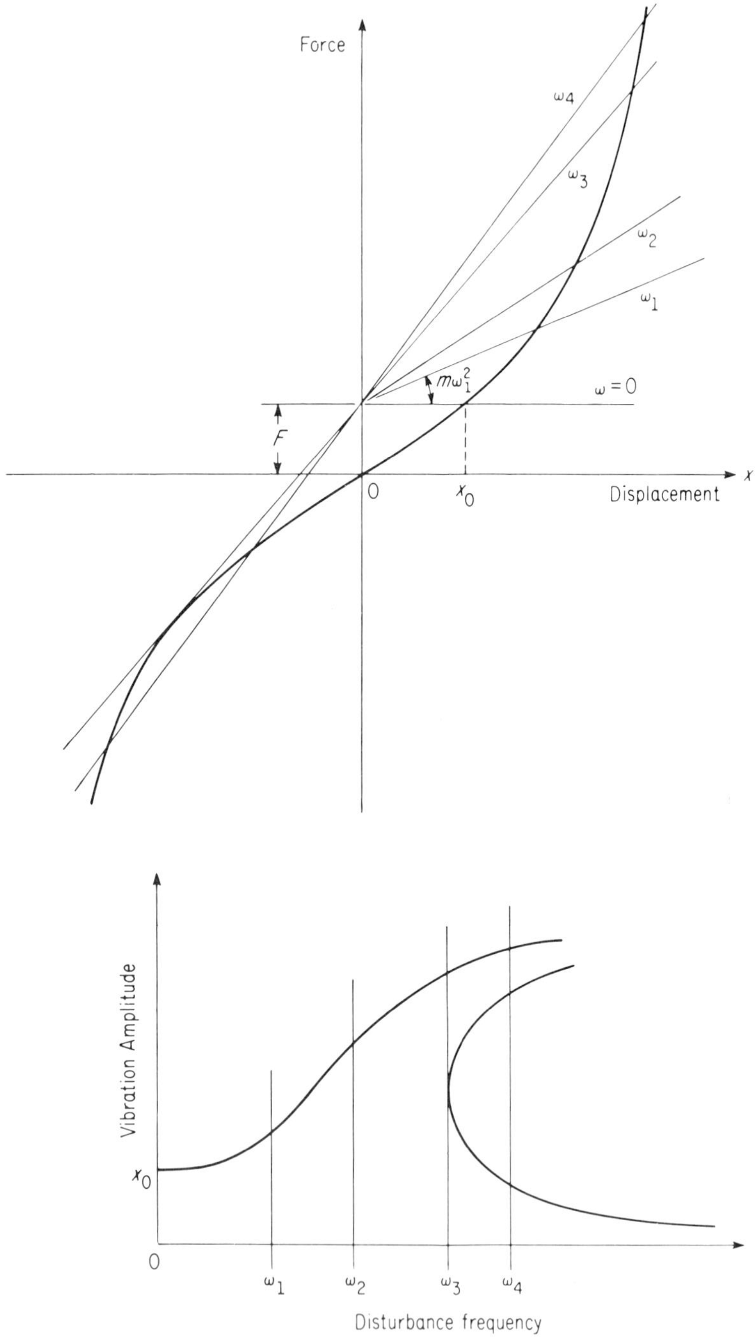

FIG. 6-33. Graphical solution for nonlinear systems.

section of the curve representing the restoring function $f(x)$ and the straight line with intercept F and slope $m\omega^2$.

The amplitude-frequency diagram (response curve) can therefore be plotted as shown in Fig. 6-33.

It is evident that the plot has two branches, one in phase and the other out of phase with the forcing function as in the linear case. However, due to the nonlinearity of the restoring function the response curves are bent and the stronger the nonlinearity the greater the bending. It should be noticed also that while one definite answer occurs in the linear problem, more than one steady-state solution is possible in the nonlinear system. For the example under consideration two solutions are possible at $\omega = \omega_3$ and three solutions at $\omega = \omega_4$ even though the parameters remain unchanged.

Some of these solutions are stable and the others unstable. The unstable solution is one which can be obtained mathematically but cannot be observed physically.

Similar graphical solutions can be obtained for the cases of foundation and unbalance excitations.

The Jump Phenomenon

An interesting phenomenon related to the stability of the motion that may occur in nonlinear vibration but never shows up in linear systems is known as the *jump phenomenon*. Let us assume that the system shown in Fig. 6-33 was under the influence of a low damping force. The effect of damping will limit the amplitude and cause a continuity in the response curve similar to the linear case. This is shown in Fig. 6-34. If the excitation is introduced at a low frequency, the response of the system will

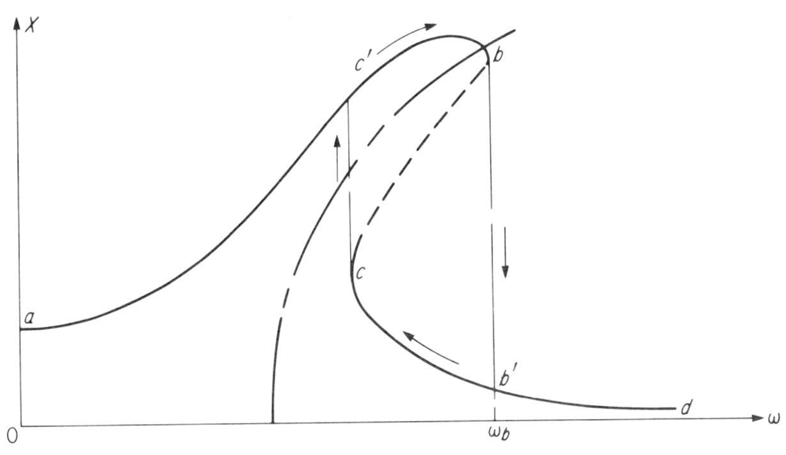

FIG. 6-34

follow the curve ab as the frequency increases. When the point b of vertical tangency is reached, the amplitude will suddenly drop to b'. The response will then follow $b'd$ as the frequency continues to increase. On the other hand, if the excitation has a frequency higher than ω_b, the response will follow the curve dc as the frequency decreases. When point c of vertical tangency is reached, the amplitude will rise suddenly from c to c' and then follows the curve $c'a$ as the frequency continues to decrease.

A vibration governed by the part bc of the response curve between the two points of vertical tangency will never physically exist, and this part of the mathematical solution is therefore unstable. It should be noted that an increase in the exciting force magnitude on the part bc will result in a decrease in the steady state amplitude. A stability criterion can therefore be stated as: If $\dfrac{\partial F}{\partial X} > 0$ the motion is stable, and if $\dfrac{\partial F}{\partial X} < 0$ the motion is unstable.

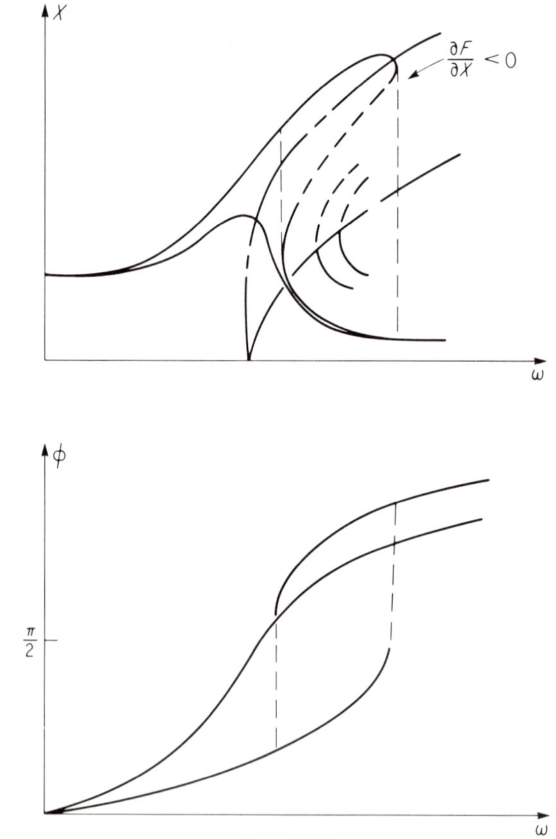

FIG. 6-35

It should also be noted that the jump phenomenon would occur simultaneously in the phase-angle curves and the response curve as shown in Fig. 6-35.

Ultraharmonic and Subharmonic Resonance

Another phenomenon which has no counterpart in linear systems is the appearance of resonances at multiples or submultiples of the forcing frequency. In ultraharmonic or superharmonic resonances, the system will vibrate at a frequency ω_n in the neighborhood of its linearized natural frequency when the forcing frequency is (ω_n/n), n being either even or odd.

In subharmonic resonance, the system vibrates at a frequency ω_n when the forcing frequency is an integral multiple of ω_n. Figure 6-36 shows

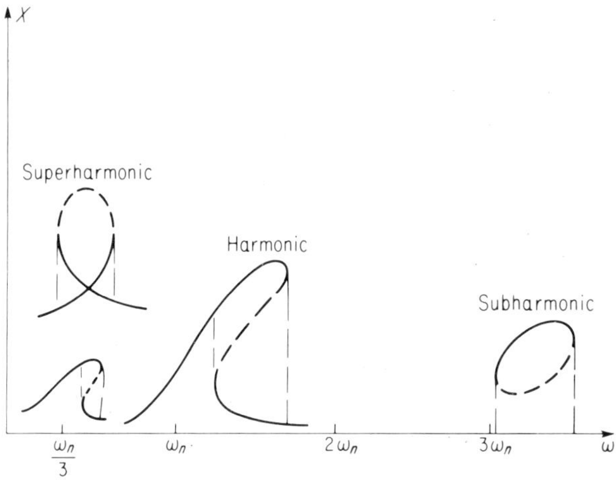

FIG. 6-36

typical superharmonic and subharmonic resonance appearing at forcing frequencies of one-third and three times the linearized natural frequency respectively.

The superharmonic resonance in systems with low damping may be of greater magnitude than the harmonic resonances and are therefore of practical importance. They also appear independent of the initial conditions in contrast to the subharmonic resonances, which will exist only for a limited combination of initial conditions and are therefore more difficult to generate. Subharmonic resonances simply cannot arise because they are unstable at the junction with the fundamental response.

Jump phenomena may also occur in both ultraharmonic and subharmonic vibrations.

6-5. SYSTEMS WITH NONLINEAR DAMPING

We have studied in a previous section the effect of frictional forces of the viscous type on the steady-state vibration of a system with a linear spring. We have found that viscous damping allows exact mathematical analysis of the vibration.

We now consider the other types of frictional forces which may not be as convenient from the analytical viewpoint, but are, however, very common in physical systems. Of all the physical characteristics of a vibratory system, damping is the hardest to evaluate accurately. The natural dissipative mechanisms are rather complex and do not usually confine themselves to one type or the other. For the sake of analysis, damping other than viscous can be classified as Coulomb, hysteretic, and hydraulic.

Coulomb Damping

In this case the frictional force is constant in magnitude and has the opposite sign of the velocity. The equation of motion for a single-degree-of-freedom system with Coulomb damping can be written as

$$m\ddot{x} + \mu N \operatorname{sgn}(\dot{x}) + kx = F \sin \omega t$$

where μ = coefficient of friction
N = normal force
$\operatorname{sgn} \dot{x} = \dot{x}/|\dot{x}|$

Since the velocity changes sign twice each cycle, the duplicity of sign in the damping term renders this equation nonlinear. A rather comprehensive analysis of this equation is given by Den Hartog in Ref. 4.

The equation can also be written in the form

$$m\ddot{x} + f(x) = F \sin \omega t$$

where $f(x)$ represents the bilinear loop shown in Fig. 6-37. Although this equation is nonlinear, it can be treated as a piecewise linear problem. However, for the sake of simplicity we shall linearize this equation by utilizing equivalent viscous damping which dissipates the same amount of energy per cycle as the Coulomb damper. We shall also assume that the solution is harmonic as an approximation.

$$x = X \sin (\omega t - \phi)$$

The energy dissipated per cycle by the Coulomb damper is

$$= 4 \int_0^X \mu N \, dx$$

$$= 4\mu N X \tag{6-48}$$

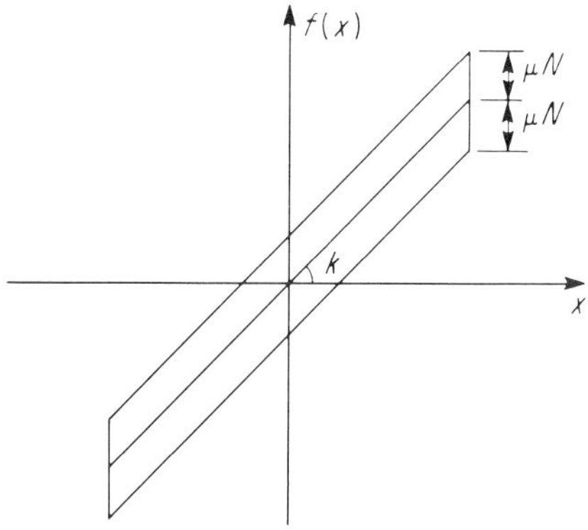

FIG. 6-37

and by equating Eqs. 6-24 and 6-48 we find

$$c_e = \text{equivalent viscous-damping coefficient} = 4\mu N/\pi X\omega$$

We also find by substituting in Eqs. 6-12 and 6-13

$$X = \frac{F}{\sqrt{(k - m\omega^2)^2 + \left(\frac{4\mu N}{\pi X}\right)^2}} \tag{6-49}$$

and

$$\tan\phi = \frac{4\mu N/\pi X}{(k - m\omega^2)} \tag{6-50}$$

Solving for X in Eq. 6-49, we get

$$X = \frac{\sqrt{F^2 - (4\mu N/\pi)^2}}{(k - m\omega^2)} \tag{6-51}$$

and

$$\text{magnification factor} = \frac{X}{F/k} = \frac{\sqrt{1 - (4\mu N/\pi F)^2}}{1 - (\omega/\omega_n)^2} \tag{6-52}$$

Equations 6-50 and 6-52 are shown in graphical form in Fig. 6-38. This approximate analysis is applicable only for small values of Coulomb friction when the amplitude of vibration is relatively high.

The Ritz averaging method can also be used to obtain an approximate solution. In this case by substituting in Eqs. 6-42 and 6-43 we have

$$H_{(x)} = \frac{1}{mX\pi}\int_0^{2\pi} kX\sin^2\sigma\,d\sigma = k/m$$

$$G(\omega, X) = \frac{4}{m\pi X}\int_0^{\frac{\pi}{2}} \mu N\cos\sigma\,d\sigma = \frac{4\mu N}{m\pi X}$$

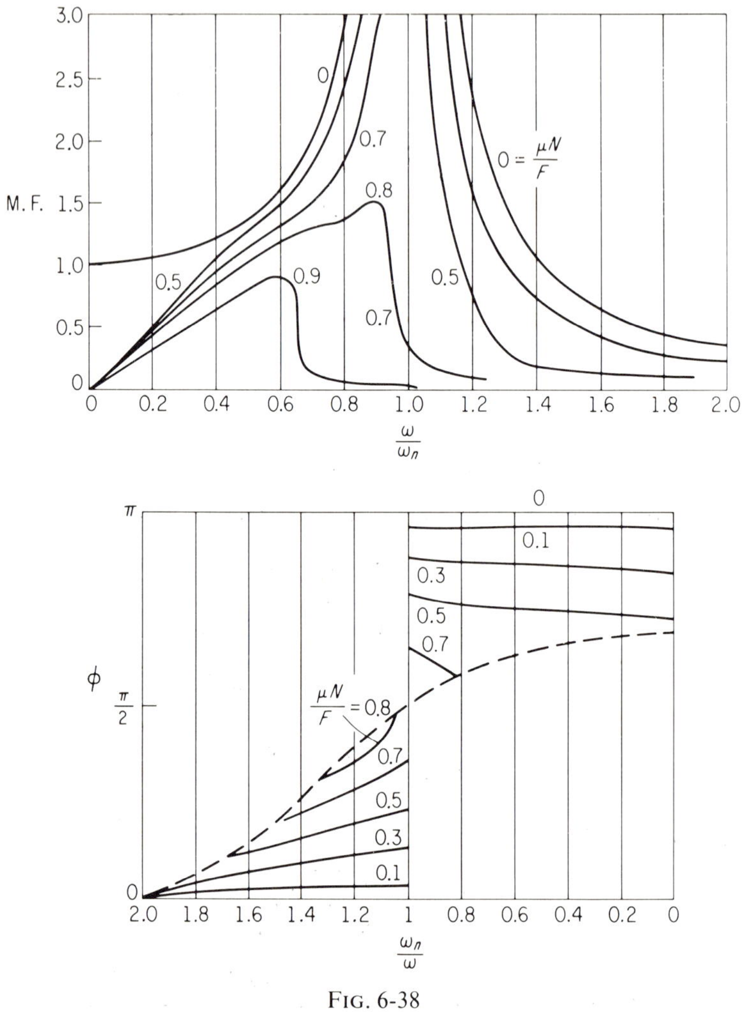

FIG. 6-38

which will give identical equations to (6-28) and (6-29) for the response curves and phase angles.

An extensive analysis of Coulomb-damped vibration isolation system is given by Ruzicka (Ref. 47).

Hysteresis Damping

When structures or structural elements are subjected to cyclic loading, there are usually some differences between the load-deformation curve which represents the loading and the curve representing the unloading.

For a complete cycle of load, the load-deformation plot takes the form of a loop known as the *hysteresis loop*. The area of this loop represents the energy dissipated per cycle. This amount of energy determines the damping characteristics of the structure.

Accurate information on the hysteresis phenomenon is scarce, due to its complexity. Investigations on vibrations involving hysteresis damping are therefore based on approximations which would allow a mathematical solution. One method of analysis is the linearization of the system as demonstrated by Bishop (Ref. 13). Knowing that the energy dissipated per cycle is of the form

$$E = nX^2 \tag{6-53}$$

where n is a material constant. The loop shown in Fig. 6-39 can therefore be approximated by a linear spring with stiffness k and an equivalent

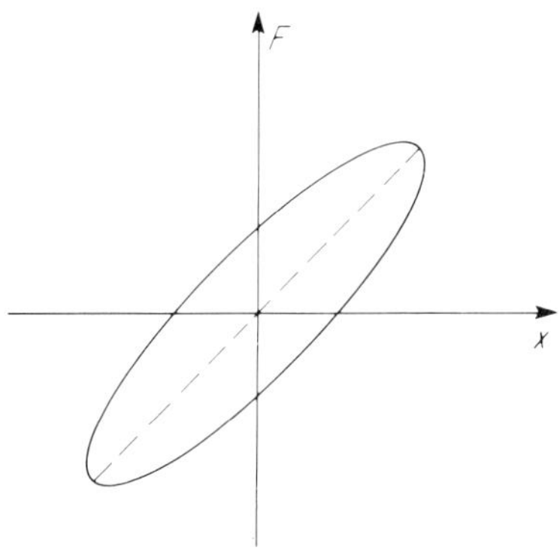

FIG. 6-39

damping coefficient which can be evaluated by equating Eqs. 6-24 and 6-53. Accordingly,

$$nX^2 = \pi c_e \omega X^2$$

where c_e = equivalent viscous damping
$= n/\pi\omega = h/\omega$
h = material constant

The equation of motion can therefore be written as

$$m\ddot{x} + \frac{h}{\omega}\dot{x} + kx = F \sin \omega t$$

which has a solution of the form

$$x = X \sin(\omega t - \phi)$$

where X is evaluated from Eq. 6-11 as

$$X = \frac{F}{\sqrt{(k - m\omega^2)^2 + h^2}}$$

$$= \frac{X_{st}}{\sqrt{(1 - (\omega/\omega_n)^2)^2 + (h/k)^2}}$$

Since h and k are constants of the loop, h/k can be considered a structural damping coefficient.

Using the notation $\eta = h/k$ we have

$$X = \frac{X_{st}}{\sqrt{(1 - (\omega/\omega_n)^2)^2 + \eta^2}} \tag{6-54}$$

and ϕ is evaluated from Eq. 6-13 as

$$\tan \phi = \frac{\eta}{1 - (\omega/\omega_n)^2} \tag{6-55}$$

Investigations by Caughey (Ref. 36) and Iwan (Ref. 38) showed that the behavior of many structures can also be approximated by bilinear hysteresis loops of the type shown in Fig. 6-40. Exact solution can be calculated for such systems. Figure 6-41 illustrates the types of response to harmonic excitation obtained from these loops. It is noticed that a series

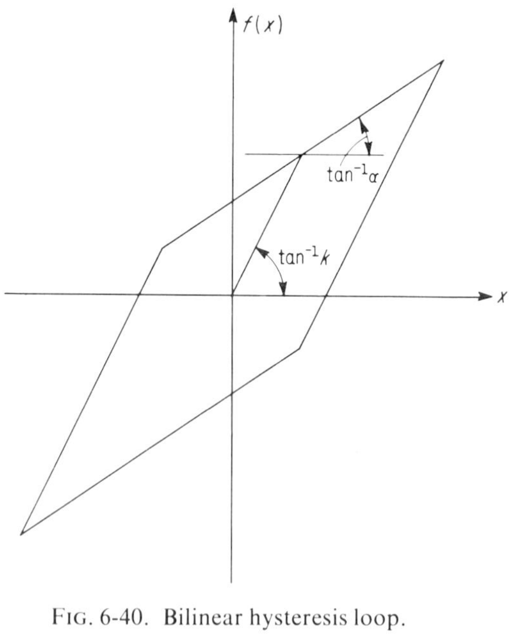

Fig. 6-40. Bilinear hysteresis loop.

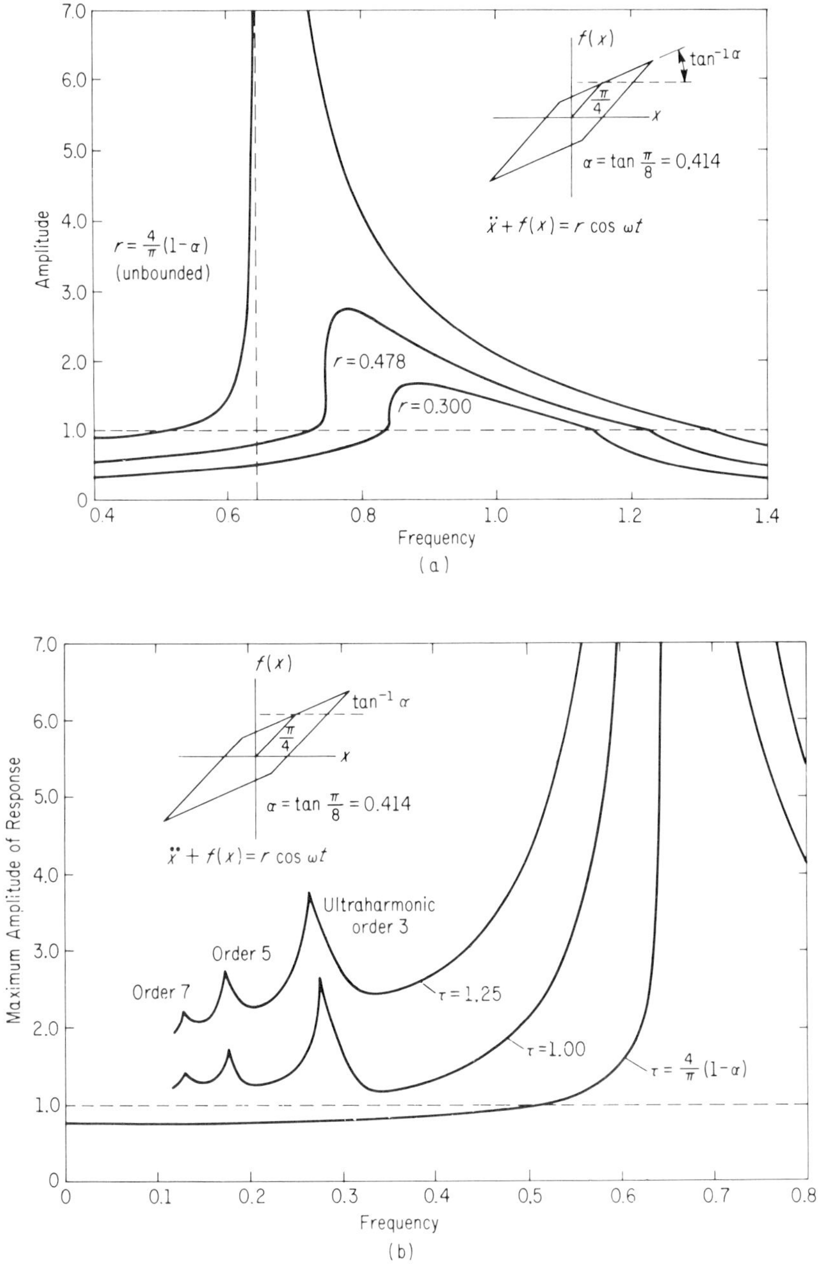

FIG. 6-41. (a) Hysteresis resonance curves, bilinear hysteresis restoring force. (b) Electric analog results, steady-state response, bilinear hysteresis restoring force.

of ultraharmonic resonances appear when the exciting force level is sufficiently high. The curves are always single-valued, and no jumps occur. It is interesting to note that when the exciting force exceeds a particular value the amplitudes become unbonded, as in the case of undamped linear systems, in spite of the hysteresis damping. The curves, however, have steep slopes on the low side of resonance.

A closer approximation of the hysteresis loop has been developed by Jennings (Ref. 40). The mathematical representation of the loops shown in Fig. 6-42 is accurate enough to simulate actual structural hysteresis,

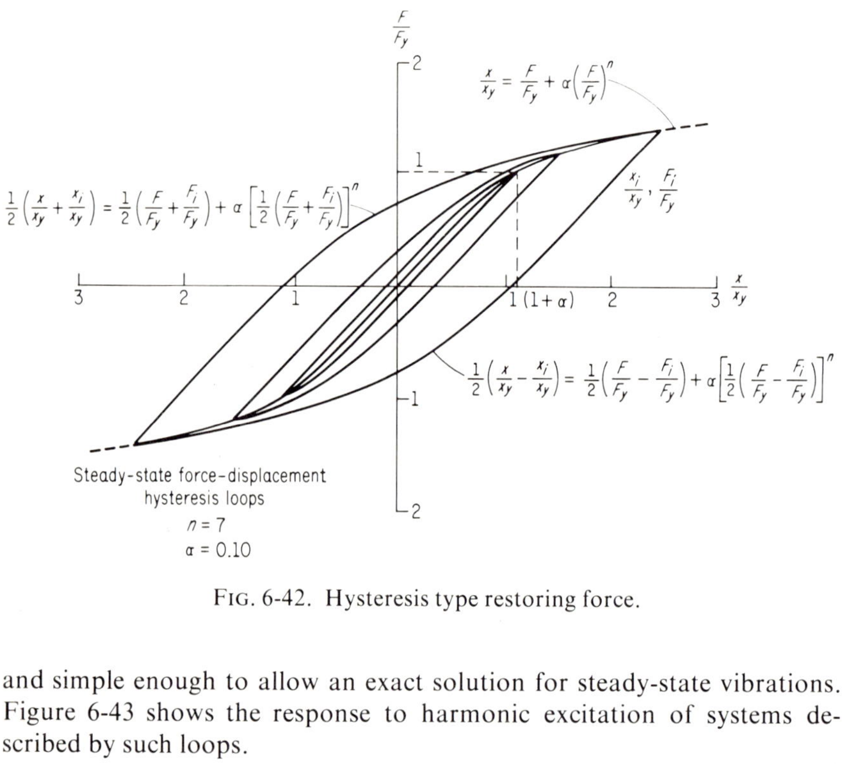

FIG. 6-42. Hysteresis type restoring force.

and simple enough to allow an exact solution for steady-state vibrations. Figure 6-43 shows the response to harmonic excitation of systems described by such loops.

Quadratic or Hydraulic Damping

This form of damping is very common in systems moving in fluids with relatively high velocities. The frictional force in this case has a magnitude which is proportional to the square of the velocity and a sign which is opposite to it.

The equation of motion in case of linear spring and quadratic damping is

$$m\ddot{x} + (\operatorname{sgn}\dot{x})c\dot{x}^2 + kx = F\sin\omega t$$

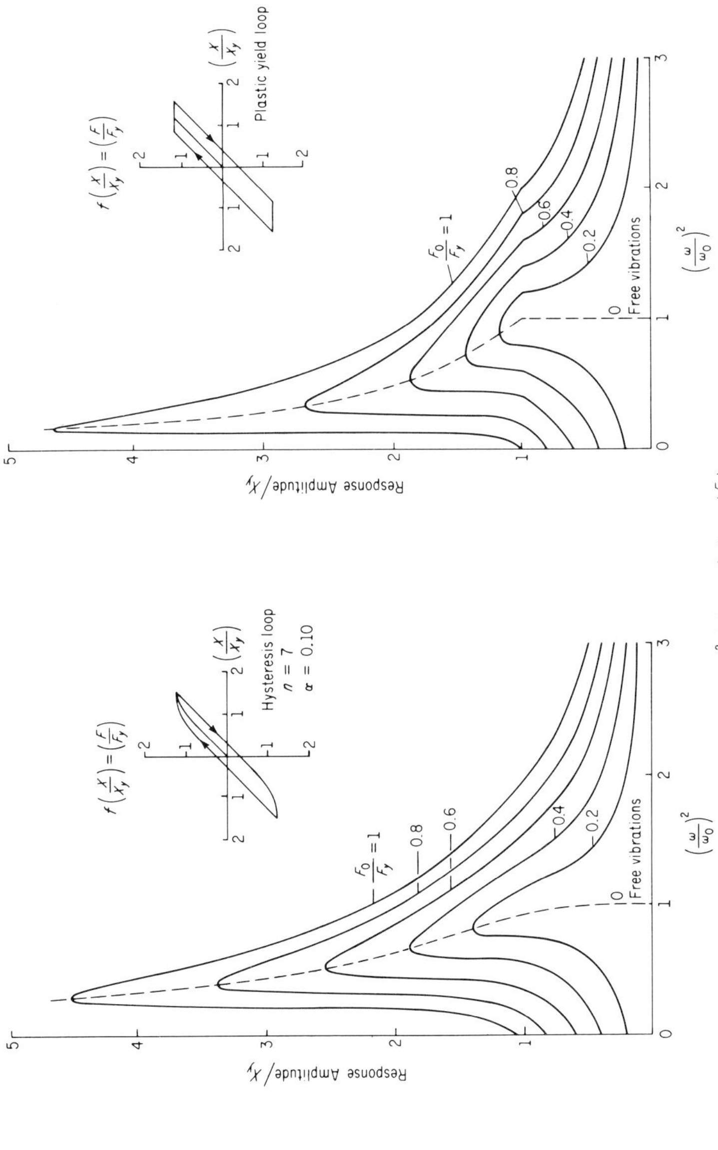

$$\frac{d^2}{dt^2}\left(\frac{X}{X_y}\right) + \omega_0^2 f\left(\frac{X}{X_y}\right) = \omega_0^2\left(\frac{F_0}{F_y}\right)\sin\omega t$$

$$\omega_0^2 = \frac{F_y}{mX_y} = (\text{natural frequency of free vibrations})^2$$

FIG. 6-43. Hysteresis type resonance curves.

Using the Ritz method, an approximate solution can be given by

$$x = X \sin(\omega t - \phi)$$

where X is determined from Eq. 6-43 as

$$(H(X) - \omega^2)^2 + G^2(X, \omega) = \left(\frac{F}{mX}\right)^2$$

and ϕ from Eq. 6-44 as

$$\tan \phi = \frac{G(X, \omega)}{H(X) - \omega^2}$$

where

$$H(X) = \frac{1}{m\pi X} \int_0^{2\pi} kx \sin^2 \sigma \, d\sigma$$

$$= \frac{k}{m}$$

and

$$G(X, \omega) = \frac{4}{m\pi X} \int_0^{\frac{\pi}{2}} (c\omega^2 X^2 \cos^2 \sigma) \cos \sigma \, d\sigma$$

$$= \frac{8}{3} \frac{c\omega^2}{m\pi} X$$

6-6. SUMMARY

In this chapter the response of systems with linear as well as nonlinear spring and damping characteristics to harmonic excitations is studied. Some of the analytical and graphical techniques useful for the solution of the equation of motion in these cases are discussed.

It has been shown that the response of linear systems to periodic excitations can be obtained by superimposing the system response to the different harmonic components of the excitation. Since the principle of superposition does not apply to nonlinear systems, the methods of this chapter with regard to these systems are applicable only when the excitations are harmonic. Numerical and analog computer methods can be used for evaluation of nonlinear system response to periodic excitations.

The reader has now completed an introductory study of free as well as forced vibrations of systems with single degree of freedom. In the following chapters we shall deal with the more complex case of systems with more than one degree of freedom. The analysis of such systems involves the solution of a set of simultaneous differential equations rather than a single equation as in this part.

PROBLEMS

6-1. A one-story building has an elastic frame structure. The mass is assumed to be concentrated at the top. What is the amplitude of the steady-state oscillation of the building if the ground vibrates horizontally in a pure harmonic motion with amplitude A and frequency ω?

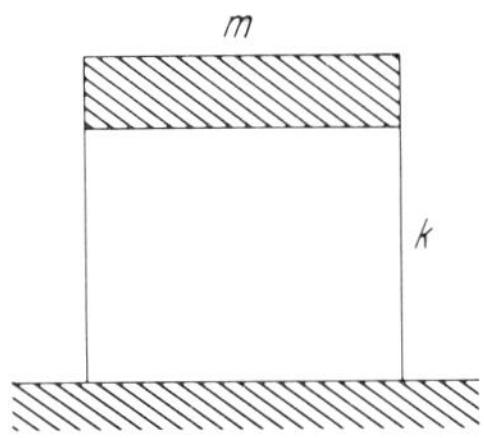

PROB. 6-1

6-2. The crankshaft of a single cylinder engine is connected to two very large flywheels at both ends. The equivalent moment of inertia of the piston, connecting rod and crank is 1 lb in sec^2. The equivalent length and diameter of the shaft at both ends are L and d inches respectively. Write an expression for the damping coefficient which limits the dynamic magnifier to 5.

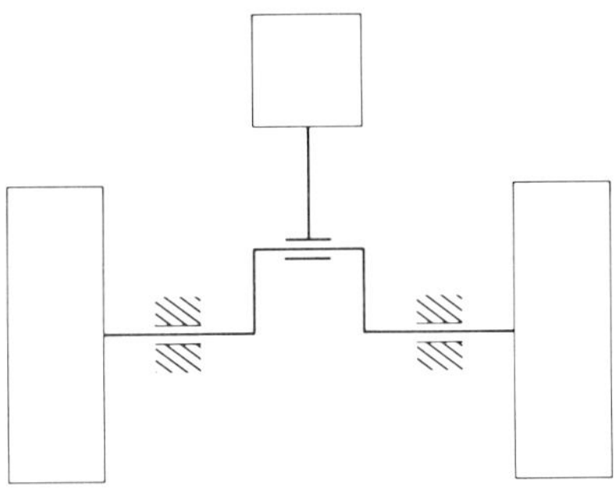

PROB. 6-2

6-3. A circular rod is clamped to a fixed horizontal base in a vertical position and carries at its free end a platform weighing 1 lb. A 1-lb force acting on the platform in a radial direction produces 0.04 in. of deflection. A small motor carrying an eccentric weight of 1 oz at $\frac{1}{4}$ in. from its axis of rotation is placed on the platform. (a) Find the motor rpm which will cause resonance. (b) If the motor speed can be increased to very high values, what amplitude of platform vibration would then be expected? (c) Find the speed range of the motor where the amplitude will not exceed $1\frac{1}{2}$ times the value calculated in (b).

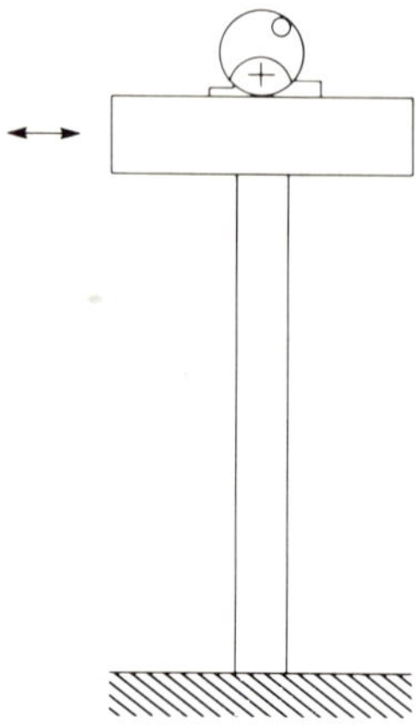

PROB. 6-3

6-4. Two frictional wheels rotate in opposite direction about fixed parallel axes as shown. A heavy bar rests on the wheels in a horizontal position. Show that if the bar is displaced from the symmetrical position, it will oscillate with a frequency $\sqrt{2\mu g/L}$. The supported length of the bar is L and the coefficient of friction between the bar and the wheels is μ.

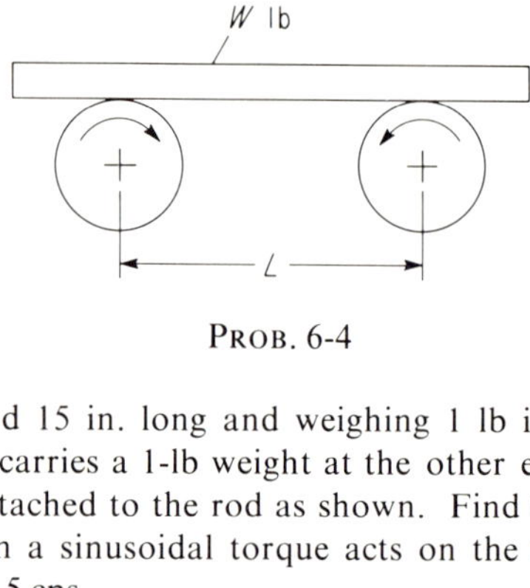

PROB. 6-4

6-5. A uniform rod 15 in. long and weighing 1 lb is hinged at a distance 6 in. from one end. It carries a 1-lb weight at the other end. Two springs with stiffness 12 lb/in. are attached to the rod as shown. Find the amplitude of vibration of the weight when a sinusoidal torque acts on the rod with an amplitude 1 lb-in. and a frequency 5 cps.

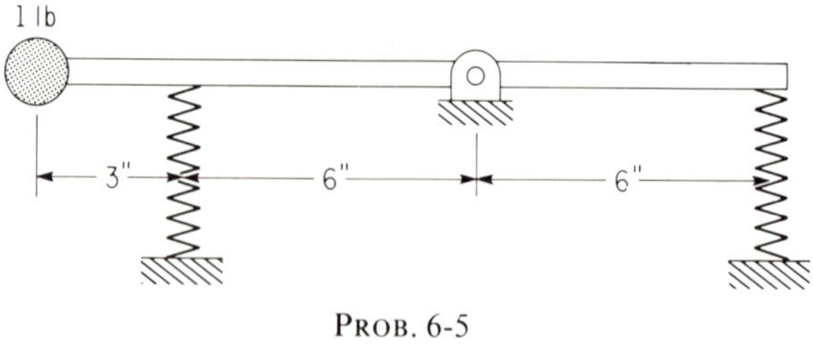

PROB. 6-5

6-6. A 4,000 rpm motor with a moment of inertia of 5 lb in. sec^2 drives a compressor with a moment of inertia of 4 lb in. sec^2. The drive system is represented schematically as shown in the figure. Neglecting the inertia and elasticity of the gears, calculate the maximum shaft torque, due to the first two harmonic components, of the compressor torque diagram shown in the figure. Assume the equivalent inertia of the motor to be very large compared to that of the compressor.

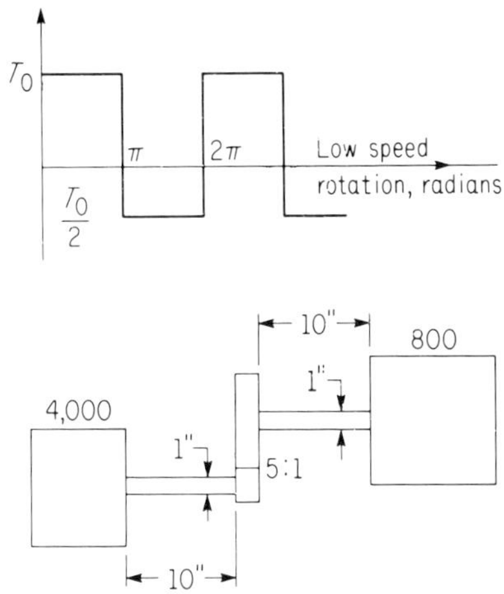

PROB. 6-6

6-7. In the preceding problem, use the phase-plane method to calculate the peak torque due to the entire periodic load.

6-8. A flywheel with a radial groove rotates with constant angular velocity ω_0 about a horizontal axis at O. A particle m slides without friction in the groove under the influence of gravity, force, centrifugal force, as well as the spring force as shown. Show that if $\omega_0^2 < k/2m$, the particle will move in space on a circle above O, and if $k/2m < \omega_0^2 < k/m$, the circle lies above O.

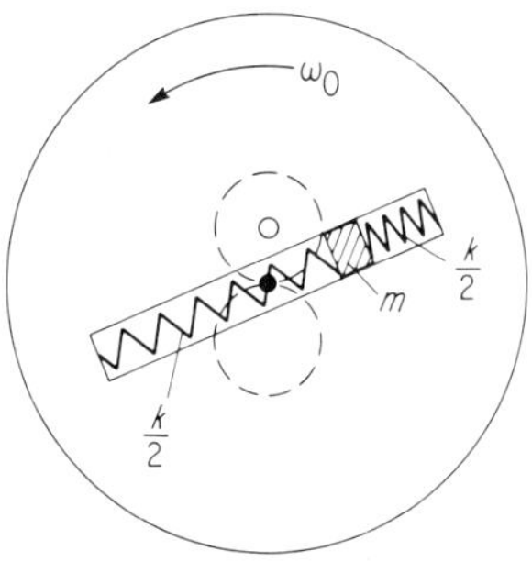

PROB. 6-8

6-9. The vibration of the system shown in the figure is damped by a viscous dashpot as well as a Coulomb damper. Write an expression for the amplitude of vibration at resonance.

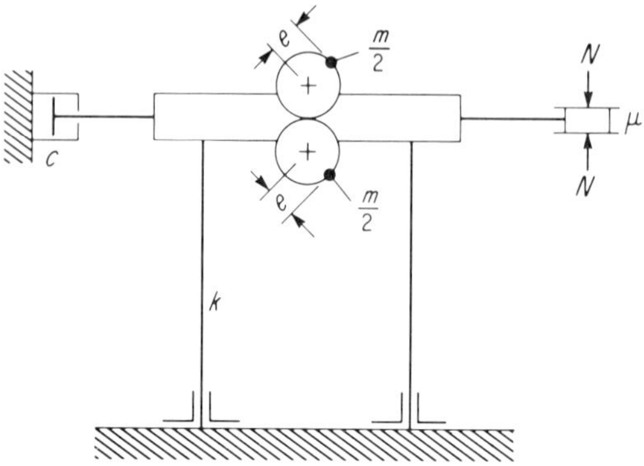

PROB. 6-9

6-10. Write an expression for the dynamic magnifier at resonance for the system shown in the figure.

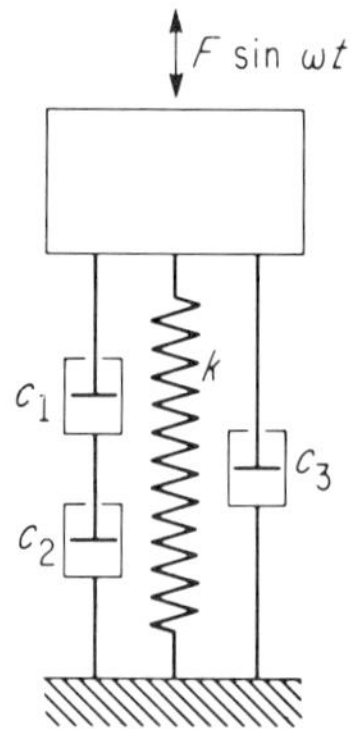

PROB. 6-10

6-11. For the manometer shown, express the damping coefficient by utilizing the pipe-friction equation.

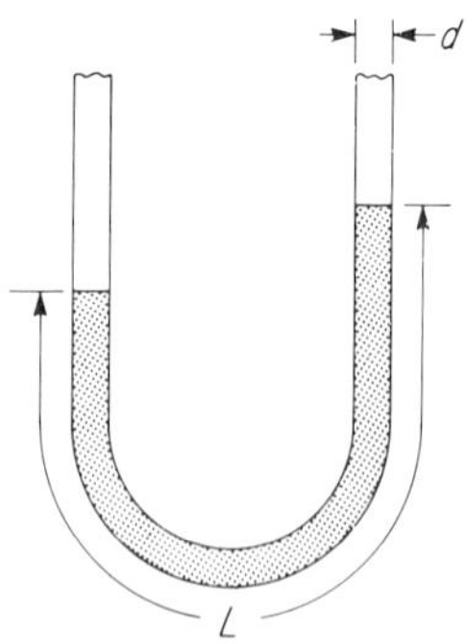

PROB. 6-11

6-12. A cantilever with a maximum deflection of 0.01 in/lb with a 1-lb weight at the end is used to detect the vibration of a platform. When in natural vibration, the amplitude of decay is 2 percent per cycle. If the observed amplitude of vibration of the mass is 0.25 in. and the frequency of platform vibration is 33 cps, calculate the amplitude δ of platform vibration.

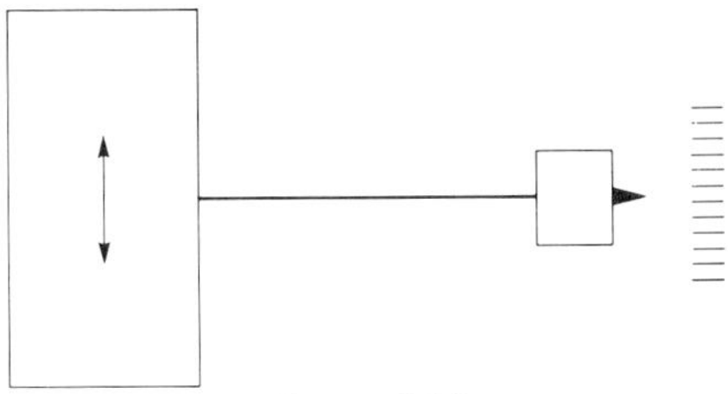

PROB. 6-12

6-13. Write an expression for the transfer function for the system shown in the figure when the input is the displacement of the foundation and the required output is the spring deflection.

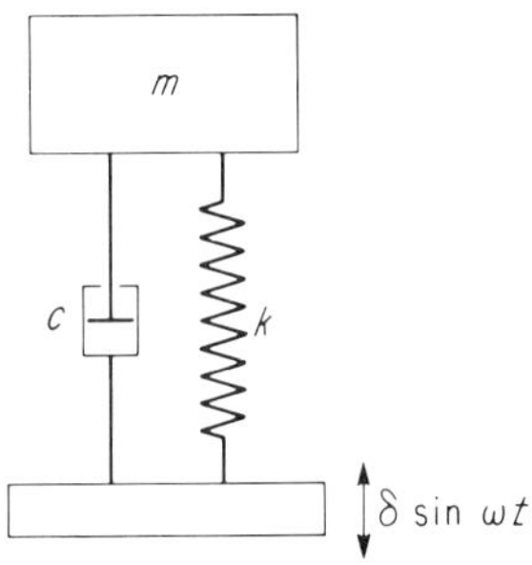

PROB. 6-13

6-14. In Prob. 6-13, find the transfer function when the input is the acceleration of the foundation and the required output is the spring deflection.

6-15. Use Fig. 6-18 to evaluate the dynamic spring force and the phase angle of the system shown here.

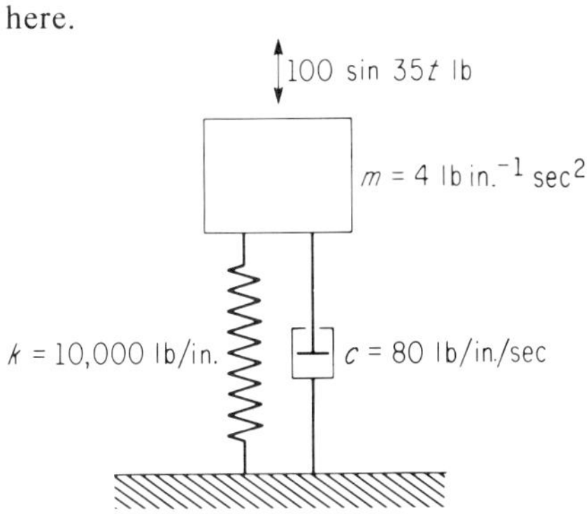

PROB. 6-15

6-16. Calculate the answers in the preceding problem by construction of a force vector diagram. Use the same diagram to evaluate the transmitted force to the foundation.

6-17. For the system shown in Prob. 6-15, calculate: (a) the undamped natural frequency of vibration, (b) the damped natural frequency of vibration, (c) the frequency at which the amplitude of forced vibration is a maximum, and (d) the frequency above which the reduction in damping coefficient will cause a reduction in the transmitted force.

6-18. For the system shown, calculate the dynamic spring force when ω is equal to (a) the undamped natural frequency, (b) the damped natural frequency. What is the maximum dynamic force for all values of ω?

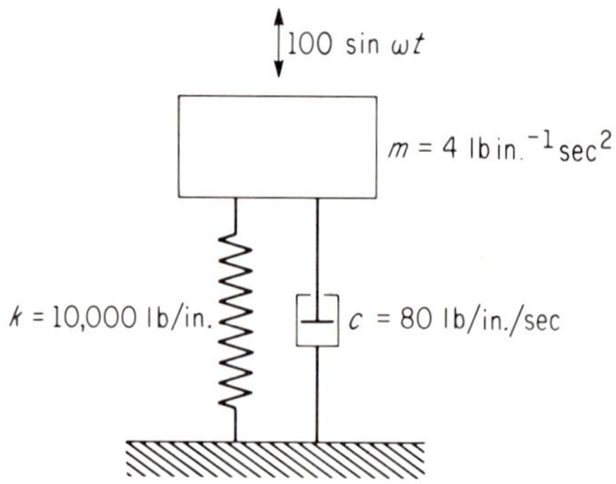

PROB. 6-18

6-19. The given single-degree-of-freedom system is subjected to a periodic force equal to

$$F_{(t)} = 10 + 25 \sin 4t + 80 \sin 6t + 200 \sin 10t$$

Write the expression for the steady-state displacement of the mass.

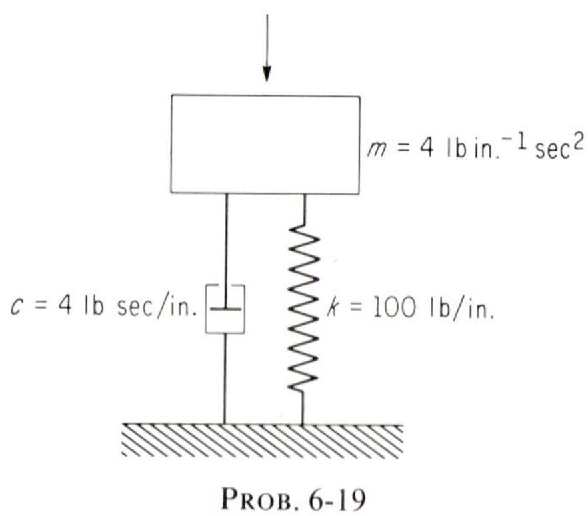

PROB. 6-19

6-20. A 100-lb rotor is centrally mounted on a simple supported shaft with 1.25-in. diameter. The distance between the supports is 12 in. If the C.G. of the rotor is 0.02 in. from the axis of rotation, calculate the shaft stress due to this imbalance when the rotor speed is 60 revolutions per second.

6-21. Use the Ritz averaging method to calculate the period of a simple pendulum when the initial release position is 135° from the equilibrium position.

6-22. The torque-deflection characteristics of a nonlinear torsional coupling are the same for both directions of twist and are determined from test as follows:

Torque, lb-in.	*Angle of twist, degrees*
20,000	0.4
40,000	0.8
60,000	1.1
80,000	1.17
100,000	1.2

Use the Ritz averaging method to calculate an equivalent stiffness factor for the coupling for a cycle with an amplitude of 1.2°.

6-23. The energy E dissipated per cycle in the hysteresis loop of a torsional coupling is given by

$$E = 270\,A^2 \text{ in. lb}$$

where A is the amplitude of vibration in degrees. Calculate the dynamic magnifier at resonance when the coupling is connecting a flywheel with moment of inertia $I = 36$ lb/in. sec^2 to a very large mass. The torsional spring constant of the coupling is 50,000 lb-in./degree.

6-24. Plot an approximate response curve (x versus ω) for the given system when the mass m is subjected to a sinusoidal force $F = 50 \sin \omega t$.

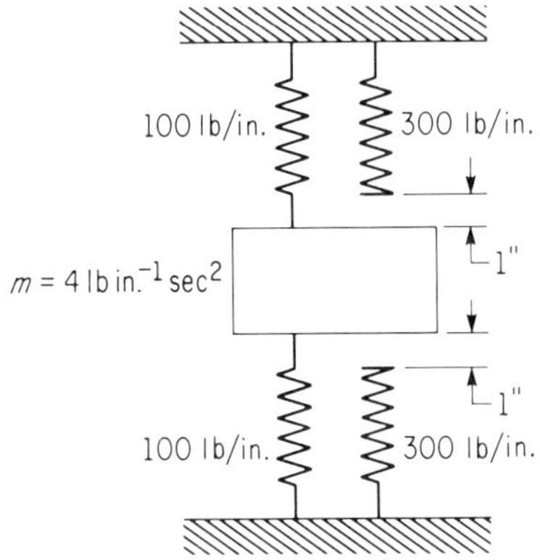

PROB. 6-24

6-25. The rod ab in the given single degree of freedom is rigid and massless. (a) Write the equation of motion for the mass m. (b) Write the expression for the natural frequency. (c) If the system is subjected to a foundation excitation at c given by $\delta = e \sin \omega t$. Write an expression for the resulting displacement at the mass.

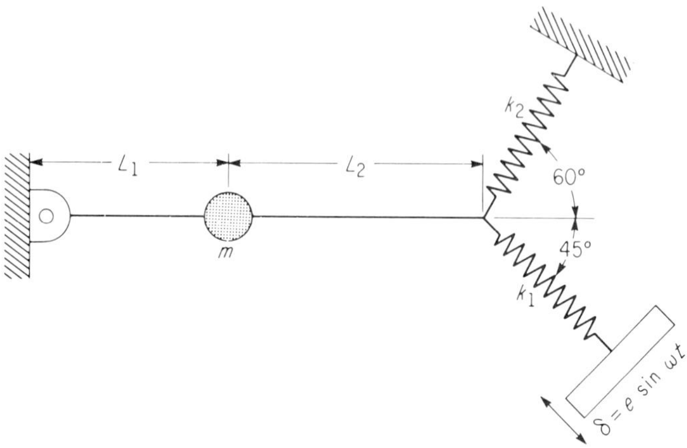

PROB. 6-25

6-26. (a) The given system is subjected to a square wave with amplitude 50 lb and frequency 7.5 rad/sec. Find the maximum dynamic spring force. (b) Compare the previous answer with the maximum dynamic spring force when the excitation is sinusoidal with an amplitude 50 lb and frequency 7.5 rad/sec. (c) If the system was critically damped and subjected to the excitation given in (b). Calculate the maximum dynamic spring force.

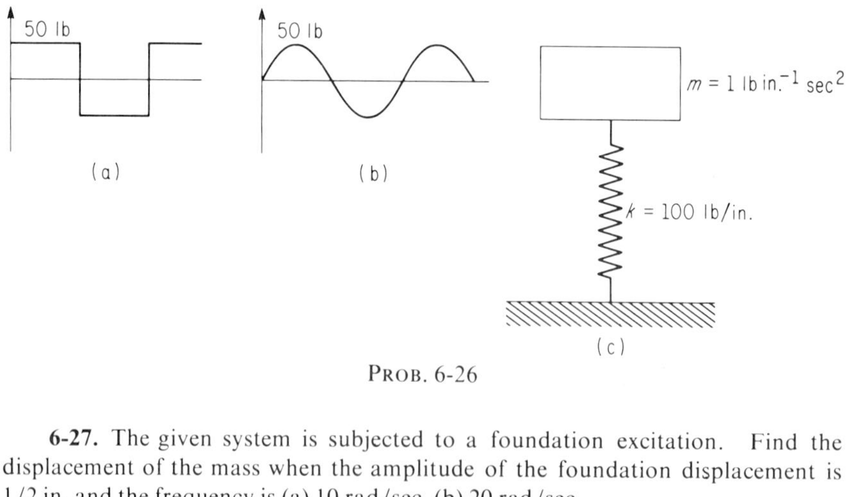

PROB. 6-26

6-27. The given system is subjected to a foundation excitation. Find the displacement of the mass when the amplitude of the foundation displacement is 1/2 in. and the frequency is (a) 10 rad/sec, (b) 20 rad/sec.

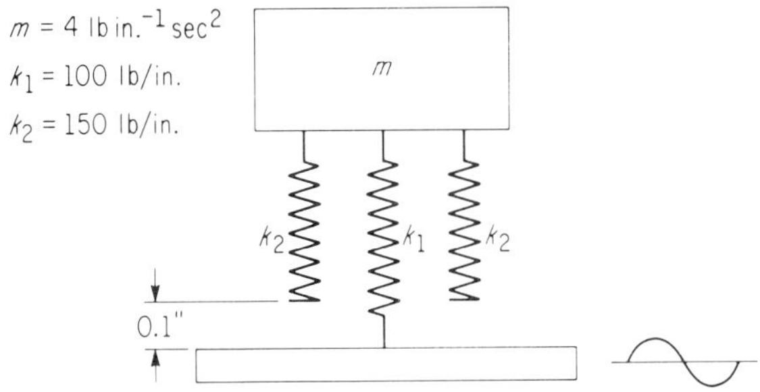

PROB. 6-27

REFERENCES

1. S. Timoshenko, *Vibration Problems in Engineering*. 3d ed. Princeton N.J.: Van Nostrand, 1955.
2. Theodore von Kármán and M. A. Biot, *Mathematical Methods in Engineering*. New York: McGraw-Hill, 1940.
3. L. S. Jacobsen and R. S. Ayre, *Engineering Vibrations*. New York: McGraw-Hill, 1958.
4. J. P. Den Hartog, *Mechanical Vibrations*. 4th ed. New York: McGraw-Hill, 1956.
5. K. N. Tong, *Theory of Mechanical Vibration*. New York: Wiley, 1960.
6. R. Burton, *Vibrations and Impact*. Reading, Mass.: Addison-Wesley, 1958.
7. C. E. Crede, *Vibration and Shock Isolation*. New York: Wiley, 1951.
8. J. W. Strutt, Lord Rayleigh, *Theory of Sound*. 2d ed. New York: Dover, 1937.
9. C. M. Harris and C. E. Crede, *Shock and Vibration Handbook*. New York: McGraw-Hill, 1961.
10. W. T. Thomson, *Mechanical Vibrations*. Englewood Cliffs, N.J.: Prentice-Hall, 1956.
11. P. Crafton, *Shock and Vibration in Linear Systems*. New York: Harper, 1961.
12. W. Jennings, *First Course in Numerical Methods*. New York: Macmillan, 1964.
13. R. E. D. Bishop and D. C. Johnson, *The Mechanics of Vibrations*. London: Cambridge U. P., 1960.
14. J. N. Warfield, *Introduction to Electronic Analog Computers*. Englewood Cliffs, N.J.: Prentice-Hall, 1959.
15. H. S. Carslaw and J. C. Jaeger, *Operational Methods in Applied Mathematics*. 2d ed. London: Oxford U.P., New York: Dover, 1948.
16. M. F. Gardner and J. L. Barnes, *Transients in Linear Systems*. New York: Wiley, 1942.
17. W. T. Thomson, *Laplace Transformations*. Englewood Cliffs, N.J.: Prentice-Hall, 1950.
18. J. A. Aseltine, *Transform Method in Linear System Analysis*. New York: McGraw-Hill, 1958.
19. H. Bateman, *Tables of Integral Transforms*. Pasadena: California Institute of Technology, 1954. Also, New York: McGraw-Hill, 1954.

20. B. J. Starkey, *Laplace Transforms for Electrical Engineers.* New York: Philosophical Library, 1955.
21. A. A. Bennett et al., *Numerical Integration of Differential Equations.* New York: Dover, 1956.
22. W. G. Bickley and A. Talbot, *Introduction to the Theory of Vibrating Systems.* London: Oxford U.P., 1961.
23. G. W. Van Santen, *Mechanical Vibrations.* Centrex Publishing Co. Eindhoven (Philips Technical Library), 1961.
24. V. V. Solodovnikov et al., *Transient Response from Frequency Response.* Infosearch, 1958.
25. J. Stoker, *Non-Linear Vibrations in Mechanical Systems.* New York: Wiley-Interscience, 1950.
26. A. Gemant, *Frictional Phenomena.* New York: Chemical Pub. Co., 1950.
27. A. Andronov et al., *Theory of Oscillators.* Reading, Mass.: Addison-Wesley, 1966.
28. N. Kryloff N. Bogoliuboff, *Introduction to Nonlinear Mechanics.* Princeton, N.J.: Princeton U.P., 1943.
29. N. Minorsky, *Introduction to Nonlinear Mechanics.* Ann Arbor, Mich.: Edwards, 1947.
30. R. Oldenburger, "Frequency Response Symposium," *Trans. ASME*, Vol. 76, No. 8 (1954). Also, New York: Macmillan, 1956.
31. W. H. Li, "Graphical Solution of Equations of Vibrations," *Proc. ASCE*, Vol. 83, No. 4 (1957).
32. W. R. Evans, *Control System Dynamics.* New York: McGraw-Hill, 1954.
33. H. S. Tsien, *Engineering Cybernetics.* New York: McGraw-Hill, 1954.
34. R. M. Rosenberg, "Nonlinear Oscillations," *Appl. Mech. Rev.*, Vol. 14, No. 11 (1961).
35. W. D. Iwan, *The Dynamic Response of Bilinear Hysteretic Systems.* Pasadena California Institute of Technology, 1961.
36. T. K. Caughey, "The Existence and Stability of Ultraharmonics and Subharmonics in Forced Nonlinear Oscillations," *J. App. Mech.*, Vol. 21, No. 4 (1954).
37. K. Klötter, "Nonlinear Vibrations Treated by the Averaging Method of Ritz," Proc. First National Cong. of Appl. Mech., *Trans. ASME*, 1952.
38. W. D. Iwan, "An Electric Analog for Systems Containing Coulomb Damping," *Exptl. Mech.*, Vol. 4, No. 8 (1964).
39. A. Seireg, "Static and Dynamic Behavior of Flexible Torsional Couplings with Nonlinear Characteristics," Ph.D. thesis, U. of Wisconsin, 1954; also *Proc. SESA*, Vol. 17, No. 2 (1960).
40. P. C. Jennings, "Response of Simple Yielding Structures to Earthquake Excitation," Ph.D. thesis, California Inst. of Technology, 1963.
41. D. E. Hudson, "Nonlinear Resonant Vibrations," ASEE Paper No. 102-3, 1962.
42. C. Hayashi, "Forced Oscillations in Nonlinear Systems." Osaka, Japan: Nippon, 1953.
43. R. Oldenburger and H. Sartorius, "The Dynamics of Automatic Controls," ASME Publication, N.Y., 1948.
44. E. Jahnke and F. Emde, *Tables of Functions with Formulae and Curves*, 4th ed. New York: Dover, 1945.

45. S. H. Crandall, *Engineering Analysis.* New York: McGraw-Hill, 1956.
46. C. Hayashi, *Nonlinear Oscillations in Physical Systems.* New York: McGraw-Hill, 1964.
47. J. Ruzicka, "Resonance Characteristics of Unidirectional Viscous and Coulomb-Damped Vibration Isolation Systems," *Trans. ASME*, Vol. 89, Ser. B, No. 4 (1967).
48. R. Beckett and J. Hurt, *Numerical Calculations and Algorithms.* New York: McGraw-Hill, 1967.

part III

Systems With a Finite Number of Degrees of Freedom: Coupled Systems

Since the fabric of the universe is most perfect, and is the work of a most wise creator, nothing whatsoever takes place in the universe in which some relation of maximum and minimum does not appear. Wherefore there is absolutely no doubt that every effect in the universe can be explained as satisfactorily from final causes, by the aid of the method of maxima and minima as it can from the effective causes themselves.

Leonhard Euler (1707–1783)

chapter 7

Problem Formulation by Lagrange's Equation

7-1. INTRODUCTION

We have discussed in Part II the behavior of systems with a single degree of freedom. In general, more than one coordinate is necessary to specify the configuration of a system. It has been shown in Chapter 1, for example, that a system of n unconstrained particles requires $3n$ coordinates to specify its configuration in space, while a system of n unconstrained rigid bodies requires $6n$ coordinates.

The equations of motion for systems with multidegrees of freedom can be obtained as previously illustrated in Part I by applying Newton's second law and Euler's equations. This method of formulation may prove tedious when complex systems are considered. By the use of more elegant techniques such as Lagrange's equation these systems can be more readily described.

The methods of solution of the preceding chapter may be extended to deal with systems with several degrees of freedom. We shall also consider in the following chapters some of the special methods which can considerably simplify the calculation procedure.

7-2. GENERALIZED COORDINATES AND LAGRANGE'S EQUATION

Considering the example of a double pendulum, Fig. 7-1, the masses m_1, m_2 will respectively undergo displacements x_1, y_1 and x_2, y_2 when their equilibrium is disturbed. These displacements are given by

$$\begin{aligned} x_1 &= l_1 \sin\theta_1 &&= f(\theta_1) \\ y_1 &= l_1 \cos\theta_1 &&= f(\theta_1) \\ x_2 &= l_1 \sin\theta_1 + l_2 \sin\theta_2 &&= f(\theta_1, \theta_2) \\ y_2 &= l_1 \cos\theta_1 + l_2 \cos\theta_2 &&= f(\theta_1, \theta_2) \end{aligned}$$

The configuration of the pendulum can be defined at any time by the coordinates x_1, y_1, x_2, y_2 or simply by the coordinates θ_1 and θ_2 which

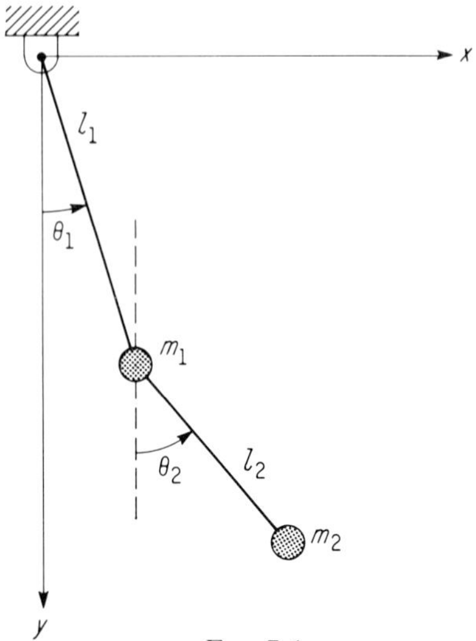

FIG. 7-1

are known as the *generalized coordinates.* Since the system has two degrees of freedom, two generalized coordinates are *all that is necessary* to describe its configuration. For a system with n degrees of freedom, there are n generalized coordinates.

The general motion of a particle in space, Fig. 7-2, is defined by its cartesian coordinates x, y, z or the polar coordinates r, θ, ϕ as

$$\begin{aligned} x &= f_x(r, \theta, \phi) \\ y &= f_y(r, \theta, \phi) \\ z &= f_z(r, \theta, \phi) \end{aligned}$$

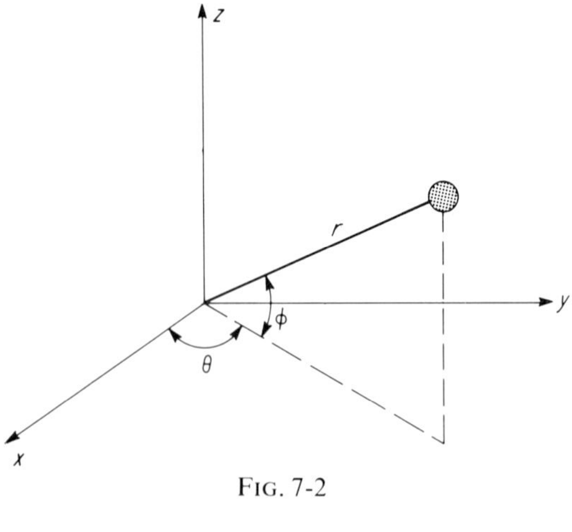

FIG. 7-2

and

$$r = f_r(x, y, z)$$
$$\theta = f_\theta(x, y, z)$$
$$\phi = f_\phi(x, y, z)$$

where $f_x, f_y, f_z, f_r, f_\theta, f_\phi$ are the proper geometrical transformation functions.

Either system can be considered the generalized coordinates for the motion of the particle. *It should be noted here that the generalized coordinates need not necessarily be variables characterizing physical measurements but can be any convenient parameters describing the motion.*

If we assume that $q_1, q_2, q_3, \ldots, q_n$ are the generalized coordinates for a system of particles with n degrees of freedom, then

$$x_i = f_{xi}(q_r)$$
$$y_i = f_{yi}(q_r)$$
$$z_i = f_{zi}(q_r)$$

where $r = 1, 2, \ldots, n$ and $i = 1, 2, 3, \ldots, n$. If we further assume that the x, y, z components of the resultant force acting on each particle are F_{xi}, F_{yi}, F_{zi}, then Newton's second law gives

$$m_i \ddot{x}_i = F_{xi}$$
$$m_i \ddot{y}_i = F_{yi}$$

and

$$m_i \ddot{z}_i = F_{zi}$$

From the x_i equation we can conclude (by applying d'Alembert's principle) that in a virtual displacement δ_{xi} the work must vanish. This requires that

$$(F_{xi} - m_i \ddot{x}_i)\,\delta x_i = 0$$

or

$$m_i \ddot{x}_i \delta x_i = F_{xi} \delta x_i$$

and similarly for y and z.

$$m_i \ddot{y}_i \,\delta y_i = F_{yi}\,\delta y_i$$
$$m_i \ddot{z}_i \,\delta z_i = F_{zi}\,\delta z_i$$

By adding for all particles we can therefore obtain the equation

$$\sum_i m_i \ddot{x}_i \,\delta x_i + m_i \ddot{y}_i \,\delta y_i + m_i \ddot{z}_i \,\delta z_i = \sum_i F_{xi} \delta x_i + F_{yi} \delta y_i + F_{zi}\,\delta z_i = \delta W$$

from which

$$\delta W = \Sigma\, m_i \ddot{x}_i \,\delta x_i + m_i \ddot{y}_i \,\delta y_i + m_i \ddot{z}_i \,\delta z_i$$

where δW represents the work done on the system in a virtual displacement $(\delta x, \delta y, \delta z)$ by all noninertial forces. Substituting,

$$\delta x_i = \frac{\partial x_i}{\partial q_r}\delta q_r \qquad \delta y_i = \frac{\partial y_i}{\partial q_r}\delta q_r$$

and

$$\delta z_i = \frac{\partial z_i}{\partial q_r}\delta q_r$$

Therefore

$$\delta W = \sum_r m_i \ddot{x}_i \frac{\partial x_i}{\partial q_r}\delta q_r$$
$$+ \text{(similar terms for } y) + \text{(similar terms for } z)$$

Since

$$\frac{d}{dt}\left(m_i \dot{x}_i \frac{\partial x_i}{\partial q_r}\right) = m_i \dot{x}_i \frac{d}{dt}\left(\frac{\partial x_i}{\partial q_r}\right) + m_i \ddot{x}_i \frac{\partial x_i}{\partial q_r}$$

then

$$\delta W = \sum_r \left[\frac{d}{dt}\left(m_i \dot{x}_i \frac{\partial x_i}{\partial q_r}\right) - m_i \dot{x}_i \frac{d}{dt}\left(\frac{\partial x_i}{\partial q_r}\right)\right]\delta q_r$$
$$+ \text{(similar terms for } y) + \text{(similar terms for } z)$$

Knowing that

$$\frac{\partial x_i}{\partial q_r} = \frac{\partial \dot{x}_i}{\partial \dot{q}_r} \qquad \text{and} \qquad \frac{d}{dt}\left(\frac{\partial x_i}{\partial q_r}\right) = \frac{\partial \dot{x}_i}{\partial q_r}$$

and substituting in the previous equation, we obtain

$$\delta W = \sum_r \left[\frac{d}{dt}\left(m_i \dot{x}_i \frac{\partial \dot{x}_i}{\partial \dot{q}_r}\right) - m_i x_i \left(\frac{\partial \dot{x}_i}{\partial q_r}\right)\right]\delta q_r$$
$$+ \text{(similar terms for } y) + \text{(similar terms for } z)$$
$$= \sum \left[\frac{d}{dt}\left\{\frac{\partial}{\partial \dot{q}_r}\left(\frac{1}{2} m_i \dot{x}_i^2\right)\right\} - \frac{\partial}{\partial q_r}\left(\frac{1}{2} m_i \dot{x}_i^2\right)\right]\delta q_r$$
$$+ \text{(similar terms for } y) + \text{(similar terms for } z) \qquad \text{(7-1)}$$

The kinetic energy T for the system can be expressed as

$$T = \sum \frac{1}{2} m_i (\dot{x}_i^2 + \dot{y}_i^2 + \dot{z}_i^2)$$

Accordingly, Eq. 7-1 takes the form

$$\delta W = \left\{\frac{d}{dt}\frac{\partial T}{\partial \dot{q}_r} - \frac{\partial T}{\partial q_r}\right\}\delta q_r \qquad \text{(7-2)}$$

The virtual work corresponding to a virtual displacement δq_r is

$$\delta W = F_r \, \delta q_r$$

where F_r is a generalized force given by

$$F_r = \frac{\partial W}{\partial q_r} \tag{7-3}$$

Equation 7-2 can therefore be written as

$$\boxed{\frac{d}{dt}\left(\frac{\partial T}{\partial \dot{q}_r}\right) - \frac{\partial T}{\partial q_r} = F_r} \tag{7-4}$$

The work done in a virtual displacement δq_r can be divided into two parts: δW_c which is the total work done by all conservative forces in the virtual displacement and δW_r which is the total work done by all the nonconservative forces.

Knowing that the change of potential energy

$$\delta U = -\delta W_c \qquad \text{and} \qquad \delta W_r = Q_r \, \delta q_r$$

where Q_r represents the general noninertial, nonconservative force, then

$$\delta W = \delta W_c + \delta W_r = -\delta U + Q_r \, \delta q_r$$

By substituting the previous result in Eq. 7-2 we obtain

$$\boxed{\frac{d}{dt}\left(\frac{\partial T}{\partial \dot{q}_r}\right) - \frac{\partial T}{\partial q_r} + \frac{\partial U}{\partial q_r} = Q_r} \tag{7-5}$$

Since the potential energy U is a function of position only and is independent of velocity $\left(\text{i.e., } \frac{\partial U}{\partial \dot{q}_r} = 0\right)$, Eq. 7-5 can be written as

$$\boxed{\frac{d}{dt}\left(\frac{\partial L}{\partial \dot{q}_r}\right) - \frac{\partial L}{\partial q_r} = Q_r} \tag{7-6}$$

where $L = T - U$.

Equations 7-4, 7-5, and 7-6 are different forms of the Lagrange equation which is a very useful tool in formulating dynamical problems, as will be shown in the following examples. It should be noted here that the equations of motion for these systems could be obtained by the use of Newton's second law and Euler's equations. However, the use of the energy concept and Lagrange's equation simplifies the formulation in many complex problems.

Example 1

In order to illustrate the derivation of the equations of motion by Lagrange's method, we shall begin with the simple pendulum, Fig. 7-3.

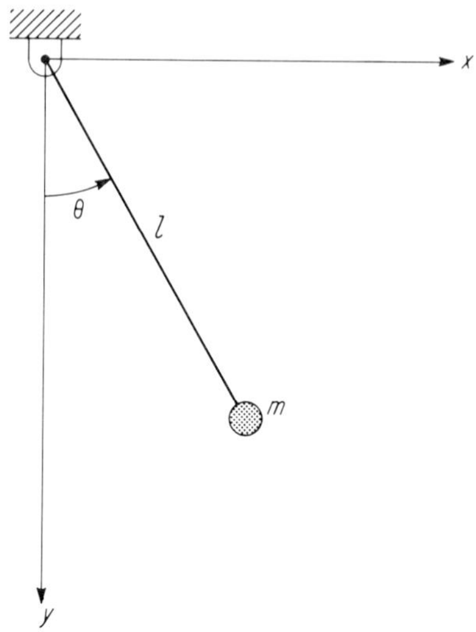

FIG. 7-3

We know that

$$
\begin{aligned}
x &= l \sin \theta \\
y &= l \cos \theta \\
\dot{x} &= l \cos \theta \dot{\theta} \\
\dot{y} &= -l \sin \theta \dot{\theta} \\
T &= \frac{1}{2} m(\dot{x}^2 + \dot{y}^2) = \frac{1}{2} m(l^2 \cos^2\theta \dot{\theta}^2 + l^2 \sin^2 \theta \dot{\theta}^2) \\
&= \frac{1}{2} ml^2 \dot{\theta}^2
\end{aligned}
$$

and

$$U = mgl\,(1 - \cos \theta)$$

In this problem θ is the generalized coordinate:

$$\frac{\partial T}{\partial \dot{\theta}} = ml^2\dot{\theta}, \quad \frac{d}{dt}\frac{\partial T}{\partial \dot{\theta}} = ml^2\ddot{\theta}, \quad \frac{\partial T}{\partial \theta} = 0$$

$$\frac{\partial U}{\partial \theta} = mgl \sin \theta$$

and $Q_r = 0$, since all forces acting on the mass are conservative, and Lagrange's equation gives

$$ml^2\ddot{\theta} + mgl \sin \theta = 0$$

$$\ddot{\theta} + \frac{g}{l} \sin \theta = 0$$

which is the same result obtained by applying Newton's second law, Eq. 4-1c.

Example 2

Considering the damped spring mass system of Fig. 7-4,

$$T = \frac{1}{2} m\dot{x}^2$$

$$U = \frac{1}{2} kx^2$$

$$\frac{d}{dt}\left(\frac{\partial T}{\partial \dot{x}}\right) = m\ddot{x}$$

$$\frac{\partial T}{\partial x} = 0$$

$$\frac{\partial U}{\partial x} = kx$$

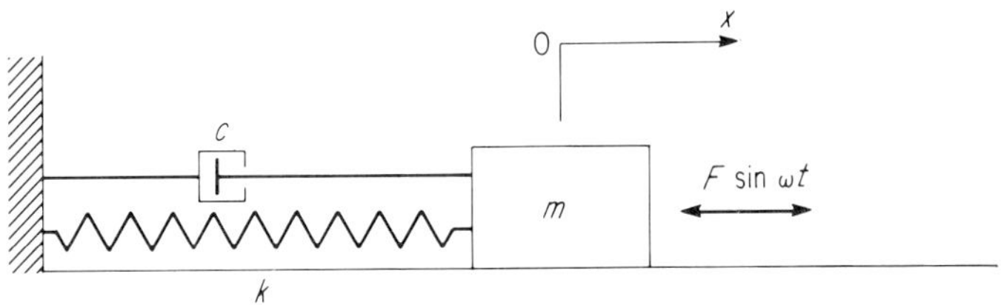

FIG. 7-4

and the nonconservative generalized force is

$$Q_r = -c\dot{x} + F \sin \omega t$$

where $-c\dot{x}$ and $F \sin \omega t$ are dissipative and exciting components respectively. Lagrange's equation therefore gives the expected result:

$$m\ddot{x} + kx = -c\dot{x} + F \sin \omega t$$

Example 3

For the double pendulum shown in Fig. 7-5,

$$x_1 = l_1 \sin \theta_1$$

$$y_1 = l_1 \cos \theta_1$$

$$x_2 = l_1 \sin \theta_1 + l_2 \sin \theta_2$$

$$y_2 = l_1 \cos \theta_1 + l_2 \cos \theta_2$$

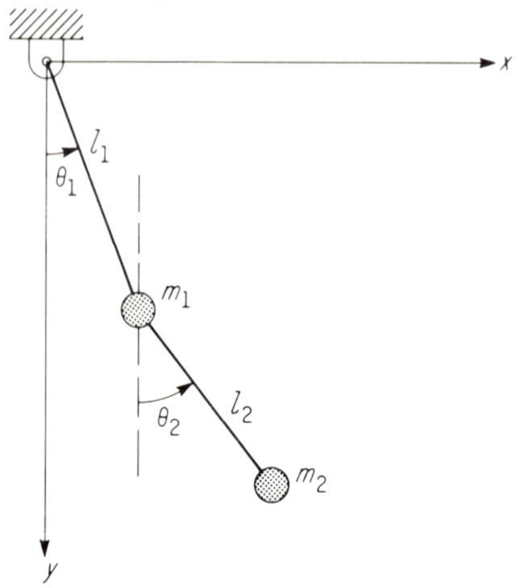

FIG. 7-5

where θ_1 and θ_2 are the generalized coordinates. Then

$$\begin{aligned}\text{K.E.} = T &= \frac{1}{2}m_1 l_1^2\dot{\theta}_1^2 \\ &\quad + \frac{1}{2}m_2[l_1^2\dot{\theta}_1^2 + l_2^2\dot{\theta}_2^2 + l_1 l_2\dot{\theta}_1\dot{\theta}_2\cos(\theta_2 - \theta_1)] \\ &= \frac{1}{2}(m_1 + m_2)l_1^2\dot{\theta}_1^2 + \frac{1}{2}m_2 l_2^2\dot{\theta}_2^2 \\ &\quad + \frac{1}{2}m_2 l_1 l_2\dot{\theta}_1\dot{\theta}_2\cos(\theta_2 - \theta_1)\end{aligned}$$

$$\text{P.E.} = U = (m_1 + m_2)gl_1(1 - \cos\theta) + m_2 g l_2(1 - \cos\theta_2)$$

and the Lagrange equations corresponding respectively to θ_1 and θ_2 are therefore

$$\begin{aligned}&(m_1 + m_2)l^2\ddot{\theta}_1 + m_2 l_1 l_2\ddot{\theta}_2\cos(\theta_2 - \theta_1) - m_2 l_1 l_2(\dot{\theta}_2^2 - \dot{\theta}_1\dot{\theta}_2) \\ &\times \sin(\theta_2 - \theta_1) - m_2 l_1 l_2\dot{\theta}_2\dot{\theta}_1\sin(\theta_2 - \theta_1) + (m_1 + m_2)gl\sin\theta_1 = 0\end{aligned}$$

and

$$\begin{aligned}m_2 l_1 l_2\cos(\ddot{\theta}_2 - \theta_1) &+ m_2 l_2^2\theta_2 \\ &+ m_2 l_1 l_2\dot{\theta}_1^2\sin(\theta_2 - \theta_1) \\ &+ m_2 g l_2\sin\theta_2 = 0\end{aligned}$$

Example 4

The three mass system shown in Fig. 7-6 has three degrees of freedom. The equations of motion of this system can be directly obtained by

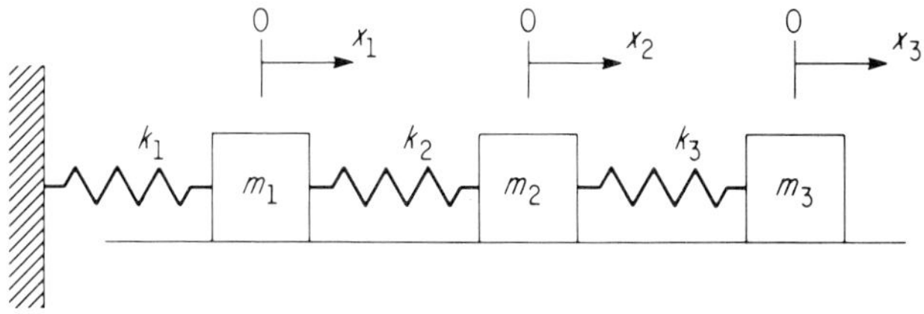

FIG. 7-6

Newton's second law as

$$\begin{aligned} m_1\ddot{x}_1 &= -k_1x_1 - k_2(x_1 - x_2) \\ m_2\ddot{x}_2 &= -k_2(x_2 - x_1) - k_3(x_2 - x_3) \\ m_2\ddot{x}_3 &= -k_3(x_3 - x_2) \end{aligned} \tag{7-7}$$

We shall, however, as an alternate method, determine these equations by the use of Lagrange's equation. In this case

$$T = \frac{1}{2}m_1\dot{x}_1^2 + \frac{1}{2}m_2\dot{x}_2^2 + \frac{1}{2}m_3\dot{x}_3^2$$

$$U = \frac{1}{2}k_1x_1^2 + \frac{1}{2}k_2(x_2 - x_1)^2 + \frac{1}{2}k_3(x_3 - x_2)^2$$

Since x_1, x_2, x_3 are the generalized coordinates in this case, the corresponding Lagrange equations are therefore

$$\begin{aligned} m_1\ddot{x}_1 + k_1x_1 - k_2(x_2 - x_1) &= 0 \\ m_2\ddot{x}_2 + k_2(x_2 - x_1) - k_3(x_3 - x_2) &= 0 \\ m_3\ddot{x}_3 + k_3(x_3 - x_2) &= 0 \end{aligned}$$

which are identical with Eq. 7-7.

Example 5

In the system, Fig. 7-7 of the mass m_1 is defined by the displacement x_1, while that of m_2 is defined by x_2 and y_2, where

$$\begin{aligned} x_2 &= x_1 + l\cos\theta \\ y_2 &= l\sin\theta \end{aligned}$$

and

$$\begin{aligned} \dot{x}_2 &= \dot{x}_1 - l\dot{\theta}\sin\theta \\ \dot{y}_2 &= \dot{\theta}l\cos\theta \end{aligned}$$

$$\begin{aligned} T &= \frac{1}{2}m_1\dot{x}_1^2 + \frac{1}{2}m_2(\dot{x}_1 - l\dot{\theta}\sin\theta)^2 + \frac{1}{2}m_2(l\dot{\theta}\cos\theta)^2 \\ &= \frac{1}{2}m_1\dot{x}_1^2 + \frac{1}{2}m_2\dot{x}_1^2 + \frac{1}{2}m_2l^2\dot{\theta}^2 - m_2l\dot{\theta}\,\dot{x}\sin\theta \end{aligned}$$

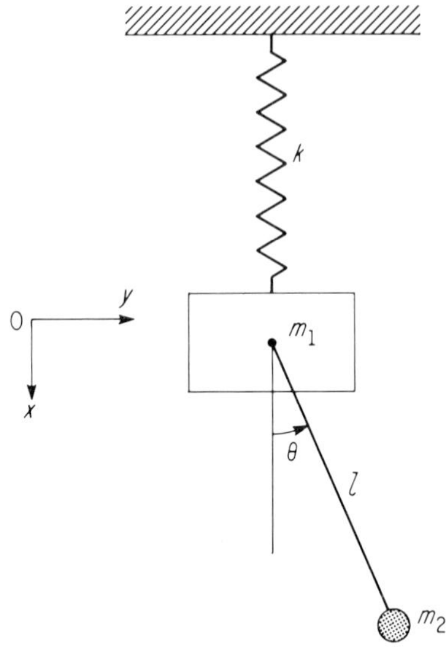

FIG. 7-7

$$U = \frac{1}{2}k(x_1 + \Delta)^2 - (m_1 + m_2)g(x_1 + \Delta) + m_2gl(1 - \cos\theta)$$

where

$$\Delta = \frac{(m_1 + m_2)g}{k}$$

Therefore

$$U = \frac{1}{2}kx_1^2 - m_2gl\cos\theta + \text{const.}$$

Since the generalized coordinates are x_1, θ,

$$\frac{d}{dt}\left(\frac{\partial T}{\partial x_1}\right) = m_1\ddot{x}_1 + m_2\ddot{x}_1 - m_2l\dot{\theta}^2\cos\theta - m_2l\ddot{\theta}\sin\theta$$

$$\frac{\partial T}{\partial x_1} = 0$$

$$\frac{\partial U}{\partial x_1} = kx_1$$

$$\frac{d}{dt}\left(\frac{\partial T}{\partial \theta}\right) = m_2l^2\ddot{\theta} - m_2l(\ddot{x}\sin\theta + \dot{x}\dot{\theta}\cos\theta)$$

$$\frac{\partial T}{\partial \theta} = -m_2\dot{x}\dot{\theta}l\cos\theta$$

$$\frac{\partial U}{\partial \theta} = m_2gl\sin\theta$$

Lagrange's equation therefore gives

$$(m_1 + m_2)\ddot{x}_1 - m_2 l\dot{\theta}^2 \cos\theta - ml\ddot{\theta}\sin\theta + kx_1 = 0$$

$$m_2 l^2\ddot{\theta} - m_2 l(\ddot{x}_1 \sin\theta + \dot{x}_1\dot{\theta}\cos\theta) + m_2 l\dot{x}_1\dot{\theta}\cos\theta + m_2 gl\sin\theta = 0$$

These equations are nonlinear. The system could, however, be linearized for small vibrations by approximating T for its value at the equilibrium position, i.e., at $x = 0$ and $\theta = 0$. In this case

$$T = \frac{1}{2}(m_1 + m_2)\dot{x}_1^2 + \frac{1}{2}m_2 l^2\dot{\theta}^2$$

and Lagrange's equations give

$$(m_1 + m_2)\ddot{x}_1 + kx_1 = 0$$

$$m_2 l^2\ddot{\theta} + m_2 gl\theta = 0 \quad (\theta \text{ substituted for } \sin\theta)$$

Example 6

This example illustrates how Lagrange's equation considerably simplifies the analysis of relatively complex systems. The system shown in Fig. 7-8 is a flexible coupling which utilized dynamical forces in the trans-

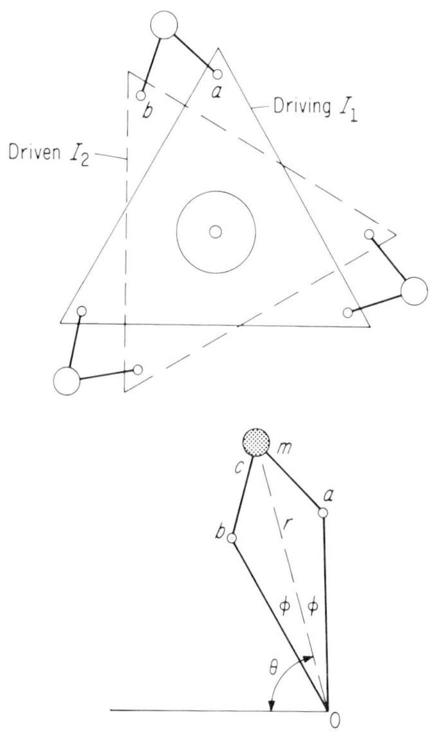

FIG. 7-8

mission of torque*. The mass m is connected through rigid links of negligible mass to both the driving and driven members with moment of inertia I_1 and I_2 respectively. If r, θ denote the polar coordinates of the mass m, its velocity will be $(\dot{r}^2 + r^2\dot{\theta}^2)^{1/2}$. The K.E. of the system can be expressed as

$$T = \frac{1}{2} I_1(\dot{\theta} + \dot{\phi})^2 + \frac{1}{2} m(r^2 + \dot{r}^2\dot{\theta}^2) + \frac{1}{2} I_2(\dot{\theta} - \dot{\phi})^2$$

where $(\theta + \phi)$ is the angular position of the drive, and $(\theta - \phi)$ is the angular position of the driven member. The virtual work accomplished by the driving torque M_1 and the load torque M_2 during virtual displacements $d\theta$ and $d\phi$ of the generalized coordinates θ and ϕ is

$$M_1(d\theta + d\phi) - M_2(d\theta - d\phi) = (M_1 - M_2)\,d\theta + (M_1 + M_2)\,d\phi$$

from which

$$Q_\theta = M_1 - M_2 \qquad \text{and} \qquad Q_\phi = M_1 + M_2$$

Applying Eq. 7-4,

$$\frac{d}{dt}\left(\frac{\partial T}{\partial q_r}\right) - \frac{\partial T}{\partial q_r} = F_r$$

The equations of motion corresponding respectively to θ and ϕ are

$$I_1(\ddot{\theta} + \ddot{\phi}) + I_2(\ddot{\theta} - \ddot{\phi}) + m\frac{d}{dt}(r^2\dot{\theta}) = M_1 - M_2$$

$$I_1(\ddot{\theta} + \ddot{\phi}) - I_2(\ddot{\theta} - \ddot{\phi}) - mr\,\dot{\theta}^2\frac{dr}{d\phi} = M_1 + M_2$$

Example 7

A mass m is resting on two springs k_1 and k_2 at distance l_1 and l_2 from its C.G., Fig. 7-9. The plane motion of the mass is determined by the translation x of the C.G. and the rotation θ about the z-axis. The K.E. in this case can be written as

$$T = \frac{1}{2} m\dot{x}^2 + \frac{1}{2} I\dot{\theta}^2$$

where I is the moment of inertia about the center of mass. The potential energy is

$$U = \frac{1}{2} k_1(x + l_1\theta)^2 + \frac{1}{2} k_2(x - l_2\theta)^2$$

*G. A. Dorton, ASME Paper No. 59 OGP-8, 1959. Also C. W. Bert, *Int. J. Mech. Sci.*, Vol. 4 (1962), p. 439.

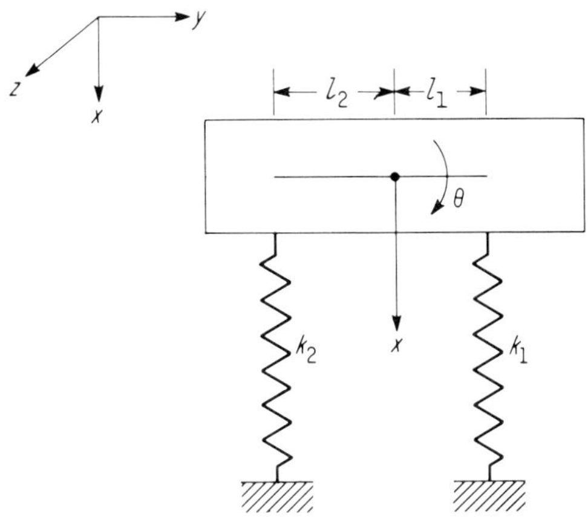

FIG. 7-9

Lagrange's equation corresponding to the generalized coordinates x_1, θ respectively are

$$m\ddot{x} + k_1(x + l_1\theta) + k_2(x - l_2\theta) = 0$$
$$I\ddot{\theta} + k_1 l_1(x + l_1\theta)k_2 l_2(x - l_2\theta) = 0$$

7-3. COUPLED SYSTEMS

One of the characteristics of systems with several degrees of freedom is that the masses are interconnected in such a way that a disturbance of one mass will cause the other masses to vibrate as well. On form of this interconnection (or coupling) is characterized by the existence of inertia terms in the equation corresponding to one coordinate which are dependent on other coordinates. This is known as *inertia or dynamic coupling* and can be found in Examples 3, 5, and 6.

The other form of coupling is known as *spring or static coupling* and is evidenced by the existence of conservative forces in the equation of motion corresponding to one coordinate which are dependent on other coordinates. This type of coupling is illustrated in Examples 4 and 7, and is the most common type in linear systems.

Coupled systems can also be classified into two groups:

1. Close-coupled or adjacently coupled systems, where the restoring forces on any mass are only dependent on the displacement of masses directly connected to it. Systems interconnected by tensile or shear type springs are of the close-coupled group.

2. Far-coupled systems, where the restoring forces acting on one mass are dependent on the displacement of any other mass in the system

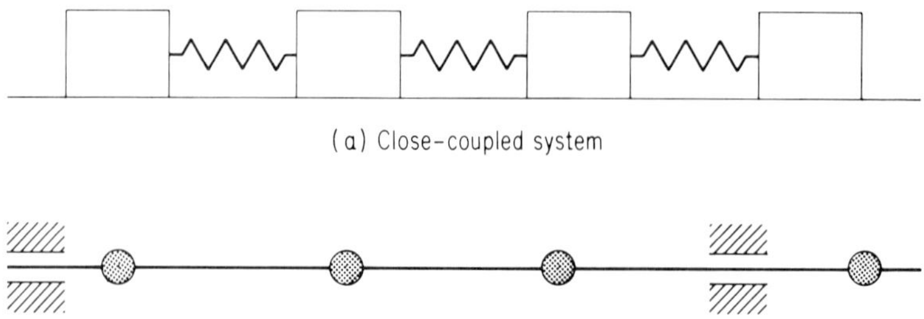

(a) Close-coupled system

(b) Far-coupled system

FIG. 7-10

even if they are not directly linked. Bending (or lateral) vibration of shafts is a case of far-coupling, Fig. 7-10b.

7-4. SUMMARY

In this chapter the equations of motion for systems with several degrees of freedom has been formulated by means of Lagrange's equation. This equation may be expressed in any of the following forms:

$$\frac{d}{dt}\left(\frac{\partial T}{\partial \dot{q}_r}\right) - \frac{\partial T}{\partial q_r} = F_r \tag{7-4}$$

$$\frac{d}{dt}\left(\frac{\partial T}{\partial \dot{q}_r}\right) - \frac{\partial T}{\partial q_r} + \frac{\partial U}{\partial q_r} = Q_r \tag{7-5}$$

$$\frac{d}{dt}\left(\frac{\partial L}{\partial \dot{q}_r}\right) - \frac{\partial L}{\partial q_r} = Q_r \tag{7-6}$$

where T = kinetic energy
U = potential energy
$L = T - U$
q_r = generalized coordinates $r = 1 \rightarrow n$
$\dot{q}_r = dq_r/dt$
F_r = generalized force corresponding to the coordinate q_r
Q_r = generalized nonconservative, noninertial force corresponding to the coordinate q_r

All these forms give identical results. The use of any of them in a particular problem depends on the convenience of expressing the energies and generalized force for the system at hand.

Lagrange's equation can have significant advantages over Newton's second law and Euler's equations in formulation of complex and nonlinear coupled systems. It is one of the most general and powerful tools of formulating physical systems.

PROBLEMS

7-1. A cylindrical roller with radius a and mass m_1 is supported by a platform which is free to oscillate as shown. If m_1 rotates without sliding on the platform, write the equations of motion for the system.

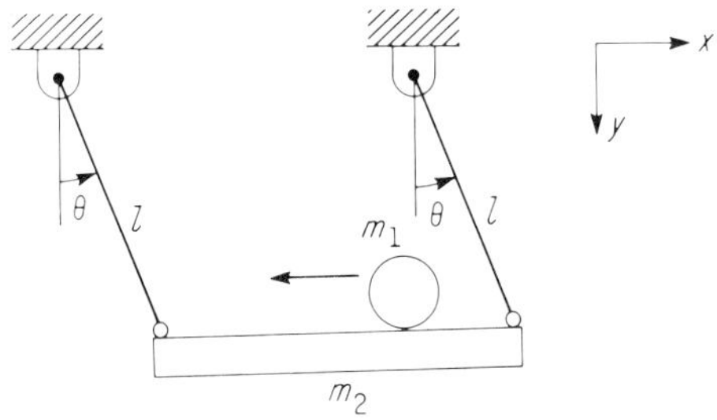

PROB. 7-1

7-2. A mass m is suspended by an elastic string having an unextended length l_0. Use Lagrange's equation to derive the equations of motion for the mass.

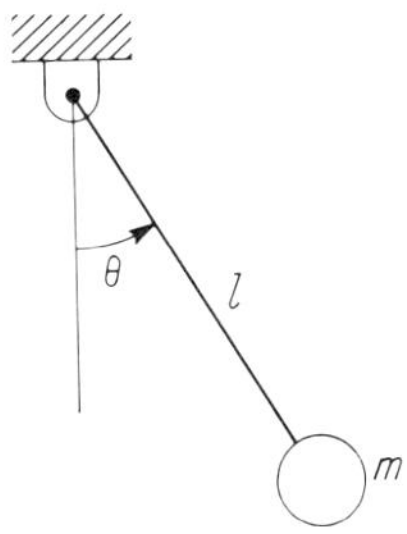

PROB. 7-2

7-3. The given system has two degrees of freedom. Write the expressions for the kinetic energy and potential energy. By the use of Lagrange's equation, derive the differential equations of motion for the system.

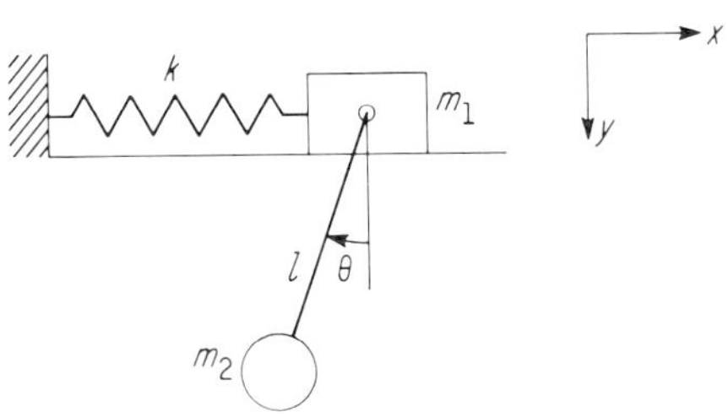

PROB. 7-3

7-4. A simple pendulum is suspended from a block with mass M which rests on a frictional surface with coefficient of friction equal to μ. The system is set in motion by an impact from another pendulum with mass m_2 and length l_2. If the coefficient of restitution between M and m_2 is e, derive the equations of motion for the system.

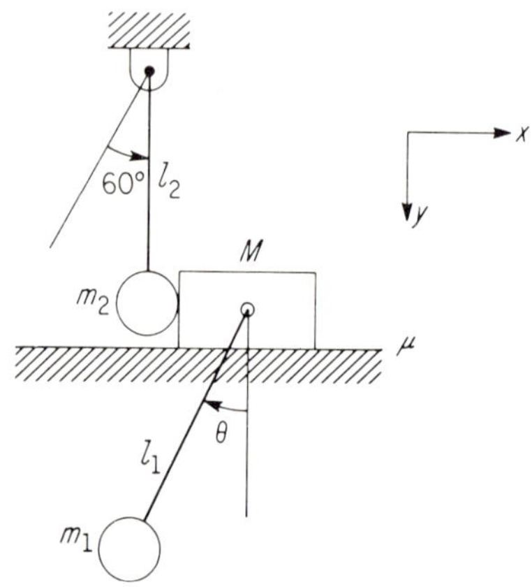

PROB. 7-4

7-5. If the hinge suspending the pendulum in Prob. 7-4 is restrained by a torsional spring with stiffness k_t and a torsional viscous damper c, write the equations of motion.

7-6. A pendulum with length l and mass m has its point of suspension restrained by two horizontal springs as shown. Derive the equations of motion.

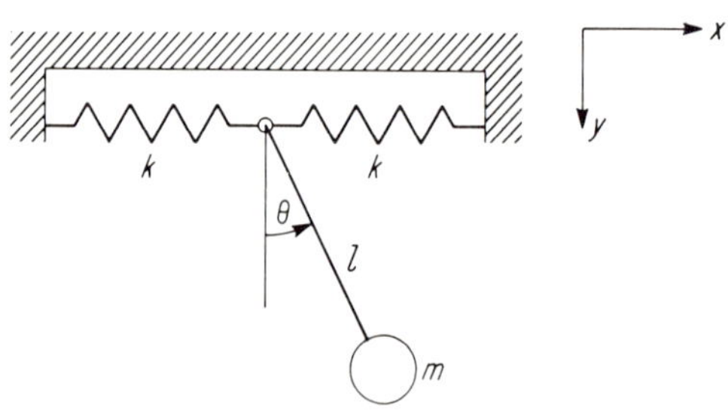

PROB. 7-6

7-7. Use Lagrange's equation to obtain the equations of motion for the given system. Assume $c \ll b$.

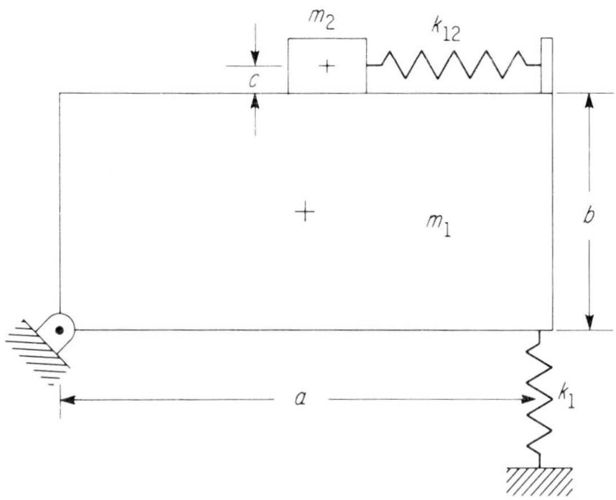

PROB. 7-7

7-8. The system shown in the sketch is a simplified model for the analysis of whip-lash during the rear impact of a car. The masses m_1, m_2 and m_3 represent the car, the seat and the passenger's body, and the passenger's head respectively. The spring k describes the compliance between the seat and the car body. The spring k_b represents the horizontal stiffness between the head and the seat. The spring k_t describes the rotational compliance of the neck. Define the generalized coordinates for the system and write the equations of motion resulting from an instantaneous change in the velocity of the car.

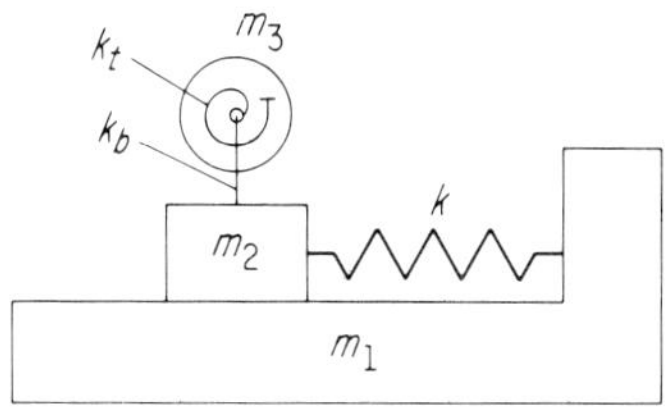

PROB. 7-8

chapter 8

Exact Solution of the Equations of Motion

8-1. INTRODUCTION

The equations of motion for systems with a finite number of degrees of freedom can be readily obtained by applying Newton's second law and Euler's equation as explained in Chapter 3. They can also be derived by means of Lagrange's equation as shown in Chapter 7.

In this chapter we shall be concerned with the exact solution of the equations of motion for linear systems whose motion can be represented by a set of simultaneous linear differential equations. The exact methods discussed in this chapter will be applied to free as well as forced vibrations and close-coupled as well as far-coupled systems. These methods are the determinant, the Laplace transform, and the impedance solutions. These exact solutions become progressively more complex as the number of the degrees of freedom increases and one must resort to approximate solutions. The latter which include both numerical and graphical techniques will be dealt with in Chapter 9. Some of these approximate methods are readily adaptable for analysis of systems with nonlinear characteristics.

8-2. DETERMINANT SOLUTIONS

Close-Coupled Systems

The following examples will be used to illustrate the application of the determinant method of solution to close-coupled linear systems.

Example 1

Free Vibration of a Two-Degree-of-Freedom Rectilinear System: In order to analyze the free vibration of the two-degree-of-freedom system shown in Fig. 8-1, the equations of motion are first written as

$$m_1\ddot{x}_1 + (k_1 + k_2)x_1 - k_2x_2 = 0$$

$$m_2\ddot{x}_2 + k_2(x_2 - x_1) = 0$$

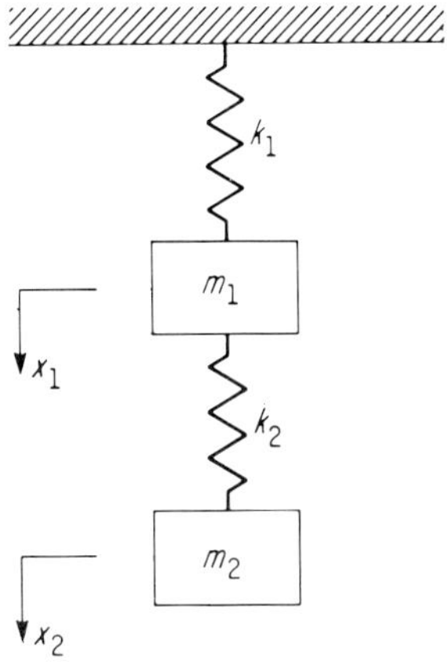

FIG. 8-1

A simple approach to the solution of these two simultaneous equations is to assume that both masses vibrate at the same frequency ω_n. Accordingly, x_1 and x_2 can be represented by

$$x_1 = ae^{i\omega_n t} \tag{8-1a}$$

$$x_2 = be^{i\omega_n t} \tag{8-1b}$$

Substituting in the equations of motion and dividing by $e^{i\omega_n t}$ we obtain

$$[-m\omega_n^2 + (k_1 + k_2)]a - [k_2]b = 0$$
$$[-k_2]a + [-m\omega_m^2 + k_2]b = 0$$

which will have a solution for a and b different from zero only if the determinant of its coefficients vanishes. This condition can be written as

$$\begin{vmatrix} -m_1\omega_n^2 + (k_1 + k_2) & -k_2 \\ -k_2 & k_2 - m_2\omega_n^2 \end{vmatrix} = 0$$

or

$$m_1 m_2 \omega_n^4 - [m_1 k_2 + m_2(k_1 + k_2)]\omega_n^2 + (k_1 + k_2)k_2 - k_2^2 = 0$$

from which

$$\omega_n^4 - \left(\frac{k_2}{m_2} + \frac{k_1 + k_2}{m_1}\right)\omega_n^2 + \frac{k_1 k_2}{m_1 m_2} = 0$$

which is a quadratic equation in ω_n^2.

Using the notation $k_2/m_2 = \omega_2^2$, $k_1/m_1 = \omega_1^2$, $m_2/m_1 = \mu$, the two solutions for ω_n^2 are

$$\omega_n^2 = \frac{1}{2}[\omega_1^2 + \omega_2^2(1 + \mu)] \pm \sqrt{\frac{1}{4}[\omega_1^2 + \omega_2^2(1 + \mu)]^2 - [\omega_1\omega_2]^2}$$

or

$$\omega_{n_1} = \pm\left\{\frac{1}{2}\,[\omega_1^2 + \omega_2^2(1 + \mu)] - \sqrt{\frac{1}{4}\,[\omega_1^2 + \omega_2^2(1 + \mu)]^2 - [\omega_1\,\omega_2]^2}\right\}^{1/2}$$

$$\omega_{n_2} = \pm\left\{\frac{1}{2}\,[\omega_1^2 + \omega_2^2(1 + \mu)] + \sqrt{\frac{1}{4}\,[\omega_1^2 + \omega_2^2(1 + \mu)]^2 - [\omega_1\,\omega_2]^2}\right\}^{1/2}$$

Accordingly, eq. 8-1 will be

$$\begin{aligned} x_1(t) &= a_1 e^{i\omega_{n_1}t} + a_2 e^{-i\omega_{n_1}t} + a_3 e^{i\omega_{n_2}} + a_4 e^{-i\omega_{n_2}t} \\ x_2(t) &= b_1 e^{i\omega_{n_1}t} + b_2 e^{-i\omega_{n_1}t} + b_3 e^{i\omega_{n_2}t} + b_4 e^{-i\omega_{n_2}t} \end{aligned}$$

where

$$\frac{b_1}{a_1} = \frac{b_2}{a_2} = \frac{[-m\omega_{n_1}^2 + (k_1 + k_2)]}{k_2} = \frac{-k_2}{[-m\omega_{n_1}^2 + k_2]} = R_1$$

and

$$\frac{b_3}{a_3} = \frac{b_4}{a_4} = \frac{[-m\omega_{n_2}^2 + (k_1 + k_2)]}{k_2} = \frac{-k_2}{[-m\omega_{n_2}^2 + k_2]} = R_2$$

Other forms of expressing $x_1(t)$ and $x_2(t)$ are

$$\begin{aligned} x_1(t) &= A_1 \sin \omega_{n_1} t + B_1 \cos \omega_{n_1} t + C_1 \sin \omega_{n_2} t + D_1 \cos \omega_{n_2} t \\ x_2(t) &= A_2 \sin \omega_{n_1} t + B_2 \cos \omega_{n_2} t + C_2 \sin \omega_{n_2} t + D_2 \cos \omega_{n_2} t \end{aligned} \tag{8-2}$$

and

$$\begin{aligned} x_1(t) &= X_{11} \sin (\omega_{n_1} t + \phi_1) + X_{12} \sin (\omega_{n_2} t + \phi_2) \\ x_2(t) &= X_{21} \sin (\omega_{n_1} t + \phi_1) + X_{22} \sin (\omega_{n_2} t + \phi_2) \end{aligned} \tag{8-3}$$

where

$$\frac{A_2}{A_1} = \frac{B_2}{B_1} = R_1 \qquad \text{and} \qquad \frac{C_2}{C_1} = \frac{D_2}{D_1} = R_2$$

also

$$\frac{X_{21}}{X_{11}} = R_1 \qquad \text{and} \qquad \frac{X_{22}}{X_{12}} = R_2$$

Equation 8-3 can therefore be written as

$$\begin{aligned} x_1(t) &= X_{11} \sin (\omega_{n_1} t + \phi_1) + X_{12} \sin (\omega_{n_2} t + \phi_2) \\ x_2(t) &= R_1 X_{11} \sin (\omega_{n_1} t + \phi_1) + R_2 X_{12} \sin (\omega_{n_2} t + \phi_2) \end{aligned} \tag{8-4}$$

which shows that the general motion for either mass is a combination of two simple harmonic oscillations at the two frequencies ω_{n_1} and ω_{n_2} which are called the *natural or normal frequencies* of the system. The integration constants X_{11}, ϕ_1, X_{12}, ϕ_2 are evaluated from the initial conditions.

Assuming at $t = 0$,

$$x_1 = X_0 \qquad \dot{x}_1 = 0$$
$$x_2 = R_1 X_0 \qquad \dot{x}_2 = 0$$

then

$$X_0 = X_{11} \sin \phi_1 + X_{12} \sin \phi_2 \tag{8-5a}$$
$$R_1 X_0 = R_1 X_{11} \sin \phi_1 + R_2 X_{12} \sin \phi_2 \tag{8-5b}$$
$$0 = \omega_{n_1} X_{11} \cos \phi_1 + \omega_{n_2} X_{12} \cos \phi_2 \tag{8-5c}$$
$$0 = \omega_{n_1} R_1 X_{11} \cos \phi_1 + \omega_{n_2} R_2 X_{12} \cos \phi_2 \tag{8-5d}$$

$$X_{11} \sin \phi_1 = X_0 \qquad \text{from Eqs. 8-5a and 8-5b}$$
$$X_{11} \cos \phi_1 = 0 \qquad \text{from Eqs. 8-5c and 8-5d}$$

By solving these equations simultaneously we find

$$X_{12} = 0$$
$$X_{11} = X_0$$
$$\phi_1 = \frac{\pi}{2}$$

and the equations of motion take the form

$$x_1(t) = X_0 \sin\left(\omega_n t + \frac{\pi}{2}\right)$$
$$x_2(t) = R_1 X_0 \sin\left(\omega_n t + \frac{\pi}{2}\right) \tag{8-6}$$

or

$$\frac{x_2(t)}{x_1(t)} = R_1$$

Similarly, if at $t = 0$,

$$x_1 = X_0 \qquad x_2 = R_2 X_0$$
$$\dot{x}_1 = 0 \qquad \dot{x}_2 = 0$$

then

$$x_1(t) = X_0 \sin\left(\omega_{n_2} t + \frac{\pi}{2}\right)$$
$$x_2(t) = R_2 X_0 \sin\left(\omega_{n_2} t + \frac{\pi}{2}\right) \tag{8-7}$$

or

$$\frac{x_2(t)}{x_1(t)} = R_2$$

These two conditions (Eqs. 8-6 and 8-7) show that if the motion is started with an initial displacement ratio equal to the amplitude ratio corresponding to one of the natural frequencies, the two masses will oscillate in a simple harmonic manner at that frequency, and in the same phase. The ratio between their displacements at any instant will be constant. The vibrations represented by Eqs. 8-6 and 8-7 respectively describe what is known as the *normal modes* or *natural modes* of the system. It can be easily seen that the general motion described by Eq. 8-4 represents superposition of natural mode vibrations.

For a system with n degrees of freedom there are n natural modes of vibration at n natural frequencies. As in the case of the two-degrees-of-freedom systems, the vibration of the system due to a general disturbance is compounded from these natural modes. Only under particular initial conditions does the system vibrate in a natural mode. In this case all particles partake in the motion at the same frequency, which is the corresponding natural frequency. They all perform simple harmonic motions which are in phase and simultaneously reach points of zero velocity.

Example 2

Free Vibration of Close Coupled Pendulums. The statically coupled system of the two identical pendulums shown in Fig. 8-2 is analyzed as follows:

$$\text{K.E.} = T = \frac{1}{2}m(l\dot{\theta}_1)^2 + \frac{1}{2}m(l\dot{\theta}_2)^2$$

$$\text{P.E.} = U = mgl(1 - \cos\theta_1) + mgl(1 - \cos\theta_2) + \frac{1}{2}kl^2(\theta_1 - \theta_2)^2$$

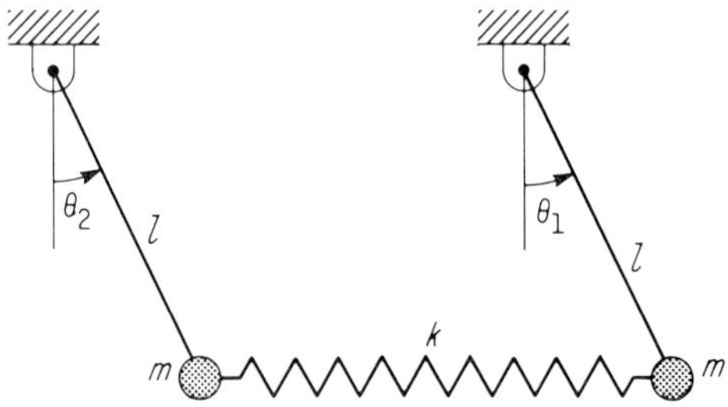

FIG. 8-2

The equations of motion can therefore be obtained by applying the Lagrange equation for θ_1 and θ_2. For small values of θ_1, θ_2 the equations reduce to

$$ml\ddot{\theta}_1 + mg\theta_1 + kl(\theta_1 - \theta_2) = 0$$
$$ml\ddot{\theta}_2 + mg\theta_2 + kl(\theta_2 - \theta_1) = 0$$

The frequency equation becomes

$$\begin{vmatrix} \left(\dfrac{k}{m} + \dfrac{g}{l} - \omega_n^2\right) & -\dfrac{k}{m} \\ -\dfrac{k}{m} & k + \dfrac{g}{l} - \omega_n^2 \end{vmatrix} = 0$$

and the solution gives the two natural frequencies

$$\omega_{n_1}^2 = \frac{g}{l}$$

$$\omega_{n_2}^2 = \frac{g}{l} + 2\frac{k}{m}$$

Also,

$$\frac{\theta_2}{\theta_1} = \frac{k/m}{k/m + g/l - \omega^2}$$

from which the amplitude ratio for the two modes of vibration are

$$R_1 = \left(\frac{\theta_2}{\theta_1}\right)_{\omega_{n_1}} = \frac{k/m}{k/m + g/l - g/l} = +1$$

$$R_2 = \left(\frac{\theta_2}{\theta_1}\right)_{\omega_{n_1}} = \frac{k/m}{k/m + g/l - g/l - 2k/m} = -1$$

The general equation for θ_1 and θ_2 can be written from Eq. 8-4 as

$$\theta_1(t) = A_1 \sin(\omega_{n_1} t + \phi_1) + B_1 \sin(\omega_{n_2} t + \phi_2)$$
$$\theta_2(t) = A_1 \sin(\omega_{n_1} t + \phi_1) - B_1 \sin(\omega_{n_2} t + \phi_2)$$

where A_1, B_1, ϕ_1, ϕ_2, are evaluated from the initial conditions of the motion.

Example 3

Three-Mass Torsional System. To study the free vibration of the three-mass torsional system shown in Fig. 8-3, the equations of motion

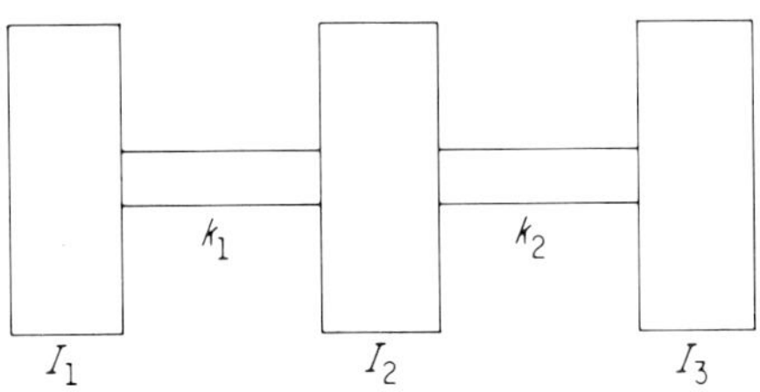

FIG. 8-3

are first obtained as:

$$I_1\ddot{\theta}_1 + k_1(\theta_1 - \theta_2) = 0$$
$$I_2\ddot{\theta}_2 + k_1(\theta_2 - \theta_1) + k_2(\theta_2 - \theta_3) = 0$$
$$I_3\ddot{\theta}_3 + k_2(\theta_3 - \theta_2) = 0$$

The frequency equation is therefore

$$\begin{vmatrix} k_1 - I_1\omega_n^2 & -k_1 & 0 \\ -k_1 & k_1 + k_2 - I_2\omega_n^2 & -k_2 \\ 0 & -k_2 & k_2 - I_3\omega_n^2 \end{vmatrix} = 0$$

which is a cubic equation in ω^2. The solution gives three roots: one is zero and the other two are real and positive.

$$\omega_n^2 = 0$$
$$\omega_{n_1}^2 = a - \sqrt{a^2 - b} \tag{8-8a}$$
$$\omega_{n_2}^2 = a + \sqrt{a^2 - b} \tag{8-8b}$$

where ω_{n_1}, ω_{n_2}, are the two natural frequencies of the system in which

$$a = \frac{1}{2}\left[\frac{k_1}{I_1} + \frac{k_1 + k_2}{I_2} + \frac{k_2}{I_3}\right]$$
$$b = \frac{k_1 k_2(I_1 + I_2 + I_3)}{I_1 I_2 I_3}$$

and the vibration modes are determined from the equations of motion as

$$\frac{\theta_2}{\theta_1} = \frac{k_1 - I_1\omega_n^2}{k_1} = 1 - \frac{\omega_n^2}{(k_1/I_1)}$$
$$\frac{\theta_3}{\theta_1} = \frac{k_2(k_1 - I_1\omega_n^2)}{k_1(k_2 - I_3\omega_n^2)} = \frac{1 - \omega_n^2/(k_1/I_1)}{1 - \omega_n^2/(k_2/I_3)}$$

when $I_1 = I_2 = I_3 = I$.

Equations 8-8a and 8-8b become

$$\omega_{n_1}^2 = \frac{k_1}{I}[(1 + c) - \sqrt{c^2 - c + 1}]$$
$$\omega_{n_2}^2 = \frac{k_1}{I}[(1 + c) + \sqrt{c^2 - c + 1}]$$

where

$$c = \frac{k_2}{k_1}$$

The values of the natural frequencies corresponding to the first and second modes respectively for a system with equal inertias are plotted in Fig. 8-4.

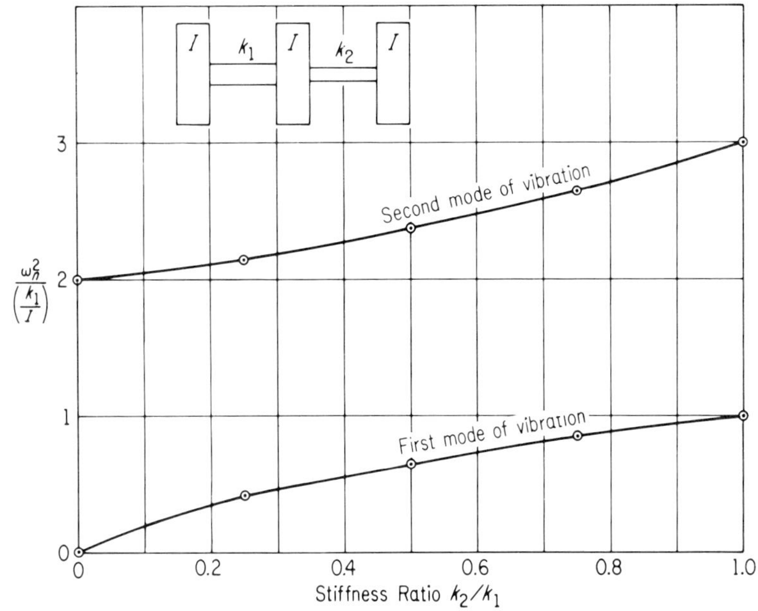

FIG. 8-4

The general solution for the system can be written as

$$\begin{aligned}\theta_1 &= \theta_0 + V_0 t + A_1 \sin \omega_{n_1} t + B_1 \cos \omega_{n_1} t \\ &\quad + C_1 \sin \omega_{n_2} t + D_1 \cos \omega_{n_2} t \\ \theta_2 &= \theta_0 + V_0 t + a_{21} A_1 \sin \omega_{n_1} t + a_{21} B_1 \cos \omega_{n_1} t \\ &\quad + a_{22} C_1 \sin \omega_{n_2} t + a_{22} D_1 \cos \omega_{n_2} t \\ \theta_3 &= \theta_0 + V_0 t + a_{31} A_1 \sin \omega_{n_1} t + a_{31} B_1 \cos \omega_{n_1} t \\ &\quad + a_{32} C_1 \sin \omega_{n_2} t + a_{32} D_1 \cos \omega_{n_2} t\end{aligned}$$

where

$$a_{21} = \left(\frac{\theta_2}{\theta_1}\right)_{\omega_{n_1}} \qquad a_{22} = \left(\frac{\theta_2}{\theta_1}\right)_{\omega_{n_2}}$$

for $\omega = \omega_{n_1}$ and $\omega = \omega_{n_2}$ respectively. It should be noted that the term $\theta_0 + V_0 t$ is the same in all three equations, since it represents the rigid body motion associated with $\omega = 0$. The constants θ_0, V_0, A_1, B_1, C_1, D_1 are to be determined from the six initial conditions of θ_1, θ_2, and θ_3.

Assuming, for example, that at $t = 0$

$$\begin{aligned}\theta_1 &= \theta_2 = \theta_3 = 0 \\ \theta_1 &= \theta_2 = 0 \\ \theta_3 &= Q\end{aligned}$$

Therefore the six conditions give the two sets of equations:

$$0 = \theta_0 + B_1 + D_1$$
$$0 = \theta_0 + a_{21}B_1 + a_{22}D_1$$
$$0 = \theta_0 + a_{31}B_1 + a_{32}D_1$$

$$0 = V_0 + \omega_{n_1}A_1 + \omega_{n_2}C_1$$
$$0 = V_0 + a_{21}\omega_{n_1}A_1 + a_{22}\omega_{n_2}C_1$$
$$Q = V_0 + a_{31}\omega_{n_1}A_1 + a_{32}\omega_{n_2}C_1$$

The simultaneous solution of the first three equations gives

$$\theta_0 = B_1 = D_1 = 0$$

while the other set of equations gives

$$V_0 = \frac{\begin{vmatrix} 0 & \omega_{n_1} & \omega_{n_2} \\ 0 & a_{21}\omega_{n_1} & a_{22}\omega_{n_2} \\ Q & a_{31}\omega_{n_1} & a_{32}\omega_{n_2} \end{vmatrix}}{|D|}$$

where

$$D = \begin{vmatrix} 1 & \omega_{n_1} & \omega_{n_2} \\ 1 & a_{21}\omega_{n_1} & a_{22}\omega_{n_2} \\ 1 & a_{31}\omega_{n_1} & a_{32}\omega_{n_2} \end{vmatrix}$$

Therefore

$$V_0 = \frac{Q[a_{22} - a_{21}]}{[(a_{22} - a_{21}) + a_{31}(1 - a_{22}) + a_{32}(a_{21} - 1)]}$$

Similarly,

$$A_1 = \frac{Q[1 - a_{22}]}{\omega_{n_1}[(a_{22} - a_{21}) + a_{31}(1 - a_{22}) + a_{32}(a_{21} - 1)]}$$

and

$$C_1 = \frac{Q[a_{21} - 1]}{\omega_{n_2}[(a_{22} - a_{21}) + a_{31}(1 - a_{22}) + a_{32}(a_{21} - 1)]}$$

The shaft torques in this case are given by

$$(T_1) = k_1(\theta_2 - \theta_1) = k_1[A_1(a_{21} - 1)\sin\omega_{n_1}t + C_1(a_{22} - 1)\sin\omega_{n_2}t]$$
$$(T_2) = k_2(\theta_3 - \theta_2) = k_2[A_1(a_{31} - a_{21})\sin\omega_{n_1}t$$
$$+ C_1(a_{32} - a_{22})\sin\omega_{n_2}t]$$

and are linear functions of Q.

Since Q represents an initial impulse E on I_3, we can write

$$Q = \frac{E}{I_3}$$

and the shaft torques due to a unit impulse $u'(t)$ at I_3 are therefore

$$(T_1)u'(t) = \frac{k_1(a_{21} - 1)(1 - a_{22}) \left[\dfrac{\sin \omega_{n_1} t}{\omega_{n_1}} + \dfrac{\sin \omega_{n_2} t}{\omega_{n_2}}\right]}{I_3[(a_{22} - a_{21}) + a_{31}(1 - a_{22}) + a_{32}(a_{21} - 1)]} \tag{8-9a}$$

$$(T_2)u'(t) = \frac{k_2 \left[(a_{31} - a_{21})(1 - a_{22}) \dfrac{\sin \omega_{n_1} t}{\omega_{n_1}} + (a_{32} - a_{22})(a_{21} - 1) \dfrac{\sin \omega_{n_2} t}{\omega_{n_2}}\right]}{I_3[(a_{22} - a_{21}) + a_{31}(1 - a_{22}) + a_{32}(a_{21} - 1)]} \tag{8-9b}$$

and consequently the torques due to any load function on I_3 can be obtained by convolution integrals as explained in Chapter 5 (dividing the load into incremental pulses and integrating the effects as in Fig. 8-5).

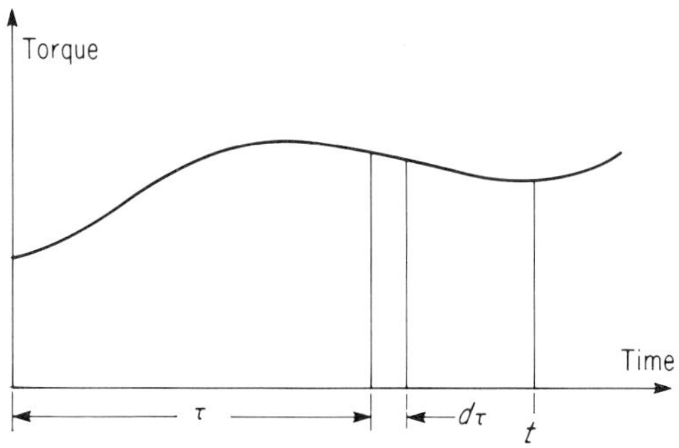

FIG. 8-5

For example, the torques due to the special case of a step function T_0 (or a suddenly applied torque T_0) at I_3 are obtained from

$$(T_1)_{T_0} = \int_0^t (T_1)_{u'(t-\tau)} T_0 \, d\tau$$

$$= T_0 \frac{k_1(a_{21} - 1)(1 - a_{22})}{I_3[D]} \int_0^t \left\{ \frac{\sin \omega_{n_1}(t - \tau)}{\omega_{n_1}} + \frac{\sin \omega_{n_2}(t - \tau)}{\omega_{n_1}} \right\} d\tau$$

By taking $(t - \tau) = \phi$,

$$(T_1)_{T_0} = T_0 \frac{k_1(a_{21} - 1)(1 - a_{22})}{I_3[D]} \int_0^t \left(\frac{\sin \omega_{n_1}\phi}{\omega_{n_1}} + \frac{\sin \omega_{n_2}\phi}{\omega_{n_2}} \right) d\phi$$

$$= T_0 \frac{k_1(a_{21} - 1)(1 - a_{22})}{I_3[D]} \left[\frac{(1 - \cos \omega_{n_1} t)}{\omega_{n_1}^2} + \frac{(1 - \cos \omega_{n_2} t)}{\omega_{n_2}^2} \right] \tag{8-10a}$$

Similarly,

$$(T_2)_{T_0} = T_0 \frac{k_2}{I_3[D]} \left[\frac{(a_{31} - a_{21})(1 - a_{22})}{\omega_{n_1}^2}(1 - \cos \omega_{n_1} t) + \frac{(a_{32} - a_{22})(a_{21} - 1)}{\omega_{n_2}^2}(1 - \cos \omega_{n_2} t) \right] \tag{8-10b}$$

where

$$[D] = (a_{33} - a_{21}) + a_{31}(1 - a_{22}) + a_{32}(a_{21} - 1)$$

Example 4

Four-Mass Torsional System. The end flywheel I_4 of the torsional system, Fig. 8-6, is subjected to a sudden change of velocity Q from an impulse. Calculate the shaft torques.

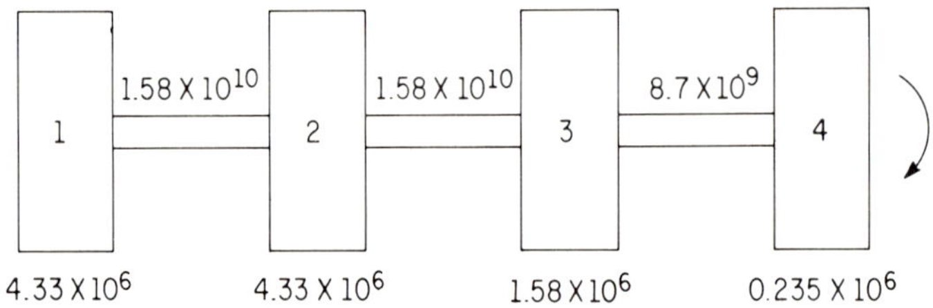

FIG. 8-6

The frequency equations for this system have the following roots:

$$\omega = 0$$
$$\omega_{n_1} = 165.4 \text{ rad/sec}$$
$$\omega_{n_2} = 201.1 \text{ rad/sec}$$
$$\omega_{n_3} = 384.3 \text{ rad/sec}$$

The three modes of vibration are given by

$$\left(\frac{\theta_2}{\theta_1}\right)_{\omega_{n_1}} = -0.499 \qquad \left(\frac{\theta_3}{\theta_1}\right)_{\omega_{n_1}} = -0.8745 \qquad \left(\frac{\theta_4}{\theta_1}\right)_{\omega_{n_1}} = -3.344$$

$$\left(\frac{\theta_2}{\theta_1}\right)_{\omega_{n_2}} = 1.2168 \qquad \left(\frac{\theta_3}{\theta_1}\right)_{\omega_{n_2}} = -0.9765 \qquad \left(\frac{\theta_4}{\theta_1}\right)_{\omega_{n_2}} = 10.56$$

$$\left(\frac{\theta_2}{\theta_1}\right)_{\omega_{n_3}} = -7.096 \qquad \left(\frac{\theta_3}{\theta_1}\right)_{\omega_{n_3}} = 17.58 \qquad \left(\frac{\theta_4}{\theta_1}\right)_{\omega_{n_3}} = -5.88$$

from which the normalized shaft twists for the three modes have the following relationships:

$$a_{11} = \left(\frac{\theta_3 - \theta_2}{\theta_2 - \theta_1}\right)_{\omega_{n_1}} = +0.25 \qquad a_{21} = \left(\frac{\theta_4 - \theta_3}{\theta_2 - \theta_1}\right)_{\omega_{n_1}} = +1.65$$

$$a_{12} = \left(\frac{\theta_3 - \theta_2}{\theta_2 - \theta_1}\right)_{\omega_{n_2}} = -0.1084 \qquad a_{22} = \left(\frac{\theta_4 - \theta_3}{\theta_2 - \theta_1}\right)_{\omega_{n_2}} = -5.20$$

$$a_{13} = \left(\frac{\theta_3 - \theta_2}{\theta_2 - \theta_1}\right)_{\omega_{n_3}} = -3.05 \qquad a_{23} = \left(\frac{\theta_4 - \theta_3}{\theta_2 - \theta_1}\right)_{\omega_{n_3}} = +2.9$$

The equations of shaft twist can be written as

$$\begin{aligned}
\theta_{21} &= (\theta_2 - \theta_1) = A_1 \sin \omega_{n_1} t + B_1 \cos \omega_{n_1} t + C_1 \sin \omega_{n_2} t \\
&\quad + D_1 \sin \omega_{n_2} t + E_1 \sin \omega_{n_3} t + F_1 \cos \omega_{n_3} t \\
\theta_{32} &= (\theta_3 - \theta_2) = a_{11} A_1 \sin \omega_{n_1} t + a_{11} B_1 \cos \omega_{n_1} t + a_{12} C_1 \sin \omega_{n_2} t \\
&\quad + a_{12} D_1 \cos \omega_{n_2} t + a_{13} E_1 \sin \omega_{n_3} t + a_{13} F_1 \cos \omega_{n_3} t \\
\theta_{43} &= (\theta_4 - \theta_3) = a_{21} A_1 \sin \omega_{n_1} t + a_{21} B_1 \cos \omega_{n_1} t + a_{22} C_1 \sin \omega_{n_2} t \\
&\quad + a_{22} D_1 \cos \omega_{n_2} t + a_{23} E_1 \sin \omega_{n_3} t + a_{23} F_1 \cos \omega_{n_3} t
\end{aligned}$$

The initial conditions are

$$\theta_{21} = \theta_{32} = \theta_{43} = 0 \quad \text{and} \quad \dot{\theta}_{43} = Q, \quad \dot{\theta}_{21} = \dot{\theta}_{32} = 0$$

Therefore, the initial displacement conditions give

$$\begin{aligned}
0 &= B_1 + D_1 + F_1 \\
0 &= a_{11} B_1 + a_{12} D_1 + a_{13} F_1 \\
0 &= a_{21} B_1 + a_{22} D_1 + a_{23} F_1
\end{aligned}$$

from which

$$B_1 = D_1 = F_1 = 0$$

and the initial conditions of the velocity give

$$\begin{aligned}
0 &= \omega_1 A_1 + \omega_2 C_1 + \omega_3 E_1 \\
0 &= a_{11} \omega_1 A_1 + a_{12} \omega_2 C_1 + a_{13} \omega_3 E_1 \\
Q &= a_{21} \omega_1 A_1 + a_{22} \omega_2 C_1 + a_{23} \omega_3 E_1
\end{aligned}$$

from which

$$\begin{aligned}
A_1 &= \frac{Q\,[a_{13} - a_{12}]}{\omega_1 [a_{21}(a_{13} - a_{12}) + a_{22}(a_{11} - a_{13}) + a_{23}(a_{12} - a_{11})]} \\
&= 7.709 \times 10^{-4}\, Q \\
C_1 &= \frac{Q\,[a_{11} - a_{13}]}{\omega_2 [a_{21}(a_{13} - a_{12}) + a_{22}(a_{11} - a_{13}) + a_{23}(a_{12} - a_{11})]} \\
&= -7.113 \times 10^{-4}\, Q \\
E_1 &= \frac{Q\,[a_{12} - a_{11}]}{\omega_3 [a_{21}(a_{13} - \alpha_{12}) + a_{22}(a_{11} - a_{13}) + a_{23}(a_{12} - a_{11})]} \\
&= 4.049 \times 10^{-5}\, Q
\end{aligned}$$

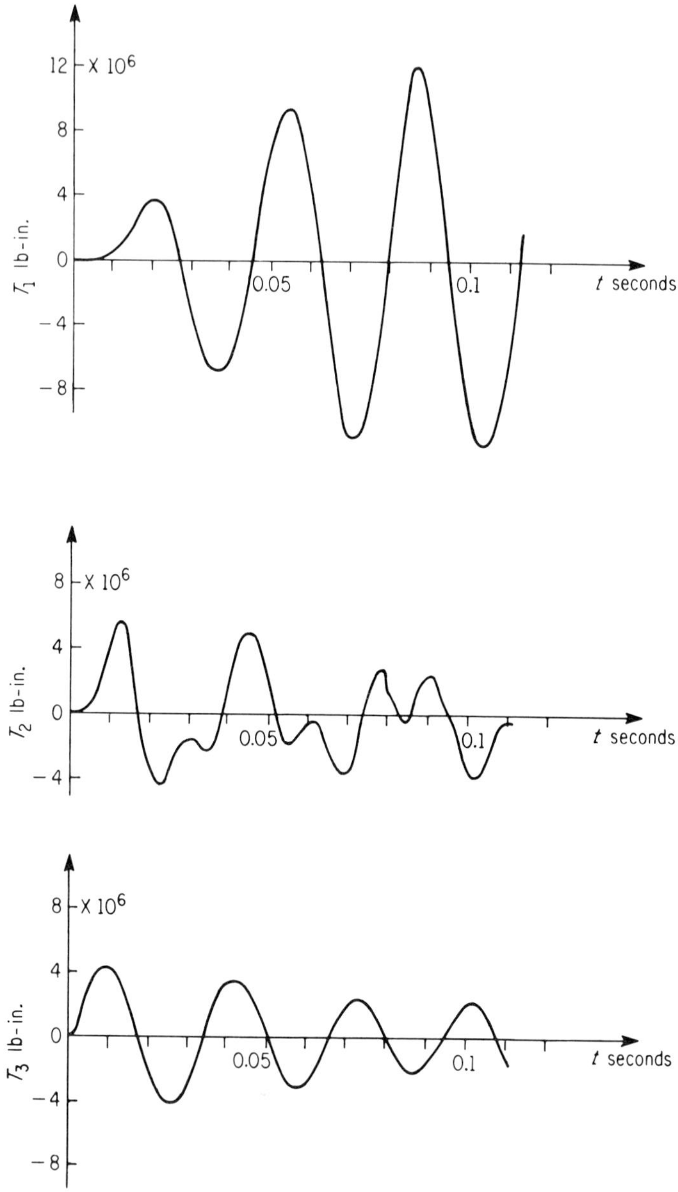

12
× 10⁶
8
4
0
−4
−8
T_1 lb-in.
0.05
0.1
t seconds
8
× 10⁶
4
0
−4
T_2 lb-in.
0.05
0.1
t seconds
8
× 10⁶
4
0
−4
−8
T_3 lb-in.
0.05
0.1
t seconds

Fig. 8-7

The shaft torques are therefore (for $Q = 1$ rad/sec)

$$\begin{aligned} T_{21} &= k_{21}\theta_{21} \\ &= 7.9 \times 10^{10}[A_1 \sin \omega_{n_1} t + B_1 \sin \omega_{n_2} t + E_1 \sin \omega_{n_3} t] \\ &= [60 \sin (165.4)t - 56 \sin (201.1)t + 3.2 \sin (384.3)t]\, 10^6 \end{aligned}$$

Similarly,

$$\begin{aligned} T_{32} &= k_{32}\theta_{32} \\ &= [30.5 \sin (165.4)t + 13.4 \sin (201.1)t - 19.5 \sin (384.3)t]\, 10^6 \end{aligned}$$

and

$$\begin{aligned} T_{43} &= k_{43}\theta_{43} \\ &= [11 \sin (165.4)t - 28.8 \sin (201.1)t + 1.09 \sin (384.3)t]\, 10^6 \end{aligned}$$

The three torque values are plotted as a function of time in Fig. 8-7. It will be left to the student to prove by using Duhamel's integral that the shaft torques in this problem due to a step load T_0 are

$$T_{21} = \frac{k_{21} T_0}{I_4 Q}\left[\frac{A_1}{\omega_{n_1}}(1 - \cos \omega_{n_1} t) + \frac{C_1}{\omega_{n_2}}(1 - \cos \omega_{n_2} t) + \frac{E_1}{\omega_{n_3}}(1 - \cos \omega_{n_3} t)\right]$$

$$T_{32} = \frac{k_{32} T_0}{I_4 Q}\left[\frac{a_{11} A_1}{\omega_{n_1}}(1 - \cos \omega_{n_1} t) + \frac{a_{12} C_1}{\omega_{n_2}}(1 - \cos \omega_{n_2} t) + \frac{a_{13} E_1}{\omega_{n_3}}(1 - \cos \omega_{n_3} t)\right]$$

$$T_{43} = \frac{k_{43} T_0}{I_4 Q}\left[\frac{a_{21} A_1}{\omega_{n_1}}(1 - \cos \omega_{n_1} t) + \frac{a_{22} C_1}{\omega_{n_2}}(1 - \cos \omega_{n_2} t) + \frac{a_{23} E_1}{\omega_{n_3}}(1 - \cos \omega_{n_3} t)\right]$$

Example 5

5-Mass Torsional System (Fig. 8-8). The same procedure can be extended to systems with any finite number of masses. The expressions for

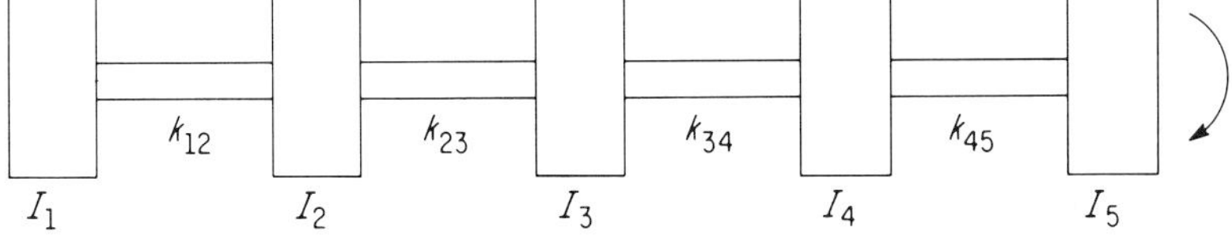

FIG. 8-8

the shaft torques become increasingly long. For example, the expressions for the shaft torques due to an impulse on mass I_5 of a five-mass system are found to be

$$
\begin{aligned}
T_{12} &= k_{12}[A_1 \sin \omega_1 t + B_1 \sin \omega_2 t + C_1 \sin \omega_3 t + D \sin \omega_4 t]\, Q \\
T_{23} &= k_{23}[a_{11} A_1 \sin \omega_1 t + a_{12} B_1 \sin \omega_2 t + a_{13} C_1 \sin \omega_3 t \\
&\qquad + a_{14} D_1 \sin \omega_4 t]\, Q \\
T_{34} &= k_{34}[a_{21} A_1 \sin \omega_1 t + a_{22} B_1 \sin \omega_2 t + a_{23} C_1 \sin \omega_3 t \\
&\qquad + a_{24} D_1 \sin \omega_4 t]\, Q \\
T_{45} &= k_{45}[a_{31} A_1 \sin \omega_1 t + a_{32} B_1 \sin \omega_2 t + a_{33} C_1 \sin \omega_3 t \\
&\qquad + a_{34} D_1 \sin \omega_4 t]\, Q
\end{aligned} \tag{8-11}
$$

where Q = sudden change in velocity of I_5 due to the impulse. The torques due to a unit step function are

$$
\begin{aligned}
T_{12} &= \frac{k_{12}}{I_5}\left[\frac{A_1}{\omega_1}(1 - \cos \omega t) + \frac{B_1}{\omega_2}(1 - \cos \omega_2 t) \right. \\
&\qquad \left. + \frac{C_1}{\omega_3}(1 - \cos \omega_3 t) + \frac{D_1}{\omega_4}(1 - \cos \omega_4 t)\right] \\
T_{23} &= \frac{k_{23}}{I_5}\left[\frac{a_{11} A_1}{\omega_1}(1 - \cos \omega_1 t) + \frac{a_{12} B_1}{\omega_2}(1 - \cos \omega_2 t) \right. \\
&\qquad \left. + \frac{a_{13} C_1}{\omega_3}(1 - \cos \omega_3 t) + \frac{a_{14} D_1}{\omega_4}(1 - \cos \omega_4 t)\right] \\
T_{34} &= \frac{k_{34}}{I_5}\left[\frac{a_{21} A_1}{\omega_1}(1 - \cos \omega_1 t) + \frac{a_{22} B_1}{\omega_2}(1 - \cos \omega_2 t) \right. \\
&\qquad \left. + \frac{a_{23} C_1}{\omega_3}(1 - \cos \omega_3 t) + \frac{a_{24} D_1}{\omega_4}(1 - \cos \omega_4 t)\right] \\
T_{45} &= \frac{k_{45}}{I_5}\left[\frac{a_{31} A_1}{\omega_1}(1 - \cos \omega_1 t) + \frac{a_{32} B_1}{\omega_2}(1 - \cos \omega_2 t) \right. \\
&\qquad \left. + \frac{a_{33} C_1}{\omega_3}(1 - \cos \omega_3 t) + \frac{a_{34} D_1}{\omega_4}(1 - \cos \omega_4 t)\right]
\end{aligned} \tag{8-12}
$$

where ω_1, ω_2, ω_3, and ω_4 are the natural frequencies of the system and (a_{11}, a_{21}, a_{31}), (a_{12}, a_{22}, a_{32}), (a_{13}, a_{23}, a_{33}) and (a_{14}, a_{24}, a_{34}) are the corresponding values of

$$\left(\frac{\theta_3 - \theta_2}{\theta_2 - \theta_1}\right), \quad \left(\frac{\theta_4 - \theta_3}{\theta_2 - \theta_1}\right), \quad \left(\frac{\theta_5 - \theta_4}{\theta_2 - \theta_1}\right)$$

for each frequency respectively.

$$A_1 = \frac{L_1}{\omega_1 L_0}$$

$$B_1 = \frac{L_2}{\omega_2 L_0}$$

$$C_1 = \frac{L_3}{\omega_3 L_0}$$

$$D_1 = \frac{L_4}{\omega_4 L_0}$$

$$L_1 = a_{22}(a_{13} - a_{14}) + a_{23}(a_{14} - a_{12}) + a_{24}(a_{12} - a_{13})$$
$$L_2 = a_{21}(a_{14} - a_{13}) + a_{23}(a_{11} - a_{14}) + a_{24}(a_{13} - a_{11})$$
$$L_3 = a_{21}(a_{12} - a_{14}) + a_{22}(a_{14} - a_{11}) + a_{24}(a_{11} - a_{12})$$
$$L_4 = a_{21}(a_{13} - a_{12}) + a_{22}(a_{11} - a_{13}) + a_{23}(a_{12} - a_{11})$$
$$L_0 = a_{31}L_1 + a_{32}L_2 + a_{33}L_3 + a_{34}L_4$$

Far-Coupled Systems

The following examples illustrate the application of the determinant method to the solution of far-coupled systems.

Example 6

Lateral Vibration of Beams Carrying Concentrated Masses. In this example the lateral vibrations of a shaft carrying two masses m_1, m_2 as shown in Fig. 8-9a and b will be considered. For such system the equations of motion can be derived by expressing the dynamic deflection at y_1 in terms of two parts y_{11} and y_{12} which are the deflections due to the inertia effects of mass m_1 and m_2 respectively. We can write the time history of motion for the system when it vibrates in a natural mode as

$$y_1(t) = Y_1 \sin(\omega t + \phi)$$
$$y_2(t) = Y_2 \sin(\omega t + \phi)$$

where ω = natural frequency

ϕ = phase angle

Y_1, Y_2 = amplitudes of vibration

Using the notation a_{ij} = the deflection at location i due to a unit force at location j, we can write the equations governing the vibration of the system from consideration of beam deflection due to inertia forces, as

$$\begin{aligned} Y_1 &= a_{11}(m_1\omega^2 Y_1) + a_{12}(m_2\omega^2 Y_2) \\ Y_2 &= a_{22}(m_2\omega^2 Y_2) + a_{21}(m_1\omega^2 Y_1) \end{aligned} \tag{8-13}$$

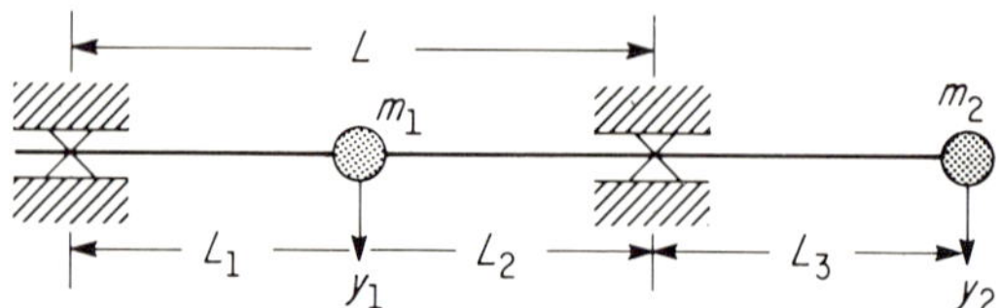

(a)

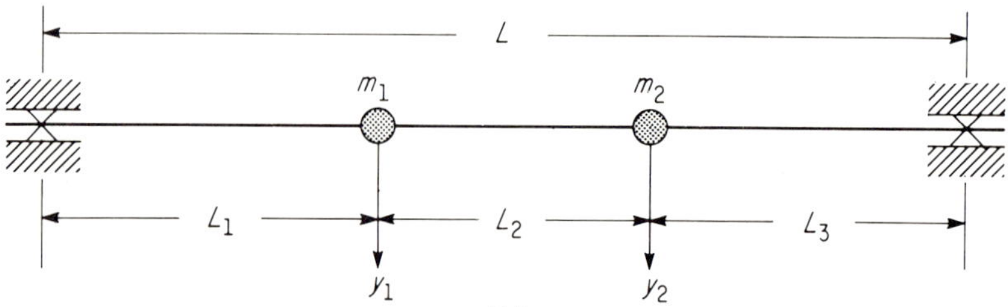

(b)

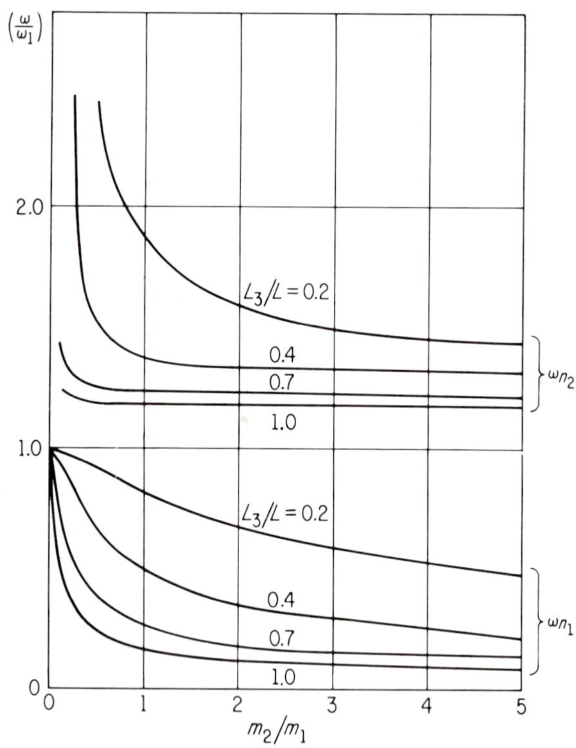

(c) Natural frequencies for system shown in case (a)

FIG. 8-9

The coefficients a_{ij} are known as the *influence coefficients* and are obtained from static deflection calculations. It should be noted here that the Maxwell reciprocal relation for linear systems gives

$$a_{ij} = a_{ji}$$

Equation 8-13 can be written in the determinant form as

$$(a_{11}m_1\omega^2 - 1)\,Y_1 + (a_{12}m_2\omega^2)\,Y_2 = 0$$
$$(a_{12}m_1\omega^2)\,Y_1 + (a_{22}m_2\omega^2 - 1)\,Y_2 = 0$$

and the frequency equation is given by

$$\begin{vmatrix} (a_{11}m_1\omega^2 - 1) & (a_{12}m_2\omega^2) \\ (a_{12}m_1\omega^2) & (a_{22}m_2\omega^2 - 1) \end{vmatrix} = 0$$

which is a quadratic equation in ω^2 resulting in two natural frequencies ω_{n_1} and ω_{n_2}. This frequency equation can be expanded as

$$\omega^4[m_1m_2(a_{11}a_{22} - a_{12}^2)] - \omega^2[a_{11}m_1 + a_{22}m_2] + 1 = 0$$

Using the notation $\omega_1^2 = \dfrac{1}{a_{11}m_1}$, where ω_1 represents the natural frequency of the system with $m_2 = 0$, the frequency equation takes the form

$$\left(\frac{\omega}{\omega_1}\right)^4\left[1 \quad \frac{a_{12}}{a_{11}}\frac{a_{12}}{a_{22}}\right] - \left(\frac{\omega}{\omega_1}\right)^2\left[\frac{m_1}{m_2}\frac{a_{11}}{a_{22}} + 1\right] + \frac{m_1}{m_2}\frac{a_{11}}{a_{22}} = 0$$

The two natural frequencies are therefore calculated from

$$\left(\frac{\omega}{\omega_1}\right)^2 = \frac{b}{2} \pm \sqrt{\left(\frac{b}{2}\right)^2 - c}$$

where

$$b = \frac{\dfrac{m_1}{m_2}\dfrac{a_{11}}{a_{22}} + 1}{\left[1 - \dfrac{a_{12}}{a_{11}}\dfrac{a_{12}}{a_{22}}\right]}$$

$$c = \frac{\dfrac{a_{11}}{a_{22}}\dfrac{m_1}{m_2}}{\left[1 - \dfrac{a_{12}}{a_{11}}\dfrac{a_{12}}{a_{22}}\right]}$$

For illustration, it will be assumed that the system shown in Fig. 8-9a has $L_1 = L_2 = L/2$.

Accordingly,

$$a_{11} = \frac{L^3}{48EI}$$

$$a_{22} = \frac{1}{EI}\left[\frac{1}{3}LL_3^2 + \frac{1}{3}L_3^3\right] = \frac{L^3}{3EI}\left[\left(\frac{L_3}{L}\right)^2 + \left(\frac{L_3}{L}\right)^3\right]$$

$$a_{12} = \frac{L^3}{16EI}\left(\frac{L_3}{L}\right) = a_{21}$$

$$\omega_1 = \sqrt{\frac{1}{a_{11}m_1}} = \sqrt{\frac{48EI}{m_1 L^3}}$$

By substituting the values of the different coefficients in the expressions for b and c and rearranging we get

$$b = \frac{1 + 16(m_2/m_1)[(L_3/L)^2 + (L_3/L)^3]}{(m_2/m_1)[7(L_3/L)^2 + 16(L_3/L)^3]}$$

and

$$c = \frac{1}{(m_2/m_1)[7(L_3/L)^2 + 16(L_3/L)^3]}$$

These expressions are used to calculate the natural frequencies for different ratios of (m_2/m_1) and (L_3/L) and the results are plotted in Fig. 8-9c.

The corresponding ratios for the amplitudes of vibration (R_1, R_2) can be obtained by substituting the values of the natural frequency in the equation

$$\frac{Y_2}{Y_1} = \frac{a_{11}m_1\omega^2 - 1}{a_{12}m_2\omega^2} = \frac{a_{12}m_1\omega^2}{a_{22}m_2\omega^2 - 1}$$

If (Y_2/Y_1) for ω_{n_1} is taken as R_1 and (Y_2/Y_1) for ω_{n_2} is taken as R_2, the general equations of motion for the system are

$$y_1(t) = A\sin(\omega_{n_1}t + \phi_1) + B\sin(\omega_{n_2}t + \phi_2) \tag{8-14a}$$

$$y_2(t) = R_1 A\sin(\omega_{n_1}t + \phi_1) + R_2 B\sin(\omega_{n_2}t + \phi_2) \tag{8-14b}$$

where A, B, ϕ_1, and ϕ_2 are calculated from the initial conditions of the motion.

Example 7

Lateral Vibrations Considering the Effect of Mass Distribution. We have assumed in the previous example that the masses carried on a shaft are concentrated at a point, and therefore the slope of the shaft at that point does not affect the result. In many mechanical systems, the dimensions of the flywheel are large compared to those of the shaft. Accordingly, due to the distribution of mass, any variation of the slope of the shaft in a vibratory motion introduces inertia couples as shown in Fig. 8-10. This effect is generally known as *gyroscopic effect* in lateral vibrations, although the word gyroscopic does not describe any true gyroscopic motion.

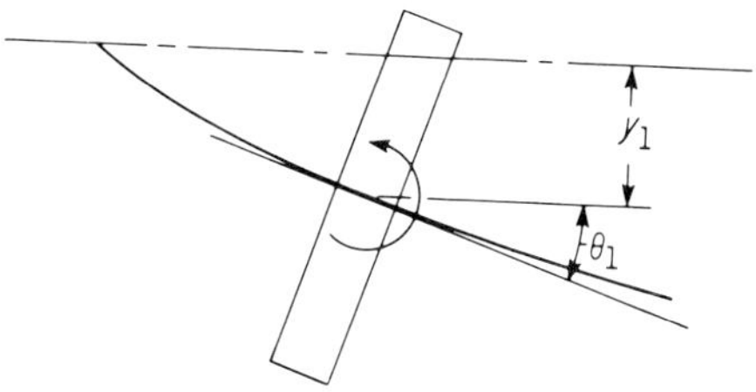

FIG. 8-10

In this example we shall study the oscillation of a cantilever beam carrying two wheels of considerable diameters as shown in Fig. 8-11.

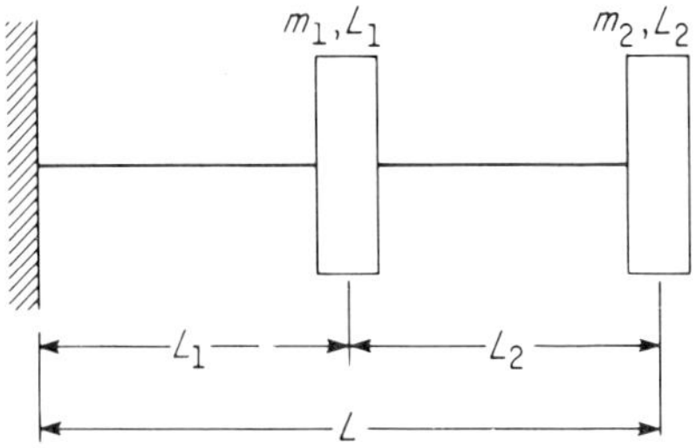

FIG. 8-11

Using the notation

m_1 = mass of flywheel at 1
m_2 = mass of flywheel at 2
I_1 = mass moment of inertia of flywheel 1 about diameter
I_2 = same as above for flywheel 2
y_1 = amplitude of dynamic displacement at 1
y_2 = amplitude of dynamic displacement at 2
θ_1 = amplitude of dynamic slope at 1
θ_2 = amplitude of dynamic slope at 2
L_1 = length of shaft between 0 and 1
L_2 = length of shaft between 1 and 2
$L = L_1 + L_2$
I = moment of inertia of shaft cross section
E = modulus of elasticity of shaft material
ω = natural frequency of vibration

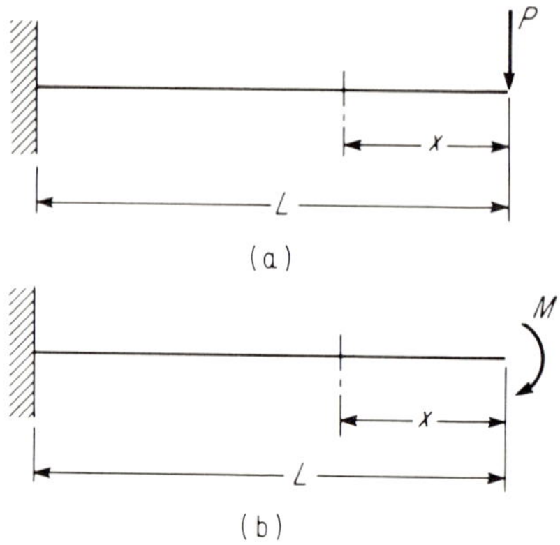

FIG. 8-12

and using the deflection relationships from Fig. 8-12a,

$$\theta_x = \frac{PL^2}{2EI}\left(1 - \left(\frac{x}{L}\right)^2\right)$$

$$y_x = \frac{PL^3}{3EI}\left[1 - \frac{3}{2}\left(\frac{x}{L}\right) + \frac{1}{2}\left(\frac{x}{L}\right)^3\right]$$

From Fig. 8-12b,

$$\theta_x = \frac{M}{EI}\left(1 - \frac{x}{L}\right)$$

$$y_x = \frac{ML^2}{2EI}\left[1 - 2\left(\frac{x}{L}\right) + \left(\frac{x}{L}\right)^2\right]$$

The equations of motion can be written as

$$y_1 = \frac{m_1\omega^2 y_1 L_1^3}{3EI} - \frac{I_1\omega^2\theta_1 L_1^2}{2EI} + \frac{m_2\omega^2 y_2 L^3}{3EI}\left[1 - \frac{3}{2}\left(\frac{L_2}{L}\right) + \left(\frac{L_2}{L}\right)^3\right]$$
$$- \frac{I_2\omega^2\theta_2 L^2}{2EI}\left[1 - 2\left(\frac{L_2}{L}\right) + \left(\frac{L_2}{L}\right)^2\right] \tag{8-12}$$

$$\theta_1 = \frac{m_1\omega^2 y_1 L_1^2}{2EI} - \frac{I_1\omega^2\theta_1 L_1}{EI} + \frac{m_2\omega^2 y_2 L^2}{2EI}\left[1 - \left(\frac{L_2}{L}\right)^2\right]$$
$$- \frac{I_2\omega^2\theta_2 L}{EI}\left[1 - \left(\frac{L_2}{L}\right)\right] \tag{8-13}$$

$$y_2 = \frac{m_2\omega^2 y_2 L^3}{3EI} - \frac{I_2\omega^2\theta_2 L^2}{2EI} + m_1\omega^2 y_1\left(\frac{L_1^3}{3EI} + \frac{L_1^2}{2EI}L_2\right)$$
$$- I_1\omega^2\theta_1\left(\frac{L_1^2}{2EI} + \frac{L_1 L_2}{EI}\right) \tag{8-14}$$

$$\theta_2 = \frac{m_2 \omega^2 y_2 L^2}{2EI} - \frac{I_2 \omega^2 \theta_2 L}{EI} + \frac{m_1 \omega^2 y_1 L_1^2}{2EI} - \frac{I_1 \omega^2 \theta_1 L_1}{EI} \tag{8-15}$$

Equations 8-12, 8-13, 8-14, and 8-15 have the form

$$\begin{aligned} A_1 y_1 + B_1 y_2 + C_1 \theta_1 + D_1 \theta_2 &= 0 \\ A_2 y_1 + B_2 y_2 + C_2 \theta_1 + D_2 \theta_2 &= 0 \\ A_3 y_1 + B_3 y_2 + C_3 \theta_1 + D_3 \theta_2 &= 0 \\ A_4 y_1 + B_4 y_2 + C_4 \theta_1 + D_4 \theta_2 &= 0 \end{aligned}$$

where y_1, y_2, θ_1, and θ_2 are the four independent coordinates of the system, and A_n, B_n, C_n, and D_n are functions of ω^2. The natural frequency for each of the four modes of vibration can be obtained from the solution of the frequency equation

$$\begin{vmatrix} A_1 & B_1 & C_1 & D_1 \\ A_2 & B_2 & C_2 & D_2 \\ A_3 & B_3 & C_3 & D_3 \\ A_4 & B_4 & C_4 & D_4 \end{vmatrix} = 0$$

where

$$A_1 = \frac{m_1 \omega^2 (L_1)^3}{3EI} - 1$$

$$A_2 = \frac{m_1 \omega^2 (L_1)^3}{3EI} \left(1 + \frac{3}{2} \frac{L_2}{L_1}\right)$$

$$A_3 = \frac{m_1 \omega^2 (L_1)^2}{2EI}$$

$$A_4 = \frac{m_1 \omega^2 (L_1)^2}{2EI}$$

$$B_1 = \frac{m_2 \omega^2 L^3}{3EI} \left[1 - \frac{3}{2}\left(\frac{L_2}{L}\right) + \frac{1}{2}\left(\frac{L_2}{L}\right)^3\right]$$

$$B_2 = \frac{m_2 \omega^2 L^3}{3EI} - 1$$

$$B_3 = \frac{m_2 \omega^2 L^2}{2EI} \left[1 - \left(\frac{L_2}{L}\right)^2\right]$$

$$B_4 = \frac{m_2 \omega^2 L^2}{2EI}$$

$$C_1 = \frac{I_1 \omega^2 (L_1)^2}{2EI}$$

$$C_2 = -\left[\frac{I_1 \omega^2 (L_1)^2}{2EI} \left(1 + 2\frac{L_2}{L_1}\right)\right]$$

$$C_3 = -\left[\frac{I_1\omega^2 L_1}{EI} + 1\right]$$

$$C_4 = -\left[\frac{I_1\omega^2 L_1}{EI}\right]$$

$$D_1 = -\left[\frac{I_2\omega^2 L^2}{2EI}\left(1 - 2\left(\frac{L_2}{L}\right) + \left(\frac{L_2}{L}\right)^2\right)\right]$$

$$D_2 = -\left[\frac{I_2\omega^2 L^2}{2EI}\right]$$

$$D_3 = -\left[\frac{I_2\omega^2 L}{EI}\left(1 - \frac{L_2}{L}\right)\right]$$

$$D_4 = -\left[\frac{I_2\omega^2 L}{EI} + 1\right]$$

and the general solution can therefore be written as

$$\begin{aligned} y_1(t) &= A\sin(\omega_{n_1}t + \phi_1) + B\sin(\omega_{n_2}t + \phi_2) \\ &\quad + C\sin(\omega_{n_3}t + \phi_3) + D\sin(\omega_{n_4}t + \phi_4) \end{aligned} \tag{8-16}$$

$$\begin{aligned} y_2(t) &= R_1 A\sin(\omega_{n_1}t + \phi_1) + R_2 B\sin(\omega_{n_2}t + \phi_2) \\ &\quad + R_3 C\sin(\omega_{n_3}t + \phi_3) + R_4 D\sin(\omega_{n_4}t + \phi_4) \end{aligned} \tag{8-17}$$

$$\theta_1(t) = R_1' A\sin(\omega_{n_1}t + \phi_1) + R_2' B\sin(\omega_{n_2}t + \phi_2) + R_3', \ldots, \text{etc.} \tag{8-18}$$

$$\theta_2(t) = R_1'' A\sin(\omega_{n_1}t + \phi_1) + R_2'' B\sin(\omega_{n_2}t + \phi_2) + R_3'', \ldots, \text{etc.} \tag{8-19}$$

where (R_1, R_1', R_1''), (R_2, R_2', R_2''), (R_3, R_3', R_3''), (R_4, R_4', R_4'') are amplitude ratios y_2/y_1, θ_1/y_1, θ_2/y_1 corresponding to ω_{n_1}, ω_{n_2}, ω_{n_3}, ω_{n_4} respectively. These ratios are obtained by the simultaneous solution of Eqs. 8-12, 8-13, 8-14, and 8-15. A, B, C, D, ϕ_1, ϕ_2, ϕ_3, and ϕ_4 are integration constants which are calculated by substituting the initial conditions of displacement and velocity in Eqs. 8-16, 8-17, 8-18, and 8-19 and their time derivatives.

8-3. ORTHOGONALITY OF PRINCIPAL MODES

Principal Modes and Normal Modes

We have shown in the previous examples that the expressions for the coordinates of the motion in a system with more than a single degree of freedom are no longer simple harmonic. Each coordinate is described by superposition of harmonic vibrations corresponding to the different natural frequencies of the motion. Taking for example a system with three degrees of freedom (q_1, q_2, q_3), the solution of the simultaneous differential equations of motion will yield the following results:

(a) Three natural frequencies of vibration ω_{n_1}, ω_{n_2}, and ω_{n_3} which are known as the *eigenvalues*.
(b) Values for the amplitude ratios (q_2/q_1), (q_3/q_1) for each of the three natural frequencies. These ratios (R_1, R_2), (R_1', R_2') and (R_1'', R_2'') are called the *eigenfunctions* for the three modes of vibration respectively.

The general equations for the three coordinates of the motion can therefore be written as

$$q_1 = A \sin(\omega_{n_1}t + \phi_1) + B \sin(\omega_{n_2}t + \phi_2) + C \sin(\omega_{n_3}t + \phi_3)$$
$$q_2 = R_1 A \sin(\omega_{n_1}t + \phi_1) + R_1' B \sin(\omega_{n_2}t + \phi_2) + R_1'' C \sin(\omega_{n_3}t + \phi_3)$$
$$q_3 = R_2 A \sin(\omega_{n_1}t + \phi_1) + R_2' B \sin(\omega_{n_2}t + \phi_2) + R_2'' C \sin(\omega_{n_3}t + \phi_3)$$

The constants A, B, and C represent the amplitudes of the different harmonic components of q_1. The constants ϕ_1, ϕ_2, and ϕ_3 represent the phase angles for the different modes of vibration. All six constants are to be evaluated from the initial conditions of the motion.

It was explained in Example 1 that only under certain specified conditions will all the coordinates of the system be described as simple harmonic functions with the same frequency and phase angle. Such motion is called a *principal mode* of vibration. For the system with coordinates q_1, q_2, q_3, these principal modes are given by

First principal mode

$$q_1 = A \sin(\omega_{n_1}t + \phi_1)$$
$$q_2 = R_1 A \sin(\omega_{n_1}t + \phi_1)$$
$$q_3 = R_2 A \sin(\omega_{n_1}t + \phi_1)$$

Second principal mode

$$q_1 = B \sin(\omega_{n_2}t + \phi_2)$$
$$q_2 = R_1' B \sin(\omega_{n_2}t + \phi_2)$$
$$q_3 = R_2' B \sin(\omega_{n_2}t + \phi_2)$$

Third principal mode

$$q_1 = C \sin(\omega_{n_3}t + \phi_3)$$
$$q_2 = R_1'' C \sin(\omega_{n_3}t + \phi_3)$$
$$q_3 = R_2'' C \sin(\omega_{n_3}t + \phi_3)$$

In a principal mode, when one of the amplitudes is made equal to 1, the motion is then referred to as a *normal mode*. The following equations therefore represent the three normal modes respectively:

First normal mode

$$q_1 = \sin(\omega_{n_1}t + \phi_1)$$
$$q_2 = R_1 \sin(\omega_{n_1}t + \phi_1)$$
$$q_3 = R_2 \sin(\omega_{n_1}t + \phi_1)$$

Second normal mode

$$q_1 = \sin(\omega_{n_2}t + \phi_2)$$
$$q_2 = R_1' \sin(\omega_{n_2}t + \phi_2)$$
$$q_3 = R_2' \sin(\omega_{n_2}t + \phi_2)$$

Third normal mode

$$q_1 = \sin(\omega_{n_3}t + \phi_3)$$
$$q_2 = R_1'' \sin(\omega_{n_3}t + \phi_3)$$
$$q_3 = R_2'' \sin(\omega_{n_3}t + \phi_3)$$

Orthogonality of Principal Modes

The oscillation of a particle in space, no matter how complicated it may appear, results from the superposition of three linear oscillations x, y, z in three perpendicular directions. In other words, if each mode is represented by a vector, the three modes will be orthogonal in three-dimensional space. This physical orthogonality of the vibrational modes of a particle leads to the concept of orthogonality of normal modes which is easily visualized in the case of particle vibration but becomes rather difficult to represent physically for systems with higher degrees of freedom. The orthogonality of the normal modes nevertheless exists for all vibrating systems and can be represented mathematically. For systems with a finite number of degrees of freedom n, the orthogonality is represented by the equation

$$\sum_{i=1}^{n} m_i q_i^{(r)} q_i^{(s)} = 0 \qquad r \neq s \tag{8-20}$$

where $q_i^{(r)}$, $q_i^{(s)}$ are the amplitudes of mass i corresponding respectively to the normal modes r and s. This relationship is known as the *orthogonality equation* for systems with n degrees of freedom and is very useful in analyzing such systems. This is detailed later in Chapter 9.

Applying this orthogonality relationship to the two-degrees-of-freedom system of Example 2, we find

$$q_1^{(1)} = 1$$
$$q_1^{(2)} = 1$$
$$q_2^{(1)} = R_1 q_1^{(1)} = 1$$
$$q_2^{(2)} = R_2 q_1^{(2)} = -1$$

Substituting in the left-hand side of Eq. 8-20 we get

$$m(1)(1) + m(1)(-1) = 0$$

which satisfies the equation

8-4. FORCED VIBRATIONS

In the previous section we have applied the determinant method of solution to the analysis of the free vibration of systems with several degrees of freedom. It has been shown that for such systems the general solution is obtained by the superposition of principal-mode oscillations. When such systems are subjected to a harmonic excitation, the solution, as in the case of single-degree-of-freedom systems, will be composed of two parts. The first part is the transient solution which is a free vibration determined by the initial conditions. The second part is a particular solution resulting from the forcing function. Due to the presence of some degree of damping in most physical systems, the transient part of the solution decays and eventually disappears. On the other hand, the second part will persist and therefore represents the steady-state motion.

The following are some cases of systems subjected to harmonic excitations.

Example 8

Harmonic Force Excitation. Studying for example the two mass system shown in Fig. 8-13 we find that in addition to the free vibration obtained from the homogeneous equations (Example 1); we have a

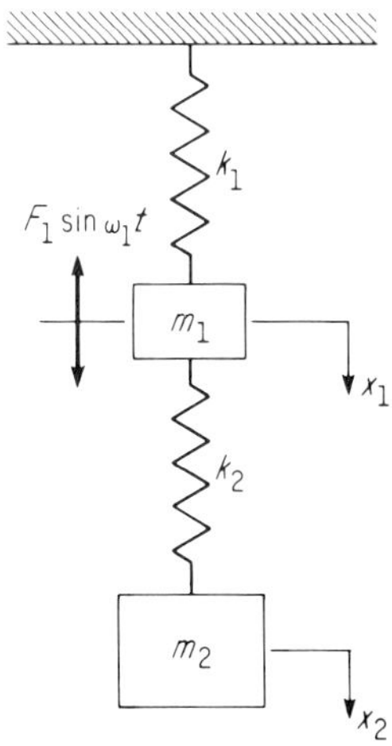

FIG. 8-13

particular integral (or a steady-state solution) which can be assumed as

$$\begin{aligned} x_1 &= X_{11} \sin \omega_1 t \\ x_2 &= X_{21} \sin \omega_1 t \end{aligned} \tag{8-21}$$

This represents a vibration having the same frequency ω_1 as the exciting force. X_{11}, X_{21} are the steady-state amplitudes of m_1, m_2 respectively. The equations of motion in this example are

$$m_1\ddot{x}_1 + k_1 x_1 + k_2(x_1 - x_2) = F_1 \sin \omega_1 t$$

$$m_2\ddot{x}_2 + k_2(x_2 - x_1) = 0$$

and by substituting the particular solution and rearranging, we have

$$[k_2 + k_1 - m_1\omega_1^2]X_{11} - k_2 X_{21} = F_1$$

$$-k_2 X_{11} + [k_2 - m_2\omega_1^2]X_{21} = 0$$

from which we obtain

$$X_{11} = \frac{\begin{vmatrix} F_1 & -k_2 \\ 0 & [k_2 - m_2\omega_1^2] \end{vmatrix}}{\begin{vmatrix} [k_2 + k_1 - m_1\omega_1^2] & -k_2 \\ -k_2 & [k_2 - m_2\omega_1^2] \end{vmatrix}}$$

$$= \frac{F_1[k_2 - m_2\omega_1^2]}{[k_2 + k_1 - m_1\omega_1^2][k_2 - m_2\omega_1^2] - k_2^2}$$

Dividing the numerator and denominator by $k_1 k_2$ and using the notation

$$\frac{k_1}{m_1} = \omega_{01}^2, \qquad \frac{k_2}{m_2} = \omega_{02}^2$$

and

$$\frac{m_2}{m_1} = \mu$$

we obtain

$$X_{11} = \frac{(F_1/k_1)\left[1 - \left(\frac{\omega_1}{\omega_{02}}\right)^2\right]}{\left[1 + \frac{k_2}{k_1} - \left(\frac{\omega_1}{\omega_{01}}\right)^2\right]\left[1 - \left(\frac{\omega_1}{\omega_{02}}\right)^2\right] - \frac{k_2}{k_1}}$$

$$= \frac{(F_1/k_1)\left[1 - \left(\frac{\omega_1}{\omega_{02}}\right)^2\right]}{1 - \left[\left(\frac{\omega_1}{\omega_{01}}\right)^2 + \mu\left(\frac{\omega_1}{\omega_{01}}\right)^2 + \left(\frac{\omega_1}{\omega_{02}}\right)^2\right] + \left(\frac{\omega_1}{\omega_{01}}\right)^2\left(\frac{\omega_1}{\omega_{02}}\right)^2}$$

$$= \frac{(F_1/k_1)\left[1 - \left(\frac{\omega_1}{\omega_{02}}\right)^2\right]}{1 - (1 + \mu)\left(\frac{\omega_1}{\omega_{01}}\right)^2 + \left(\frac{\omega_1}{\omega_{02}}\right)^2\left[\left(\frac{\omega_1}{\omega_{01}}\right)^2 - 1\right]} \tag{8-22}$$

Similarly,

$$X_{21} = \frac{\begin{vmatrix} k_2 + k_1 - m_1\omega_1^2 & F_1 \\ -k_2 & 0 \end{vmatrix}}{\begin{vmatrix} k_1 + k_2 - m_1\omega_1^2 & -k_2 \\ -k_2 & k_2 - m_2\omega_1^2 \end{vmatrix}}$$

$$= \frac{F_1/k_1}{1 - (1+\mu)\left(\dfrac{\omega_1}{\omega_{01}}\right)^2 + \left(\dfrac{\omega_1}{\omega_{02}}\right)^2\left[\left(\dfrac{\omega_1}{\omega_{01}}\right)^2 - 1\right]} \tag{8-23}$$

and the general solution combines the particular integral of Eq. 8-21 and the free vibration given in Eq. 8-1. It should be noted that X_{11} and X_{21} will reach infinitely large values when

$$1 - (1+\mu)\left(\frac{\omega_1}{\omega_{01}}\right)^2 + \left(\frac{\omega_1}{\omega_{02}}\right)^2\left[\left(\frac{\omega_1}{\omega_{01}}\right)^2 - 1\right] = 0 \tag{8-24}$$

which gives the resonant frequencies of the system.

If the sinusoidal force is acting on mass m_2 (Fig. 8-14), the particu-

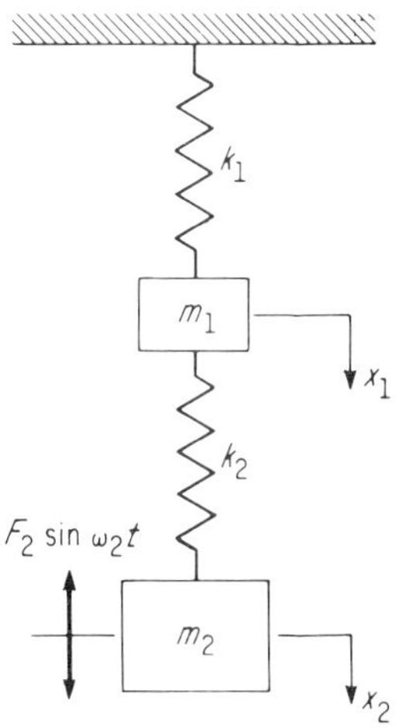

FIG. 8-14

lar solution will be

$$x_1 = X_{12} \sin \omega_2 t$$

$$x_2 = X_{22} \sin \omega_2 t$$

and following the same procedure as before, we get

$$X_{12} = \frac{\begin{vmatrix} 0 & -k_2 \\ F_2 & k_2 - m_2\omega_2^2 \end{vmatrix}}{\begin{vmatrix} k_2 + k_1 - m_1\omega_2^2 & -k_2 \\ -k_2 & k_2 - m_2\omega_2^2 \end{vmatrix}}$$

$$= \frac{F_2/k_1}{1 - (1 + \mu)\left(\dfrac{\omega_2}{\omega_{01}}\right)^2 + \left(\dfrac{\omega_2}{\omega_{02}}\right)^2\left[\left(\dfrac{\omega_2}{\omega_{01}}\right)^2 - 1\right]}$$

$$X_{22} = \frac{\begin{vmatrix} k_2 + k_1 - m_1\omega_2^2 & 0 \\ -k_2 & F_2 \end{vmatrix}}{\begin{vmatrix} k_2 + k_1 - m_1\omega_2^2 & -k_2 \\ -k_2 & k_2 - m_2\omega_2^2 \end{vmatrix}}$$

$$= \frac{(F_2/k_1)\left[1 + \dfrac{k_1}{k_2} - \left(\dfrac{\omega_2}{\omega_{02}}\right)^2\right]}{1 + (1 + \mu)\left(\dfrac{\omega_2}{\omega_{01}}\right)^2 + \left(\dfrac{\omega_2}{\omega_{02}}\right)^2\left[\left(\dfrac{\omega_2}{\omega_{01}}\right)^2 - 1\right]}$$

In general, for a linear system with n degrees of freedom (Fig. 8-15) particular solutions for $x_1, x_2, \ldots, x_n$ are obtained for each force excitation independently. Since superposition applies in linear systems, the summation of these solutions at any mass gives the resultant forced vibration of that mass. For example, the forced vibration $x_3(t)$ at mass 3 due to sinusoidal forces on all masses is given by

$$x_3(t) = \sum_{i=1}^{i=n} X_{3i} \sin \omega_i t$$

where

$$X_{31} = \frac{\begin{vmatrix} k_1 + k_2 - m_1\omega_1^2 & -k_2 & F_1 & 0 \ldots \\ -k_2 & k_2 + k_3 - m_2\omega_1^2 & 0 & 0 \ldots \\ 0 & -k_3 & 0 & -k_4 \ldots \\ 0 & 0 & 0 & k_4 + k_5 - m_4\omega_1^2 \\ 0 & 0 & 0 & \ldots \end{vmatrix}}{\begin{vmatrix} k_1 + k_2 - m_1\omega_1^2 & -k_2 & 0 & 0 \ldots \\ -k_2 & k_2 + k_3 - m_2\omega_1^2 & -k_3 & 0 \ldots \\ 0 & -k_3 & k_3 + k_4 - m_3\omega_1 & -k_4 \ldots \\ 0 & 0 & -k_4 & k_4 + k_5 - m_4\omega_1^2 \\ 0 & 0 & \ldots & \ldots \end{vmatrix}}$$

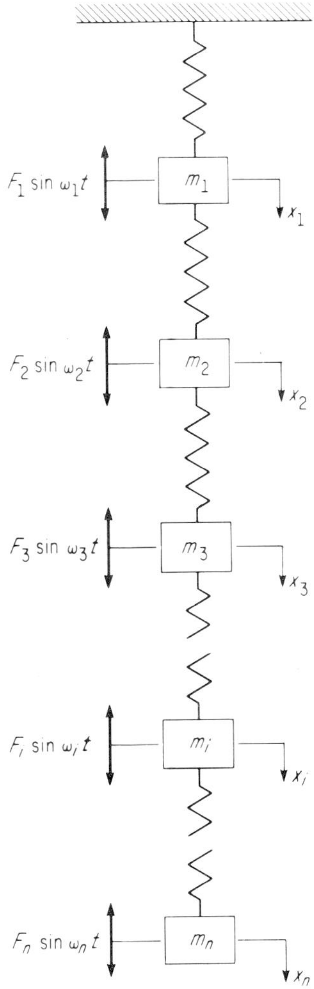

FIG. 8-15

and

$$X_{32} = \frac{\begin{vmatrix} k_1 + k_2 - m_1\omega_2^2 & -k_2 & 0 & 0 \ldots \\ -k_2 & k_2 + k_3 - m_2\omega_2^2 & F_2 & 0 \ldots \\ 0 & -k_3 & 0 & -k_4 \ \ldots \\ 0 & 0 & 0 & k_4 + k_5 - m_4\omega_2^2 \\ 0 & 0 & 0 & \ldots \end{vmatrix}}{|\text{ Determinant of coefficients }|}$$

and so on. The resonant frequencies which are also the natural frequencies of the system are the values of ω which make the determinant of coefficients vanish.

Example 9

Harmonic Foundation Excitation. The equation of motion for a two-mass system subjected to platform excitation (Fig. 8-16) can be written as

$$m_1\ddot{x}_1 + k_1(x_1 - x_f) + k_2(x_1 - x_2) = 0$$
$$m_2\ddot{x}_2 + k_2(x_2 - x_1) = 0$$

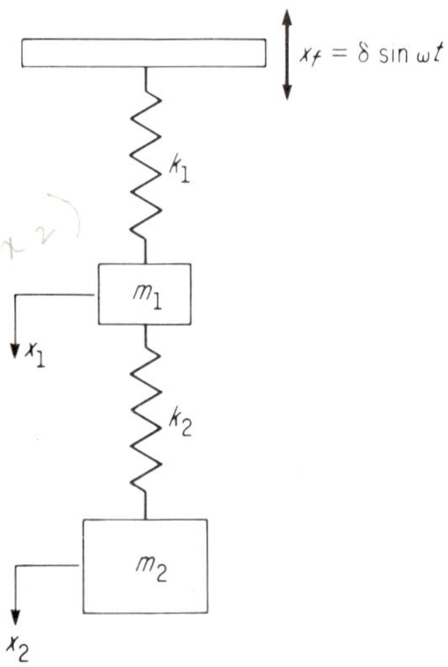

FIG. 8-16

Taking harmonic functions for the particular solution

$$x_1 = X_{1f} \sin \omega t$$
$$x_2 = X_{2f} \sin \omega t$$

and by substitution in the equations of motion and rearranging,

$$(k_1 + k_2 - m_1\omega^2)X_{1f} - k_2 X_{2f} = k_1 \delta$$
$$-k_2 X_{1f} + k_2 - m_2\omega^2 X_{2f} = 0$$

we obtain a solution similar to Eqs. 8-22 and 8-23 as

$$X_{1f} = \frac{\delta\left[1 - \left(\dfrac{\omega}{\omega_{02}}\right)^2\right]}{1 - (1 + \mu)\left(\dfrac{\omega}{\omega_{01}}\right)^2 + \left(\dfrac{\omega}{\omega_{02}}\right)^2\left[\left(\dfrac{\omega}{\omega_{02}}\right)^2 - 1\right]} \tag{8-25}$$

and

$$X_{2f} = \frac{\delta}{1 - (1 + \mu)\left(\frac{\omega}{\omega_{01}}\right)^2 + \left(\frac{\omega}{\omega_{02}}\right)^2 \left[\left(\frac{\omega}{\omega_{01}}\right)^2 - 1\right]} \tag{8-26}$$

and resonance occurs when ω makes the denominator equal zero.

Example 10

Unbalance Excitation. If the mass m_1 is subjected to an excitation due to an eccentric mass m rotating at a frequency ω_1, the equations of motion for the two-mass system shown in Fig. 8-17 become

$$m_1\ddot{x}_1 + k_1x_1 + k_2(x_1 - x_2) = (m\omega_1^2 e)\sin\omega_1 t$$
$$m_2\ddot{x}_2 + k_2(x_2 - x_1) = 0$$

Using the same notation as in Example 1, the solution takes the form

$$x_1 = X_1 \sin\omega_1 t$$
$$x_2 = X_2 \sin\omega_1 t$$

where

$$X_1 = \frac{\left[1 - \left(\frac{\omega_1}{\omega_{02}}\right)^2\right]\frac{m}{m_1}e\left(\frac{\omega_1}{\omega_{01}}\right)^2}{\left[1 + \frac{k_2}{k_1} - \left(\frac{\omega_1}{\omega_{01}}\right)^2\right]\left[1 - \left(\frac{\omega_1}{\omega_{02}}\right)^2\right] - \frac{k_2}{k_1}}$$

$$= \frac{\left[1 - \left(\frac{\omega_1}{\omega_{02}}\right)^2\right]\frac{m}{m_1}e\left(\frac{\omega_1}{\omega_{01}}\right)^2}{1 - (1 + \mu)\left(\frac{\omega_1}{\omega_{01}}\right)^2 + \left(\frac{\omega_1}{\omega_{02}}\right)^2\left[\frac{\omega_1}{\omega_{01}}^2 - 1\right]} \tag{8-27}$$

and

$$X_2 = \frac{\frac{m}{m_1}e\left(\frac{\omega_1}{\omega_{01}}\right)^2}{\left[1 + \frac{k_2}{k_1} - \left(\frac{\omega_1}{\omega_{01}}\right)^2\right]\left[1 - \left(\frac{\omega_1}{\omega_{02}}\right)^2\right] - \frac{k_2}{k_1}}$$

$$= \frac{\frac{m}{m_1}e\left(\frac{\omega_1}{\omega_{01}}\right)^2}{1 - (1 + \mu)\left(\frac{\omega_1}{\omega_{01}}\right)^2 + \left(\frac{\omega_1}{\omega_{02}}\right)^2\left[\left(\frac{\omega_1}{\omega_{01}}\right)^2 - 1\right]} \tag{8-28}$$

and resonance occurs when ω_1 makes the denominator equal zero.

If the unbalance is acting on the second mass at a frequency ω_2, the

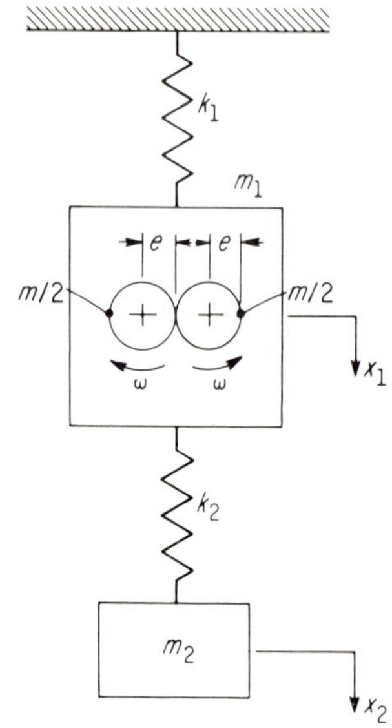

FIG. 8-17

equations of motion will be

$$m_1\ddot{x}_1 + k_1x_1 + k_2(x_1 - x_2) = 0$$
$$m_2\ddot{x}_2 + k_2(x_2 - x_1) = m\omega_2^2 e \sin \omega_2 t$$

and the solution is given by

$$x_1 = X_1 \sin \omega_2 t$$
$$x_2 = X_2 \sin \omega_2 t$$

where

$$X_1 = \frac{\dfrac{k_2}{k_1}\dfrac{m}{m_2}\left(\dfrac{\omega_2}{\omega_{02}}\right)^2 e}{1 - (1+\mu)\left(\dfrac{\omega_2}{\omega_{01}}\right)^2 + \left(\dfrac{\omega_2}{\omega_{02}}\right)^2\left[\left(\dfrac{\omega_2}{\omega_{01}}\right)^2 - 1\right]}$$

and

$$X_2 = \frac{\left[1 + \dfrac{k_2}{k_1} - \left(\dfrac{\omega_2}{\omega_{01}}\right)^2\right]\dfrac{m}{m_2}\left(\dfrac{\omega_2}{\omega_{02}}\right)^2 e}{1 - (1+\mu)\left(\dfrac{\omega_2}{\omega_{01}}\right)^2 + \left(\dfrac{\omega_2}{\omega_{02}}\right)^2\left[\left(\dfrac{\omega_2}{\omega_{01}}\right)^2 - 1\right]}$$

Example 11

The Undamped Dynamic Absorber. An important practical conclusion can be drawn from the analysis of the systems of Figs. 8-13, 8-16, and 8-17. In all these cases, if the secondary spring-mass system is selected in such a way that its natural frequency ω_{02} is equal to the exciting frequency ω_1 then the vibration of main mass will vanish. This can be easily deduced from Eqs. 8-22, 8-25, 8-27 respectively. This result is independent of the natural frequency of the main system. It can also be seen in all three cases that if the secondary spring-mass system did not exist, the displacement amplitude X_1 of mass m_1 would become infinitely large as ω_1 approaches ω_{01} which is the natural frequency for this single-degree-of-freedom system. However, for the same condition, if the secondary system is chosen such that $k_1/m_1 = k_2/m_2$ (which is equivalent to $\omega_{01} = \omega_{02}$), the amplitude of the mass m_1 will be zero at $\omega_1 = \omega_{01}$. In other words, if the mass m_1 represents an elastically supported structure or machine which is in resonance with a sinusoidal excitation, its vibration can be made to vanish by introducing a secondary system having the same natural frequency as the main system. This will be theoretically valid independent of the size of the secondary mass or the spring as long as their ratio remains the same as that of the main system. The secondary system is generally known as the *dynamic damper*. The size of the dynamic damper is determined by the single parameter μ which in this case determines the mass ratio as well as the spring ratio.

$$\mu = \frac{m_2}{m_1} = \frac{k_2}{k_1}$$

Two factors, however, have to be considered in choosing the value of μ:

(a) Since μ controls the two resonant frequencies of the two mass system by virtue of Eq. 8-24, the value of μ will be important in determining the range of speeds in which the absorber remains effective. When Eq. 8-24 is rewritten for $\omega_{01} = \omega_{02}$ it takes the form

$$1 - (1 + \mu)\left(\frac{\omega_1}{\omega_{01}}\right)^2 + \left(\frac{\omega_1}{\omega_{01}}\right)^2 \left[\left(\frac{\omega_1}{\omega_{01}}\right)^2 - 1\right] = 0$$

$$1 - (2 + \mu)\left(\frac{\omega_1}{\omega_{01}}\right)^2 + \left(\frac{\omega_1}{\omega_{01}}\right)^4 = 0 \tag{8-29}$$

The graph of (ω_1/ω_{01}) versus μ is plotted in Fig. 8-18 which shows that the increase in absorber mass spreads the range between the two resonant frequencies.

(b) It can also be seen from Eq. 8-23 that μ controls the amplitude of vibration of the secondary system. When $\omega_1 = \omega_{01} = \omega_{02}$, Eqs. 8-23, 8-26, 8-28 respectively take the form

$$|X_{21}| = \frac{F_1}{\mu k_1} = \frac{F_1}{k_2} \qquad \left(\text{since } \mu = \frac{k_2}{k_1}\right)$$

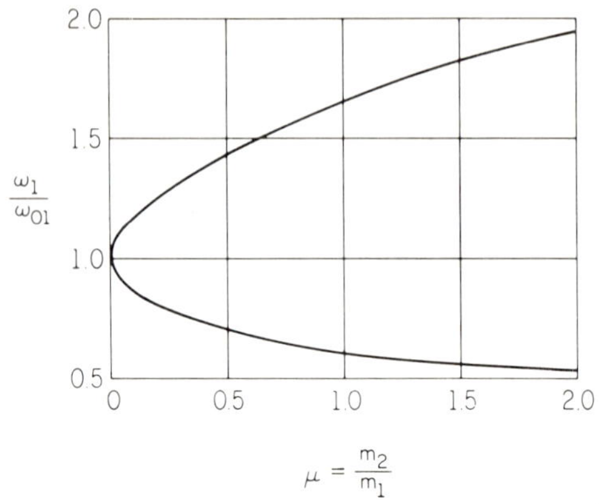

FIG. 8-18

$$|X_{2f}| = \frac{\delta}{\mu}$$

$$|X_2| = \frac{(m/m_1)e}{\mu}$$

which all indicate that the larger μ is, the lower will be the amplitude of the secondary system. The previous conditions indicate that a dynamic absorber can be designed to limit the resonant amplitude of the main mass within a specified range by means of a secondary system chosen such that

$$\frac{k_2}{m_2} = \frac{k_1}{m_1}$$

The size of the secondary system is determined by the parameter

$$\mu = \frac{m_2}{m_1} = \frac{k_2}{k_1}$$

which satisfies the frequency range and strength consideration. This may prove impractical when the exciting frequency is expected over a wide range, as in the case of variable-speed drives. The introduction of damping the secondary system renders a feasible solution for these cases.

Example 12

The Damped Dynamic Absorber. Complete response of the system with damped absorber, Fig. 8-19, has been analyzed by J. Ormondroyd and J. P. Den Hartog (Ref. 4). A typical frequency response of the main mass is illustrated in Fig. 8-20 for a particular value of μ and different damping coefficients. It shows that independent of the magnitude of damping, there are two points common to all response curves. These points can be obtained from the intersection of the undamped response

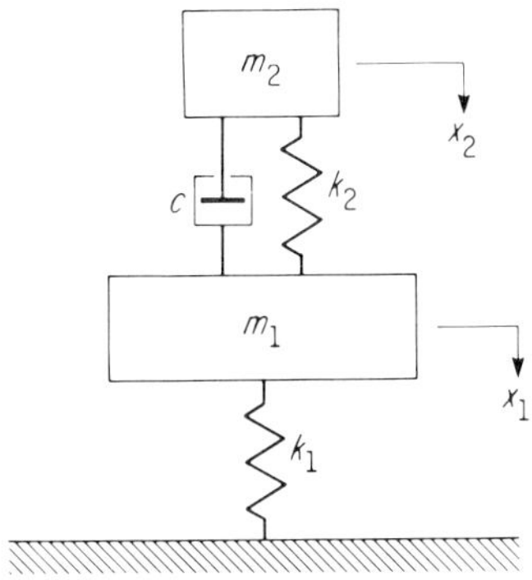

FIG. 8-19

with the response with infinite damping. The latter corresponds to a single-mass system with mass equal to the sum of the masses $(m_1 + m_2)$ and with stiffness k_1. A limiting condition, therefore, exists when these two points are on the same horizontal line. This condition is satisfied when

$$\frac{\sqrt{k_2/m_2}}{\sqrt{k_1/m_1}} = \frac{1}{1 + \mu} \tag{8-30}$$

The amplitude of vibration (X_1) of mass m_1 for these two points is given by

$$(X_1)_{a \text{ or } b} = \frac{F_1}{k_1}\sqrt{1 + \frac{2}{\mu}} \tag{8-31}$$

The optimal damping coefficient for which the response curves have a horizontal tangent close to a and b (and consequently give minimum

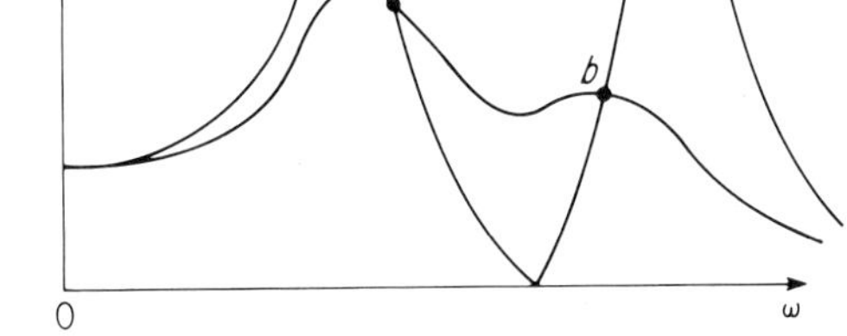

FIG. 8-20

amplitudes over the frequency spectrum) is found from

$$c^2 = \frac{3\mu^2 k_1 m_2}{2(1+\mu)^3} \tag{8-32}$$

A special case of the damped absorber is the Lanchester damper, (Fig. 8-21a). In this type $k_2 = 0$ and the damper c is the only connection

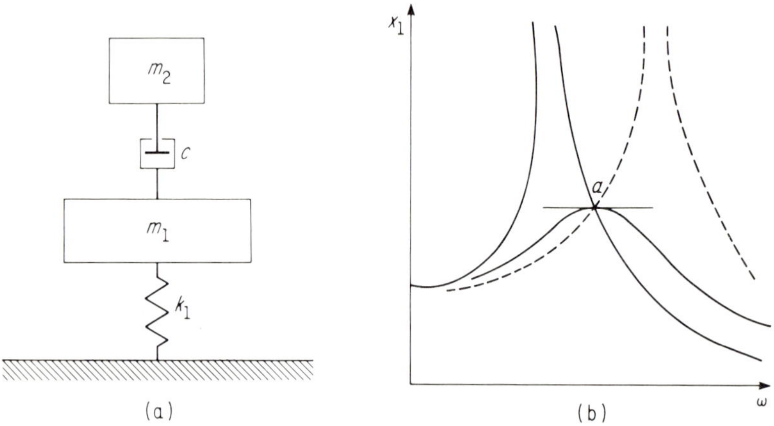

FIG. 8-21

between the main and secondary masses. The optimal condition for this damper is obtained as in the previous case from the intersection of the response curves having $c = 0$ and $c = \infty$, (Fig. 8-21b). These are equivalent to the response for the two undamped single-mass systems with stiffness k_1 and mass m_1 or $(m_1 + m_2)$ respectively. The maximum amplitude at the point of intersection a is

$$(X_1)_a = 1 + \frac{2}{\mu} \tag{8-33}$$

where $\mu = m_2/m_1$, and the optimal damping coefficient is

$$c = \frac{2\, m_2 k_1\, (\mu)}{(2+\mu)(1+\mu)} \tag{8-34}$$

Another type of absorber which has received attention in recent years is the *impact type absorber* shown in Fig. 8-22. In this case the motion of the main mass is damped by the impact of a mass m_2 which is free to move within a gap d.

Several studies are available in the literature which deal with the free vibration, forced vibration and stability of this absorber. (Refs. 7–16.)

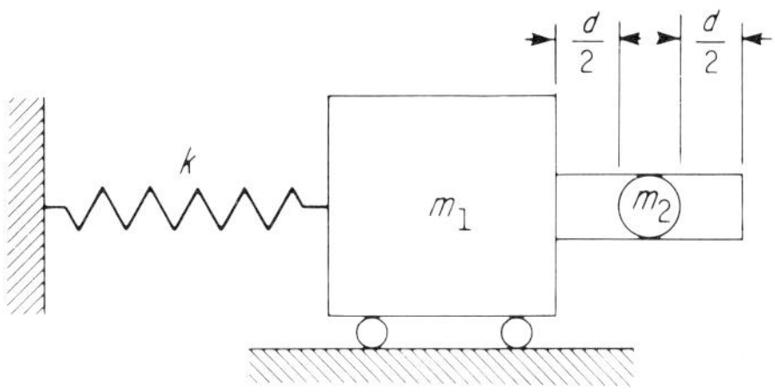

FIG. 8-22

8-5. LAPLACE TRANSFORMS

The use of Laplace transforms has been previously illustrated in solution of single-degree-of-freedom problems. The method can also be utilized for linear systems with multiple degrees of freedom, as will be shown in the case of the two-mass system given in Fig. 8-23. The

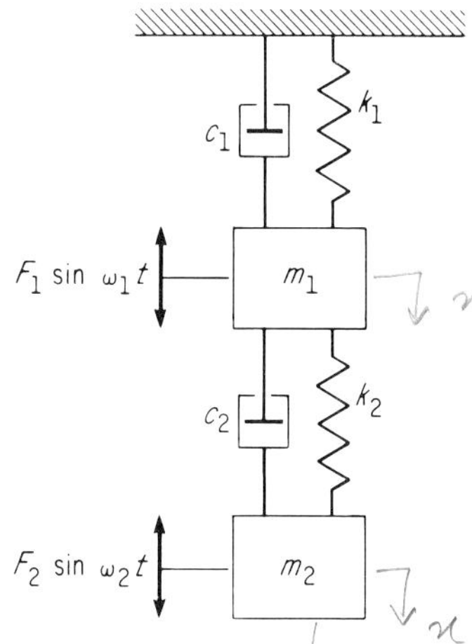

FIG. 8-23

equations of motion in this case are

$$\begin{aligned} m_1\ddot{x}_1 + c_1\dot{x}_1 + c_2(\dot{x}_1 - \dot{x}_2) + k_1x_1 + k_2(x_1 - x_2) &= F_1 \sin \omega_1 t \\ m_2\ddot{x}_2 + c_2(\dot{x}_2 - \dot{x}_1) + k_2(x_2 - x_1) &= F_2 \sin \omega_2 t \end{aligned} \tag{8-35}$$

Using the notation

$$\begin{aligned} X_1, X_2 &= \text{Laplace transforms of } x_1, x_2 \\ (x_1)_0, (x_2)_0 &= \text{initial displacements of the two masses} \\ (\dot{x}_1)_0, (\dot{x}_2)_0 &= \text{initial velocities of the two masses} \end{aligned}$$

the transformation of Eqs. 8-35 can be obtained by utilizing the tables in the appendix:

$$m_1p^2X_1 - m_1p(x_1)_0 - m_1(\dot{x}_1)_0 + (c_1 + c_2)pX_1 - (c_1 + c_2)(x_1)_0$$
$$-c_2pX_2 + c_2(x_2)_0 + (k_1 + k_2)X_1 - k_2X_2 = \frac{F_1\omega_1}{p^2 + \omega_1^2}$$
$$m_2p^2X_2 - m_2p(x_2)_0 - m_2(\dot{x}_2)_0 + c_2pX_2 - C_2(x_2)_0 - c_2pX_1$$
$$+ c_2(x_1)_0 + k_2X_2 - k_2X_1 = \frac{F_2\omega_2}{p^2 + \omega_2^2}$$

Rearranging, we get

$$[m_1p^2 + (c_1 + c_2)p + k_1 + k_2]\,X_1 - [c_2p + k_2]\,X_2$$
$$= \frac{F_1\omega_1}{p^2 + \omega_1^2} + (c_1 + c_2 + m_1p)(x_1)_0 + m_1(\dot{x}_1)_0 - c_2(x_2)_0 \qquad (8\text{-}36)$$
$$-c_2p + k_2X_1 + m_2p^2 + c_2p + k_2X_2$$
$$= \frac{F_2\omega_2}{p^2 + \omega_2^2} - c_2(x_1)_0 + (c_2 + m_2p)(x_2)_0 + m_2(\dot{x}_2)_0 \qquad (8\text{-}37)$$

from which X_1 and X_2 are obtained by the simultaneous solution of the two algebraic equations. The inverse transform for X_1, X_2 gives $x_1(t)$ and $x_2(t)$, respectively. The expressions for X_1 and X_2 are usually expanded in partial fractions before the inverse transform is carried out.

Example 13

In the previous problem take

$$F_1 = F_2 = 0 \qquad m_1 = m_2 = m$$
$$k_1 = k_2 = k \qquad c_1 = c_2 = 0$$

and the initial conditions

$$\text{at } t = 0, \quad (\dot{x}_1)_0 = V_0$$
$$(\dot{x}_2)_0 = (x_2)_0 = (x_2)_0 = 0$$

Equations 8-36 and 8-37 become

$$[mp^2 + 2k]X_1 - kX_2 = mV_0$$
$$-kX_1 + [mp^2 + k]X_2 = 0$$

Taking $k/m = \omega_0^2$,

$$X_1 = \frac{\begin{vmatrix} V_0 & -\omega_0^2 \\ 0 & p^2 + \omega_0^2 \end{vmatrix}}{\begin{vmatrix} (p)^2 + 2\omega_0^2 & -\omega_0^2 \\ -\omega_0^2 & p^2 + \omega_0^2 \end{vmatrix}}$$
$$= \frac{V_0[(p)^2 + \omega_0^2]}{p^4 + 3\omega_0^2p^2 + \omega_0^4},$$

and

$$X_2 = \frac{V_0 \omega_0^2}{p^4 + 3\omega_n^2 p^2 + \omega_0^4}$$

for the inverse transform, the expression for X_1 and X_2 should be expanded in partial fractions. The equation $[p^4 + 3\omega_0^2 p^2 + \omega_0^4] = 0$ will have no positive roots for a stable vibration. Any real roots have to be zero or negative and the imaginary part of the complex roots represents the natural frequency of the vibration.

Assuming in this case that $p^2 = -a^2$ and $p^2 = -b^2$,

$$\omega_{n_1} = a \quad \text{and} \quad \omega_{n_2} = b$$

and

$$X_1 = \frac{V_0[p^2 + \omega_0^2]}{[p^2 + a^2][p^2 + b^2]}$$

$$= \frac{V_0(-a^2 + \omega_0^2)/(b^2 - a^2)}{p^2 + a^2} + \frac{V_0(-b^2 + \omega_0^2)/(a^2 - b^2)}{p^2 + b^2}$$

and the inverse transform gives

$$x_1(t) = \frac{V_0(\omega_0^2 - a^2)}{a(b^2 - a^2)}(\sin at) + \frac{V_0(\omega_0^2 - b^2)}{b(a^2 - b^2)}(\sin bt)$$

Similarly,

$$x_2(t) = \frac{V_0 \omega_0^2}{a(b^2 - a^2)}(\sin at) + \frac{V_0 \omega_0^2}{b(b^2 - a^2)}(\sin bt)$$

where

$$a = \omega_0 \sqrt{1.5 - \sqrt{(1.5)^2 - 1}} = 0.525\,\omega_0 = \omega_{n_1}$$

$$b = \omega_0 \sqrt{1.5 = \sqrt{1.5 - 1}} = 1.65\,\omega_0 = \omega_{n_2}$$

and consequently the solution is given by

$$x_1(t) = \frac{V_0}{\omega_0}[0.564 \sin(0.525)\omega_0 t + 0.427 \sin(1.65)\omega_0 t]$$

$$x_2(t) = \frac{V_0}{\omega_0}[0.778 \sin(0.525)\omega_0 t + 0.242 \sin(1.65)\omega_0 t]$$

Example 14

In the previous problem take

$$F_2 = 0$$

$$k_1 = k_2 = k$$

$$m_1 = m_2 = m$$

$$c_1 = c_2 = 0$$

and the initial conditions

$$\text{at } t = 0 \quad \dot{x}_1 = \dot{x}_2 = 0$$

$$x_1 = x_2 = 0$$

Then

$$(mp^2 + 2k)X_1 - kX_2 = \frac{F_1\omega_1}{p^2 + \omega_1^2}$$

$$-kX_1 + (mp^2 + k)X_2 = 0$$

Dividing through by k and using the notation $\omega_0^2 = k/m$ and $X_s = F_1/k$, we obtain

$$X_1 = \frac{\begin{vmatrix} \dfrac{X_s\omega_1}{p^2 + \omega_1^2} & (-1) \\ 0 & \left(\dfrac{p}{\omega_0}\right)^2 + 1 \end{vmatrix}}{\begin{vmatrix} \left(\dfrac{p}{\omega_0}\right)^2 + 2 & (-1) \\ -1 & \left(\dfrac{p}{\omega_0}\right)^2 + 1 \end{vmatrix}}$$

$$= \frac{X_s\left[\left(\dfrac{p}{\omega_0}\right) + 1\right]\left[\dfrac{\omega_1}{p^2 + \omega_1^2}\right]}{\left(\dfrac{p}{\omega_0}\right)^4 + 3\left(\dfrac{p}{\omega_0}\right)^2 + 1}$$

$$X_2 = \frac{X_s\left[\dfrac{\omega_1}{p^2 + \omega_1^2}\right]\omega_0^4}{p^4 + 3p^2\omega_0^2 + \omega_0^4}$$

Following the same procedure as in the previous example and knowing that the roots of the equation obtained by making the denominator equal zero are

$$p^2 = -a^2 \quad \text{and} \quad p^2 = -b^2$$

then

$$X_1 = X_s\omega_0^4\left\{\frac{\left[\left(1 - \dfrac{a^2}{\omega_0^2}\right)/(b^2 - a^2)\right]\omega_1/(\omega_1^2 - a^2)}{(p^2 + a^2)} + \frac{\left[\left(1 - \dfrac{b^2}{\omega_0^2}\right)/(a^2 - b^2)\right]\omega_1/(\omega_1^2 - b^2)}{(p^2 + b^2)}\right\}$$

and

$$X_2 = X_s\omega_0^4\left[\frac{\omega_1/(\omega_1^2 - a^2)}{p^2 + a^2} + \frac{\omega_1/(\omega_1^2 - b^2)}{p^2 + b^2}\right]$$

8-6. IMPEDANCE METHOD

The impedance represents a measure of dynamic resistance. In the electrical system, the impedance represents the ratio of voltage to current and by classical mechanical analogy, it should be defined as *the ratio of force to velocity*. Impedance can also be used in mechanical systems to indicate the ratio of applied force to displacement or acceleration as well.

Mobility and admittance are the inverse of impedance and represent the ratio of velocity to force.

The use of impedance and mobility techniques considerably simplifies the analysis of a certain class of vibration problems. One example is the case of the response of a certain point in a complex structure to an excitation at a different point. An attempt will be made here briefly to familiarize the reader with the impedance concept by means of some illustrative examples.

Considering the simple system of Fig. 8-24, the impedance at point 1

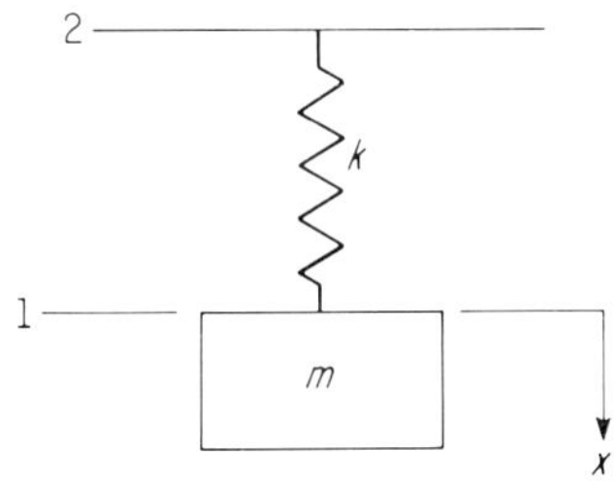

FIG. 8-24

can be derived by the help of the complex plane (Fig. 8-25) as

$$Z_1 = \frac{F_1}{V_1} = \frac{ma_1}{V_1} = \frac{-m\omega^2 x_1}{i\omega x_1} = m\omega i$$

where F_1 = force
Z_1 = impedance
x_1 = displacement
V_1 = velocity
a_1 = acceleration
ω = frequency of the vibration

and the subscript 1 refers to point 1. Similarly,

$$V_2 = V_1 + V_{12}$$
$$F_2 = kx_{12} = F_1 = F$$

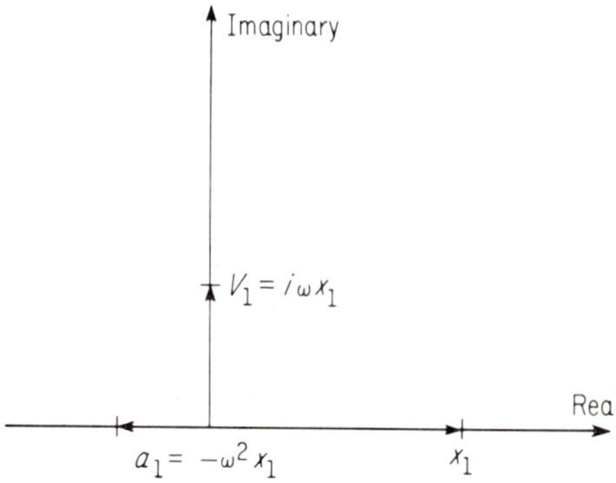

FIG. 8-25

where x_{12} = relative displacement between points 1 and 2

$V_{12} = dx_{12}/dt$

= relative velocity between points 1 and 2

The impedance Z_2 at point 2 can therefore be calculated from

$$\frac{1}{Z_2} = \frac{V_2}{F_2} = \frac{V_1}{F_1} + \frac{V_{12}}{F} = \frac{1}{Z_1} + \frac{i\omega x_{12}}{kx_{12}}$$

or

$$Z_2 = \frac{1}{\dfrac{1}{m\omega i} + \dfrac{i\omega}{k}}$$

If the system of Fig. 8-24 includes damping as shown in Fig. 8-26, the impedance at point 1 will be

$$Z_1 = \frac{F_1}{V_1} = \frac{-m\omega^2 x_1}{i\omega x_1} = im\omega$$

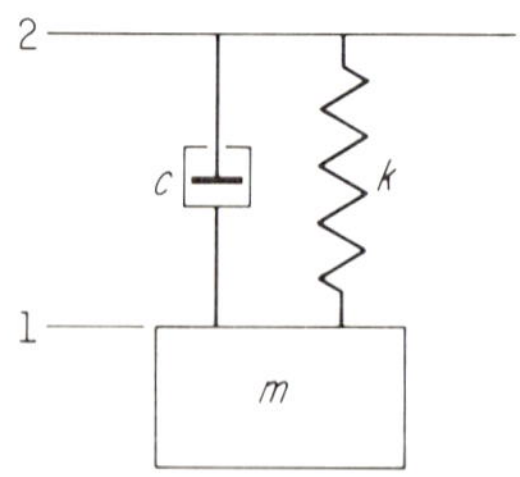

FIG. 8-26

and Z_2 can be obtained as follows:

$$F_1 = F_2 = kx_{12} + ic\omega x_{12} = F$$

$$V_2 = V_1 + V_{12}$$

from which

$$\frac{V_2}{F_2} = \frac{V_1 + V_{12}}{F} = \frac{V_{12}}{F} + \frac{V_1}{F}$$

$$= \frac{1}{Z_1} + \frac{i\omega x_{12}}{kx_{12} + ic\omega x_{12}}$$

$$= \frac{1}{Z_1} + \frac{i\omega}{k + ic\omega}$$

Therefore

$$Z_2 = \frac{1}{\dfrac{1}{im\omega} + \dfrac{i\omega}{k + ic\omega}}$$

For the case of the dynamic absorber (Fig. 8-27), it can be easily seen that Z_1 and Z_2 are the same as in the previous case:

$$Z_1 = \frac{F_1}{V_1} = i\omega m_1$$

$$Z_2 = \frac{1}{\dfrac{1}{im_1\omega} + \dfrac{i\omega}{k_1 + ic\omega}}$$

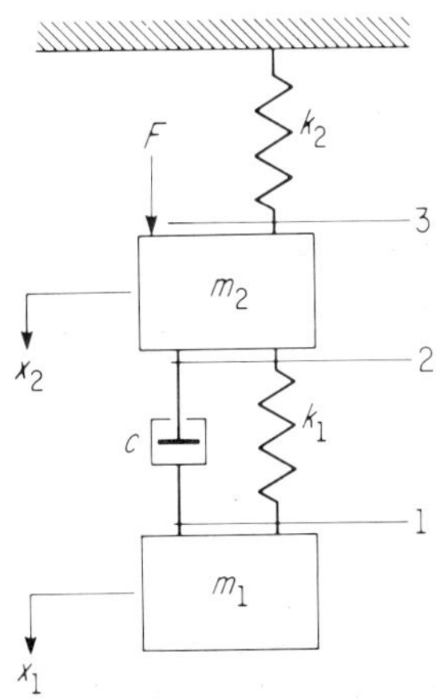

FIG. 8-27

The impedance Z_3 at point 3 can be calculated from

$$Z_3 = \frac{F_3}{V_3}$$

Since

$$V_3 = V_2$$

and

$$F_3 = F_2 - m_2\omega^2 x_2$$

Therefore

$$Z_3 = Z_2 - \frac{m_2\omega^2 x_2}{i\omega x_2} = i\omega m_2 + Z_2$$

$$= i\omega m_2 + \cfrac{1}{\cfrac{1}{im_1\omega} + \cfrac{i\omega}{k_1 + ic\omega}} \tag{8-38}$$

On the other hand, Z_3 can also be expressed as

$$Z_3 = \frac{k_2 x_2}{i\omega x_2} = \frac{k_2}{i\omega} \tag{8-39}$$

In the case of free vibration the impedance as calculated from Eqs. 8-38 and 8-39 should have the same value. This requires that

$$\frac{k_2}{i\omega} = i\omega m_2 + \cfrac{1}{\cfrac{1}{im_1\omega} + \cfrac{i\omega}{k_1 + ic\omega}}$$

which is the characteristic equation for the system under consideration.

The impedance Z for a dynamic absorber can similarly be written as

$$Z = \frac{F}{V_2}$$

where F = applied force on mass 2

$= F_3 + k_2 x_2$

and the impedance is calculated as

$$Z = \frac{F_3}{V_2} + \frac{k_2 x_2}{V_2} = Z_3 + Z_3'$$

$$= \frac{k_2}{i\omega} + i\omega m_2 + \cfrac{1}{\cfrac{1}{im_1\omega} + \cfrac{i\omega}{k_1 + ic\omega}} \tag{8-40}$$

In order for the vibration absorber to be effective, Z must be as large as possible. Equation 8-40 can therefore be used to compare different absorber designs.

The impedance method is also effectively used for evaluation of the natural frequencies of multimass systems. Taking as an example the undamped four-mass system shown in Fig. 8-28, the impedance at differ-

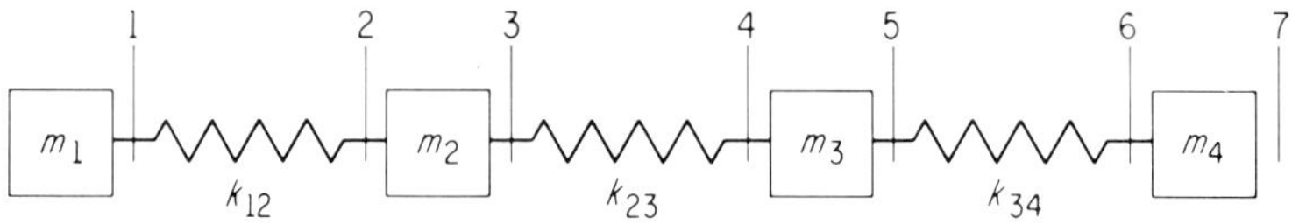

FIG. 8-28

ent locations is written as

$$Z_1 = \frac{F_1}{V_1} = -\frac{m_1\omega^2 x_1}{i\omega x_1} = im_1\omega$$

$$Z_2 = \frac{F_2}{V_2}$$

$$V_2 = V_1 + V_{12}, \quad F_2 = F_1 = k_{12}x_{12}$$

$$\frac{1}{Z_2} = \frac{V_1}{F_1} + \frac{V_{12}}{k_{12}x_{12}} = \frac{1}{Z_1} + \frac{i\omega x_{12}}{k_{12}x_{12}}$$

$$Z_2 = \frac{1}{\dfrac{1}{im_1\omega} + \dfrac{i\omega}{k_{12}}}$$

$$Z_3 = \frac{F_3}{V_3}$$

$$= \frac{F_2 - m_2\omega^2 x_2}{V_2}$$

$$= Z_2 - \frac{m_2\omega^2 x_2}{i\omega x_2} = im_2\omega + Z_2$$

$$= im_2\omega + \frac{1}{\dfrac{1}{im_1\omega} + \dfrac{i\omega}{k_{12}}}$$

Similarly,

$$Z_4 = \frac{1}{\dfrac{i\omega}{k_{23}} + \dfrac{1}{Z_3}}$$

$$Z_5 = im_3\omega + Z_4$$

$$Z_6 = \frac{1}{\dfrac{i\omega}{k_{34}} + \dfrac{1}{Z_5}}$$

and

$$Z_7 = im_4\omega + Z_6$$

For free vibration ($Z_7 = 0$) which gives the characteristic equation for evaluation of the natural frequencies for the system.

The impedance concept can also be utilized in conjunction with the differential equations of motion to evaluate the response of multimass linear systems to general disturbances. This is illustrated in the following case. A comprehensive account of application of mechanical impedance methods is given in Ref. 18.

8-7. TWO-MASS TORSIONAL SYSTEM UNDER GENERAL CONDITION OF VISCOUS DAMPING

The following is a comprehensive analysis of the damping influence on the vibratory behavior of the two-mass torsional system of Fig. 8-29.

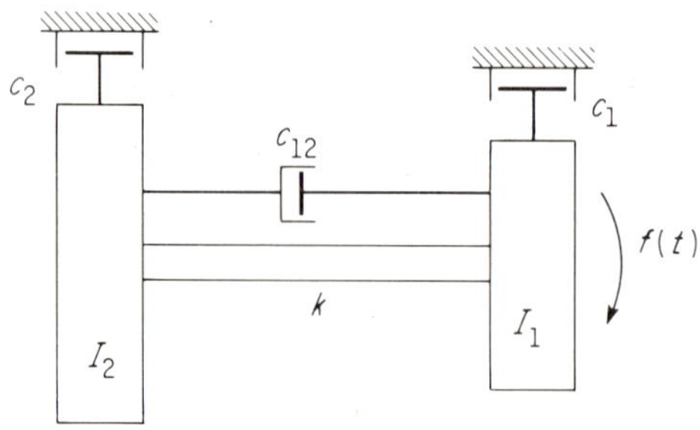

FIG. 8-29. Two-mass torsional system with general damping condition.

The damping is considered viscous (or proportional to the velocity) to simplify the analysis. The damping coefficients which are defined in single-degree-of-freedom systems by their ratio to the critical coefficient will be defined here by a different nondimensional coefficient to simplify the formulation. The frequency equation is also given in nondimensional terms to facilitate computations.

Using the notation

I_1, I_2 = moment of inertia of driving and driven masses respectively.

k = stiffness of connecting shafts and couplings

θ_1, θ_2 = angular displacement of driving and driven masses respectively

c_1, c_2 = damping coefficients at driving and driven masses respectively

c_{12} = damping coefficient for relative motion between the two masses

$f(t)$ = exciting force function.

$$p = \frac{d}{dt}$$

$$p^2 = \frac{d^2}{dt^2}$$

$$c_1 = n_1 \sqrt{k(I_1 + I_2)}$$

$$c_2 = n_2 \sqrt{k(I_1 + I_2)}$$

$$c_{12} = n_{12} \sqrt{k(I_1 + I_2)}$$

n_1, n_2, n_{12} = nondimensional damping coefficients

I_e = equivalent moment of inertia of system

$$= \frac{I_1 I_2}{I_1 + I_2}$$

ω_0 = natural frequency for undamped system

$$= \sqrt{\frac{k}{I_e}}$$

b = natural frequency for damped system

$$R = \frac{I_1}{I_2}$$

$(T_s)_{T_0}$ = shaft torque due to a step load of magnitude T_0

$(T_s)_E$ = shaft torque due to an impulse of strength E

c_{crit} = critical damping coefficient for system with $c_1 = c_2 = 0$

$$= 2I_e\omega_0$$

$$= (n_{12})_{\text{crit}} \sqrt{k(I_1 + I_2)}$$

$$(n_{12})_{\text{crit}} = 2/\sqrt{R + \frac{1}{R} + 2}$$

The differential equations describing the motion of the system (Fig. 8-29) are given by

$$I_1 p^2\theta_1 + k(\theta_1 - \theta_2) + c_1 p\theta_1 + c_{12}p(\theta_1 - \theta_2) = f(t) \tag{8-41}$$

$$I_2 p^2\theta_2^2 + k(\theta_2 - \theta_1) + c_2 p\theta_2 + c_{12}p(\theta_2 - \theta_1) = 0 \tag{8-42}$$

from which

$$\theta_2 = \frac{k + c_{12}p}{I_2 p^2 + k + (c_2 + c_{12})p}\,\theta_1 \tag{8-43}$$

$$\theta_1 = f(t)/\left[I_1 p^2 + k + (c_1 + c_{12})p - \frac{(k + c_{12}p)^2}{I_2 p^2 + k + (c_2 + c_{12})p}\right] \tag{8-44}$$

Note that

Z_1 = the impedance of the system at I_1 which will be taken as the ratio between force and displacement

$$= \frac{f(t)}{\theta_1}$$

$$= \frac{[I_1 p^2 + k + (c_1 + c_{12})p][I_2 p^2 + k + (c_2 + c_{12})p] - [k + c_{12}p]^2}{I_2 p^2 + k + (c_2 + c_{12})p} \tag{8-45}$$

and for free vibration

$$Z_1 = 0$$

from which

$$p^4 + \left[\frac{c_1}{I_1} + \frac{c_2}{I_2} + \frac{c_{12}(I_1 + I_2)}{I_1 I_2}\right] p^3 + \left[\frac{c_{12}}{I_1 I_2}(c_1 + c_2) + \frac{c_1}{I_1}\frac{c_2}{I_2} + k\frac{(I_1 + I_2)}{I_1 I_2}\right] p^2 + \left[\frac{k}{I_1 I_2}(c_1 + c_2)\right] p = 0 \tag{8-46}$$

which can be written in terms of the nondimensional quantities n_1, n_{12}, n_2, and R as follows:

$$\left(\frac{p}{\omega_0}\right)^4 + \left[n_1\sqrt{\frac{1}{R}} + n_2\sqrt{R} + n_{12}\sqrt{R + \frac{1}{R} + 2}\right]\left(\frac{p}{\omega_0}\right)^3 + [n_{12}(n_1 + n_2) + n_1 n_2 + 1]\left(\frac{p}{\omega_0}\right)^2 + \left[(n_1 + n_2)/\sqrt{R + \frac{1}{R} + 2}\right]\left(\frac{p}{\omega_0}\right) = 0 \tag{8-47}$$

This represents the characteristic equation and is identical with the corresponding equation obtained by Laplace transform. Solving for p gives the following roots:

$$p = 0, p_1, p_2, p_3$$

A stable vibration will occur when the following conditions are satisfied:

$$\begin{aligned} p_1 &= \text{real root} = \text{zero or negative} \\ p_2, p_3 &= \text{complex roots} = a \pm ib \\ a &= \text{zero or negative} \end{aligned}$$

and the natural frequency of the system is given by the imaginary part b of the complex roots.

The system under consideration is of practical interest since it approximately describes many mechanical drives. Solution of the characteristic Eq. 8-47 has been obtained for different values of R and for different damping conditions. The values of the natural frequency are given in Appendix.

Transient Response

To obtain the transient response of the system under consideration, the force function $f(t)$ in Eq. 8-41 will be taken as the unit step function $u(t)$. The angular displacements of the two masses due to this unit step function can be written as

$$\theta_1 = f_1(t)\,u(t) = \frac{1}{Z_1}u(t) = \frac{I_2 p^2 + k + (c_2 + c_{12})p}{(I_1 I_2)p(p - p_1)(p - p_2)(p - p_3)}u(t) \tag{8-48}$$

and

$$\theta_2 = f_2(t)\,u(t) = \frac{1}{Z_2}u(t) = \frac{k + c_{12}p}{(I_1 I_2)p(p - p_1)(p - p_2)(p - p_3)}u(t) \tag{8-49}$$

where Z_1, Z_2 are the impedances of the system with regard to θ_1 and θ_2 respectively, Eqs. 8-48 and 8-49 can be expanded into partial fractions and solved by operational calculus. The shaft torque due to the unit step function $u(t)$ will be

$$(T_s)_{u(t)} = k(\theta_1 - \theta_2) = k[f_1(t) - f_2(t)]u(t)$$

The shaft torque due to a suddenly applied load T_0 can be written as

$$(T_s)_{T_0} = k[f_1(t) - f_2(t)]T_0 \tag{8-50}$$

and the shaft torque due to an impulse E will be

$$(T_s)_E = E\frac{d(T_s)\,u(t)}{dt} = Ek\left[\frac{df_1(t)}{dt} - \frac{df_2(t)}{dt}\right] \tag{8-51}$$

This procedure is applied, as an example, to the system under consideration with four different conditions of damping. The resulting equations for shaft torques when a vibration occurs are given as follows:

Condition 1

$$n_1 = n_{12} = n_2 = 0$$
$$p = 0, \quad p = \pm i\omega_0$$

Natural frequency $= \omega_0$

The equations for the shaft torque due to a step load and an impulse are respectively

$$(T_s)_{T_0} = \frac{T_0}{R+1}[1 - \cos \omega_0 t]$$

$$(T_s)_E = \frac{E}{R+1}\omega_0[\sin \omega_0 t]$$

Condition 2

$$n_1 = n_2 = 0$$

$$p = 0, \quad p = a + ib, \quad p = a - ib$$

$$a = -\frac{n_{12}}{2}\left[\sqrt{R + \frac{1}{R} + 2}\right]\omega_0$$

Nondimensional values of a are given in Fig. 8-30.

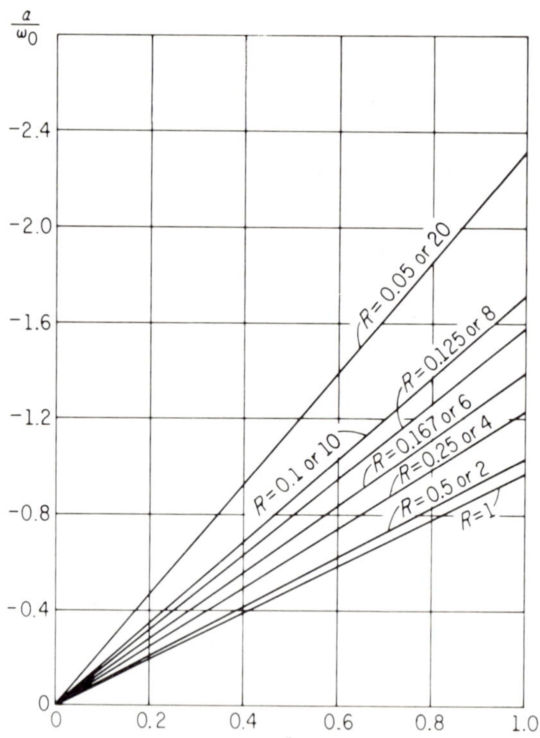

FIG. 8-30. Values of a for condition (2) $n_1 = n_2 = 0, n_{12} = n$.

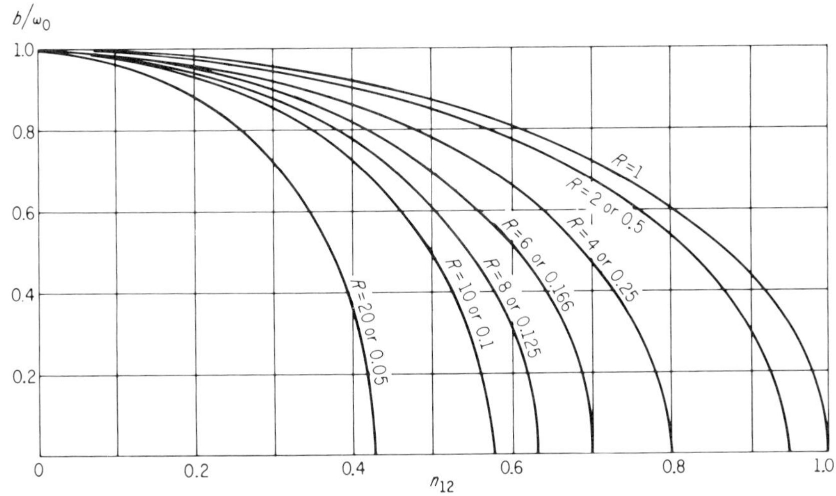

FIG. 8-31. $n_1 = n_2 = 0$.

$$b = \left[\sqrt{1 - \left(\frac{n_{12}}{2}\right)^2 \left(R + \frac{1}{R}\right) + 2}\right]\omega_0 = \text{natural frequency}$$

Nondimensional values of b are plotted in Fig. 8-31.

It can be easily seen from the previous equation that no vibration will occur in this case when

$$b \geq 0$$

from which the critical condition $b = 0$ is given by

$$n_{12} = \frac{2}{\sqrt{R + \frac{1}{R} + 2}}$$

which represents the critical damping coefficient.

The equations for the shaft torque due to a step load and a pulse are respectively

$$(T_s)_{T_0} = \frac{T_0}{R + 1}\left[1 - \frac{\omega_0}{b} e^{at} \cos(bt + \psi)\right]$$

$$(T_s)_E = -E\left(\frac{\omega_0}{b}\right)\omega_0 e^{at} \cos(bt + \psi + \theta)$$

where

$$\sin\psi = \frac{a}{\omega_0} \qquad \cos\psi = \frac{b}{\omega_0}$$

and

$$\sin\theta = \frac{b}{\omega_0} \qquad \cos\theta = \frac{a}{\omega_0}$$

Therefore

$$\theta + \psi = 90^\circ \qquad \text{or} \qquad \theta = 90^\circ - \psi$$

and the equation for $(T_s)_E$ reduces to

$$(T_s)_E = E\left(\frac{\omega_0}{b}\right)\omega_0 e^{at}\sin(bt)$$

Condition 3

$$n_1 = n_{12} = 0, \quad n_2 = n$$

In this case

$$p = 0, \quad p = D_1, \quad p = a + ib, \quad p = a - ib$$

For a vibration to occur D_1 and a should be negative. Their values are given respectively in Figs. 8-32 and 8-33 for different values of R and n.

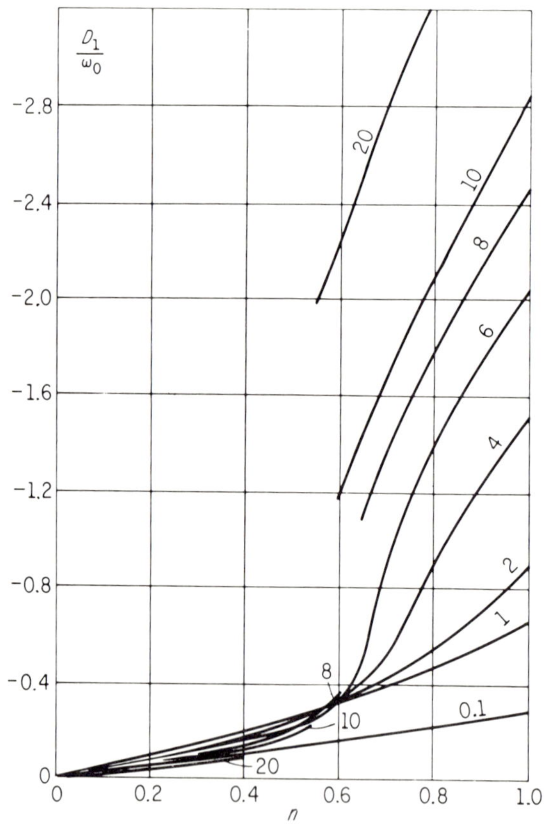

FIG. 8-32. Values of D_1 for conditions 3 and 4. Numbers on curves designate R for condition 3; $1/R$ for condition 4.

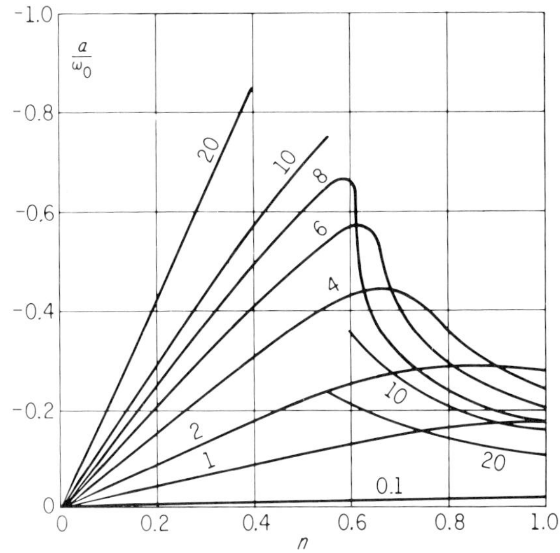

FIG. 8-33. Values of a for conditions 3 and 4. Numbers on curves designate R for condition 3; $1/R$ for condition 4.

b = natural frequency of the vibration whose value is plotted in a nondimensional form in Fig. 8-34.

The equations for the shaft torque due to a step load and an impulse are in this case

$$(T_s)_{T_0} = T_0[1 - Se^{D_1 t} - Ae^{at}\cos(bt + \psi)]$$
$$(T_s)_E = -E[D_1 Se^{D_1 t} + \omega_0 Ae^{at}\cos(bt + \psi + \theta)]$$

where

$$S = \left[\left(\frac{D_1}{\omega_0}\right)^2 + n_2\sqrt{R}\left(\frac{D_1}{\omega_0}\right) + \frac{1}{1+R}\right] \Big/ \left[\left(\frac{D_1}{\omega_0} - \frac{a}{\omega_0}\right)^2 + \left(\frac{b}{\omega_0}\right)^2\right]$$

$$A = 2\sqrt{M^2 + N^2}$$

$$M = \frac{-\left(\frac{a}{\omega_0} - \frac{D_1}{\omega_0}\right)\left(\frac{D_1}{\omega_0}\right) - \left[\left(\frac{a}{\omega_0}\right)^2 - \left(\frac{b}{\omega_0}\right)^2 + n_2\sqrt{R}\left(\frac{a}{\omega_0}\right) + \frac{1}{1+R}\right]}{2\left[\left(\frac{D_1}{\omega_0} - \frac{a}{\omega_0}\right)^2 + \left(\frac{b}{\omega_0}\right)^2\right]}$$

$$N = \frac{\left(\frac{b}{\omega_0}\right)^2\left(\frac{D_1}{\omega_0}\right) - \left(\frac{a}{\omega_0} - \frac{D_1}{\omega_0}\right)\left[\left(\frac{a}{\omega_0}\right)^2 - \left(\frac{b}{\omega_0}\right)^2 + n_2\sqrt{R}\left(\frac{a}{\omega_0}\right) + \frac{1}{1+R}\right]}{2\left(\frac{b}{\omega_0}\right)\left[\left(\frac{D_1}{\omega_0} - \frac{a}{\omega_0}\right)^2 + \left(\frac{b}{\omega_0}\right)^2\right]}$$

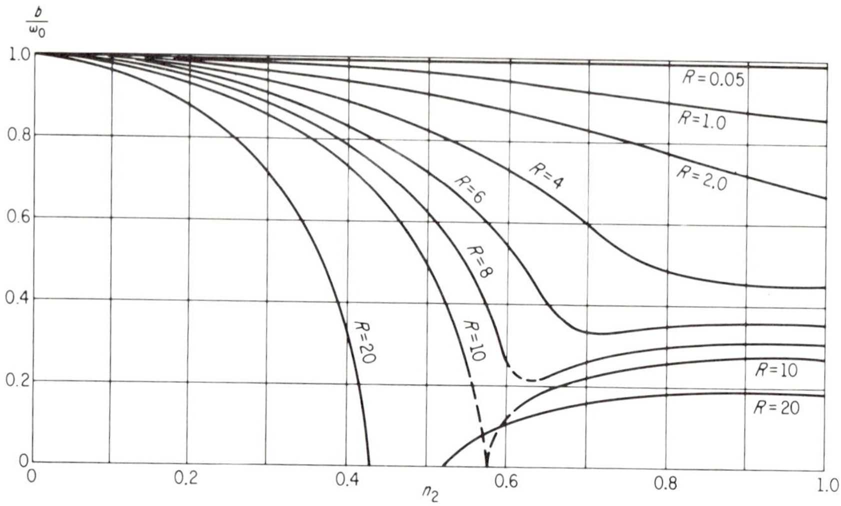

FIG. 8-34. $n_1 = n_{12} = 0$.

ψ is calculated from

$$\sin \psi = 2N/A \qquad \cos \psi = 2M/A$$

and θ is calculated from

$$\sin \theta = b/\sqrt{a^2 + b^2} \qquad \cos \theta = a/\sqrt{a^2 + b^2}$$

Condition 4

$$n_{12} = n_2 = 0, \quad n_1 = n$$

$$p = 0, \quad p = D_1, \quad p = a + ib, \quad p = a - ib$$

For a vibration to exist, D_1 and a should be negative. Figures 8-32 and 8-33 show plots of these values for different values of R and n.

b = natural frequency of the vibration whose value is plotted in Fig. 8-34.

The torque equations in this case are

$$(T_s)_{T_0} = \frac{T_0 \left[e^{D_1 t} + \sqrt{1 + \left(\frac{a}{b} - \frac{D_1}{b} \right)^2} e^{at} \cos (bt + \psi) \right]}{(R + 1) \left[\left(\frac{b}{\omega_0} \right)^2 + \left(\frac{a}{\omega_0} - \frac{D_1}{\omega_0} \right)^2 \right]}$$

$$(T_s)_E = \frac{E \left[D_1 e^{D_1 t} + \sqrt{a^2 + b^2} \sqrt{1 + \left(\frac{a}{b} - \frac{D_1}{b} \right)^2} e^{at} \cos (bt + \psi + \theta) \right]}{(R + 1) \left[\left(\frac{b}{\omega_0} \right)^2 + \left(\frac{a}{\omega_0} - \frac{D_1}{\omega_0} \right)^2 \right]}$$

where ψ is given by

$$\sin\psi = \frac{-\left(\frac{\omega_0}{b}\right)^2\left(\frac{a - D_1}{1 + R}\right)}{2\left[1 + \left(\frac{a}{b} - \frac{D_1}{b}\right)^2\right]} \qquad \cos\psi = \frac{-\left(\frac{\omega_0}{b}\right)^2\left(\frac{1}{R + 1}\right)}{2\left[1 + \left(\frac{a}{b} - \frac{D_1}{b}\right)^2\right]}$$

and θ is given by

$$\sin\theta = b/\sqrt{a^2 + b^2} \qquad \cos\theta = a/\sqrt{a^2 + b^2}$$

Examples—Numerical

The following are numerical examples for the shaft torque in the previously discussed systems due to step loads and impulse applied on the mass I_1.

Example 1

We shall assume in this case that

$$R = 0.1, \quad n_1 = n_{12} = n_2 = 0$$

and

$$\omega_0 = 100$$

This example is the same as Condition 1, and the equations for the shaft torques due to a step load T_0 and a pulse E are therefore

$$(T_s)_{T_0} = \frac{T_0}{1.1}[1 - \cos 100t]$$

$$(T_s)_E = 90.9E[\sin 100t]$$

These equations are plotted as a function of time in Fig. 8-35 and represent a continuous oscillation due to the absence of damping.

Example 2

Taking in this case

$$R = 0.1, \quad n_{12} = 0.4, \quad \omega_0 = 100 \text{ rad/sec}, \quad n_1 = n_2 = 0$$

the values of a and b can be obtained from Figs. 8-30 and 8-31 as

$$a = -69.6, \quad b = 71.8 = \text{natural frequency}$$

from which

$$\psi = -44°$$

The shaft torques are therefore

$$(T_s)_{T_0} = \frac{T_0}{1.1} 1 - 1.39\, e^{-69.6t} \cos(71.8t - 44°)$$

$$(T_s)_E = E\,(139)\, e^{-69.6t} \sin 71.8t$$

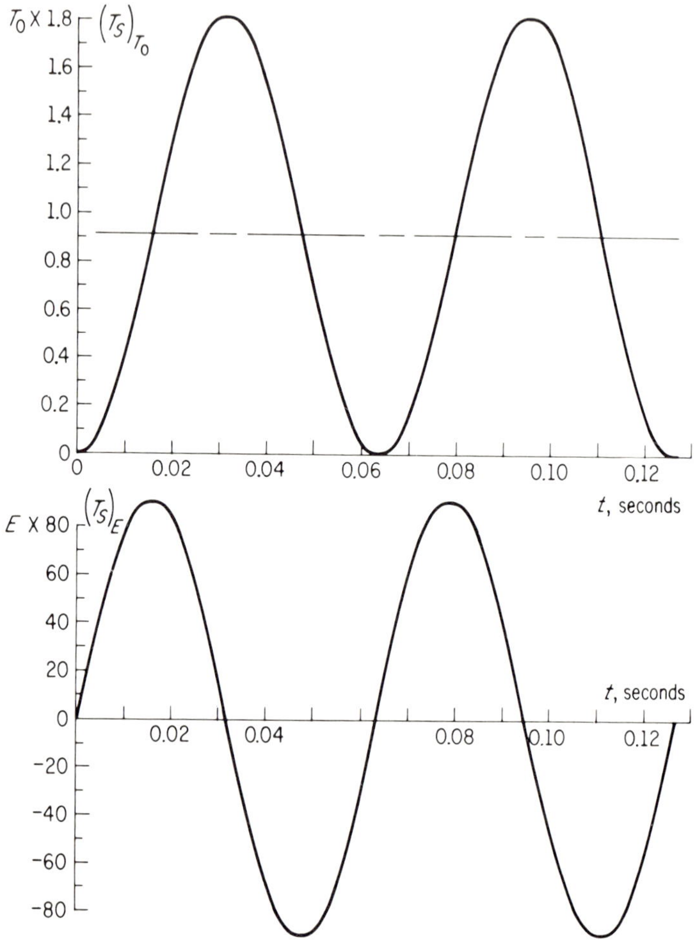

FIG. 8-35. Shaft torques for Example 1. $R = 0.1$, $n_1 = n_2 = n_{12} = 0$, $\omega_0 = 100$ rad/sec.

These equations are plotted in Fig. 8-36 and represent a highly damped oscillation.

Example 3

Assuming that

$$R = 0.1, \quad n_2 = 0.4, \quad \omega_0 = 100, \quad n_1 = n_{12} = 0$$

the values of D_1, a, and b can be obtained from Figs. 8-32, 8-33, and 8-34.

$$D_1 = -11.514$$

$$a = -0.5674$$

$$b = 99.93 = \text{natural frequency}$$

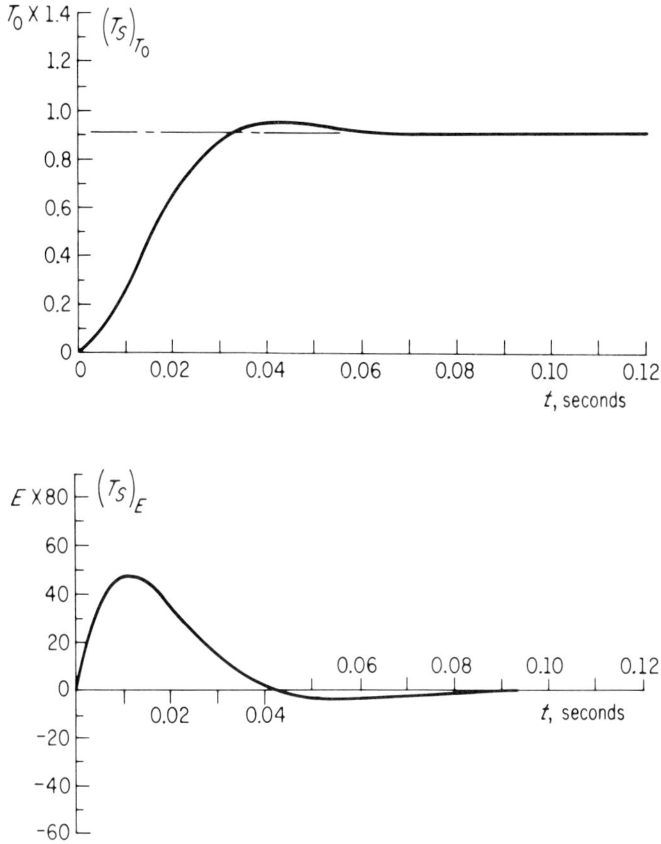

FIG. 8-36. Shaft torques for Example 2. $R = 0.1$, $n_2 = n_1 = 0$, $n_{12} = 0.4$, $\omega_0 = 100$ rad/sec.

The values of S, M, N, A, ψ, and θ can therefore be calculated as shown in Condition 3:

$$S = 0.898$$
$$M = 0.0503$$
$$N = -0.0527$$
$$A = 2\sqrt{M^2 + N^2} = 0.146$$
$$\psi = -46^\circ$$
$$\theta = 90^\circ 20'$$

The equations for the torques are

$$(T_s)_{T_0} = [1 - 0.898\, e^{-11.514t} - 0.146\, e^{-0.5674t} \cos(bt - 46^\circ)]$$

$$(T_s)_E = -E[-10.35\,e^{-11.514t} + 14.6\,e^{-0.5674t}\cos(bt - 46^\circ + 90^\circ\,20'))]$$
$$= E[+\,10.35\,e^{-11.514t} + 14.6\,e^{-0.5674t}\sin(bt - 45^\circ 40')]$$

The previous equations are plotted in Fig. 8-37.

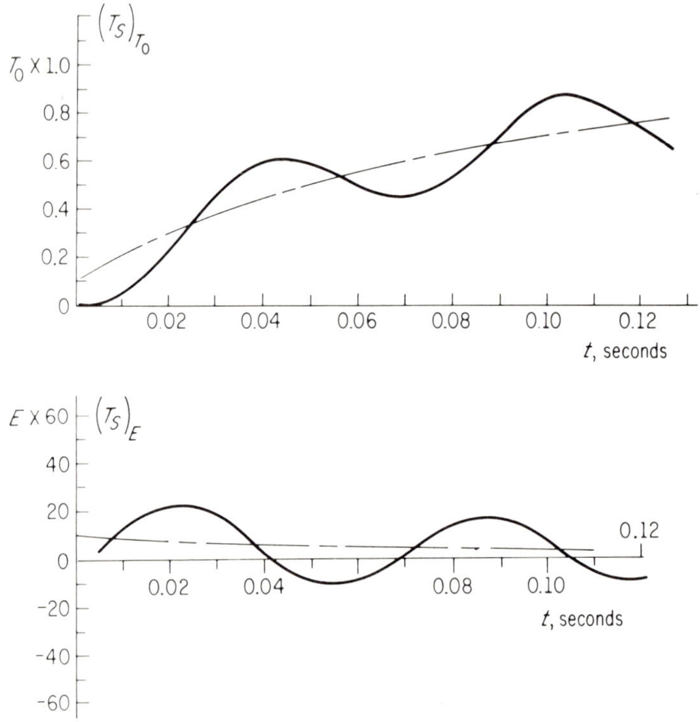

FIG. 8-37. Shaft torques for Example 3. $R = 0.1$, $n_1 = n_{12} = 0, n_2 = 0.4, \omega_0 = 100$ rad/sec.

Example 4

We shall assume in this example that

$$R = 0.1, \quad n_1 = 0.4, \quad \omega_0 = 100, \quad n_2 = n_{12} = 0$$

This is the same as for Condition 4. Proceeding as shown in that case, we get

$$D_1 = -13.582$$
$$a = -56.45$$
$$b = 72.659 = \text{natural frequency}$$
$$\psi = 180^\circ - 29.5^\circ$$
$$\theta = 180^\circ - 53^\circ$$

The shaft-torque equations are therefore

$$(T_s)_{T_0} = T_0[0.413\, e^{-13.582t} - 0.475\, e^{-56.45t} \cos(72.659t - 29.5°)]$$
$$(T_s)_E = E[-5.61\, e^{-13.582t} + 43\, e^{-56.45t} \cos(bt - 82.5°)]$$

Figure 8-38 shows the plot of these equations.

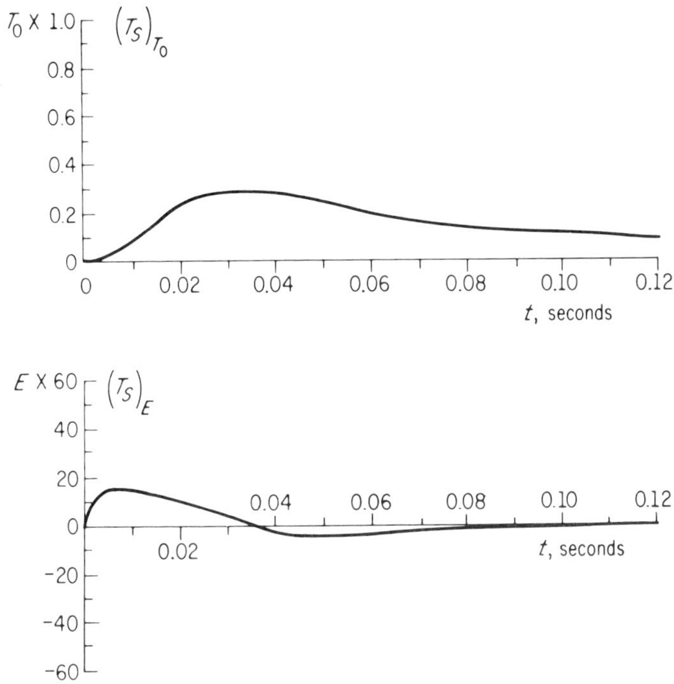

FIG. 8-38. Shaft torques for Example 4. $R = 0.1$, $n_2 = n_{12} = 0, n = 0.4, \omega_0 = 100$ rad/sec.

8-8. SUMMARY

Several techniques have been presented in this chapter for exact solution of the linear system of differential equations describing systems with a finite number of degrees of freedom. The determinant method has been applied to close-coupled and far-coupled systems. The special case of the dynamic absorber is discussed. The Laplace transforms and impedance methods are briefly demonstrated. The example of a two-mass torsional system with general condition of viscous damping is treated, and because of its practical importance extensive charts for its analysis are given.

In chapter 9, some of the numerical and graphical methods solutions are considered.

PROBLEMS

8-1. The given system is constrained for motion in the x-y plane only. The system parameters are

$$\begin{aligned} m &= 0.02 \quad \text{lb-in.}^{-1}\text{-sec}^2 \\ k_1 &= 100 \quad \text{lb/in.} \\ k_2 &= 60 \quad \text{lb/in.} \\ k_3 &= 50 \quad \text{lb/in.} \end{aligned}$$

Calculate the natural frequencies and the normal modes. Check the orthogonality of the normal modes.

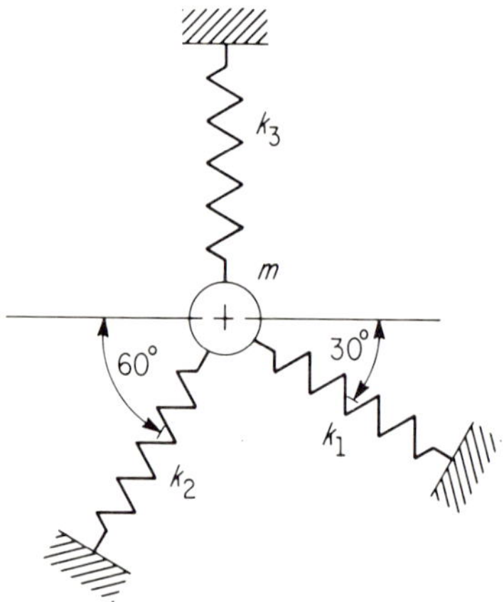

PROB. 8-1

8-2. The mass in Prob. 8-1 is subjected to a short duration impulse with strength E lb-sec in the direction of the x-axis. Derive the equations of the resulting motion.

8-3. A two-rotor torsional system is subjected to a short duration impulse with strength E lb-in.-sec at I_1. Derive the equations of motion for the two rotors and obtain an expression for the shaft torque due to this impulse.

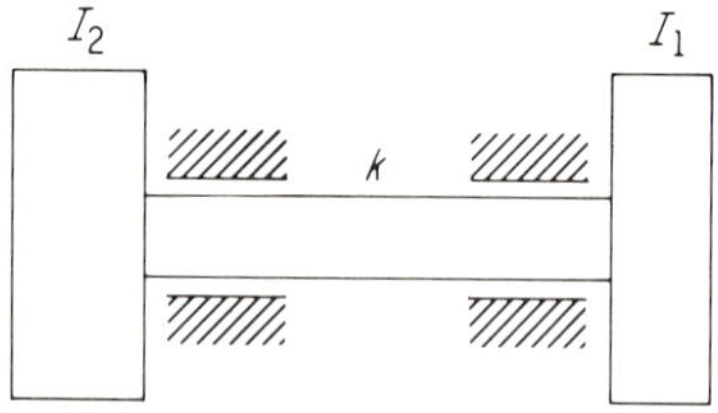

PROB. 8-3

8-4. The rotor I_1 of the system shown in Prob. 8-3 is subjected to a rectangular torque pulse with magnitude T_0 and duration t_0. Write the general expression for the dynamic torque in the shaft.

8-5. A ramp-type torque which increases linearly from 0 to T_0 after a time t_0 is applied to the rotor I_1 of Prob. 8-3. Find an expression for the shaft torque.

8-6. Derive an expression for the dynamic torque in the shaft of the system shown in Prob. 8-3 when rotors I_1 and I_2 are subjected to sinusoidal torques $T_1 \sin \omega_1 t$ and $T_2 \sin \omega_2 t$ respectively.

8-7. The given three-rotor system is subjected to a periodic torque input at the end rotor given by

$$f(t) = F_1 \sin \omega t + F_2 \cos 2\omega t$$

Obtain the expressions for the dynamic torques in the shafts.

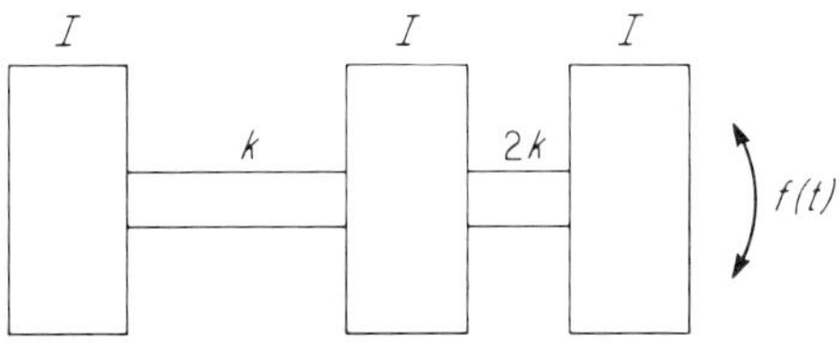

PROB. 8-7

8-8. A three-car train is moving with a velocity V_0 when a constant braking force F_0 is suddenly applied to the middle car. Derive expressions for the resulting motion. Assume all cars and their elastic coupling elements to be identical.

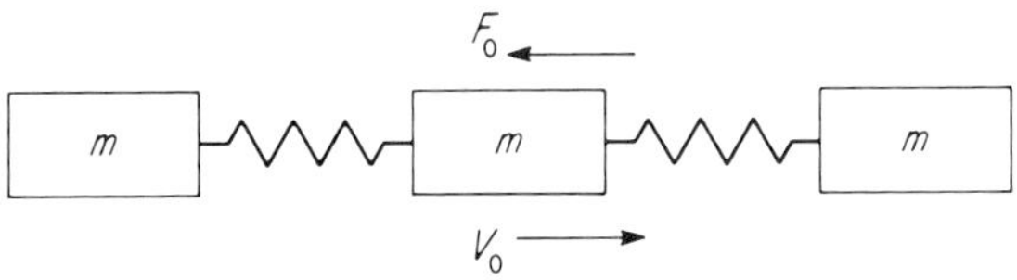

PROB. 8-8

8-9. A step load with magnitude F_0 is applied at mass m_2 of the given two-mass rectilinear system. Write the expression for the dynamic force in spring k_1.

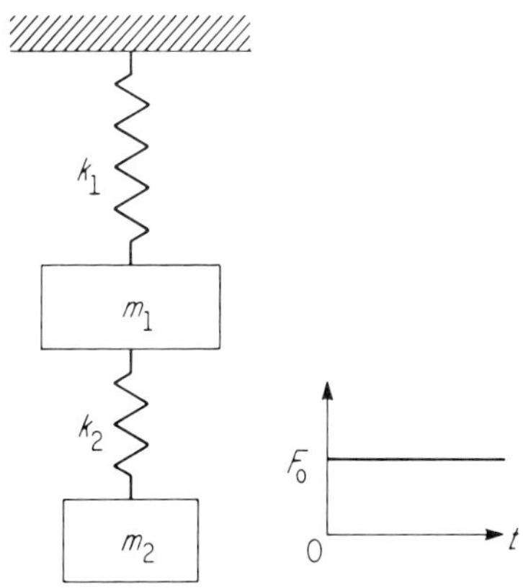

PROB. 8-9

8-10. A car which is diagramatically represented in the given sketch experiences a sudden drop in the road level while traveling at a speed V_0. Derive the expressions for the resulting oscillations of the car body. Assume plane motion. The car mass is m and its moment of inertia about the center of mass is I.

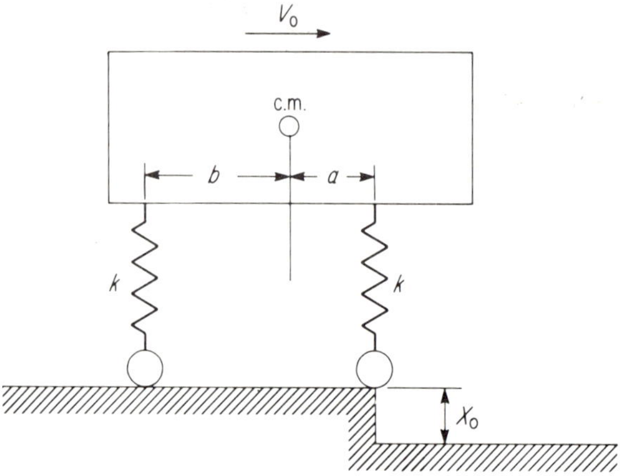

PROB. 8-10

8-11. The given sketch represents a cantilevered uniform steel shaft carrying two weights W_1 and W_2. Given the following values

$$W_1 = 200 \text{ lb}$$
$$W_2 = 100 \text{ lb}$$
$$l_1 = 6 \text{ in.}$$
$$l_2 = 4 \text{ in.}$$
$$(EI)_1 = (EI)_2 = 30 \times 10^6 \text{ lb in.}^2$$

find the natural frequencies for the latteral vibrations of the system and the corresponding modes of vibration.

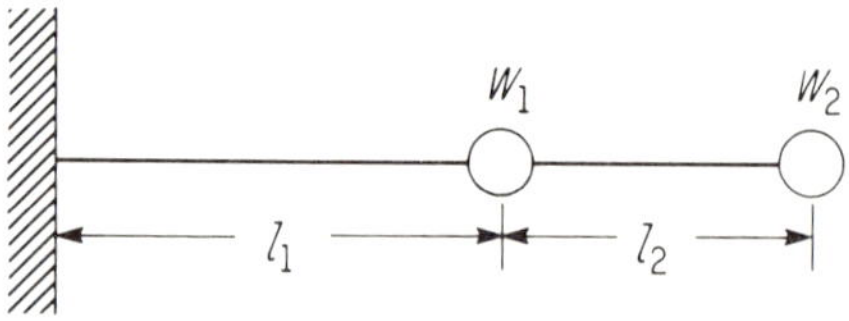

PROB. 8-11

8-12. The weight W_1 of Prob. 8-11 is subjected to a short duration impulse of strength E lb-sec. Write the expressions for the resulting motion.

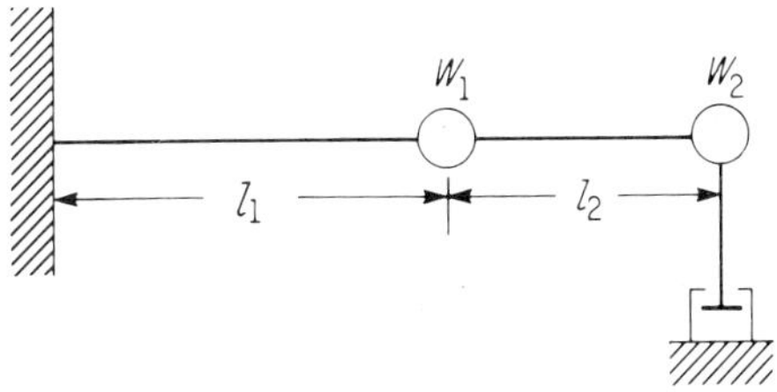

PROB. 8-13

8-13. If the weight W_2 of Prob. 8-11 is resisted by a viscous dash pot with a damping coefficient $c = 100$ lb-in.$^{-1}$-sec, find the natural frequencies and the normal modes for the system.

chapter 9

Approximate Solutions

9-1. INTRODUCTION

The reader can realize from the previous chapter that exact methods of solution become progressively tedious as the number of degrees of freedom increases. This is why numerical, graphical, and analog methods become advantageous.

In this chapter some of these techniques are briefly discussed.

9-2. TABULAR METHODS

The Holzer Tabulation for Close-Coupled Systems

This method is basically a systematic tabulation of the numerical results from the equations of motion under certain assumed conditions. The solution is obtained when the assumed values satisfy all the boundary conditions of the system. This in general may require several trials, and can be easily done with the help of desk calculators or digital computers. The Holzer method is discussed in the following section for the analysis of the free vibration of multidegree-of-freedom systems with and without damping.

Conservative Systems

In the free vibration of the system given in Fig. 9-1, the boundary conditions are such that there are no external torques acting on the system. Assuming harmonic oscillations of all masses, the equations of motion can be written as

$$I_1 \ddot{\theta}_1 = -k_{12}(\theta_1 - \theta_2)$$

from which

$$(\theta_1 - \theta_2) = (\theta_{12}) = \frac{I_1 \omega^2 \theta_1}{k_{12}}$$

where ω is the frequency of the oscillation. Similarly,

$$(\theta_2 - \theta_3) = (\theta_{23}) = \frac{I_1 \omega^2 \theta_1 + I_2 \omega^2 \theta_2}{k_{23}}$$

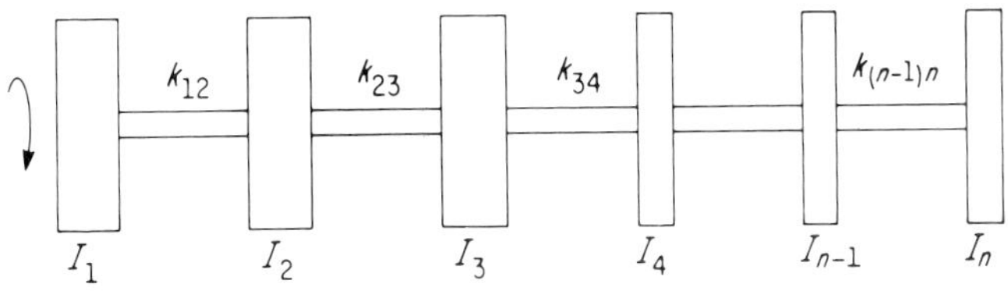

FIG. 9-1

and in general terms,

$$(\theta_{n-1} - \theta_n) = \frac{\sum_{i=1}^{n-1} I_i \omega^2 \theta_i}{k_{(n-1)n}}$$

Since all shaft torques are internal torques in the system, the condition for zero external torque on I_n can therefore be written as

$$\sum_{i=1}^{n-1} I_i \omega^2 \theta_i = 0$$

The Holzer method is based on successive assumption of a value ω for the natural frequency. Each assumption is followed by a calculation of the configuration corresponding to that frequency. The degree of inaccuracy of the assumption is detected by the magnitude of the resulting external torque:

$$\sum_{i=1}^{n} I_i \omega^2 \theta_i$$

A natural frequency should satisfy the boundary conditions of the system and accordingly should give (Fig. 9-2)

$$\sum_{i=1}^{n} I_i \omega^2 \theta_i = 0$$

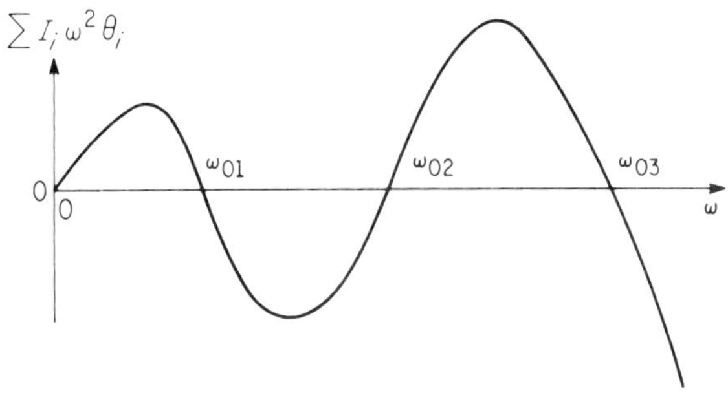

FIG. 9-2

The values of θ_i corresponding to that frequency represent the corresponding natural mode. Since θ_1 (which is chosen arbitrarily) is usually taken equal to one radian, the θ_i values in the table constitute the normal mode for the system.

The system under consideration has n degrees of freedom and we should therefore expect n natural frequencies and n normal modes. The first value of ω which satisfies the boundary condition is $\omega = 0$ and the corresponding mode is

$$\theta_1 = \theta_2 = \theta_3 = \cdots = \theta_n = 1$$

This represents rotation of the system as a rigid body. The successive assumption of ω should give the natural frequencies $\omega_1, \omega_2, \ldots, \omega_{(n-1)}$ and the corresponding modes of vibration.

The same analysis is applicable to the close-coupled rectilinear system shown in Fig. 9-3 where m_i and x_i would be used instead of I_i and θ_i.

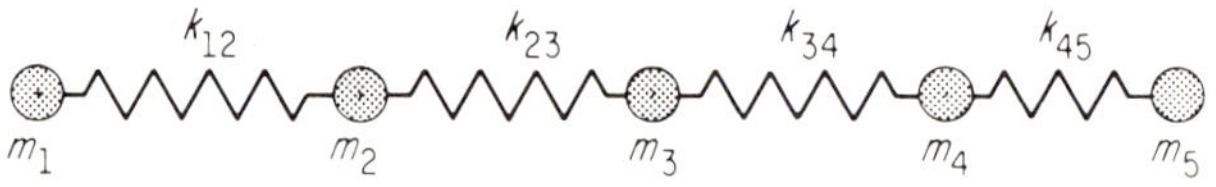

FIG. 9-3

It is also useful to use the effective inertia at station n for detecting the natural frequencies by the Holzer method. The effective inertia at station n is calculated from

$$I_e \ddot{\theta}_n = \sum_i^n I_i \omega^2 \theta_i$$

since

$$|\ddot{\theta}_n| = \omega^2 \theta_n$$

therefore

$$I_e = \frac{\sum_i^n I_i \theta_i}{\theta_n}$$

and when I_e is plotted for successive values of ω the natural frequencies are found when $I_e = 0$. The advantage of this type of plot (Fig. 9-4) is that it can also be used to detect frequencies at which θ_n becomes infinitely small. These are determined when I_e becomes infinite, and are known as antiresonance frequencies.

The boundary conditions for the system shown in Fig. 9-5 are

$$\theta_1 = 0 \quad \text{and} \quad \theta_6 = 0$$

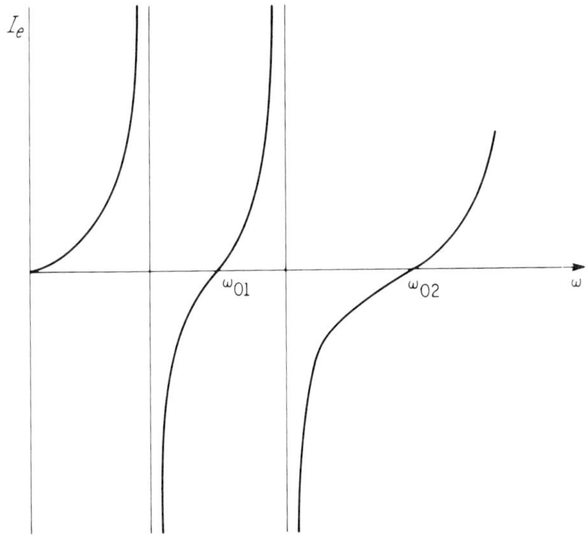

FIG. 9-4

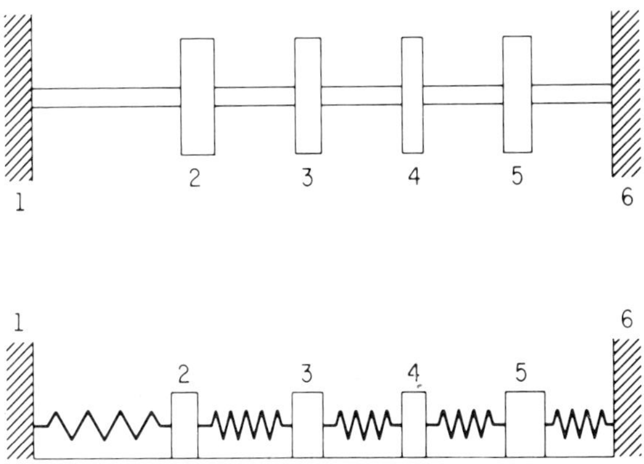

FIG. 9-5

We can start the tabulation with $\theta_1 = 0$ and $\theta_2 = 1$. The natural frequency in this case should satisfy the other condition $\theta_6 = 0$. We can also start with $\theta_6 = 0$ and $\theta_5 = 1$. The natural frequency should satisfy the other condition, $\theta_1 = 0$, in this case.

The system given in Fig. 9-6 has the boundary conditions $\theta_5 = 0$ and the external torque $(T_e)_1 = 0$. Here we can either start with $(T_e)_1 = 0$, $\theta_1 = 1$ and satisfy $\theta_5 = 0$ or start with $\theta_5 = 0$, $\theta_4 = 1$ and satisfy $(T_e)_1 = 0$.

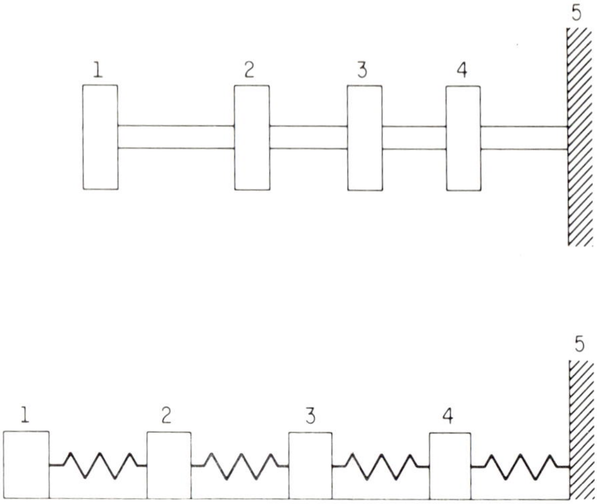

FIG. 9-6

Example

Holzer tables (Tables 9-1 to 9-4) for the four vibratory modes of a free five-mass torsional system are given here as an example. The different moments of inertia and torsional stiffnesses for the system are as tabulated. Notice that for the first vibratory mode θ changes sign once. This occurs between stations 2 and 3. The point on the connecting shaft where $\theta = 0$ is called a *node*. The tables for the second, third, and fourth modes show the existence of 2, 3, and 4 nodes respectively. The number of nodes can be taken to define the order of the vibratory mode.

Damped Systems

The system shown in Fig. 9-7 represents a torsional system with general condition of damping. We shall denote c_j for the damping coefficient between mass j and the fixed frame. We shall also denote $d_{j,(j+1)}$ for the damping coefficient between mass j and mass $j + 1$. The equations

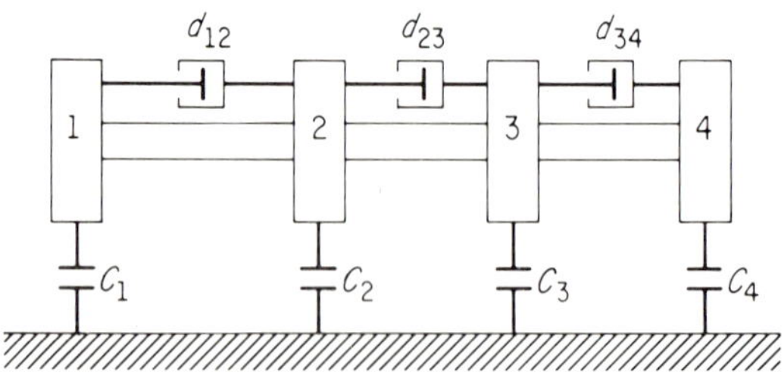

FIG. 9-7

TABLE 9-1
Holzer Tables for Undamped System
First Mode: $\omega_{n_1} = 69.416$ rad/sec

Station	I, lb_m/sec^2	θ, rad	$\Sigma I\omega^2\theta$	k, lb_m/rad	$\Delta\theta = \frac{\Sigma I\omega^2\theta}{k}$
1	0.1644×10^6	1.0	0.792385×10^9	0.10788×10^6	0.7345
2	0.1644×10^6	0.2655	0.10028×10^{10}	0.350×10^9	2.8651
3	0.7111×10^6	−2.5996	0.112041×10^9	0.1322×10^9	0.8474
4	0.1333×10^6	−3.447	0.89895×10^8	0.16322×10^9	0.551
5	0.4666×10^6	−3.998	$-0.49655 \approx 0$		

TABLE 9-2
Second Mode: $\omega_{n_2} = 119.515$ rad/sec

Station	I, lb_m/sec^2	θ, rad	$\Sigma I\omega^2\theta$	k, lb_m/rad	$\Delta\theta = \frac{\Sigma I\omega^2\theta}{k}$
1	0.1644×10^6	1.0	0.234889×10^{10}	0.10788×10^{10}	2.1772
2	0.1644×10^6	−1.1772	-0.416115×10^9	0.350×10^9	−1.1889
3	0.7111×10^6	0.0117	-0.404185×10^9	0.1322×10^9	−3.0569
4	0.1333×10^6	3.0686	-0.345743×10^9	0.16322×10^9	−2.1182
5	0.4666×10^6	5.1868	$2.4087 \approx 0$		

TABLE 9-3
Third Mode: $\omega_{n_3} = 128.595$ rad/sec

Station	I, lb_m/sec^2	θ, rad	$\Sigma I\omega^2\theta$	k, lb_m/rad	$\Delta\theta = \frac{\Sigma I\omega^2\theta}{k}$
1	0.1644×10^6	1.0	0.27193×10^{10}	0.10788×10^{10}	2.5205
2	0.1644×10^6	−1.5205	-0.141556×10^{10}	0.350×10^9	−4.0444
3	0.7111×10^6	2.5239	0.15524×10^{10}	0.1322×10^9	11.7410
4	0.1333×10^6	−9.2171	0.13492×10^{10}	0.16322×10^9	8.2660
5	0.4666×10^6	−17.4831	$-2.3473 \approx 0$		

TABLE 9-4
Fourth Mode: $\omega_{n_4} = 492.9$ rad/sec

Station	I, lb_m/sec^2	θ, rad	$\Sigma I\omega^2\theta$	k, lb_m/rad	$\Delta\theta = \frac{\Sigma I\omega^2\theta}{k}$
1	0.1644×10^6	1.0	0.39952×10^{11}	0.10788×10^{10}	37.031
2	0.1644×10^6	−36.031	-0.13995×10^{13}	0.350×10^9	−3998.731
3	0.7111×10^6	3962.7	0.67062×10^{14}	0.1322×10^9	0.507163×10^6
4	0.1333×10^6	-0.5032×10^6	-0.9595×10^{14}	0.16322×10^9	-0.5878×10^6
5	0.4666×10^6	0.846×10^5	$20.6 \approx 0$		

of motion in this case are the same as in the undamped case except for the complex nature of the j inertia and stiffness terms. By assuming sinusoidal oscillations, the general form for these equations will be

$$-I_j\omega^2\theta_j + k_{j,(j+1)}(\theta_j - \theta_{j,j+1}) + ic_j\omega\theta_j + d_{j,j+1}\omega(\theta_j - \theta_{j,(j+1)}) = 0$$

$$-(I_j\omega^2 - ic_j\omega)\theta_j + (k_{j,(j+1)} + id_{j,j+1}\omega)(\theta_j - \theta_{j,j+1)}) = 0$$

where ω = the frequency of the vibration. The corresponding Holzer tables will have complex numbers. The natural frequencies are determined by the condition that the magnitude of the residual torque T_e is a minimum. Notice that the angular displacement will also be complex which indicates that in the damped case, there are phase differences between the oscillations of the different masses.

Example

Tables 9-5 and 9-6 give sample Holzer results for the damped system shown in Fig. 9-8 at the two natural frequencies. These frequencies are

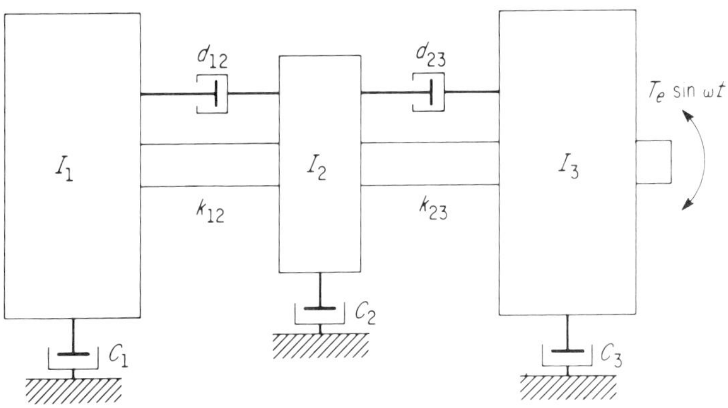

FIG. 9-8

evaluated from a series of Holzer calculations where the magnitude of the residual torque is plotted versus assumed frequencies. The plot exhibits minimum values at the natural frequencies. The system characteristics in this case are

$$I_1 = 3.0 \text{ lb-in. sec}^2$$
$$I_2 = 1.5 \text{ lb-in. sec}^2$$
$$I_3 = 8 \text{ lb-in. sec}^2$$
$$k_{12} = 0.4 \times 10^6 \text{lb-in./rad}$$
$$k_{23} = 8 \times 10^6 \text{lb-in./rad}$$
$$c_1 = c_2 = c_3 = 200 \text{ lb-in./(rad/sec)}$$
$$d_{12} = d_{23} = 200 \text{ lb-in./(rad/sec)}$$

TABLE 9-5

$\omega_{n_1} = 370$ rad/sec

Station	$\frac{(I\omega^2 - ic\omega)}{10^6}$	θ	$\frac{(I\omega^2 - ic\omega)\theta}{10^6}$	$\frac{\Sigma(I\omega^2 - ic\omega)\theta}{10^6}$	$\frac{k + id\omega}{10^6}$	$\Delta\theta$
1	$0.4107 - 0.074i$	1.0	$0.4107 - 0.074i$	$0.4107 - 0.074i$	$0.4 + 0.074i$	$0.957 - 0.362i$
2	$0.2053 - 0.074i$	$0.0403 + 0.3625i$	$0.0351 + 0.0715i$	$0.446 - 0.0025i$	$0.8 + 0.074i$	$0.557 - 0.0008i$
3	$1.10 - 0.074i$	$-0.0154 + 0.3633i$	$0.01 + 0.399i$	$0.456 + 0.396i$		

$$T_e = 10^6\sqrt{(0.456)^2 + (0.396)^2} = 0.604 \times 10^6$$

TABLE 9-6

$\omega_{n_2} = 2480$ rad/sec

Station	$\frac{(I\omega^2 - ic\omega)}{10^8}$	θ	$\frac{(I\omega^2 - ic\omega)\theta}{10^8}$	$\frac{\Sigma(I\omega^2 - ic\omega)\theta}{10^8}$	$\frac{k + id\omega}{10^6}$	$\Delta\theta$
1	$0.1845 - 0.005i$	1.0	$0.1845 - 0.005i$	$0.1845 - 0.005i$	$0.4 + 0.456i$	$17.57 - 23i$
2	$0.0923 - 0.005i$	$-16.57 + 23i$	$-1.415 + 2.2i$	$-1.23 + 2.2i$	$0.4 + 0.456i$	$-13.62 + 28.34i$
3	$0.492 - 0.005i$	$-2.95 - 5.34i$	$-1.48 - 2.61i$	$-2.71 - 0.41i$	$8 + 0.456i$	

$$T_e = 10^8\sqrt{(2.71)^2 + (0.41)^2} = 2.741 \times 10^8$$

Tabular Method for the Far-Coupled Systems

A procedure similar to that of the Holzer method has been devised for the far-coupled systems by Myklestad (Ref. 25) and Prohl (Ref. 26). The tabulation has been further simplified by W. Thomson (Ref. 19). The procedure can be explained as follows:

Considering the beam Fig. 9-9 carrying masses $m_1, m_2, \ldots, m_n, \ldots, m_k$ at spans $l_1, l_2, \ldots, l_k$ to be oscillating with a frequency ω; the equations

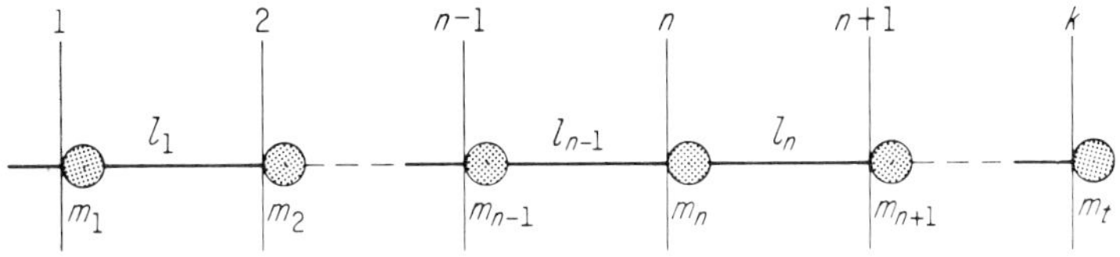

FIG. 9-9

for the shearing forces, bending moments, slopes and deflections at stations $1, 2, \ldots, n, \ldots, k$ which are located immediately to the left of the masses, are given by

$$\begin{aligned} V_n &= V_{n-1} + m_{n-1}\omega^2 y_{n-1} \\ M_n &= M_{n-1} + V_n l_{n-1} \\ \theta_n &= \theta_{n-1} - V_n \alpha_{n-1} - M_{n-1}\beta_{n-1} \\ y_n &= y_{n-1} - \theta_{n-1} l_{n-1} + V_n \alpha'_{n-1} + M_{n-1}\beta'_{n-1} \end{aligned} \tag{9-1}$$

where V, M, θ, and y are the shearing force, bending moment slope and deflection respectively. α, α', β and β' are the influence coefficients and are defined as follows:

$\alpha_{n-1}, \beta_{n-1}$ = slope of beam at station n measured with respect to tangent at $n - 1$ due to a unit force and unit moment at station $n - 1$.

$\alpha'_{n-1}, \beta'_{n-1}$ = deflections at station n measured from a tangent at $n - 1$ due to a unit force and unit moment at $n - 1$.

When the beam segments have uniform cross section, the influence coefficients can be obtained by the moment area method (Fig. 9-10) as

$$\alpha_{n-1} = \left[\frac{l^2}{2EI}\right]_{n-1}$$

$$\beta_{n-1} = \left[\frac{l}{EI}\right]_{n-1}$$

$$\alpha'_{n-1} = \left[\frac{l^3}{6EI}\right]_{n-1}$$

$$\beta'_{n-1} = \left[\frac{l^2}{2EI}\right]_{n-1}$$

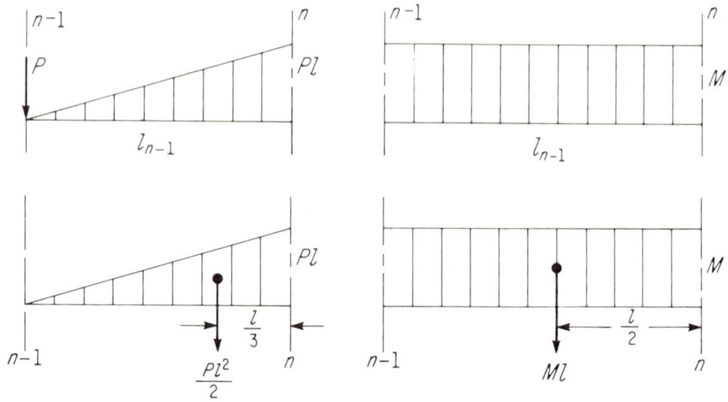

FIG. 9-10

Boundary Conditions

The boundary conditions for different types of beam supports are

	V	M	θ	y
Simple support	V	0	θ	0
Free	0	0	θ	y
Fixed	V	M	0	0

In the analysis of the beam, the starting station will be one of the previous types and accordingly will have two nonzero boundary values. Two tabulations will therefore be necessary to accommodate them. Successive values of ω are assumed as in the Holzer method. The two tabulations are carried out by utilizing Eq. 9-1. The value of ω which simultaneously satisfies the two final boundary conditions will be a natural frequency.

Simply Supported Beams

Taking, for example, a simply supported beam, the two initial boundary values are $V = V_i$, $\theta = \theta_i$. The final boundary values for any ω are obtained from the two tables and will have the form

$$\begin{aligned} M_f &= aV_i + b\theta_i \\ y_f &= cV_i + d\theta_i \end{aligned} \tag{9-2}$$

where a, c are the moment and deflection values respectively from the V-table for $V_i = 1$ and $\theta_i = 0$. Similarly, b, d are the moment and deflection values respectively from the θ-table for $\theta_i = 1$ and $V_i = 0$.

We can arbitrarily satisfy one of the two final boundary values, for example, take $M_f = 0$ in Eq. 9-2 from which

$$\theta_i = -\frac{a}{b} V_i$$

by substituting in the second boundary conditions

$$y_f = \left[c - \frac{a}{b}d\right] V_i$$
$$= RV_i$$

which will be satisfied when $R = 0$. Therefore, by successive assumption of ω, the corresponding R is evaluated. The values of ω giving $R = 0$ are the natural frequencies of the beam.

Cantilever Beam (Fig. 9-11)

The nonzero initial boundary values in this case are $\theta = \theta_i$ and $y = y_i$. From the corresponding tables we can find the final boundary

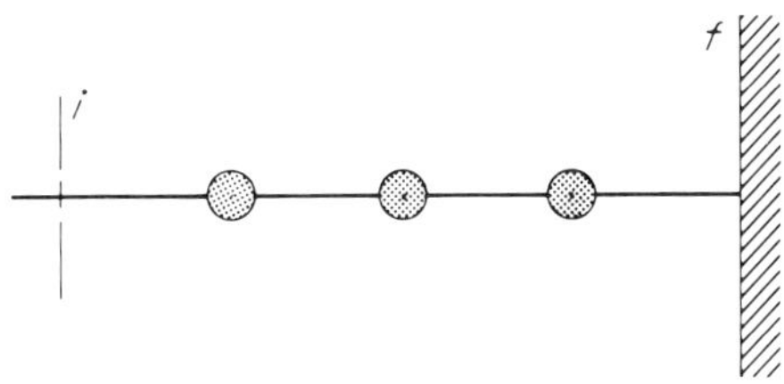

FIG. 9-11

values as

$$\theta_f = a\theta_i + by_i$$
$$y_f = c\theta_i + dy_i$$

where a, c are the values of the slope and deflection respectively from the θ-table and b, d are the corresponding values from the y-table.

Satisfying the $\theta_f = 0$ we obtain $y_i = -(a/b)\theta_i$ and by substitution in the y_f equation we get

$$y_f = \left[d - \frac{b}{a}c\right] y_i$$
$$= Ry_i$$

and all boundary conditions are satisfied when $R = 0$.

Example

The beam shown in Fig. 9-12 has

$$l = 10 \text{ in.}$$
$$W = 386 \text{ lb}$$
$$l_1 = 0.6l$$
$$l_2 = 0.2l = l_3$$
$$(EI_1) = (EI_2) = (EI_3) = 10^7 \text{ lb-in.}^2$$

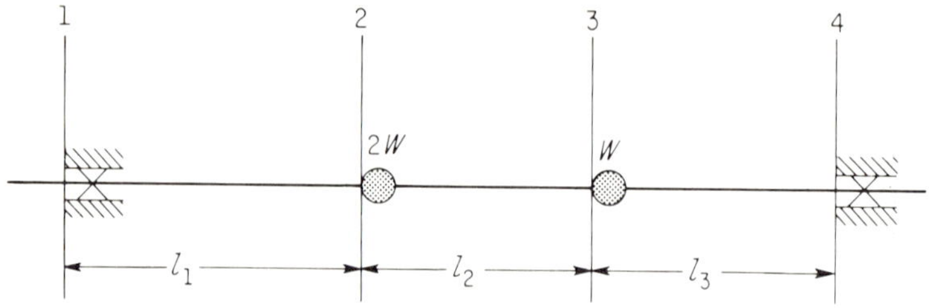

FIG. 9-12

Prove by the use of the tabulation method that $\omega = 465$ rad/sec is a natural frequency.

Solution

$$\alpha_1 = 18 \times 10^{-7}, \quad \alpha_2 = 2 \times 10^{-7} = \alpha_3, \quad \beta_1 = 6 \times 10^{-7},$$
$$\beta_2 = 2 \times 10^{-7} = \beta_3$$
$$\alpha_1' = 36 \times 10^{-7}, \quad \alpha_2' = 1.33 \times 10^{-7} = \alpha_3', \quad \beta_1' = 18 \times 10^{-7},$$
$$\beta_2' = 2 \times 10^{-7} = \beta_3'$$
$$\omega = 465 \text{ rad/sec}, \quad \omega^2 = 21.7 \times 10^4$$

The initial boundary values are

$$V_1 = V_i, \quad \theta_1 = \theta_i, \quad y_1 = 0, \quad M_1 = 0$$

The V-table will be as follows:

Station	$m\omega^2$	V	M	θ	y
1	0	1	0	0	0
2	$(43.4)10^4$	1	6	$(-18)10^{-7}$	$(36)10^{-7}$
3	$(21.7)10^4$	2.56	11.12	$(-35.12)10^{-7}$	$(87.3)10^{-7}$
4	0	4.46	20.04		$(186.54)10^{-7}$

from which $a = 20.04$ and $c = (186.54)10^{-7}$.

Similarly, *the θ-table* will be

Station	$m\omega^2$	V	M	θ	y
1	0	0	0	1	0
2	$(43.4)10^4$	0	0	1	-6
3	$(21.7)10^4$	$(-2.6)10^6$	$(-5.2)10^6$	(1.52)	-8.346
4	0	4.415	$(-14.03)10^6$		-12.91

from which $b = (-14.03)10^6$ and $d = -12.91$.

Therefore

$$(R)_{(\omega=465)} = c - \frac{a}{b} d = (186.54)10^{-7} - \frac{(20.04)}{(-14.03)}(-12.9)10^{-6}$$
$$= (186.54)10^{-7} - (185)10^{-7} \approx 0$$

which shows that $\omega = 465$ rad/sec is a natural frequency. Have we chosen a different value for ω, the remainder would have been considerably different from zero. The natural frequencies would be obtained by the successive assumption of ω and searching for the values giving $R = 0$.

The so-called "gyroscopic effects" resulting from consideration of the rotors as rigid bodies rather than particles can be easily incorporated in this tabulation procedure. An additional term should be included at each mass to account for the inertial moment due to the angular vibration of the rotor. This requires a modification of Eq. 9-1 by substituting M_{n-1} whenever it appears in the right-hand side by $[M_{n-1} + I_{n-1}\omega^2\theta_{n-1}]$. The same procedure can then be followed to calculate the natural frequency.

9-3. NUMERICAL METHODS

Matrix Iteration

The matrix-iteration procedure is a convenient numerical method which enables us (especially with the help of high-speed digital computers) to solve the simultaneous equations of motion by successive approximations. When the equations are written in the matrix form, arbitrary values are assigned to the different coordinate amplitudes. By substituting these values in the equations, new values will result. These are in turn substituted in the equations and so on, until the substituted and resulting values are approximately the same and therefore constitute a solution. The matrix-iteration procedure provides a very useful tool for solution of coupled linear conservative systems or systems where the equations of motion are conveniently written in a matrix form. For detailed discussions on matrix methods the reader is referred to texts on numerical analysis (Ref. 20, for example).

We shall try a solution for the problem of Fig. 9-12 by means of matrix iteration for illustration. The system is redrawn in Fig. 9-13. The first step is to calculate the necessary influence coefficients for the system. These represent the deflection at one station due to a unit force at another station. For the simply supported uniform beam under consideration,

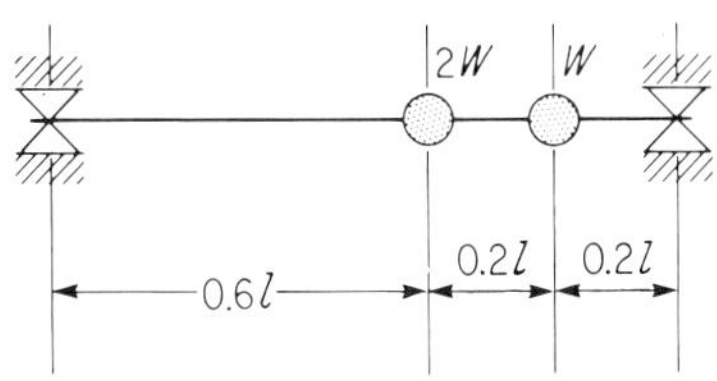

FIG. 9-13

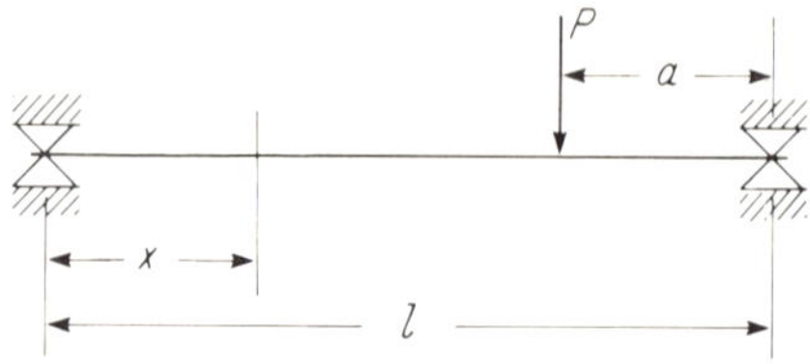

FIG. 9-14

these coefficients can be calculated from the equation (see Fig. 9-14)

$$y_x = \frac{Pax}{6EI\,l}(l^2 - x^2 - a^2)$$

where P is the load, EI is the flexural rigidity of the beam, and l, x, and a are the dimensions defined in Fig. 9-14. The influence coefficients are therefore

$$\begin{aligned}
a_{11} &= \text{deflection at station 1 due to a unit load at station 1} \\
&= \frac{0.4 \times 0.6l^2}{6EIl}(l^2 - (0.6)^2l^2 - (0.4)^2l^2) \\
&= \frac{0.0192l^3}{EI} \\
a_{12} &= \text{deflection at station 1 due to a unit load at station 2} \\
&= \frac{0.2 \times 0.6l^2[l^2 - (0.6)^2l^2 - (0.2)^2l^2]}{6EIl} \\
&= \frac{(0.012)l^3}{EI} \\
a_{21} &= \text{deflection at station 2 due to a unit load at station 1} \\
&= \frac{0.6 \times 0.2l^2[l^2 - (0.2)^2l^2 - (0.6)^2l^2]}{6EIl} \\
&= \frac{(0.012)l^3}{EI} \\
a_{22} &= \text{deflection at station 2 due to a unit load at station 2} \\
&= \frac{0.2 \times 0.8l^2(l^2 - (0.8)^2l^2 - (0.2)^2l^2)}{6EIl} \\
&= \frac{(0.00853)l^3}{EI}
\end{aligned}$$

Notice that $a_{12} = a_{21}$, which is expected according to *Maxwell's reciprocity principle.* This principle states that in linear systems the deflection at a station i due to a unit load at station j is equal to the deflection at station j due to a unit load at station i. This can be written as $a_{ij} = a_{ji}$. Now we can proceed to write the equation of motion for the system based on consideration of the beam deflection under the inertia forces. These equations are

$$y_1 = a_{11}\omega^2 m_1 y_1 + a_{12}\omega^2 m_2 y_2$$
$$y_2 = a_{21}\omega^2 m_1 y_1 + a_{22}\omega^2 m_2 y_2$$

and when written in the matrix form will be

$$\begin{vmatrix} y_1 \\ y_2 \end{vmatrix} = \omega^2 \begin{vmatrix} a_{11}m_1 & a_{12}m_2 \\ a_{12}m_1 & a_{22}m_2 \end{vmatrix} \begin{vmatrix} y_1 \\ y_2 \end{vmatrix}$$

which is a convenient notation for

$$\begin{matrix} y_1 \searrow \\ y_2 \nearrow \end{matrix} = \rightrightarrows \omega^2 \begin{matrix} \nearrow (a_{11}m_1 y_1 + a_{12}m_2 y_2) \\ \searrow (a_{22}m_1 y_1 + a_{22}m_2 y_2) \end{matrix}$$

substituting the influence coefficients in the matrix we obtain

$$\begin{vmatrix} y_1 \\ y_2 \end{vmatrix} = \frac{\omega^2 l^3 W}{EIg} \begin{vmatrix} 0.0384 & 0.012 \\ 0.024 & 0.00853 \end{vmatrix} \begin{vmatrix} y_1 \\ y_2 \end{vmatrix} \tag{9-3}$$

We are now ready to start the iteration process. We can assume any values for y_1, y_2 as a first approximation. The closer these values are to the first mode of vibration, the faster the convergence toward the first mode. For lack of better values, let us take $y_1 = y_2 = 1$ as a first approximation. By substitution in the matrix we get

$$\begin{vmatrix} y_1 \\ y_2 \end{vmatrix} = \frac{\omega^2 l^3 W}{EIg} \begin{vmatrix} 0.0384 + 0.012 \\ 0.024 + 0.00853 \end{vmatrix} = \frac{\omega^2 l^3 W}{EIg} \begin{vmatrix} 0.0504 \\ 0.03253 \end{vmatrix}$$
$$= \frac{\omega^2 l^3 W}{EIg} (0.0504) \begin{vmatrix} 1 \\ 0.647 \end{vmatrix}$$

This last step is called *normalizing the matrix*, which means keeping one of the amplitudes equal to unity. We can now use $y_1 = 1$ and $y_2 = 0.647$ as a second approximation which gives

$$\begin{vmatrix} y_1 \\ y_2 \end{vmatrix} = \frac{\omega^2 l^3 W}{EIg} \begin{vmatrix} 0.0384 & 0.012 \\ 0.024 & 0.00853 \end{vmatrix} \begin{vmatrix} 1 \\ 0.647 \end{vmatrix}$$

$$= \frac{\omega^2 l^3 W}{EIg} \begin{vmatrix} 0.0384 + 0.00776 \\ 0.024 + 0.00555 \end{vmatrix}$$

$$= \frac{\omega^2 l^3 W}{EIg} (0.04616) \begin{vmatrix} 1 \\ 0.64 \end{vmatrix}$$

Using the resulting values for the third iteration we get

$$\begin{vmatrix} y_1 \\ \\ y_2 \end{vmatrix} = \frac{\omega^2 l^3 W}{EIg} \begin{vmatrix} 0.0384 + 0.00769 \\ \\ 0.024 \quad + 0.00546 \end{vmatrix} = \frac{\omega^2 l^3 W}{EIg}(0.04609) \begin{vmatrix} 1 \\ \\ 0.64 \end{vmatrix}$$

This indicates that we now have the solution for the first mode. The normal mode is given by $y_2/y_1 = 0.64$ and the corresponding natural frequency is obtained from

$$\frac{\omega^2 l^3 W}{EIg}(0.04609) = 1 \quad \text{which gives } \omega \approx 4.65\sqrt{\frac{EIg}{Wl^3}}$$

It should be noted here that in matrix solutions, the iteration always converges to the first mode except for the unlikely condition when the first assumed values represent exactly one of the higher modes. In order to obtain the higher modes by iteration, the matrix of Eq. 9-3 has to be modified for each mode by utilizing the orthogonality relationship Eq. 8-23 between the mode under consideration and the previous modes. Therefore in this case:

$$\sum_{i=1}^{i=2} m_i y_i^{(1)} y_i^{(2)} = 0$$

where $y_i^{(1)}$ and $y_i^{(2)}$ are the first- and second-mode amplitudes respectively.

$$\frac{2W}{g} y_1^{(1)} y_1^{(2)} + \frac{W}{g} y_2^{(1)} y_2^{(2)} = 0$$

$$2 \times 1 \times y_1^{(2)} + 1 \times 0.64\, y_2^{(2)} = 0$$

$$y_1^{(2)} = -0.32\, y_2^{(2)}$$

We can also write

$$y_2^{(2)} = y_1^{(2)} + y_2^{(2)} \tag{9-4}$$

We can now obtain a new matrix:

$$\begin{vmatrix} y_1^{(2)} \\ y_2^{(2)} \end{vmatrix} = \begin{vmatrix} 0 & -0.32 \\ 0 & 1 \end{vmatrix} \begin{vmatrix} y_1^{(2)} \\ y_2^{(2)} \end{vmatrix}$$

and the matrix for the second mode will be

$$\begin{vmatrix} y_1 \\ \\ y_2 \end{vmatrix} = \left[\frac{\omega^2 l^3 W}{EIg}\right] \begin{vmatrix} 0.0384 & 0.012 \\ \\ 0.024 & 0.00853 \end{vmatrix} \begin{vmatrix} 0 & -0.32 \\ \\ 0 & 1 \end{vmatrix} \begin{vmatrix} y_1 \\ \\ y_2 \end{vmatrix}$$

and the matrix of coefficients is obtained by multiplying the rows of the first matrix by the columns of the second.

$$\begin{vmatrix} y_1 \\ \\ y_2 \end{vmatrix} = \frac{\omega^2 l^3 W}{EIg} \begin{vmatrix} A_1 & A_2 \\ \\ B_1 & B_2 \end{vmatrix} \begin{vmatrix} y_1 \\ \\ y_2 \end{vmatrix}$$

where $A_1 = 0.384 \times 0 + 0.012 \times 0 = 0$

$$A_2 = -0.0386 \times 0.32 + 0.012 \times 1 = -0.000288$$

$$B_1 = 0.024 \times 0 + 0.00853 \times 0 = 0$$

$$B_2 = -0.024 \times 0.32 + 0.00853 \times 1 = +0.00085$$

$$\begin{vmatrix} y_1 \\ y_2 \end{vmatrix} = \frac{\omega^2 l^3 W}{EIg} \begin{vmatrix} 0 & -0.000288 \\ 0 & +0.00085 \end{vmatrix} \begin{vmatrix} y_1 \\ y_2 \end{vmatrix}$$

which gives the second mode directly as

$$\begin{vmatrix} y_1 \\ y_2 \end{vmatrix} = \frac{\omega^2 l^3 W}{EIg}(0.00085) \begin{vmatrix} -0.34 \\ 1 \end{vmatrix}$$

The normal mode is given by $y_2/y_1 = 1/-0.34 = -2.94$ and the corresponding natural frequency is

$$\omega_n = 34.2\sqrt{\frac{EIg}{Wl^3}} \quad \text{rad/sec}$$

In general, for any number of degrees of freedom the "reducing" or "sweeping" matrix (which is the matrix necessary to extract the influence of the lower modes from the original matrix equation) can be obtained from the orthogonality relationships between the mode under consideration and the preceding ones. In order to complete a square matrix, the rest of the coordinates are set equal to themselves.

The matrix method which is briefly introduced here is a powerful tool of analysis of linear systems. Programs for matrix iteration are usually available in most computer-program libraries.

For rigorous mathematical generalization of this technique, the reader is referred to texts on numerical analysis or matrix algebra dealing with the eigenvalue problem. (Refs. 24, 29, for example).

Runge-Kutta Method

There are many numerical methods of solution for systems of differential equations such as the Euler Predictor-Corrector, Adams-Bashforth, and Runge-Kutta methods (Refs. 29, 30). The latter is extensively used and is generally available in most computer-program libraries.

The Runge-Kutta method will be first described without proof for the solution of a general second-order differential equation. It will then be extended for the solution of a coupled system of general second-order equations.

Assuming a general equation of the form

$$\ddot{x} = f(t, x, \dot{x})$$

a step-by-step solution for this equation can be obtained starting at any step n where t_n, x_n, and $\dot{x}_n$ are known. First, the following quantities are calculated:

$$k_1 = hf(t_n, x_n, \dot{x}_n)$$

$$k_2 = hf\left[\left(t_n + \frac{h}{2}\right), \left(x_n + \frac{h}{2}\dot{x}_n + \frac{h}{8}k_1\right), \left(\dot{x}_n + \frac{k_1}{2}\right)\right]$$

$$k_3 = hf\left[\left(t_n + \frac{h}{2}\right), \left(x_n + \frac{h}{2}\dot{x}_n + \frac{h}{8}k_1\right), \left(\dot{x}_n + \frac{k_2}{2}\right)\right]$$

$$k_4 = hf\left[(t_n + h), \left(x_n + h\dot{x}_n + \frac{h}{2}k_3\right), (\dot{x}_n + k_3)\right]$$

where h = a selected small increment of time

$x_n, \dot{x}_n$ = displacement and velocity at any time t_n

The change in displacement Δx and velocity $\Delta \dot{x}$ can then be calculated as

$$\Delta x = h\left[\dot{x}_n + \frac{1}{6}(k_1 + k_2 + k_3)\right]$$

$$\Delta \dot{x} = \frac{1}{6}[k_1 + 2k_2 + 2k_3 + k_4]$$

and the new values of displacement and velocity at $t_{n+1} = t_n + h$ are given by

$$x_{n+1} = x_n + \Delta x$$

$$\dot{x}_{n+1} = x_n + \Delta \dot{x}$$

Therefore, knowing the initial conditions of displacement and velocity, the complete history of the motion can be numerically obtained.

The same procedure can be extended for the solution of a general system of differential equations. Assuming these equations to be

$$\ddot{q}_r = f_r[t, (q_1, q_N), (\dot{q}_1, \dot{q}_N)]$$

$$= f_r[t, (q_r)_{r=1\to N}, (\dot{q}_r)_{r=1\to N}]$$

where r = 1 to N

N = number of degrees of freedom

The solution in this case can be calculated as follows:

$$k_{1r} = hf_r[t_n, q_{rn}, \dot{q}_{rn}]$$

$$k_{2r} = hf_r\left[\left(t_n + \frac{h}{2}\right), \left(q_{rn} + \frac{h}{2}\dot{q}_{rn} + \frac{h}{8}k_{1r}\right)_{r=1\to N'}\left(\dot{q}_{rn} + \frac{k_{1r}}{2}\right)_{r=1\to N}\right]$$

$$k_{3r} = hf_r\left[\left(t_n + \frac{h}{2}\right), \left(q_{rn} + \frac{h}{2}\dot{q}_{rn} + \frac{h}{8}k_{1r}\right)_{r=1\to N}, \left(\dot{q}_{rn} + \frac{k_{2r}}{2}\right)_{r=1\to N}\right]$$

$$k_{4r} = hf_r\left[(t_n + h), \left(q_{rn} + h\dot{q}_{rn} + \frac{h}{2}k_{3r}\right)_{r=1\to N}, (\dot{q}_{rn} + k_{3r})_{r=1\to N}\right]$$

where h = a selected increment of time

$q_{rn}, \dot{q}_{rn}$ = generalized coordinates and their time derivatives

The change in displacements and velocities are calculated as

$$\Delta q_r = h\left[\dot{q}_r + \frac{1}{6}(k_{1r} + k_{2r} + k_{3r})\right]$$

$$\Delta q_r = \frac{1}{6}[k_{1r} + 2k_{2r} + 2k_{3r} + k_{4r}]$$

The new values of q_r and $\dot{q}_r$ at time $t_{(n+1)} = t_n + h$ are therefore

$$q_{r(n+1)} = q_{rn} + \Delta q_r$$
$$\dot{q}_{r(n+1)} = \dot{q}_{rn} + \Delta\dot{q}_r$$

The complete history of motion can accordingly be obtained in a step-by-step fashion.

9-4. GRAPHICAL SOLUTIONS

The Generalized Phase-Plane- Method

The method explained here is due to Abramson and Chu (Ref. 28) and is applicable to the general case of multidegree-of-freedom systems. It yields directly the steady-state as well as the transient response. In order to increase the practicality and accuracy of this technique, the method is formulated for solution by high-speed digital computers.

The equations of motion for a system with N degrees of freedom can be written as

$$\begin{aligned}\ddot{q}_n &= f_n(q_1, q_2, \ldots, q_n, \dot{q}_1, \dot{q}_2, \ldots, \dot{q}_N, t)\\ &= -p_0^2 q_n + G_n(q_1, q_2, \ldots, q_N, \dot{q}_1, \ldots, \dot{q}_N, t)\end{aligned}$$

where $n = 1, 2, \ldots, N$ and p_0 = a single constant which may be taken as a convenient velocity scale factor.

This equation can be rewritten in terms of the phase-plane variables $\dot{q}_n, q_n$ by substituting $\ddot{q}_n = \dot{q}_n d\dot{q}_n/dq_n$, and therefore

$$\frac{dq_n}{d\left(\dfrac{\dot{q}_n}{p_0}\right)} = \frac{-\left(\dfrac{\dot{q}_n}{p_0}\right)}{q_n - \dfrac{G_n}{p_0^2}} = \tan\phi_n \tag{9-5}$$

where $(\tan \phi_n)$ is the slope of the plot in the nth phase plane as shown in Fig. 9-14.

A step-by-step solution can therefore be obtained by simultaneous plots in the N phase planes.

Knowing the initial conditions for $q_n, \dot{q}_n$, points A_n can be plotted in the different phase planes and the corresponding ϕ_n can be calculated from Eq. 9-5. The line T-T can therefore be plotted to represent the tangent at A_n. The normal to T-T at A_n will intersect the q_n axis at O_n. Now the arc plotted with center O_n and radius $\rho_n = O_nA_n$ will determine the new points B_n after a suitable incremental time $p_0 \delta t$. The conditions at the new points B_n are then used in Eq. 9-5 to find the new slopes, and so on.

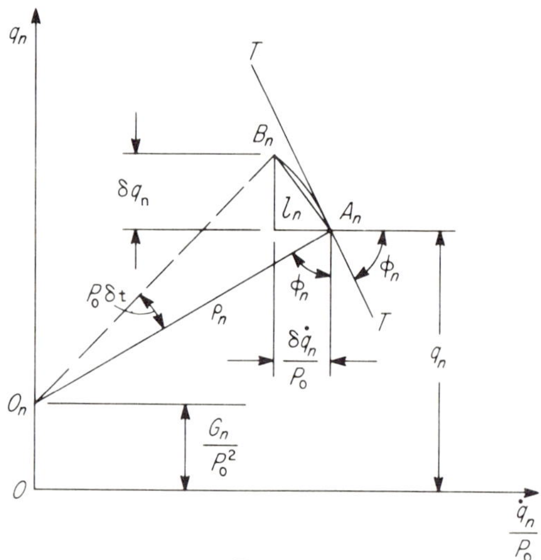

FIG. 9-15

It can be easily seen from Fig. 9-15 that

$$\delta q_n = l_n \sin\left(\phi_n - \frac{p_0 \delta t}{2}\right)$$

$$\delta\left(\frac{\dot{q}_n}{p_0}\right) = -l_n \cos\left(\phi_n - \frac{p_0 \delta t}{2}\right)$$

where $l_n = 2\rho_n \sin \dfrac{p_0 \delta t}{2}$

$$\rho_n = \sqrt{(q_n - \delta_n)^2 + (\dot{q}_n/p_0)^2}$$

$$\phi_n = \tan^{-1} \frac{\dot{q}_n/p_0}{q_n - \delta_n}$$

$$\delta_n = \frac{G_n}{p_0^2}$$

It should be noted here that several iterations may be necessary at each point to determine the point O_n which is applicable for the phase plots between A_n and B_n.

Graphical solutions by the phase-plane method are convenient for certain classes of systems with a small number of degrees of freedom. As a general rule such graphical constructions become tedious for complex systems with many degrees of freedom. The formulation of the procedure as given before is suitable for programming for computer solutions.

9-5. USE OF THE ANALOG COMPUTER FOR SOLUTION OF SYSTEMS WITH FINITE DEGREES OF FREEDOM

It has been shown previously that the analog computer can be effectively utilized for the solution of single degree of freedom systems with linear or nonlinear characteristics. When the system has more than one degree of freedom, the computer solution is performed by coupling the different loops representing the different coordinates as shown in the case of the two-degrees-of-freedom system of Fig. 9-16. The equations

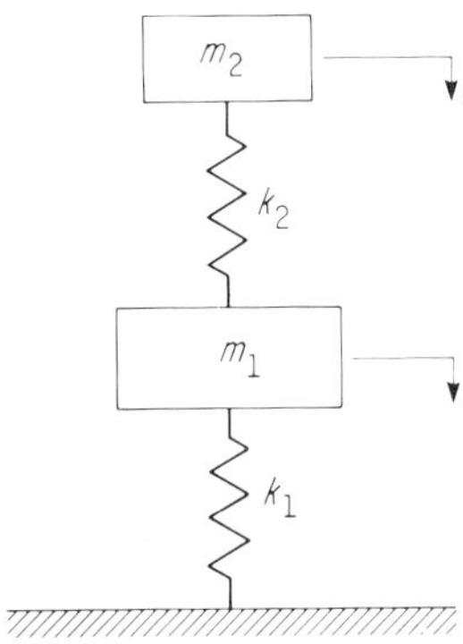

FIG. 9-16

describing the system are

$$\ddot{x}_1 = -\frac{k_1}{m}x_1 - \frac{k_2}{m}(x_1) + \frac{k_2}{m}x_2$$

$$\ddot{x}_2 = -\frac{k_2}{m}x_2 + \frac{k_2}{m}x_1$$

and the computer diagram will be as shown in Fig. 9-17.

Example

A system simulating a car with shock absorbers (Fig. 9-18) will be considered as an illustration of the use of analog computers in systems

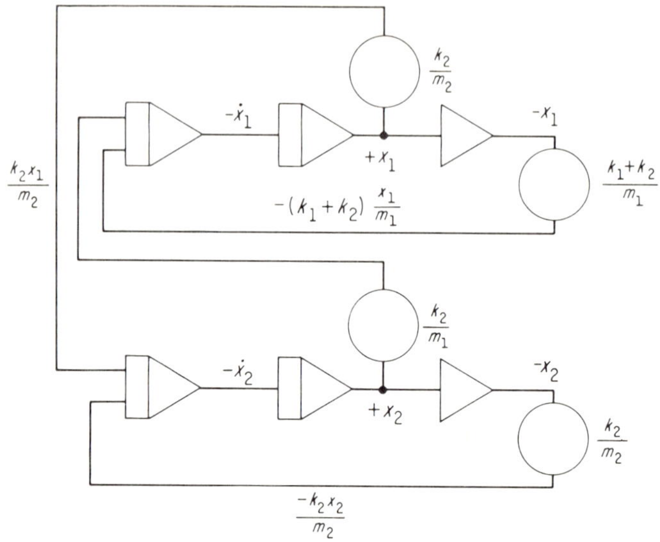

FIG. 9-17

with finite degrees of freedom. Using the notation

m_1 = one quarter of mass of automobile
m_2 = mass of the wheel and axle
k_1 = spring constant of main auto spring
k_2 = spring constant of tire (assumed linear)
c_1 = shock-absorber damping constant
x_1 = displacement of auto body
x_2 = displacement of wheel
x_3 = roadway profile displacement

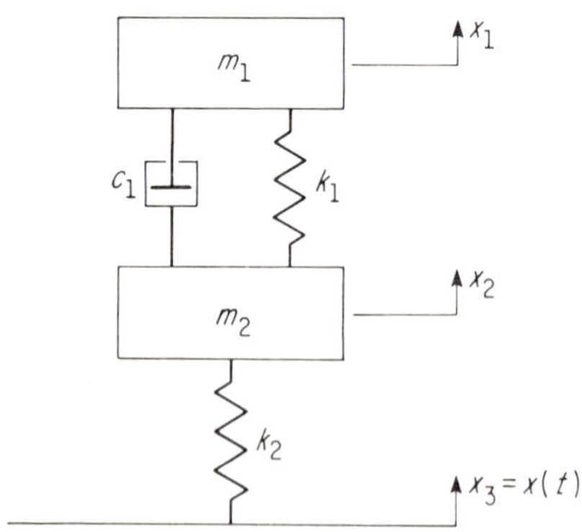

FIG. 9-18

Taking

$$\begin{aligned} m_1 &= 25 \text{ slugs} \\ m_2 &= 2 \text{ slugs} \\ k_1 &= 1000 \text{ lb/ft} \\ k_2 &= 4500 \text{ lb/ft} \\ c_1 &= \text{variable} \end{aligned}$$

and the initial conditions

$$x_1 = x_2 = \frac{dx_1}{dt} = \frac{dx_2}{dt} = 0 \quad \text{at} \quad t = 0$$

System Equations

The differential equations of motion of the system are

$$\frac{d^2x_1}{dt^2} = -\frac{c_1}{m_1}\left(\frac{dx_1}{dt} - \frac{dx_2}{dt}\right) - \frac{k_1}{m_1}(x_1 - x_2)$$

$$\frac{d^2x_2}{dt^2} = -\frac{c_1}{m_2}\left(\frac{dx_2}{dt} - \frac{dx_1}{dt}\right) - \frac{k_1}{m_2}(x_2 - x_1) - \frac{k_2}{m_2}(x_2 - x_3)$$

Scaled Equations

Assume the following maximum values of the scaled variables to assure that the computer problem voltages are as high as possible but less than ± 10 volts.

$$\left[5\,\frac{d^2x_1}{dt^2}\right] \quad \left[0.8\,\frac{d^2x_2}{dt^2}\right]$$

$$\left[5\,\frac{dx_1}{dt}\right] \quad \left[0.8\,\frac{dx_2}{dt}\right]$$

$$[50x_1] \quad [50x_2]$$

The scaled equations then become

$$[5\ddot{x}_1] = -\left(\frac{c_1}{25}\right)[5\dot{x}_1] + \left(\frac{c_1}{4}\right)[0.8\dot{x}_2] - (4.0)[50x_1] + (4.0)[50x_2]$$

$$[0.8\ddot{x}_2] = -\left(\frac{c_1}{2}\right)[0.8\dot{x}_2] + (0.08c_1)\,[5\dot{x}_1] - (44)[50x_2] + (8)[50x_1] + (1800)[x_3]$$

( ) = Combination potentiometer settings and amplifier gains.
[ ] = Voltage outputs from computing components.

Computer Diagram (Fig. 9-19)

Each input to every integrator is multiplied by a factor $1/\beta$ to change the speed with which the system behavior is obtained on the computer.

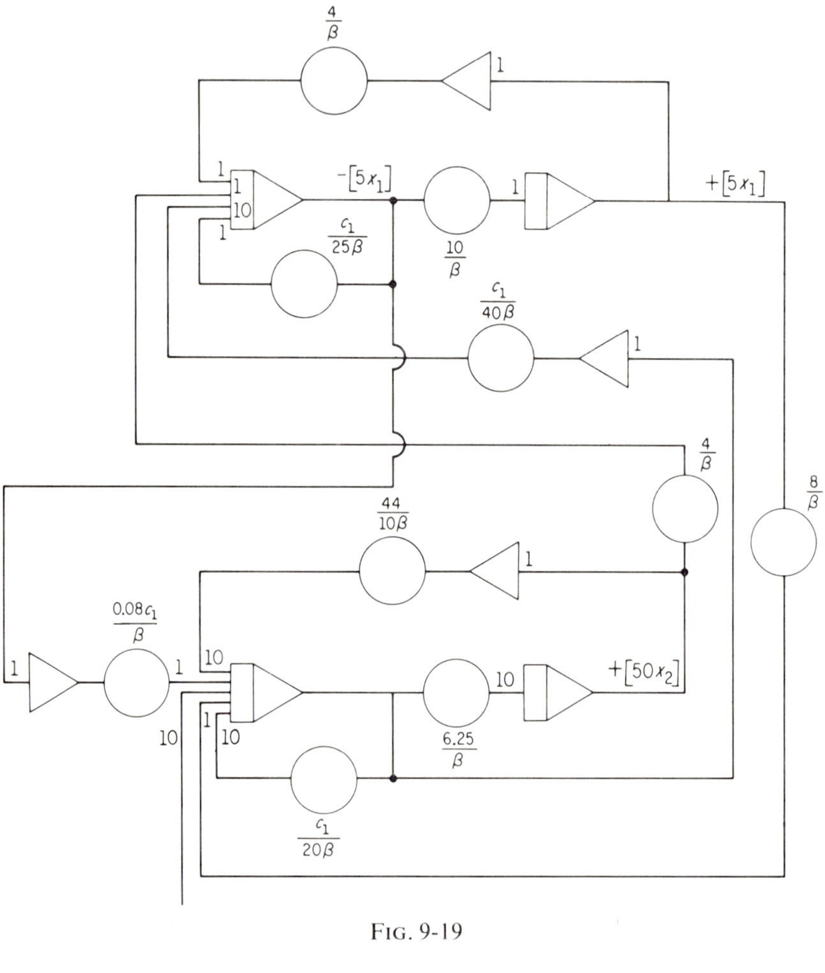

FIG. 9-19

In this example $\beta = 10$ appears appropriate and the computer-circuit behavior will be ten times slower than that of the suspension system. Fig. 9-20 shows the computer solution when x_3 is a square pulse of 35 millisec duration. c_1 is taken as 20, 10, and 100 lb-ft^{-1} sec. respectively.

In many cases of dynamic systems the use of "hybrid computers" becomes advantageous. Such computers combine the logic of the digital computer and the versatility of the analog systems. An extensive account of analog and hybrid computing can be found in Ref. 31.

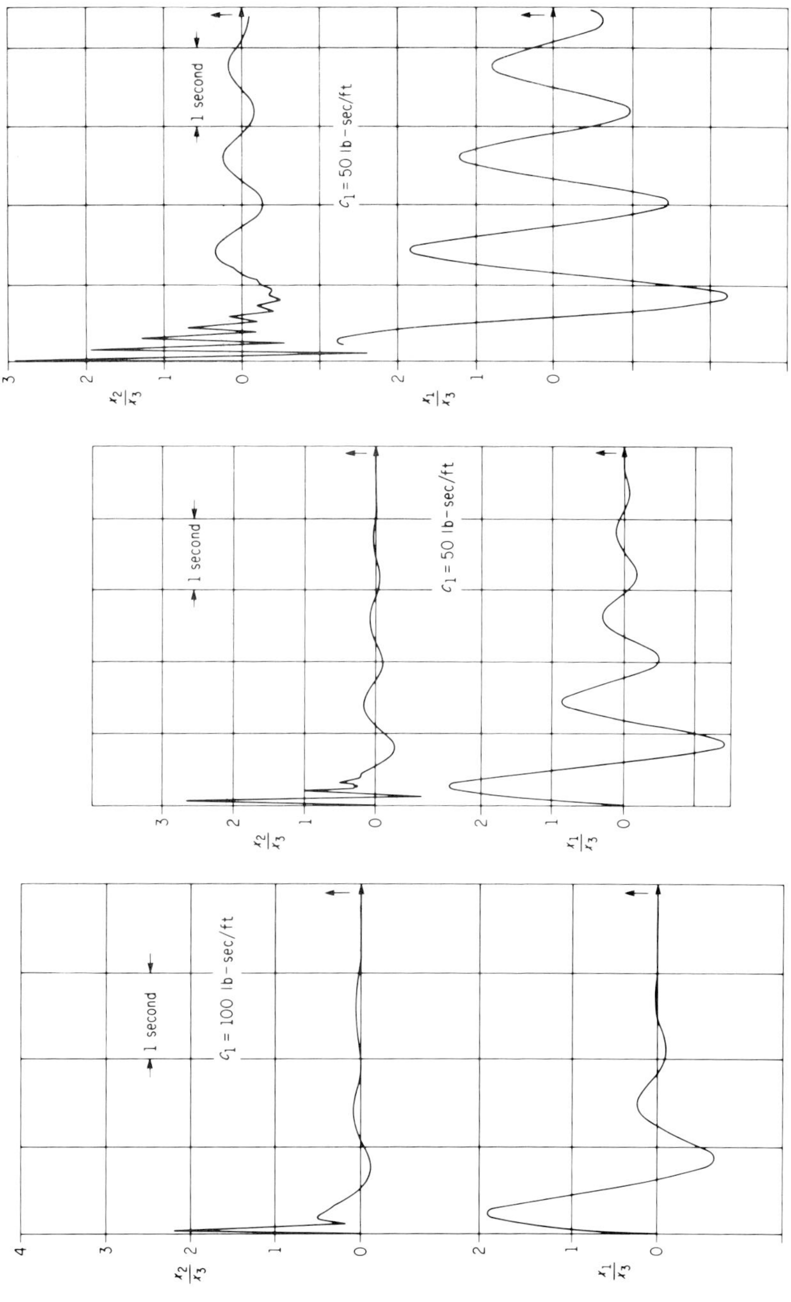
1 second
$c_1 = 50$ lb-sec/ft
$\frac{x_2}{x_3}$
$\frac{x_1}{x_3}$
1 second
$c_1 = 50$ lb-sec/ft
$\frac{x_2}{x_3}$
$\frac{x_1}{x_3}$
1 second
$c_1 = 100$ lb-sec/ft
$\frac{x_2}{x_3}$
$\frac{x_1}{x_3}$

FIG. 9-20

9-6. SUMMARY

In this chapter several approximate methods of solution have been presented for the systems with a finite number of degrees of freedom. Holzer tabulation for close-coupled systems and the Mykelstad-Prohl method for the far-coupled systems has been discussed. The matrix-iteration technique for linear systems has been briefly illustrated. The powerful Runge-Kutta method for the solution of the general case of a system of second-order differential equations has been described. The application of the analog computer to the solution of such systems has also been demonstrated.

PROBLEMS

The systems in the following problems have natural frequencies which are approximately evaluated as given in each case. Use tabular methods to check the accuracy of these frequencies and to determine the corresponding normal modes.

9-1. $\omega_{n_1} = 0,$ $\omega_{n_2} = 0.518\sqrt{k/I},$
$\omega_{n_3} = 1.0\sqrt{k/I},$ $\omega_{n_4} = 1.41\sqrt{k/I},$
$\omega_{n_5} = 1.73\sqrt{k/I}$ and $\omega_{n_6} = 1.93\sqrt{k/I}$ rad/sec.

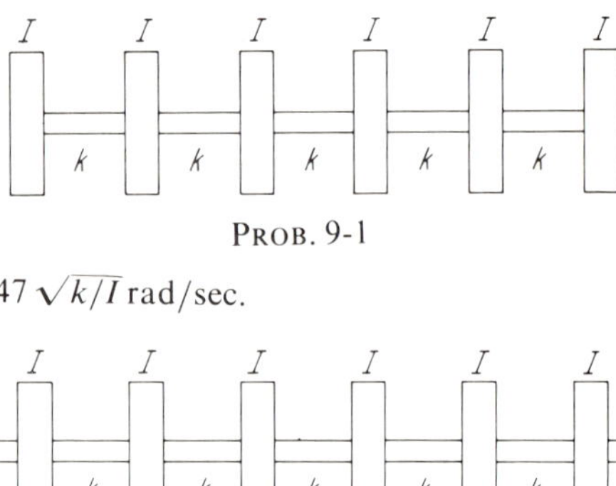

PROB. 9-1

9-2. $\omega_{n_1} = 0.347\sqrt{k/I}$ rad/sec.

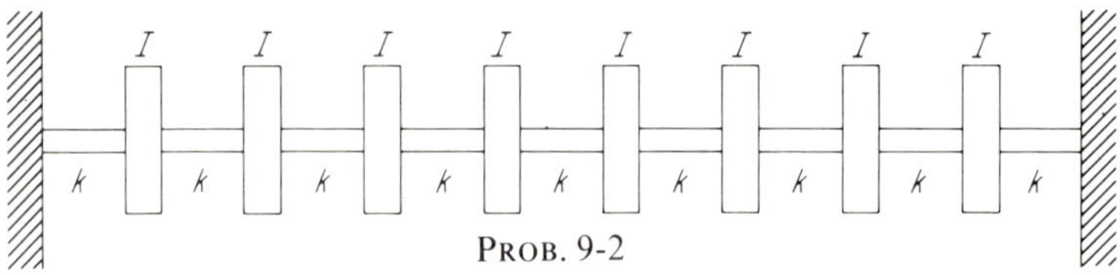

PROB. 9-2

9-3. $\omega_{n_1} = 0.3\sqrt{k/I},$ $\omega_{n_2} = 0.81\sqrt{k/I},$
$\omega_{n_3} = 1.45\sqrt{k/I}$ and $\omega_{n_4} = 2.83\sqrt{k/I}$ rad/sec.

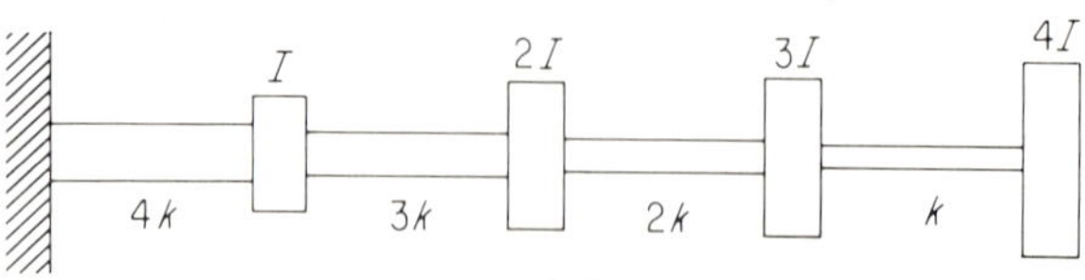

PROB. 9-3

9-4. $m_1 = 1.0$ lb-in.$^{-1}$-sec^2
$m_2 = 10$ lb-in.$^{-1}$-sec^2
$k = 9000$ lb/in.
$c_1 = 126$ lb-in.$^{-1}$-sec
$\omega_n = 72.5$ rad/sec.

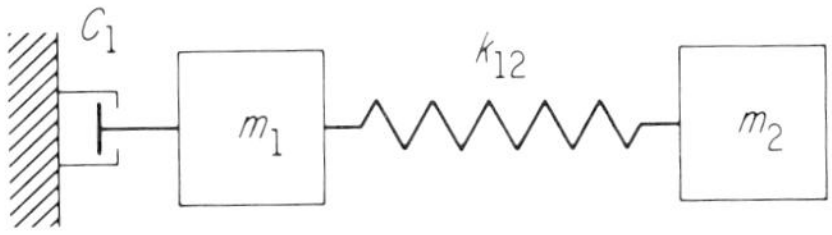

PROB. 9-4

9-5. $\omega_{n_1} = 0, \quad \omega_{n_2} = \sqrt{k/m},$
$\omega_{n_3} = 1.732\sqrt{k/m}$ rad/sec.

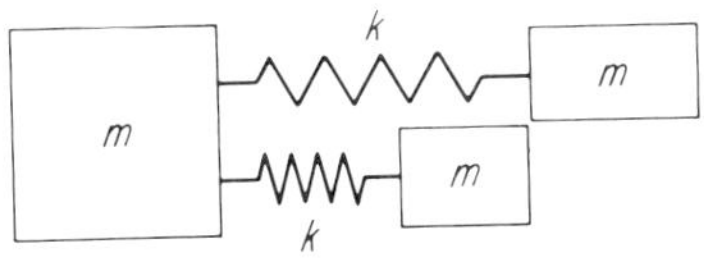

PROB. 9-5

9-6.

$m_1 = 10$ lb-in.$^{-1}$-sec^2		$k_{12} = 100$ lb/in.	
$m_2 = 15$	"	$k_{23} = 200$	"
$m_3 = 20$	"	$k_{34} = 200$	"
$m_4 = 10$	"	$k_{45} = 100$	"
$m_5 = 10$	"	$k_{56} = 150$	"
$m_6 = 20$	"		

$\omega_{n_1} = 1.64$ rad/sec.

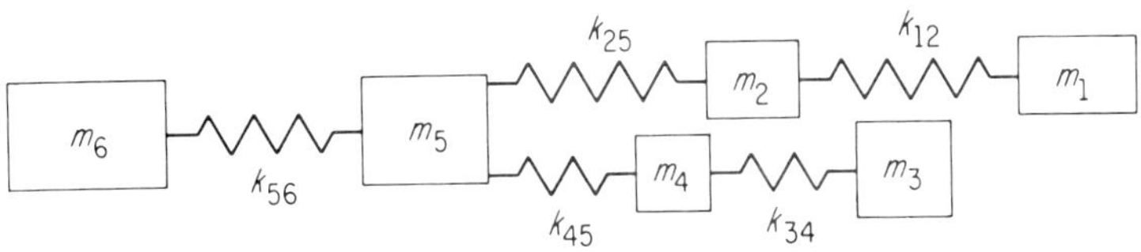

PROB. 9-6

9-7. Uniform beam with constant EI

$$\omega_{n_1} = 0.125\sqrt{EIg/WL^3} \text{ rad/sec}$$

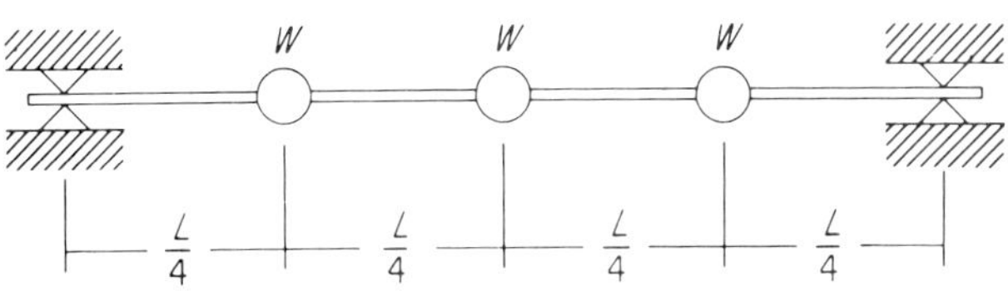

PROB. 9-7

9-8. Uniform beam with constant $EI = 30 \times 10^6$ lb/in.^{2}

$$W_1 = 200 \text{ lb}, \quad W_2 = 100 \text{ lb}$$
$$l_1 = 6 \text{ in.}, \quad l_2 = 4 \text{ in.}$$
$$\omega_n = 500 \text{ rad/sec}$$

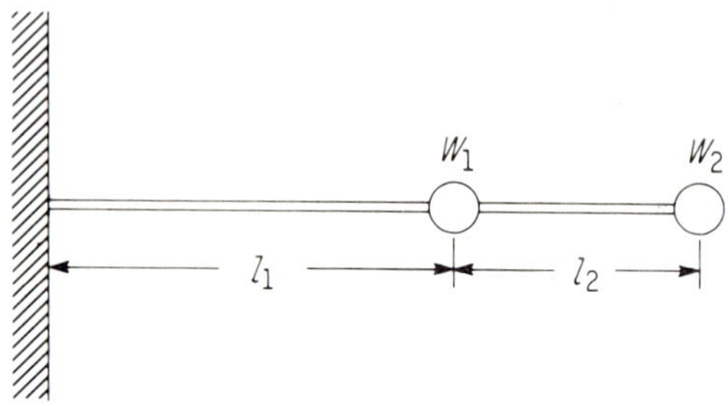

PROB. 9-8

9-9. Uniform beam with constant EI

$$W_1 = 500 \text{ lb}, \quad W_2 = 100 \text{ lb}$$
$$\omega_{n_1} = 4.57\sqrt{EI/L^3} \text{ rad/sec}$$

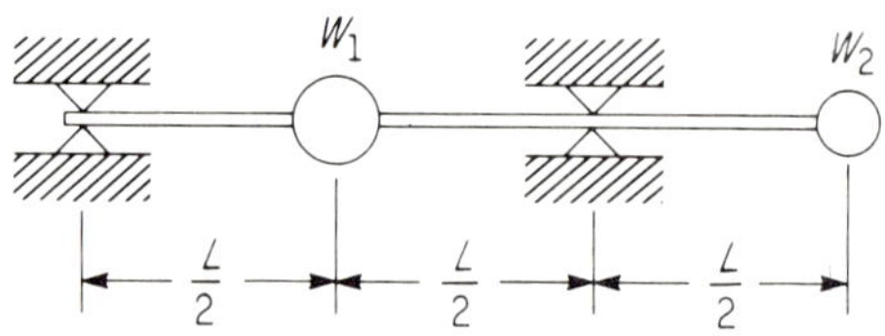

PROB. 9-9

chapter 10

Normal-Mode Solutions for Lumped Parameter Systems

10-1. INTRODUCTION

It has been shown in the previous chapters that a system of n masses can be readily described by n differential equations. Because of the physical coupling between the masses, the equations are not independent and must be solved simultaneously. This may be prohibitively difficult for a system with many degrees of freedom and complex forcing functions.

However, it is possible that a linear conservative system can be represented in a more simple manner by means of the normal-mode method.

This technique enables the differential equations of motion to be uncoupled by a simple transformation of coordinates. The uncoupled equations are independent single-coordinate equations and therefore can be solved singly. Each equation represents one of the vibrational modes for the system. A summation of the independent modal solutions produces the solution in the original system coordinates. The use of the normal-mode method greatly simplifies the analysis of linear systems, since the response of such systems to any forcing function can be easily obtained by convolution integrals. Normal-mode techniques also provide physical insight into the contribution of each mode to the total response of the system. The concept of normal modes has been discussed briefly in Chapter 8.

This chapter gives a summary of the normal-mode method and its application to undamped and damped systems. A technique is also presented which permits an approximate mathematical model for a damped multimass system to be constructed entirely from experimentally obtained sinusoidal frequency-response data.

10-2. NORMAL-MODE EQUATIONS FOR CONSERVATIVE SYSTEMS

The systems shown in Figs. 10-1 and 10-2 are conservative n-degree of freedom systems with far and close coupling respectively. The differ-

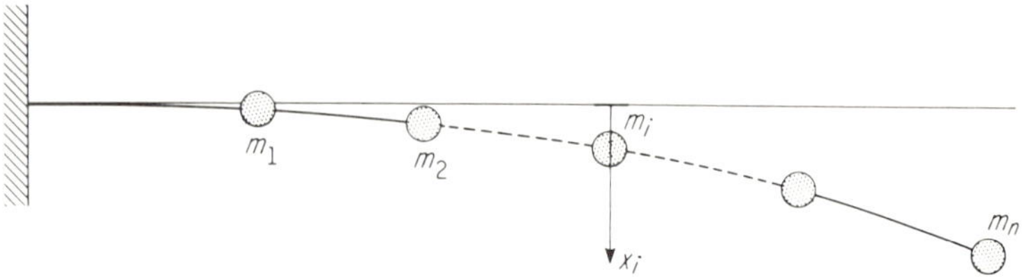

FIG. 10-1

ential equations describing the free oscillation of these systems can be derived from Lagrange's equation

$$\frac{d}{dt}\left(\frac{\partial T}{\partial \dot{x}_i}\right) - \frac{\partial T}{\partial x_i} + \frac{\partial U}{\partial x_i} = 0$$

where x_i = displacement of the ith mass in an inertial reference,

$i = 1 \rightarrow n$

T = kinetic energy of system

$= \frac{1}{2} \sum_{i=1}^{n} \sum_{j=1}^{n} m_i x_i^2$

U = potential energy of system

$= \frac{1}{2} \sum_{i=1}^{n} \sum_{j=1}^{n} K_{ij} x_i x_j$

K_{ij} = array of coefficients describing stiffness of system, $K_{ij} = K_{ji}$

n = number of degrees of freedom

The resulting n equations can be simultaneously solved to produce the eigenvalues (or natural frequencies) ω_k for $k = 1$ to n, and the corresponding eigenvectors ϕ_{ik} which describe the normal modes of vibration.

The coupled differential equations of motion can be transformed (see Ref. 41, Chapter V) into a set of uncoupled equations of the form

$$\ddot{\eta}_k + \omega_k \eta_k = 0 \qquad k = 1 \rightarrow n \tag{10-1}$$

where η_k represents the normal coordinates.

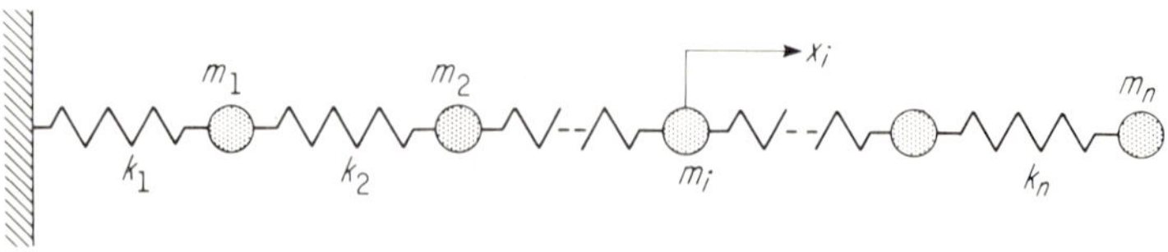

FIG. 10-2

The solution in the original system coordinates can be found from

$$x_i = \sum_{k=1}^{n} \phi_{ik}\, \eta_k \tag{10-2}$$

which defines the coordinate transformation necessary to uncouple the equations of motion of the system.

As was discussed in Chapter 8, an important property of normal modes is their orthogonality. The orthogonality relations for any two modes r and s are stated as

$$\begin{aligned} \sum_i m_i \phi_{ir} \phi_{is} &= 0 \qquad \text{for } r \neq s \\ &= M_k \qquad \text{for } r = s = k \end{aligned} \tag{10-3}$$

and

$$\begin{aligned} \sum_{ij} K_{ij} \phi_{is} \phi_{jr} &= 0 \qquad \text{for } r \neq s \\ &= M_k \omega_k \qquad \text{for } r = s = k \end{aligned} \tag{10-4}$$

where

$$M_k = \sum_i m_i \phi_{ik}^2 \tag{10-5}$$

and

$$\omega_k = \text{eigenvalues}$$

10-3. SYSTEMS SUBJECTED TO GENERAL FORCING FUNCTIONS

The solution of this problem is simplified considerably by transforming to normal coordinates. This technique will be applied to systems subjected to excitations at the masses as well as to base excitations.

Mass Excitation

The system shown in Fig. 10-3 is subjected to general force excitations at the different mass. Lagrange's equation in normal coordinates for this system can be written as

$$\frac{d}{dt}\left(\frac{\partial T}{\partial \dot{\eta}_k}\right) - \frac{\partial T}{\partial \eta_k} + \frac{\partial U}{\partial \eta_k} = \frac{\partial W}{\partial \eta_k}$$

where W = work done on the system by the different forces

$$= \sum_i F_i x_i$$

$$= \sum_i F_i \phi_{ik}\, \eta_k$$

$$\frac{\partial W}{\partial \eta_k} = \sum_i F_i \phi_{ik}$$

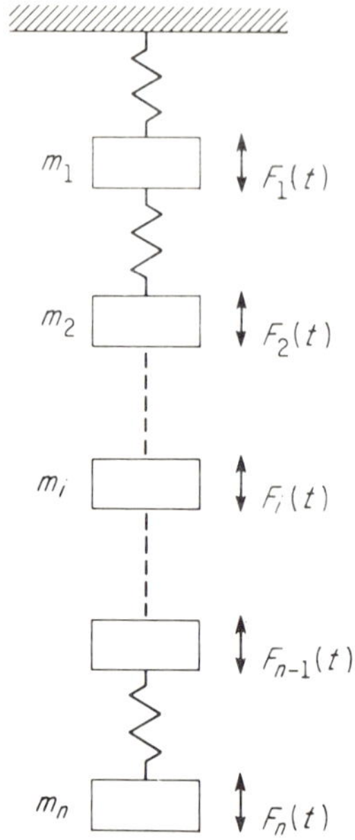

FIG. 10-3

By substituting T, U, and W in Lagrange's equation, the result is the set of uncoupled equations

$$\begin{aligned} \ddot{\eta}_k + \omega_k^2 \eta_k &= \frac{\Sigma F_i \phi_{ik}}{M_k} \\ &= \frac{F_k}{M_k} \end{aligned} \tag{10-6}$$

This equation has the same form as that of a single-degree-of-freedom system subjected to a force excitation. Assuming zero initial conditions for the displacement and velocity, the general solution of this equation can be obtained as shown in Chapter 5 by a convolution (Duhamel's) integral. This gives

$$\eta_k = \frac{1}{M_k \omega_k} \int_0^t F_k(\tau) \sin \omega_k (t - \tau) d\tau$$

and the solution in the original system coordinates is then calculated from

$$x_i = \sum_k \phi_{ik} \eta_k$$

Base Excitation

The response of the system shown in Fig. 10-4 to a ground excitation can be obtained in a similar manner to the previous case. Assuming that

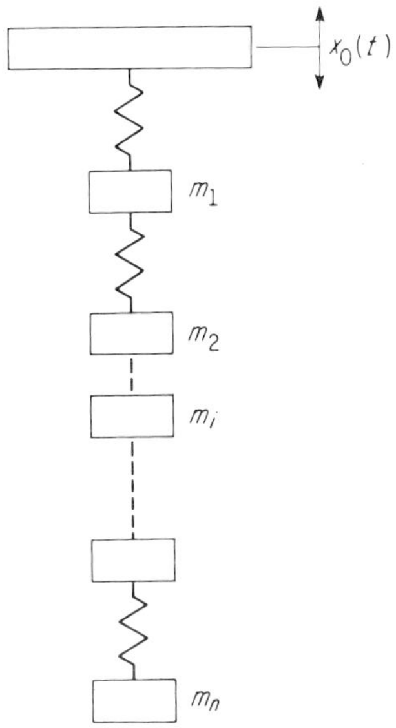

FIG. 10-4

the excitation is given by $x_0(t)$, the displacement coordinates y_i relative to the base are therefore

$$y_i = x_i - x_0$$

The coordinates y_i can be transformed to normal coordinates by

$$y_i = \Sigma \phi_{ik} \eta_k$$

Therefore

$$\begin{aligned} T &= \frac{1}{2} \sum_i m_i (\dot{y}_i + \dot{x}_0)^2 \\ &= \frac{1}{2} \sum_i m_i \left(\sum_k \phi_{ik} \dot{\eta}_k + \dot{x}_0 \right)^2 \\ &= \frac{1}{2} \sum_k M_k \dot{\eta}_k^2 + \dot{x}_0 \sum_i m_i \sum_k \phi_{ik} \dot{\eta}_k + \frac{\dot{x}_0^2}{2} \sum_i m_i \end{aligned}$$

$$\begin{aligned} U &= \frac{1}{2} \sum_{ij} K_{ij} x_i x_j \\ &= \frac{1}{2} \sum_{i,j} K_{ij} \phi_{ir} \phi_{is} \eta_{ir} \eta_{is} \\ &= \frac{1}{2} \sum_k M_k \omega_k^2 \eta_k^2 \end{aligned}$$

Substitution in Lagrange's equation gives

$$\frac{d}{dt}\left(\Sigma M_k \dot{\eta}_k + \dot{x}_0 \sum_i m_i \phi_{ik}\right) + \Sigma M_k \omega_k^2 \eta_k = 0$$

from which

$$\ddot{\eta}_k + \omega_k^2 \eta_k = -\ddot{x}_0 \sum_i m_i \phi_{ik} / M_k$$

$$= -a_k \ddot{x}_0 \tag{10-7}$$

where

$$a_k = \frac{\Sigma m_i \phi_{ik}}{M_k} \tag{10-8}$$

is known as the *modal-participation factor.*

The solution of this equation for zero initial conditions can also be obtained by convolution integration as

$$\eta_k = -\frac{a_k}{\omega_k} \int_0^t \ddot{x}_0(\tau) \sin \omega_k (t - \tau) d\tau$$

and the solution in the original system coordinates is therefore

$$y_i = \sum_k \phi_{ik} \eta_k$$

10-4. SYSTEMS WITH RAYLEIGH DAMPING

Free Vibration

It has been shown that uncoupled differential equations in normal coordinates exist for conservative systems. Such formulation simplifies the analysis of these systems, especially when they are subjected to complex excitations.

The normal-mode method does not generally apply to damped systems. Only a particular class of damped systems originally defined by Rayleigh (Ref. 32) and later generalized by Caughey (Ref. 33) can be uncoupled by the same transformation which uncouples conservative systems.

Lagrange's equation for a damped system can be written as

$$\frac{d}{dt}\left(\frac{\partial T}{\partial \dot{\eta}_k}\right) - \frac{\partial T}{\partial \eta_k} + \frac{\partial U}{\partial \eta_k} = -\frac{\partial D}{\partial \eta_k}$$

where D = dissipated energy

$$= \frac{1}{2} \Sigma c_{ij} \dot{x}_i \dot{x}_j$$

Using the same transformation as in conservative systems,

$$x_i = \phi_{ik} \eta_k$$

Therefore

$$D = \frac{1}{2} \Sigma c_{ij} \phi_{ik} \phi_{jk} \dot{\eta}_k^2$$

$$\frac{\partial D}{\partial \dot{\eta}_k} = \sum_{ij} c_{ij} \phi_{ik} \phi_{jk} \dot{\eta}_k$$

and Lagrange's equation yields

$$\ddot{\eta}_k + \omega_k^2 \eta_k = -\frac{\Sigma c_{ij} \phi_{ik} \phi_{jk} \dot{\eta}_k}{\Sigma m_{ij} \phi_{ik} \phi_{jk}} \tag{10-9}$$

This equation is uncoupled when $c_{ij}/m_{ij} = c_i/m_i$ is constant. This condition which is known as uniform mass damping occurs when the damping coefficient at each mass is proportional to the magnitude of the mass. The normal-mode equations in this case become

$$\ddot{\eta}_k + R_1 \dot{\eta}_k + \omega_n^2 \eta_k = 0 \tag{10-10}$$

where $R_1 = c_i/m_i$.

Another damping condition which uncouples the modal equation can be obtained by substituting the orthogonality condition Eq. 10-4 in Eq. 10-9. This gives

$$\ddot{\eta}_k + \omega_k^2 \eta_k = -\frac{\Sigma c_{ij} \phi_{ik} \phi_{jk} \dot{\eta}_k}{\Sigma K_{ij} \phi_{ik} \phi_{jk} / \omega_k^2}$$

which is uncoupled when $c_{ij} = K_{ij}$ = constant. This condition is known as *uniform viscous damping* and occurs when each spring has a damper acting in parallel with it with a coefficient proportional to the spring constant. Uniform viscous damping is an appropriate model for systems in which the damping effect is an inherent property the spring material. The normal-mode equations are therefore given by

$$\ddot{\eta}_k + R_2 \omega^2 \dot{\eta}_k + \omega_k^2 \eta_k = 0 \tag{10-11}$$

where $R_2 = c_{ij}/K_{ij} = c_i/k_i$.

Linear damped systems can be described in uncoupled modal equations if the damping matrix is a linear function of the stiffness and the inertia matrix.

In general this is expressed as

$$c_{ij} = R_1 m_{ij} + R_2 K_{ij} \tag{10-12}$$

Such systems are known as *Rayleigh systems* and have the same normal modes (eigenvectors) as the corresponding undamped systems regardless of the magnitude of the damping.

Using the notation ζ_k = normal-mode damping ratio, Rayleigh systems can be described by

$$\ddot{\eta}_k + 2\zeta_k\omega_k\dot{\eta}_k + \omega_k^2\eta_k = 0 \tag{10-13}$$

For close-doupled systems, ζ_k can be expressed in terms of system parameters as follows:

(a) For uniform mass damping

$$2\zeta_k\omega_k = R_1$$

$$\zeta_k = \frac{R_1}{2\omega_k} \tag{10-14}$$

(b) For uniform viscous damping

$$2\zeta_k\omega_k = R_2\omega_k^2$$

$$\zeta_k = \frac{R_2}{2}\omega_k \tag{10-15}$$

(c) For both uniform mass and spring damping

$$2\zeta_k\omega_k = R_1 + R_2\omega_k^2$$

$$\zeta_k = \frac{1}{2}\left[\frac{R_1}{\omega_k} + R_2\omega_k\right] \tag{10-16}$$

10-5. RAYLEIGH SYSTEMS SUBJECTED TO GENERAL FORCING FUNCTIONS

It can be easily seen from Eqs. 10-6 and 10-13 that the normal-mode equations for a system subjected to excitations F_i at the mass can be derived as

$$\ddot{\eta}_k + 2\zeta_k\omega_k\dot{\eta}_k + \omega_k^2\eta_k = \frac{\Sigma F_i\phi_{ik}}{M_k} \tag{10-17}$$

and the uncoupled coordinate is evaluated by a convolution integral of the form

$$\eta_k = \frac{1}{M_k\omega_d}\int_0^t F_k(\tau)e^{-\delta_k\omega_d(t-\tau)}\sin\omega_d(t-\tau)\,d\tau$$

where

$$F_k = \sum_i F_i\phi_{ik}$$

$$\omega_d = \omega_k\sqrt{1-\zeta_k^2}$$

$$\delta_k = \zeta_k/\sqrt{1-\zeta_k^2}$$

and the solution is calculated from $x_i = \sum_k \phi_{ik} \eta_k$. Similarly from Eqs. 10-6 and 10-13, the normal-mode equations for the case of a system with Rayleigh damping subjected to base excitation $\ddot{x}_0(t)$ can be derived as

$$\ddot{\eta}_k + 2\zeta_k \omega_k \dot{\eta}_k + \omega_k^2 \eta_k = -a_k \ddot{x}_0(t) \tag{10-18}$$

where a_k is the modal participation factor (Eq. 10-7). The uncoupled coordinates are therefore similarly evaluated from

$$\eta_k = \frac{-a_k}{\omega_d} \int_0^t \ddot{x}_0(\tau) e^{-\delta_k \omega_d (t-\tau)} \sin \omega_d (t - \tau)\, d\tau$$

and the solution for the motion relative to the base is obtained from

$$y_i = \sum_k \phi_{ik} \eta_k$$

10-6. AN APPROXIMATE NORMAL-MODE METHOD FOR DAMPED LUMPED PARAMETER SYSTEMS

It has been shown in the previous section that a particular class of damped systems (known as *Rayleigh systems*) can be uncoupled by the same transformation which uncouple conservative systems. Certain other damped systems can also be uncoupled but only in complex coordinates. Foss (Ref. 34), O'Kelly (Ref. 35), and Halfman (Ref. 36) have described the complex transformations that are required.

Damping in physical structures is usually of a complex nature. It is usually convenient to associate damping in multidegree-of-freedom systems with the different normal modes and assume it to be viscous (Refs. 38, 39). The method described here (Ref. 40) allows any lumped-parameter linear system subjected to any forcing function to be approximately represented by uncoupled equations of the same form as those which represent the Rayleigh system.

In this method, the equations describing the system response will be written in a form similar to the classical normal-mode equations. Fictitious uncoupled coordinates η_{ik} for each mass i at each mode k and corresponding damping ratios ζ_{ik} will be used in these equations. The system response will be obtained by superposition of these fictitious coordinates whose value can be calculated by equating the derived solution to experimental data (or known response data) at the natural frequencies.

The method can be used within practical limitations to construct a mathematical model of an arbitrary excited system entirely from experimentally obtained sinusoidal frequency-response data. Components of mass, stiffness, and damping need not be known if the experimental data consist of displacement and phase information for all masses over a frequency range that includes all resonances.

The Approximate Normal-Mode Method

The equations representing the kth mode of a linear damped lumped-parameter system, Fig. 10-5, subjected to ground excitation $x_0(t)$ will be

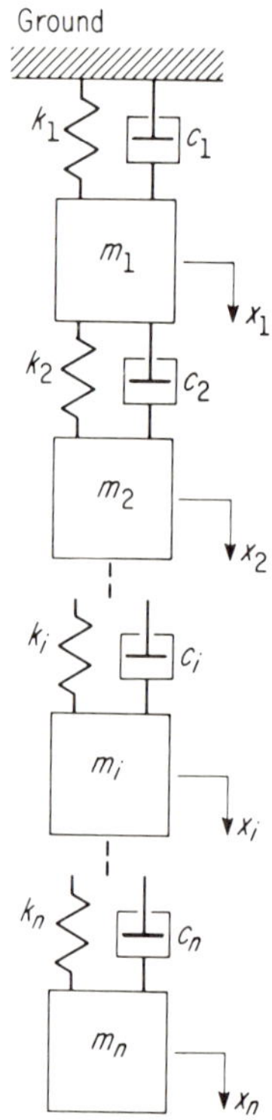

FIG. 10-5. Idealized multimass system.

approximately written as

$$\ddot{\eta}_{ik} + 2\xi_{ik}\omega_k\dot{\eta}_{ik} + \omega_k^2\eta_{ik} = -a_k\ddot{x}_0(t) \tag{10-19}$$

Accordingly, the fictitious uncoupled coordinates η_{ik} are evaluated from

$$\eta_{ik} = \frac{-a_k}{\omega_{id}}\int_0^t \ddot{x}_0(\tau)e^{-\delta_{ik}\omega_{id}(t-\tau)}\sin\omega_{id}(t-\tau)\,d\tau \tag{10-20}$$

where ξ_{ik} = the fictitious damping ratios

$\omega_{id} = \omega_k(1 - \xi_{ik}^2)^{1/2}$

$\delta_{ik} = \xi_{ik}/(1 - \xi_{ik})^{1/2}$

$$a_k = \frac{\Sigma m_i \phi_{ik}}{M_k} = \text{model participation factor} \tag{10-21}$$

ϕ_{ik} = eigenvectors

ω_k = eigenvalues = natural frequencies for undamped systems

m_i = different masses in system

$$M_k = \sum_i m_i \phi_{ik}^2 \tag{10-22}$$

Similarly, for general force excitations $f_i(t)$, the equations will be

$$\ddot{\eta}_{ik} + 2\xi_{ik}\omega_k\dot{\eta}_k + \omega_k^2\eta_{ik} = \frac{\Sigma \phi_{ik} f_i(t)}{M_k} \tag{10-23}$$

and the response can be evaluated from

$$\eta_{ik} = \frac{1}{M_k \omega_{id}} \int_0^t \phi_{ik} f_i(\tau) e^{-\delta_{ik}\omega_{id}(t-\tau)} \sin \omega_{id}(t - \tau)\, d\tau \tag{10-24}$$

The response of the system relative to the base for both conditions respectively can be written as

$$x_i = \sum_k \phi_{ik}\eta_{ik}, \quad y_i = \sum_k \phi_{ik}\eta_{ik} \tag{10-25}$$

The accuracy of this method requires careful selection of the fictitious damping ratios ξ_{ik}. For Rayleigh systems the value of ξ_{ik} will be the same for all masses at any particular mode and will be equal to the modal damping ratio ξ_k. Accordingly, the fictitious coordinates η_{ik} will be the same for all masses at any particular mode and will be equal to the normal coordinate η_k. In the proposed analysis ξ_{ik} is evaluated from known steady-state response values for the system at the different resonant frequencies.

Considering the case of a harmonic ground excitation $x_0(t) = X_0 e^{j\omega t}$, Eq. 10-19 becomes

$$\ddot{\eta}_{ik} + 2\xi_{ik}\omega_k\dot{\eta}_k + \omega_k^2\eta_k = a_k\omega^2 x_0(t)$$

Assuming the steady-state solution to be

$$\eta_{ik} = N_{ik} e^{j(\omega t - \theta_{ik})}$$

and solving, we obtain

$$N_{ik} = \frac{a_k \left(\dfrac{\omega}{\omega_k}\right)^2 X_0}{\sqrt{\left[1 - \left(\dfrac{\omega}{\omega_k}\right)^2\right]^2 + \left[2\xi_{ik}\dfrac{\omega}{\omega_k}\right]^2}} \tag{10-26}$$

and

$$\theta_{ik} = \arctan \frac{2\xi_{ik} \dfrac{\omega}{\omega_k}}{1 - \left(\dfrac{\omega}{\omega_k}\right)^2} \tag{10-27}$$

From Eq. 10-25 the transfer functions for the relative displacements become

$$(hy_i)_\omega = \frac{Y_i}{X_0} = \vec{\sum_{k=1}^{n}} \frac{\phi_{ik} a_k \omega^2/\omega_k^2}{\sqrt{\left[1 - \dfrac{\omega^2}{\omega_k^2}\right]^2 + \left[2\xi_{ik} \dfrac{\omega}{\omega_k}\right]^2}} = \vec{\sum_k} hy_{ik} \tag{10-28}$$

If the absolute coordinates x_i (relative to a fixed frame) are used, the transfer functions in this case take the form

$$(hx_i)_\omega = \frac{X_i}{X_0} = \vec{\sum_{k=1}^{n}} \frac{\phi_{ik} a_k}{\sqrt{\left[1 - \dfrac{\omega^2}{\omega_k^2}\right]^2 + \left[2\xi_{ik} \dfrac{\omega}{\omega_k}\right]^2}} \tag{10-29}$$

When exciting at a particular natural frequency ω_K Eq. 10-28 becomes

$$(hy_i)_K = Q_{iK} = \vec{\sum_k} \frac{\phi_{ik} a_k \omega_K^2/\omega_k^2}{\sqrt{\left[1 - \left(\dfrac{\omega_K}{\omega_k}\right)^2\right]^2 + \left[2\xi_{ik} \dfrac{\omega_K}{\omega_k}\right]^2}} \tag{10-30}$$

where $\vec{\sum_k}$ represents vector summation according to the phase relations of Eq. 10-27.

Now assuming that Q_{ik} are known from experimental data, then ξ_{iK} can be obtained from Eq. 10-30 by simultaneous solution. This is extremely laborious and can be avoided by recognizing that each ξ_{iK} is approximately independent of the remaining ξ_{ik}. This results in two possible simplifications.

A. The 90° Relative-Phase Assumption

In this case it will be assumed that $\xi_{ik} = 0$ when $k \neq K$, and therefore the relative phase between the resonant component and the nonresonant components is 90°. Accordingly, Eq. 10-30 takes the form

$$Q_{iK}^2 = \left[\frac{\phi_{iK} a_K}{2\xi_{iK}}\right]^2 + \left[\sum_{k \neq K} \frac{\phi_{ik} a_k \left(\dfrac{\omega_K^2}{\omega_k^2}\right)}{\left(1 - \dfrac{\omega_K^2}{\omega_k^2}\right)}\right]^2$$

where $\sum_{k \neq K}$ represents algebraic summation. This equation contains a single unknown ξ_{iK} which can be found directly from

$$\xi_{iK} = \frac{\phi_{iK} a_K / 2}{\sqrt{Q_{iK}^2 - \left[\sum_{k \neq K} \frac{\phi_{ik} a_k \omega_K^2 / \omega_k^2}{1 - \frac{\omega_K^2}{\omega_k^2}}\right]^2}} \tag{10-31}$$

B. The Single-Resonance Assumption

If the nonresonant components are assumed to have negligible effect at ω_K, Eq. 10-31 will reduce to

$$\xi_{iK} = \frac{\phi_{iK} a_K}{2Q_{iK}} \tag{10-32}$$

For the case of mass excitations the ξ_{ik} ratios are obtained by analogous procedure. For the particular case where $F_i = 0$ when $i \neq 1$, the transfer function becomes

$$(hx_i)_\omega = \frac{k_1 x_i}{F_1} = \vec{\sum_k} \frac{k_1 \phi_{ik} / M_k}{\omega_k^2 \sqrt{\left[1 - \frac{\omega^2}{\omega_k^2}\right]^2 + \left[2\xi_{ik} \frac{\omega}{\omega_k}\right]^2}}$$

Assuming no damping associated with the spring k_1, then

$$\sum_i m_i \omega_k^2 \phi_{ik} = k_1$$

from which

$$a_k = \frac{\Sigma m_i \phi_{ik}}{M_k} = \frac{k_1}{\omega_k^2 M_k}$$

and the transfer function can be rewritten as

$$(hx_i)_\omega = \frac{k_1 x_i}{F_1} = \vec{\sum_k} \frac{\phi_{ik} a_k}{\sqrt{\left[1 - \frac{\omega^2}{\omega_k^2}\right]^2 + \left[2\xi_{ik} \frac{\omega}{\omega_k}\right]^2}} \tag{10-33}$$

The equations for the 90°-phase and single-resonance assumptions respectively are

$$\xi_{iK} = \frac{\phi_{iK} a_K}{2\sqrt{Q_{iK}^2 - \left[\sum_{k \neq K} \frac{\phi_{iK} a_K}{1 - \frac{\omega_K^2}{\omega_k^2}}^2\right]^2}} \tag{10-34}$$

and

$$\xi_{iK} = \frac{\phi_{iK} a_K}{2Q_{iK}} \qquad [10\text{-}32]$$

It can be seen that the single-resonance assumption yields the same equation for both types of excitation.

10-7. DERIVATION OF MATHEMATICAL MODEL FROM EXPERIMENTAL DATA

The approximate normal-mode method is particularly useful in the mathematical representation of nonconservative systems from experimentally obtained data. The mathematical model can then be used to predict the system response to any input. Furthermore, it can be utilized with the aid of digital or analog computers for design of damping devices which are suitable for a particular excitation.

Most physical structure can be approximately represented by a linear lumped-parameter system. The magnitudes of the different masses, stiffnesses, and damping coefficients are usually difficult to evaluate directly. However, the different transfer functions Q_{ik} and phase relations can be experimentally obtained by applying a harmonic excitation with known magnitudes through a frequency range which includes all resonances, and then recording the oscillations of the different masses. The linearity of the system is easily checked by repeating the test for several input amplitudes.

As shown by Eqs. 10-19, 10-20, 10-23, 10-24, and 10-25, the system can be completely described and its response to any function evaluated once the different values of ω_k, ϕ_{ik}, a_k and ξ_{ik} are known.

Theoretically ω_k are the undamped natural frequencies of the system. However, except for large magnitudes of damping, ω_k can be considered as the frequency corresponding to the response peaks. The eigenvectors ϕ_{ik} are the amplitude ratios when the undamped system vibrates freely at ω_k. It is reasonable to expect that the resonant-amplitude ratios (except for large damping) will provide a close approximation to ϕ_{ik}. Accordingly,

$$|\phi_{ik}| \approx \frac{Q_{ik}}{Q_{1k}} \qquad (10\text{-}35)$$

where mass 1 is the reference mass. ϕ_{ik} is taken as positive or negative when the recorded $(x_i)_{\omega_k}$ is approximately in phase or out of phase with $(x_1)_{\omega_k}$ respectively.

The values of a_k can be obtained from the experimental data by noting that when $\omega = 0$ both Eqs. 10-29 and 10-33 reduce to

$$\sum_k \phi_{ik} a_k = 1 \qquad (10\text{-}36)$$

Since the approximate values of ϕ_{ik} are already known, Eq. 10-36 can be easily written for the different masses. The different values of a_k are obtained from simultaneous solution of these equations. The fictitious damping ratios ξ_{ik} are then evaluated from Eqs. 10-31, 10-32, or 10-34. Notice that when Eq. 10-32 is used together with Eq. 10-35, this approximation will yield equal fictitious damping ratios for all masses at any particular resonant frequency given by

$$\xi_k = \xi_{ik} = \frac{a_k}{2Q_{1k}} \qquad \text{for } i = 1 \rightarrow n \tag{10-37}$$

In cases when this approximation is permissible, it will follow that the fictitious coordinates will also be the same for any particular ω_k and Eq. 10-19 can be represented as an approximate normal-mode equation

$$\ddot{\eta}_k + 2\zeta_k\dot{\eta}_k + \omega_k^2\eta_k = -a_k\ddot{x}_0(t)$$

where ζ_k is calculated from Eq. (10-37).

Example

The classical response solution shown in Fig. 10-7 will be assumed to represent the experimental response of a two-mass system subjected to harmonic ground excitations, Fig. 10-6. The response curves are for a

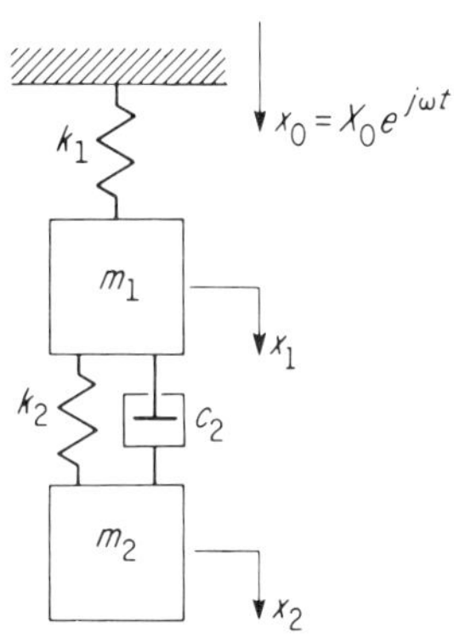

FIG. 10-6. Two-mass system subjected to harmonic ground excitation.

system with mass ratio $\mu = 0.2$ and a frequency ratio satisfying Den Hartog's optimum relationship for a dynamic absorber ($f = 1/1 + \mu = 0.833$). The natural frequency for the main system (ω_{11}) is chosen as 10 cps and the damping ratio for the absorber system ($\zeta_2 = c_2/2m_2\omega_{22}$) is taken as 0.05.

It will now be assumed that the only description available for the

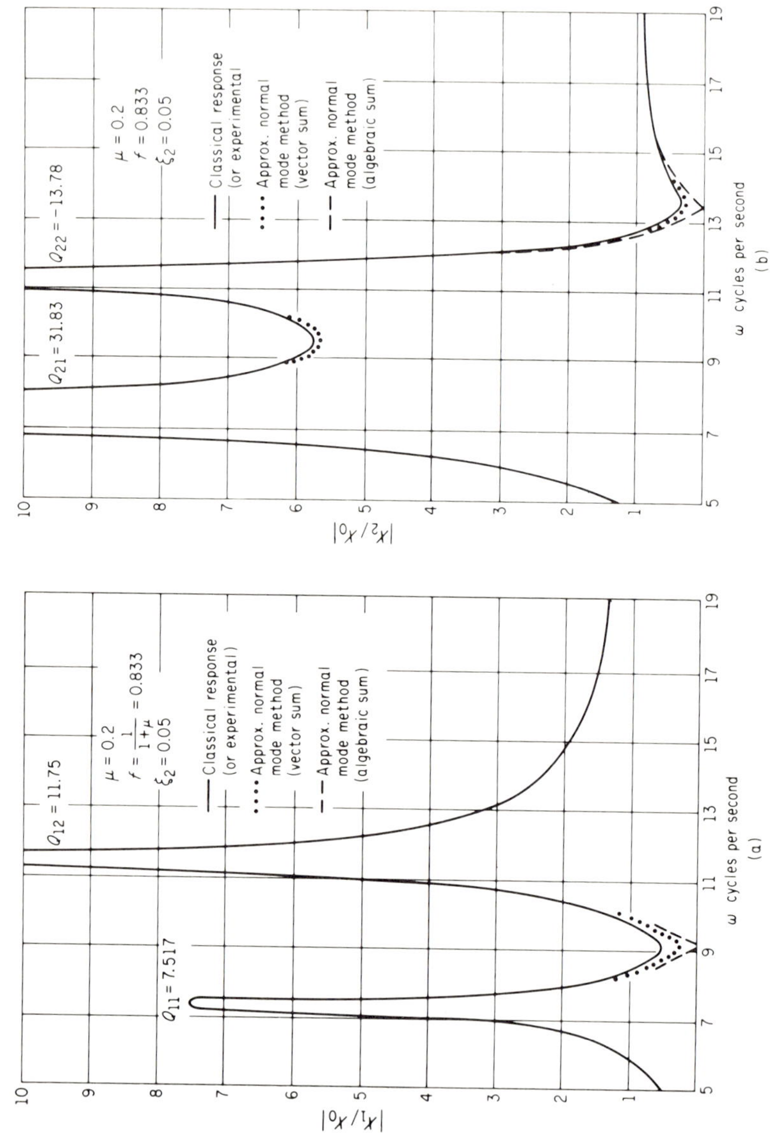
$Q_{12} = 11.75$
$Q_{11} = 7.517$
$\mu = 0.2$
$f = \frac{1}{1+\mu} = 0.833$
$\xi_2 = 0.05$
Classical response (or experimental)
Approx. normal mode method (vector sum)
Approx. normal mode method (algebraic sum)
$|X_1/X_0|$
ω cycles per second
(a)
$Q_{21} = 31.83$
$Q_{22} = -13.78$
$\mu = 0.2$
$f = 0.833$
$\xi_2 = 0.05$
Classical response (or experimental)
Approx. normal mode method (vector sum)
Approx. normal mode method (algebraic sum)
$|X_2/X_0|$
ω cycles per second
(b)

FIG. 10-7

system is its response to a harmonic ground excitation and that the mass, stiffness, and damping parameter are unknown. The response curves give

$$
\begin{aligned}
Q_{11} &= 7.517 \quad \text{at} \quad \omega = 7.3 \text{ cps} \\
Q_{12} &= 11.75 \quad \text{at} \quad \omega = 11.4 \text{ cps} \\
Q_{21} &= 31.83 \quad \text{at} \quad \omega = 7.31 \text{ cps} \\
Q_{22} &= -13.78 \quad \text{at} \quad \omega = 11.4 \text{ cps}
\end{aligned}
$$

Therefore, the two natural frequencies will be taken as

$$
\begin{aligned}
\omega_1 &= 7.31 \text{ cps} \\
\omega_2 &= 11.4 \text{ cps}
\end{aligned}
$$

and the eigenvectors will be approximately taken as

$$
\begin{aligned}
\phi_{11} &= \phi_{12} = 1 \\
\phi_{21} &= +31.83/7.517 = +4.235 \\
\phi_{22} &= -13.78/11.75 = -1.17
\end{aligned}
$$

Equation 10-36 gives

$$
\begin{aligned}
\phi_{11}a_1 + \phi_{12}a_2 &= 1 \\
\phi_{21}a_1 + \phi_{22}a_2 &= 1
\end{aligned}
$$

and by substituting the eigenvectors we obtain

$$
\begin{aligned}
a_1 + a_2 &= 1 \\
+4.235a_1 - 1.17a_2 &= 1
\end{aligned}
$$

from which

$$
\begin{aligned}
a_1 &= 0.4 \\
a_2 &= 0.6
\end{aligned}
$$

Using the single-resonance assumption (Eq. 10-32), we get

$$
\begin{aligned}
\xi_{11} &= 0.4/2 \times 7.517 = 0.0265 \\
\xi_{12} &= 0.6/2 \times 11.75 = 0.0255 \\
\xi_{21} &= 0.4 \times 4.235/2 \times 31.83 = 0.0265 \\
\xi_{22} &= 0.6 \times 1.17/2 \times 13.78 = 0.0255
\end{aligned}
$$

Now the response curves are calculated by the vector summation of Eq. 10-28 and the results are shown in Fig. 10-7. The figure also shows response curves where the summation of Eq. 10-28 is done algebraically. The same procedure is repeated for the system discussed above when ξ_2 is changed to 0.1. The results are shown in Fig. 10-8.

When a digital computer was utilized to compare approximate

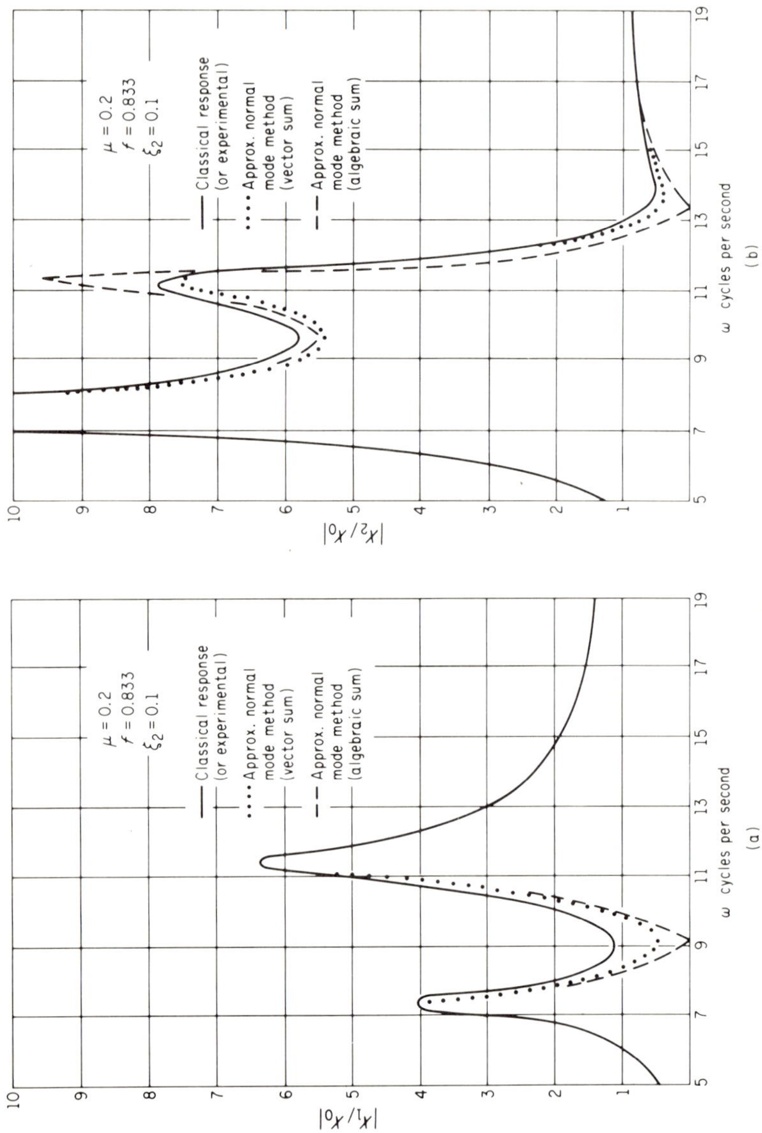
$\mu = 0.2$
$f = 0.833$
$\xi_2 = 0.1$
Classical response (or experimental)
Approx. normal mode method (vector sum)
Approx. normal mode method (algebraic sum)
$|X_1/X_0|$
$|X_2/X_0|$
ω cycles per second
(a)
(b)

FIG. 10-8

normal-mode responses with classical responses for all combinations of the following parameters for a two-mass system.

$$\mu = 0.2, 0.5, 1, 2$$
$$k_2/k_1 = 0.05, 0.1388, 0.5, 1.0$$
$$\xi_2 = 0.005, 0.01, 0.03, 0.05, 0.1, 0.2$$

The results show that for systems with ξ_2 values below 0.1 the agreement between the classical and the approximate normal-mode solutions is acceptable for all practical purposes. The calculated fictitious damping ratios ξ_{ik} (Eq. 10-31) and ξ_k (Eq. 10-37) are equivalent within 6 percent for all the computed systems when the corresponding transfer function Q_{ik} is greater than 3. The error in the eigenvectors obtained from the peak responses is less than 10 percent for systems with ξ_2 smaller than 0.1.

10-8. SUMMARY

The normal-mode method for undamped systems and systems with classical damping has been presented in this chapter. An approximate method has also been introduced for linear damped lumped-parameter systems. The approximate method permits a mathematical model to be constructed for a multidegree-of-freedom system from experimentally obtained frequency-response data.

REFERENCES

1. S. Timoshenko, *Vibration Problems in Engineering*. 3d ed. Princeton, N.J.: Van Nostrand, 1955.
2. W. T. Thomson, *Vibration Theory and Applications*. Englewood Cliffs, N.J.: Prentice-Hall, 1964.
3. J. P. Den Hartog, *Mechanical Vibrations*. 4th ed. New York: McGraw-Hill, 1956.
4. J. Ormondroyd and J. P. Den Hartog, "Theory of Dynamic Vibration Absorbers," *Trans. ASME*, Vol. 50 (1928), p. APM-241.
5. J. E. Brock, "A Note on the Damped Vibration Absorber," *Trans. ASME*, Vol. 68 (1946), p. A284.
6. *Trans. ASME*, Special Vibration Issue, Vol. 89, Ser. B, No. 4 (November 1967).
7. P. Lieber and D. P. Jensen, "An Acceleration Damper Development, Design, and Some Applications," *Trans. ASME*, Vol. 67 (1945), pp. 523–530.
8. R. T. McGoldrick, "Experiments with an Impact Vibration Damper," David Taylor Model Basin Report No. 816, Washington, D. C., 1952.
9. I. Vigness, "Bouncing and Rotation Excited by Vibration," Naval Research Laboratory Report No. 4358, Washington, D.C., May 1954.

10. C. Grubin, "On the Theory of the Acceleration Damper," *Trans. ASME, J. Appl. Mech.*, Vol. 78 (1956), pp. 373–378.
11. R. N. Arnold, "Response of an Impact Vibration Absorber to Forced Vibration," 9th Int. Congress on Applied Mechanics, U. of Brussels, 1957.
12. G. B. Warburton, discussion of C. Grubin, "On the Theory of the Acceleration Damper," *Trans. ASME, J. Appl. Mech.*, Vol. 79 (1957), pp. 322–324.
13. H. G. Kaper, "The Behavior of a Mass-Spring System Provided with a Discontinuous Dynamic Absorber," *App. Sci. Res.*, Ser. A, Vol. 10 (1961).
14. S. F. Masri and T. K. Caughey, "On the Stability of the Impact Damper," *Trans. ASME, J. Appl. Mech.*, Ser. E, Vol. 88 (1966), pp. 586–592.
15. D. M. Egle, "An Investigation of an Impact Vibration Absorber," M.S. thesis, Tulane University, New Orleans, La., 1962.
16. D. M. Egle, "An Investigation of an Impact Vibration Absorber," *Trans. ASME*, Vol. 89, Ser. B, No. 4 (1967).
17. D. L. Cheng, Analysis of Linear Systems. Reading, Mass.: Addison-Wesley, 1959.
18. R. Plunkett (ed.), "Mechanical Impedance Methods for Mechanical Vibrations," *ASME*, N.Y. (1958).
19. W. T. Thomson, "A Note on Tabular Methods for Flexural Vibration," *J. Aeron. Sci.*, (January 1953), pp. 62–63.
20. W. T. Thomson, "Shock Spectra of a Nonlinear System," *J. Appl. Mech.*, Vol. 27 (1960), pp. 528–534.
21. R. V. Churchill, *Operational Mathematics.* 2d ed. New York: McGraw-Hill, 1963.
22. A. H. Church, *Mechanical Vibrations.* New York: Wiley, 1957.
23. L. A. Pipes, *Applied Mathematics for Engineers and Physicists.* 2d ed. New York: McGraw-Hill, 1958.
24. E. Pestel and F. Leckie, *Matrix Methods in Elastomechanics.* New York: McGraw-Hill, 1963.
25. N. O. Myklestad, "A New Method of Calculating Natural Modes of Uncoupled Bending Vibrations of Airplane Wings and Other Types of Beams," *J. Aeron. Sci.* (April 1944), pp. 153–162.
26. M. A. Prohl, "A General Method for Calculating Critical Speeds of Flexible Rotors," *Trans. ASME*, Vol. 66 (1945), p. A-142.
27. W. T. Thomson, "Matrix Solution for the Vibration of Nonuniform Beams," *J. Appl. Mech.*, Vol. 17, No. 2 (1950).
28. H. N. Abramson and Wen-Hwa Chu, "Application of the Generalized Phase-plane-δ Method to Multidegree-of-Freedom Vibrating Systems," *Trans. ASME, J. Appl. Mech.*, Vol. 29, No. 3 (1962).
29. R. Beckett and J. Hurt, *Numerical Calculations and Algorithms.* New York: McGraw-Hill, 1967.
30. A. I. Katz and J. A. Carlson, "Analog and Digital Computers," in H. A. Rothbart (ed.), *Mechanical Design and Systems Handbook.* New York: McGraw-Hill, 1964.
31. G. A. Korn and T. M. Korn, *Electronic Analog and Hybrid Computers.* New York: McGraw-Hill, 1964.

32. J. W. Strutt, Lord Rayleigh, *Theory of Sound.* 2d ed. New York: Dover, 1937.
33. T. K. Caughey, "Classical Normal Modes in Damped Linear Dynamic Systems," *ASME Trans., J. Appl. Mech.,* Vol. 82, Ser. E (June 1960), pp. 640–643.
34. K. A. Foss, "Coordinates Which Uncouple the Equations of Motion of Damped Linear Dynamic Systems, *Trans. ASME, J. Appl. Mech.,* Vol. 25 (1958), pp. 361–364.
35. M. E. J. O'Kelly, "Vibration of Viscously Damped Linear Dynamic Systems," California Institute of Technology, Publication (May 1964).
36. R. L. Halfman, *Dynamics*, Vol. 2. Reading, Mass.: Addison-Wesley, 1962.
37. ASME Colloquium on Structural Damping, December 1959.
38. D. Young, "Response of Structural Systems to Ground Shock," ASME Colloquium on Shock, M. V. Barton (ed.) *ASME*, N.Y. (1960).
39. C. E. Crede, "Failure Resulting from Vibration," in S. H. Crandall (ed.), *Random Vibration*, Vol. 2. Cambridge, Mass.: M.I.T. Press, 1963.
40. A. Seireg and L. Howard, "An Approximate Normal-Mode Method for Damped Lumped-Parameter Systems," *Trans. ASME*, Vol. 89, Ser. B, No. 4 (November 1967).
41. Theodore von Kármán and M. A. Biot, *Mathematical Methods in Engineering.* New York: McGraw-Hill, 1940.
42. S. H. Crandall (ed.), *Random Vibration.* 2 vols. Cambridge, Mass.: M.I.T. Press, 1963.
43. C. M. Harris and C. E. Crede, *Shock and Vibration Handbook.* New York: McGraw-Hill, 1961.
44. I. Vigness, Navy High-Impact Shock Machines for Lightweight and Medium-Weight Equipment, Naval Research Laboratory Report No. 5618.
45. L. S. Pechter et al., "Technique for Determining the Frequency Response of Linear Mechanical Systems," *J. Acous. Soc. Amer.*, Vol. 37, No. 1 (January 1965).
46. T. K. Caughey and M. E. J. O'Kelly, "Classical Normal Modes in Damped Linear Dynamic Systems," *ASME Trans., J. Appl. Mech.*, Vol. 32, Ser. E, No. 3 (September 1965).
47. L. K. Howard, "Some Considerations on the Vibration of Linear, Viscosity Damped, Multimass Systems, M.S. thesis, Marquette U., Milwaukee, Wis., 1965.

part IV

Continuous Systems

It was characteristic of Saint-Venant's work (1797–1886) that he was never satisfied by giving only a general solution. He always wanted to present his results in such a form that engineers could use them without difficulty in practical applications. Using Boussinesq's solution, Saint-Venant, in collaboration with Flamant, prepared diagrams which illustrate various phases of longitudinal impact for various values of the ratio r of the mass of the struck bar to that of the striking mass.

S. Timoshenko, History of Strength of Materials,
McGraw-Hill, New York, 1953

chapter 11

Wave Propagation and Vibration in Continuous Systems

11-1. INTRODUCTION

It has been stated in Part I that a continuum has the following characteristics:

(a) It has an infinite number of degrees of freedom.

(b) Its particles are capable of propagating a disturbance without departing from the neighborhood of their equilibrium position.

(c) For the one-dimensional case, the equation of motion takes the form

$$\frac{\partial^2 q}{\partial t^2} = c^2 \frac{\partial^2 q}{\partial x^2} \tag{11-1}$$

which is the well-known wave equation.

In this chapter the propagation of disturbance in a continuum will be studied in relation to its infinite modes of vibration. Methods of evaluating these modes for one- and two-dimensional problems will be considered. Simple cases of the three-dimensional problems will also be discussed.

11-2. PROPAGATION OF DISTURBANCE—WAVE MOTION

We shall study in this section the behavior of a continuum whose particle motion is confined to a single direction. Some examples of this type are the plane vibration of a string, the longitudinal or torsional vibration of beams, and the vibration of a fluid in a tube. In all these cases the motion can be described by Eq. 11-1 as shown in Part I. q represents the parameter of the motion and x is the direction in which the disturbance is propagated. It is not difficult to find that a general solution for this equation can be given as

$$q = f_1(x - ct) + f_2(x + ct) \tag{11-2}$$

where f_1 and f_2 are any twice-differentiable functions.

The first term on the right in Eq. 11-2 represents a function $f_1(x)$ at $t = 0$ and moving in the positive x direction with a velocity c. Similarly, the second part represents another function $f_2(x)$ at $t = 0$ and moving in the negative x-direction with the same velocity c. The general solution therefore represents two waves traveling along the x-axis in two opposite directions. The functions f_1 and f_2 are determined by the initial disturbance.

We can now investigate the motion in view of the general solution. We shall consider for the sake of illustration the longitudinal wave propagation in rods. In this case, the velocity c of wave propagation is given by

$$c = \sqrt{E/\rho}$$

and the general expression for the displacement of any cross section is

$$u = f_1(x - ct) + f_2(x + ct)$$

The velocity $= v = du/dt = -c[f_1'(x - ct) - f_2'(x + ct)]$ and the transmitted force is

$$F = -EA\frac{\partial u}{\partial x} = -EA[f_1'(x - ct) + f_2'(x + ct)]$$

If we consider only the wave propagated in the positive x-direction therefore

$$u = f_1(x - ct)$$

$$v = -cf_1'(x - ct)$$

and

$$F = -EAf_1'(x - ct)$$

$$= \frac{EA}{c}v \tag{11-3}$$

Equation 11-3 shows that the transmitted force is proportional to the particle velocity and not the acceleration as would have been expected.

Knowing that

$$F = \sigma A$$

where σ = stress
A = cross section of bar

we therefore find from Eq. 11-3

$$\sigma = \frac{E}{c}v \tag{11-4}$$

Now considering the energy of the wave, we find that the strain energy (or potential energy) per unit length is given by

$$\text{P.E.} = \frac{1}{2} A \frac{\sigma^2}{E} \tag{11-5}$$

and the kinetic energy per unit length is

$$\begin{aligned} \text{K.E.} &= \frac{1}{2} A \rho v^2 \\ &= \frac{1}{2} A \frac{E}{c^2} v^2 \\ &= \frac{1}{2} A \frac{E}{c^2} \left(\frac{\sigma^2}{E^2} c^2 \right) \\ &= \frac{1}{2} A \frac{\sigma^2}{E} \end{aligned} \tag{11-6}$$

The work done by the external force per unit length is

$$W = A \sigma \left(\frac{\sigma}{E} \right) = A \frac{\sigma^2}{E} \tag{11-7}$$

Equations 11-5, 11-6, and 11-7 show that

$$W = P.E. + K.E.$$

from which it can be concluded that the total energy is half potential and half kinetic. Equation 11-3 can be written in the form

$$\frac{F}{v} = \frac{EA}{c} = \rho A c = Z \tag{11-8}$$

where Z represents the impedance of the system.

Example

Find the transmitted force, and the energy per unit length in a simple harmonic wave.

In this case we shall assume the displacement to be

$$u = f(x - ct)$$

from which

$$u = -\frac{1}{c} f\left(t - \frac{x}{c} \right)$$

or

$$u = g\left(t - \frac{x}{c} \right) \tag{11-9}$$

which gives u primarily as a function of time. The formulation given in Eq. 11-9 is more descriptive in cases such as the one under consideration where u is a harmonic function of time. Taking ω as the wave frequency we can write

$$u = u_0 \sin \omega \left(t - \frac{x}{c}\right)$$

from which

$$v = u' = \omega u_0 \cos \omega \left(t - \frac{x}{c}\right)$$

The transmitted force is therefore

$$F = \frac{EA}{c} v$$

$$= \frac{EA}{c} \omega u_0 \cos \omega \left(t - \frac{x}{c}\right)$$

The stress is

$$\sigma = \frac{E}{c} \omega u_0 \cos \omega \left(t - \frac{x}{c}\right)$$

The impedance is

$$Z = \frac{EA}{c} = \rho A c$$

The potential and kinetic energy per unit length are

$$\text{K.E.} = \text{P.E.} = \frac{1}{2} A \frac{E}{c^2} \omega^2 u_0^2 \cos^2 \omega \left(t - \frac{x}{c}\right)$$

$$= \frac{1}{2} A \rho \omega^2 u_0^2 \cos^2 \omega \left(t - \frac{x}{c}\right)$$

Total work done by transmitted force per unit length is

$$W = A \rho \omega^2 u_0^2 \cos^2 \omega \left(t - \frac{x}{c}\right)$$

and

$$\text{Time average of work done} = {}^1\!/_2 A \rho \omega^2 u_0^2$$

$$\text{Transmitted power} = \text{average} (F\, du/dt)$$

$$= \text{average} \frac{EA}{c} \omega^2 u_0^2 \cos^2 \omega \left(t - \frac{x}{c}\right)$$

$$= \frac{1}{2} \frac{EA}{c} \omega^2 u_0^2$$

$$= {}^1\!/_2 A \rho c \omega^2 u_0^2$$

The Reflection of Waves

There are cases where a discontinuity exists in the medium due to change in geometry or physical characteristics. In such cases the wave will not be fully transmitted and partial reflection is expected. We shall consider here, for the sake of illustration, the case of a harmonic wave traveling from a medium with characteristics A_1, ρ_1, E_1 to another medium with A_2, ρ_2, E_2 as shown in Fig. 11-1. A, ρ, and E respectively

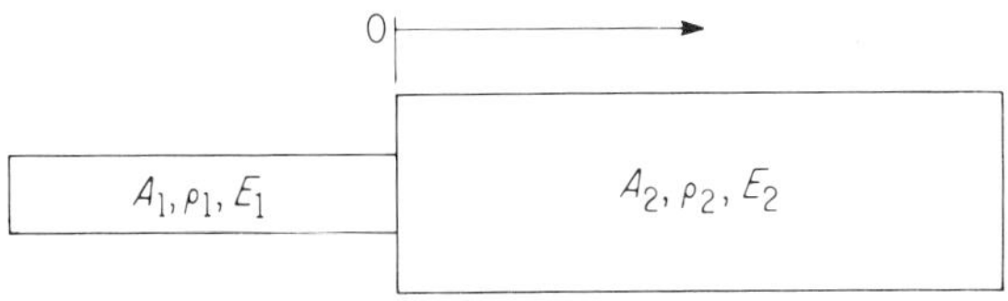

FIG. 11-1

represent the cross-sectional area, the mass density and the modulus of elasticity. The velocity of wave propagation in each media is

$$c_1 = \sqrt{E_1/\rho_1} \qquad c_2 = \sqrt{E_2/\rho_2}$$

We shall consider a complex exponential form for the harmonic wave to accommodate any change in phase. The incident wave will therefore be

$$u_{1i} = g_{1i}\left(t - \frac{x}{c}\right) = U_{1i}e^{i\omega(t - x/c_1)}$$

which represents a harmonic wave with frequency ω traveling in medium 1 in the positive x-direction. The reflected wave can similarly be represented by

$$u_{1r} = g_{1r}(t + x/c_1) = U_{1r}e^{i\omega(t+x/c_1)}$$

which represents a harmonic wave traveling in medium 1 in the negative x-direction. Also the transmitted wave will take the form

$$u_2 = g_2(t - x/c_2) = U_2e^{i\omega(t-x/c_2)}$$

which represents a wave traveling in medium 2 in the positive x-direction. The resulting waves in the two media therefore can be written as

$$\begin{aligned} u_1 &= u_{1i} + u_{1r} \\ &= U_{1i}e^{i\omega(t-x/c_1)} + U_{1r}e^{i\omega(t+x/c_1)} \end{aligned}$$

and

$$u_2 = U_2e^{i\omega(t-x/c_2)}$$

The values of U_{1i}. U_{1r}, and U_2 can be obtained by satisfying the conditions at the junction $x = 0$. These conditions are when

$$x = 0$$

$$u_1 = u_2 \tag{11-10}$$

and

$$F_1 = F_2 \tag{11-11}$$

Equation 11-10 gives

$$U_{1i}e^{i\omega t} + U_{1r}e^{i\omega t} = U_2 e^{i\omega t}$$

or

$$U_{1i} + U_{1r} = U_2 \tag{11-12}$$

and knowing that $F = -EA\partial u/\partial x$ Eq. 11-11 gives

$$\frac{E_1 A_1}{c_1} U_{1i} - U_{1r} = \frac{E_2 A_2}{c_2} U_2$$

$$Z_1 U_{1i} - U_{1r} = Z_2 U_2 \tag{11-13}$$

Now solving Eqs. 11-12 and 11-13 simultaneously, we get

$$U_{1r}/U_{1i} = (Z_1 - Z_2)/(Z_1 + Z_2) \tag{11-14}$$

and

$$U_2/U_{1i} = 2Z_1/(Z_1 + Z_2) \tag{11-15}$$

These results show that when $Z_1 = Z_2$ there will be no reflection and all the wave will be transmitted. On the other hand, if $Z_2 = \infty$, which represents a fixed end, then $U_{1r}/U_{1i} = -1$. which indicates reflection without change of sign. For the case of a free end (i.e. $Z_2 = 0$), $U_{1r}/U_{1i} = +1$, which indicates reflection with change of sign.

From consideration of the energy at the junction, the following conclusion can be reached:

$$\frac{\text{Reflected energy}}{\text{Incident energy}} = \left(\frac{Z_1 - Z_2}{Z_1 + Z_2}\right)^2 \tag{11-16}$$

$$\frac{\text{Transmitted energy}}{\text{Incident energy}} = \frac{4Z_1 Z_2}{(Z_1 + Z_2)^2} \tag{11-17}$$

which shows that when the impedances Z_1 and Z_2 are equal, all the energy is transmitted and no energy is reflected.

Example

Considering the case (Fig. 11-2) of impact between two bars with equal cross section and length having velocity v_{1i} and v_{2i} respectively the particle velocity v at the instant when A_1 contacts A_2 assumes an average value between v_{1i} and v_{2i}, i.e.,

$$v = \frac{v_{1i} + v_{2i}}{2}$$

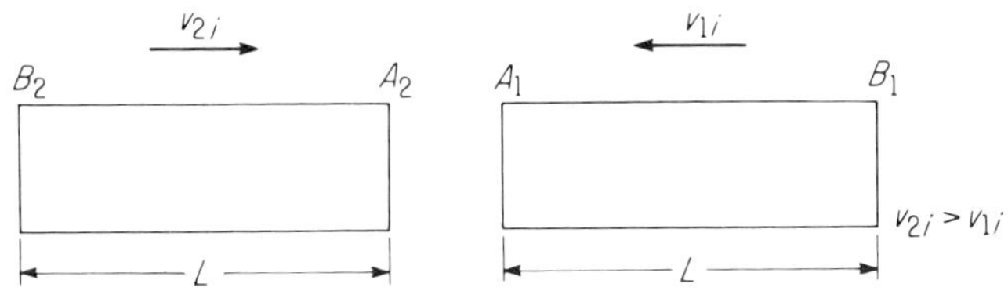

FIG. 11-2

A compressive wave is then propagated through bar 1 to the right with velocity c until it reaches the free end after a time L/c where it is reflected as a tensile wave. By superposition the particle velocities are given by all particles in bar 1 at $t = L/c$ will have a velocity

$$v_1 = \frac{v_{1i} + v_{2i}}{2} - v_{1i}$$
$$= \frac{v_{2i} - v_{1i}}{2}$$

to the right and the bar is uniformly stress with a compressive stress = Ev/c at $t = 2L/c$. All particles in the bar will have a velocity

$$v_1 = \frac{v_{2i} - v_{1i}}{2} + \frac{v_{1i} + v_{2i}}{2}$$
$$= v_{2i}$$

to the right and the bar at this instant is completely relieved of stress.

Similarly at $t = L/c$, all particles in bar 2 will have a velocity

$$v_2 = \frac{v_{1i} + v_{2i}}{2} - v_{2i}$$
$$= \frac{v_{1i} - v_{2i}}{2}$$

to the left and the entire bar is subjected to a compressive stress = Ev/c at $t = 2L/c$. All particles in the bar will have a velocity

$$v_2 = \frac{v_{1i} - v_{2i}}{2} + \frac{v_{1i} + v_{2i}}{2}$$
$$= v_{1i}$$

to the left and the bar is completely relieved of stress.

It can therefore be seen that at $t = 2L/c$ when the wave comes back to sections A_1, A_2 separation will occur and the impact is accordingly terminated with the two bars exchanging their initial velocities.

Similar illustrations can be made for more complex cases of bars with different lengths and cross sections (Ref. 1). In problems where the impacting bars have many variations in cross section, the accounting of all transmitted and reflected waves becomes a tedious task. The utilization of digital computers in such cases is a necessity. A procedure for computer utilization for stress calculations in this type of problem is explained in Ref. 2.

11-3. PROPAGATION OF WAVES IN A TWO-DIMENSIONAL SOLID

It has been shown in the previous section that the wave velocity in bars subjected to longitudinal impact is given by

$$c = \sqrt{\frac{E}{\rho}} \tag{11-18}$$

Similarly, in the case of torsional impact, the waves propagate with a velocity

$$c_s = \sqrt{\frac{G}{\rho}} \tag{11-19}$$

In plane problems, two types of waves exist. The first type propagates in the direction of the plane and are called *dilatational waves.* The other propagates normal to the plane and are called *transverse* or *shear waves.* The velocity of propagation of both waves can be calculated by analysis of the two-dimensional solid (Ref. 1) and their values are given by

$$c_d = \text{dilatation wave velocity}$$

$$= \sqrt{\frac{E(1-\nu)}{(1+\nu)(1-2\nu)\rho}} = c\sqrt{\frac{(1-\nu)}{(1+\nu)(1-2\nu)}} \tag{11-20}$$

where ν = Poisson's ratio

$$c_t = \text{transverse wave velocity}$$

$$= c_s$$

$$= \sqrt{G/\rho} \tag{11-21}$$

It can be seen from Eq. 11-20 that dilatational waves (compressive or tensile waves in two dimensional solids) propagate faster than longi-

Material	Poisson's Ratio	Wave-Propagation Velocity	
		Dilatational c_d fps	Transverse c_s fps
Aluminum	0.34	20,900	10,200
Brass	0.35	14,000	6,700
Glass	0.34	22,300	10,700
Iron	0.28	19,500	10,500
Lead	0.45	7,100	2,300
Lucite	0.35	8,700	4,200
Polystyrene	0.32	7,500	3,900
Steel	0.31	19,500	10,200

tudinal waves in bars. For example, $c_d = 1.16c$ when $\nu = 0.3$. This can be explained by the fact that lateral displacement in the two-dimensional case is suppressed while lateral contraction and expansion is assumed to exist in bars under impact.

Dilatational and transverse wave velocities for some materials are given in the accompanying table.

11-4. PROPAGATION OF WAVES OVER THE SURFACE OF ELASTIC SOLID BODIES (RAYLEIGH WAVES)

Lord Rayleigh was the first to investigate these waves which propagate over the surface with very little penetration in the body itself. In this sense they are similar to waves produced by a stone on a water surface and are important in earthquake detection as well as in the collision of solids.

The analysis of the problem (Refs. 3, 4) produces values for the velocity of propagation for these waves as

$$c_R = 0.9194\sqrt{G/\rho} \qquad \text{for } \nu = 0.25$$

and for the extreme case of $\nu = 0.5$,

$$c_R = 0.9553\sqrt{G/\rho}$$

The ratio of the amplitudes of the horizontal to vertical displacements at the surface for these waves is 0.681 for $\nu = 0.25$.

11-5. EFFECT OF BOUNDARY CONDITIONS ON VIBRATION OF CONTINUUM

The previous section dealt with the solution of simple continuum problems with regard to the propagation, reflection, and transmission of disturbances in the continuum. In this section we shall investigate the vibration resulting from a disturbance of a finite continuum with known boundary conditions. The emphasis will be on determining the modes of vibration (or the so-called *eigenfunctions*) and the corresponding natural frequencies (or *eigenvalues*). The equation of motion for the one-dimensional continuum has been found to take the form

$$\frac{\partial^2 q}{\partial t^2} = c^2 \frac{\partial^2 q}{\partial x^2}$$

where c is the velocity of wave propagation in the continuum. In the case of a free vibration at a particular mode, the solution for this equation can be written as

$$q = q_x \sin \omega t$$

where q_x is a function of x representing the mode of vibration and ω is the natural frequency corresponding to this mode. It would be expected that q_x gives a constant ratio for the displacement at all points of the continuum at all times. The function q_x should therefore satisfy the geometrical conditions at the boundary of the finite continuum.

Example 1

Let us now, for example, consider the case (Fig. 11-3) of a string with length L subjected to a disturbance. We shall assume that the tension in

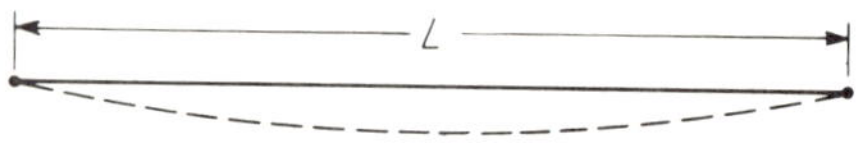

FIG. 11-3

the string is P and the mass per unit length is m. The equation of motion is

$$\frac{\partial^2 y}{\partial x^2} = \frac{m}{P}\frac{\partial^2 y}{\partial t^2} = \frac{1}{c^2}\frac{\partial^2 y}{\partial t^2} \qquad [3\text{-}5]$$

We shall now assume that

$$y = y_x \sin \omega t$$

from which

$$\frac{\partial^2 y}{\partial t^2} = -\omega^2 y_x \sin \omega t, \quad \frac{\partial^2 y}{\partial x^2} = \frac{\partial^2 y_x}{\partial x^2} \sin \omega t$$

and by substitution in Eq. 3-5 we get

$$\frac{\partial^2 y_x}{\partial x^2} = -\frac{\omega^2}{c^2} y_x$$

from which

$$y_x = A \sin \frac{\omega}{c} x + B \cos \frac{\omega}{c} x$$

where A and B are integration constants to be determined from the conditions at the boundary. In this case $y_x = 0$ at $x = 0$, $x = L$. Therefore

$$0 = B$$

and

$$0 = A \sin \frac{\omega L}{c}$$

From the latter equation, either $A = 0$, which means no vibration, or

$$\frac{\omega L}{c} = \pi, \quad 2\pi, \ldots, n\pi, \ldots, \qquad n \text{ an integer}$$

$$\omega = \frac{n\pi c}{L}, \qquad n = 1 \rightarrow \infty$$

which are the natural frequencies corresponding to the modes

$$y_x = A_n \sin \frac{n\pi x}{L}$$

The solution in this case for any mode can be given by

$$y_n = A_n \sin \frac{n\pi x}{L} \sin \frac{n\pi c}{L} t$$

where $n = 1 \rightarrow \infty$, and for a general disturbance the solution can be written as the summation of all modes:

$$y = \sum_{n=1}^{\infty} A_n \sin \frac{n\pi x}{L} \sin \left(\frac{n\pi c}{L} t + \phi_n \right)$$

where the coefficients A_n and the phase angle ϕ_n for each mode are evaluated from the initial conditions of the disturbance.

Example 2

Considering now the case (Fig. 11-4) of the longitudinal vibration of a beam with length L, cross section A, and mass per unit volume ρ. The

FIG. 11-4

equation of motion is given by

$$\frac{\partial^2 u}{\partial t^2} = c^2 \frac{\partial^2 u}{\partial x^2} \qquad [3\text{-}7]$$

The boundary conditions in this case are such that the forces are zero at the boundary, i.e., $(EA\,\partial u/\partial x) = 0 = \partial u_x/\partial x$ at $x = 0$ and $x = L$, which gives

$$0 = \frac{\omega}{c} A \qquad \text{and} \qquad 0 = \frac{\omega}{c} B \sin \frac{\omega}{c} L$$

from which

$$\frac{\omega L}{c} = n\pi$$

where n is an integer. The natural frequencies are, therefore,

$$\omega_n = \frac{n\pi c}{L}$$

corresponding to the modes $u_x = B_n \cos(n\pi x/L)$ and the general solution is

$$u = \sum_{n=1}^{\infty} B_n \cos\frac{n\pi x}{L} \sin\left(\frac{n\pi c}{L}t + \phi_n\right)$$

It has been shown in the previous examples that the solution of the wave equation for simple cases of finite continuum together with the conditions at the boundary give the modes of vibration and the corresponding natural frequencies. It has also been shown that the fundamental frequencies can be determined directly from the time taken for a wave to travel between the two boundaries.

$$\tau = 2\pi/\omega_{n_1} = 2\pi/(\pi c/L) = 2L/c$$

The higher natural frequencies are integer multiples of the fundamental.

Orthogonality of Modes

In the given examples, it was found that the modes of vibration are given by a set of harmonic functions. It was also found that these functions $(q_x)_n$ correspond to a definite set of characteristic values ω_n known as the natural frequencies of the system or the eigenvalues of the problem.

It is well known from integral calculus that over a complete cycle

$$\int \sin\frac{i\pi x}{L}\sin\frac{k\pi x}{L}\,dx = 0$$

and

$$\int \cos\frac{i\pi x}{L}\cos\frac{k\pi x}{L}\,dx = 0$$

for $i \neq k$. This can be used as a proof that the characteristic functions (or eigenfunctions) are mutually orthogonal in the given examples. This orthogonality condition is similar to that given for systems with several degrees of freedom. It applies to any type of eigenfunctions and is not only restricted to harmonic functions. It can also be stated in its general form for a one-dimensional continuum as

$$\int (q_x)_i (q_x)_k\,dx = 0$$

where the integration is performed over the boundary for $i \neq k$. It can also be extended to two-dimensional continua as

$$\int\int (Z_{xy})_i (Z_{xy})_k \, dx dy = 0$$

where the integration is performed over the surface for $i \neq k$.

The Rayleigh-Ritz Method

In more complex problems, an exact mathematical solution will be very difficult to obtain and one must resort to approximate solutions. One such procedure is the energy method which can be used in conjunction with the so-called *Ritz functions* to obtain solutions for a wide class of continuum problems.

In the Rayleigh-Ritz method, a finite series function for the displacement is assumed to satisfy the geometrical boundary conditions of the problem. The series could be one of the Fourier type or any other class of functions. The series includes a finite number of coefficients A_n which may be considered as the coordinates of the system. Each term in the series must be chosen to satisfy the boundary conditions independent of the variation of the coefficient.

Expressions for the strain energy U in the maximum deflected position and for the maximum kinetic energy T are then obtained. A set of simultaneous equations can be derived from

$$\frac{\partial (T - U)}{\partial A_n} = 0 \tag{11-22}$$

The natural frequencies of the system are the values which make the determinant of the coefficients of the set of equations 11-22 equal to zero. The number of terms in the displacement expression depends on the shape of the continuum, the boundary conditions, and the mode of vibration under consideration.

Example 3

We shall again consider (Fig. 11-5) the vibration of the string in Example 1 as an illustration for the Ritz method. In this case the displacement function for the fundamental frequency will be assumed as

$$y = y_x \sin \omega t$$

and y_x can be assumed as a single term

$$y_x = A \sin \pi x / L$$

which satisfies the boundary conditions $y = 0$ at $x = 0$ and $x = L$. The expression for the strain energy in the maximum deflected position can be

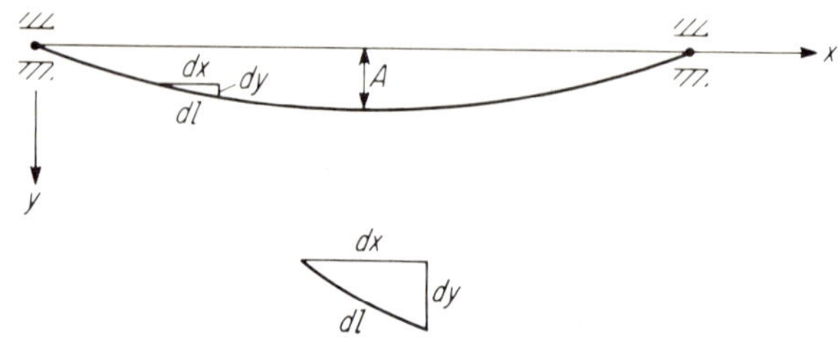

$$dl^2 = dx^2 + dy^2$$

$$dl = \sqrt{1 + (dy/dx)^2}\, dx$$

$$(dl - dx) \approx \tfrac{1}{2}(dy/dx)^2\, dx$$

FIG. 11-5

given by

$$\begin{aligned} U &= \int_0^L P(dl - dx) \\ &= \frac{1}{2} P \int_0^L (dy/dx)^2\, dx \\ &= \frac{1}{2} PA^2(\pi/L)^2 \int_0^L \cos^2 \frac{\pi x}{L}\, dx \\ &= \frac{\pi^2}{2L} A^2 P \end{aligned}$$

where P is the tension in the string. The expression for the kinetic energy is

$$\begin{aligned} T &= \int_0^L \frac{1}{2} m(dy/dt)^2_{\max}\, dx \\ &= \int_0^L \frac{1}{2} m\omega^2 A^2 \sin^2 \frac{\pi x}{L}\, dx \\ &= \frac{1}{2} m\omega^2 A^2 L \end{aligned}$$

Accordingly, Eq. 11-22 gives

$$\omega = \frac{\pi}{L}\sqrt{P/m} = \frac{\pi c}{L}$$

Example 4

The Rayleigh-Ritz Method will be applied to the case of a simple supported beam. The coordinate system is assumed to be the same as in Fig. 11-5.

We shall consider for this case a Ritz function of the form

$$y = y_x \sin \omega t$$

y_x will be assumed as the harmonic series

$$y_x = \sum_{i=1}^{n} A_i \sin \frac{i\pi x}{L} \tag{11-23}$$

where each term in the series satisfies the boundary conditions. In this case the flexural strain energy can be given by

$$U = \int_0^L \frac{1}{2} M \, d\theta$$

where

$$M = EI \frac{d^2y}{dx^2}$$

$$\theta = dy/dx$$

$$d\theta = \frac{d^2y}{dx^2} dx$$

therefore

$$U = \int_0^L \frac{1}{2} EI \left(\frac{d^2y}{dx^2}\right)^2 dx \tag{11-24}$$

and

$$T = \int_0^L \frac{1}{2} \rho \, dx (\omega y_x)^2 \tag{11-25}$$

from Eq. 11-23

$$\frac{d^2y_x}{dx^2} = -\sum_{i=1}^{n} A_i \left(\frac{i\pi}{L}\right)^2 \sin \frac{i\pi x}{L}$$

$$U = \frac{EI}{2} \int_0^L \left[\sum_{i=1}^{n} A_i \left(\frac{i\pi}{L}\right)^2 \sin \frac{i\pi x}{L}\right]^2 dx$$

$$= \frac{EI\pi^4}{2L^4} \int_0^L \left[\sum_{i=1}^{n} A_i i^2 \sin \frac{i\pi x}{L}\right]^2 dx$$

$$= \frac{EI\pi^4}{2L^4} \left[\frac{L}{2} \sum_{i=1}^{n} A_i i^4\right]$$

$$= \frac{EI^4}{4L^3} \sum_{i=1}^{n} A_i^2 i^4$$

$$T = \frac{1}{2}\rho\omega^2 \int_0^L \left[\sum_{i=1}^{n} A_i \sin\frac{i\pi x}{L}\right]^2 dx$$

$$= \frac{1}{2}\rho\omega^2\left[\frac{L}{2}\sum_{i=1}^{n} A_i^2\right]$$

$$= \frac{1}{4}\rho\omega^2 L \sum_{i=1}^{n} A_i^2$$

Therefore

$$\frac{\partial(T - U)}{\partial A_i} = \frac{L}{2}\rho\omega^2 A_i - L^4\frac{EI\pi^4}{2L^3}A_i = 0$$

from which

$$\omega^2 = \left(\frac{i\pi}{L}\right)^4 \frac{EI}{\rho}$$

or

$$\omega = i^2\left(\frac{\pi}{L}\right)^2 \sqrt{\frac{EI}{\rho}}$$

If the reference coordinate systems are taken as shown in Fig. 11-6, a

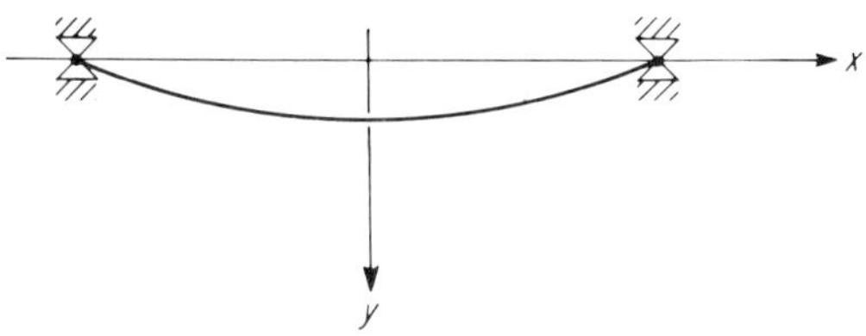

FIG. 11-6

power-series function can be used to satisfy the boundaries as

$$y_x = [f(x)]\Sigma A_n(2x/L)^n$$

where $f(x) = [1 - (2x/L)^2]$ satisfies both boundaries in this case. The same procedure as before can be followed to determine the natural frequencies of vibration.

Example 5

To find the fundamental frequency for the flexural vibration (Fig. 11-7) of a cantilever beam, we shall consider a Ritz function of the Fourier type as

$$y = y_x \sin\omega t$$

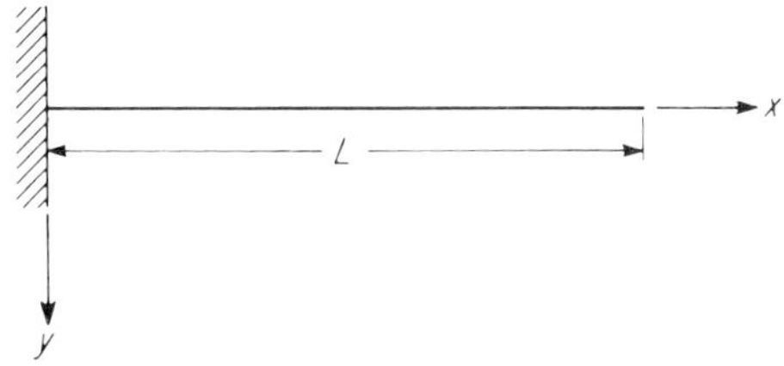

FIG. 11-7

where

$$y_x = A_0 + \sum_{n=1,2,3}^{n} A_n \cos \frac{n\pi x}{2L}$$

The conditions at the boundary are

$$\begin{array}{llll} 1 & \text{at} & x = 0, & y = 0 \\ 2 & \text{at} & x = 0, & dy/dx = 0 \\ 3 & \text{at} & x = L, & d^2y/dx^2 = 0 \\ 4 & \text{at} & x = L, & d^3y/dx^3 = 0 \end{array}$$

The assumed y_x satisfies condition 2 and 3. In order to satisfy conditions 1 and 4 respectively,

$$A_0 + \Sigma A_n = 0 \tag{11-26}$$

and

$$\Sigma(-1)^{(n-1)/2} A_n \left(\frac{n\pi}{2L}\right)^3 = 0 \tag{11-27}$$

Accordingly, for one-term approximation three terms have to be considered, since Eqs. 11-26 and 11-27 will reduce the number of the coefficients in the y_x expression to one. In this case

$$y_x = A_0 + A_1 \cos \frac{\pi x}{2L} + A_3 \cos \frac{3\pi x}{2L}$$

To satisfy the boundary conditions 1 and 4,

$$A_1 = 27A_3, \quad A_0 = -28A_3$$

Therefore

$$y_x = -28A_3 + 27A_3 \cos \frac{\pi x}{2L} + A_3 \cos \frac{3\pi x}{2L}$$

The maximum kinetic energy is therefore

$$T = \tfrac{1}{2}\omega^2 \rho \int_0^L y_x^2 \, dx$$

$$= \omega^2 \rho L A_3^2 [\{28^2 + 27^2 + \tfrac{1}{8}\} - (28)(54\tfrac{1}{3})]$$

And the strain energy in the maximum deflected position is

$$U = \tfrac{1}{2} EI \int_0^L (d^2 y_x/dx^2)^2\, dx$$

$$= \tfrac{1}{2} EI \int_0^L A_3^2[-27(\pi/2L)^2 \cos(\pi x/2L)$$

$$- (3\pi/2L)^2 \cos(3\pi x/2L)]^2 dx$$

$$= \tfrac{1}{2} EILA_3^2[(27)^2(\pi/2L)^4(\tfrac{1}{4}) + 27(\pi/2L)^4(\tfrac{1}{4})]$$

From

$$\frac{\delta(T - U)}{\partial A_3} = 0$$

we find

$$\omega = 0.4089(\pi/L)^2 \sqrt{EI/\rho}$$

The solution by exact theory in this case is

$$\omega = 0.357(\pi/L)^2 \sqrt{EI/\rho}$$

which indicates that the one-term approximation is not sufficient and more terms will be necessary in this case for good approximation.

As an alternative method we can consider a power series for the y_x function as

$$y_x = f(x) \sum_n A_n (x/L)^n$$

where $f(x)$ is chosen to satisfy the geometrical boundary. In this case we can use

$$f(x) = (x/L)^2$$

The same procedure is followed to evaluate the fundamental frequency. The number of terms utilized in the solution should be chosen to give the required accuracy.

Example 6

Find a suitable y_x function for the beam fixed at both ends as shown in Fig. 11-8.

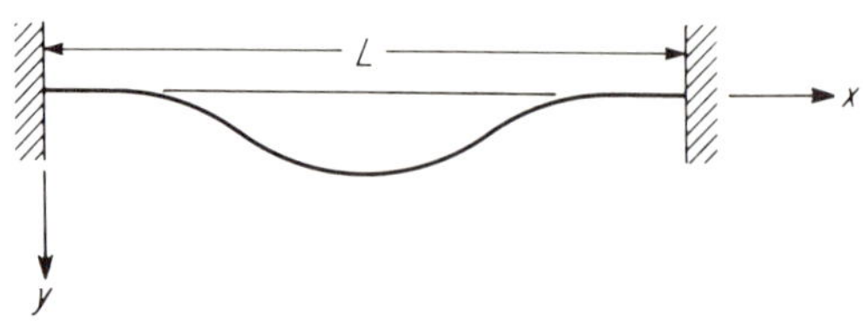

FIG. 11-8

The geometrical boundary conditions in this case are

$$\text{at } x = 0, \quad y = 0, \quad \text{and} \quad dy/dx = 0$$
$$\text{at } x = L, \quad y = 0, \quad \text{and} \quad dy/dx = 0$$

A series of harmonic functions can be given as

$$y_x = A_0 + \sum_{n=1}^{\infty} A_n \cos(2n\pi x/L)$$

A power series can be alternatively used with coordinates as shown in Fig. 11-9:

$$y_x = [1 - (2x/L)^2]^2 \Sigma A_n (2x/L)^n$$

where $[1 - (2x/L)^2]^2$ satisfies the fixation at boundaries.

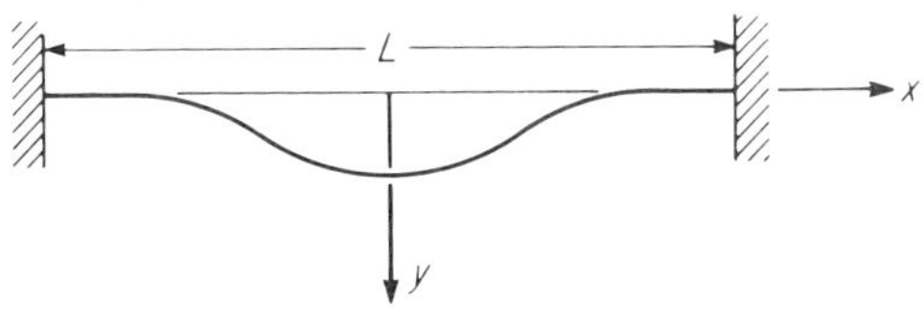

FIG. 11-9

Example 7

To find the fundamental frequency for the simple supported beam with variable cross section shown in Fig. 11-10 we can assume the same

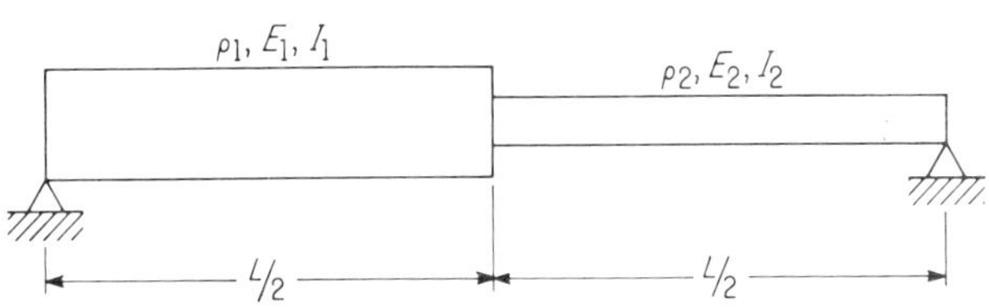

FIG. 11-10

y_x as in Example 4 since the geometrical boundary conditions are not changed.

In this case,

$$y = \Sigma A_n \sin \frac{n\pi x}{L} \sin \omega t$$

$$T = \frac{1}{2} \rho_1 \omega^2 \int_0^{l_1} y_x^2 \, dx + \frac{1}{2} \rho_2 \omega^2 \int_{l_1}^{l_2} y_x^2 \, dx$$

$$U = \frac{EI}{2} \int_0^{l_1} (d^2 y_x/dx^2)^2 \, dx + \frac{EI}{2} \int_{l_1}^{l_2} (d^2 y_x/dx^2)^2 \, dx$$

and the natural frequency can be evaluated by lettering the determinant of coefficients equal zero for the set of equations:

$$\frac{\partial(T - U)}{\partial A_n} = 0$$

Example 8

For the case of a simple supported beam with a concentrated mass (see Fig. 11-11), the y_x function will again be the same as in Example 4.

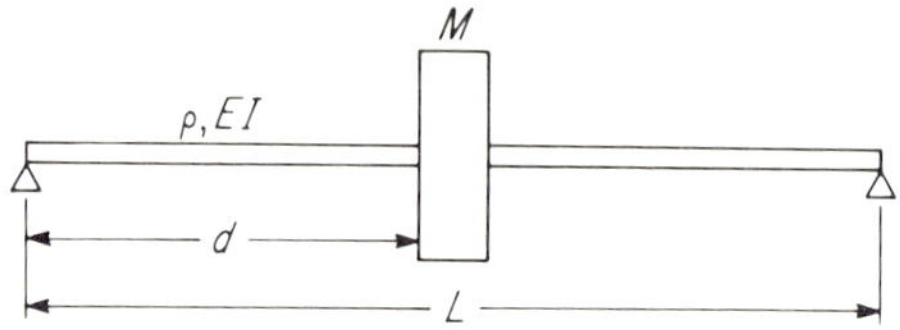

FIG. 11-11

The expression for the kinetic energy will be

$$T = \frac{1}{2}\,\omega^2 \int_0^l y_x^2\,dx + \frac{1}{2}\,M\omega^2 y_d^2$$

where y_d is the y_x function at $x = d$. The strain energy function will be the same as in Example 4:

$$U = \frac{EI}{2} \int_0^l (d^2 y_x/dx^2)^2\,dx$$

The natural frequencies for different cases of beams are given in Table 11-1. (Ref. 6).

TABLE 11-1
Lateral Vibrations of Beams

$$f_n = K\,\frac{r}{L^2} \times 10^4 \times \frac{c}{c_{\text{steel}}}$$

where f_n = natural frequency for nth mode cps
K = constant of mode and boundary condition
r = radius of gyration = $\sqrt{I/A}$, inches (use maximum r for variable-section beams)
L = beam length, inches
c = $\sqrt{E/\rho}$ = wave propagation velocity for material
c_{steel} = $\sqrt{E_{\text{steel}}/\rho_{\text{steel}}}$ = 196,320 in./sec

Condition		K				
		$n = 1$	$n = 2$	$n = 3$	$n = 4$	$n = 5$
 Clamped-clamped	 Free-free	71.95	198.29	388.73	642.73	959.94

TABLE 11-1 (continued)

Condition	K				
	$n = 1$	$n = 2$	$n = 3$	$n = 4$	$n = 5$
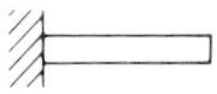 Clamped-free	11.30	70.85	198.30	388.73	642.60
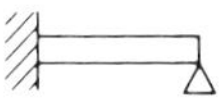 Clamped-hinged 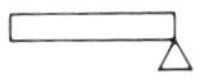 Free-hinged	49.57	160.65	335.17	573.20	874.65
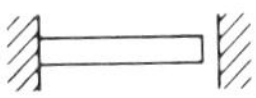 Clamped-guided  Free-guided	17.98	97.18	239.98	446.25	715.98
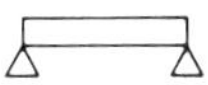 Hinged-hinged 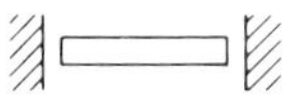 Guided-guided	31.73	126.93	285.60	507.73	793.33
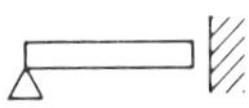 Hinged-guided	7.93	71.40	198.33	388.73	642.60
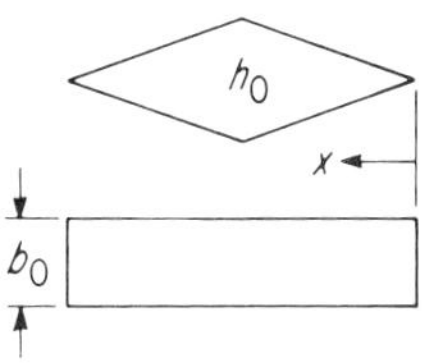 $b = b_0$, $h = h_0 \frac{x}{L}$ Free-free	21.21	56.97			
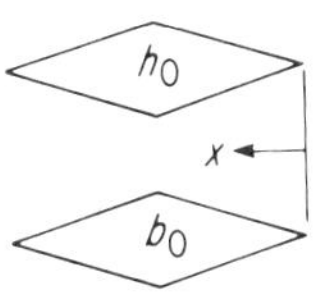 $h = h_0 \frac{x}{L}$, $b = b_0 \frac{x}{L}$ Free-free	32.73	76.57			
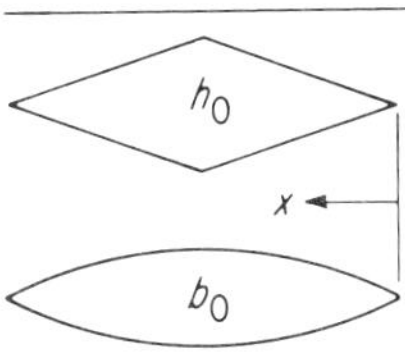 $h = h_0 \frac{x}{L}$, $b = b_0 \sqrt{\frac{x}{L}}$ Free-free	25.66	66.06			

TABLE 11-1 (continued)

Condition		K				
		$n = 1$	$n = 2$	$n = 3$	$n = 4$	$n = 5$
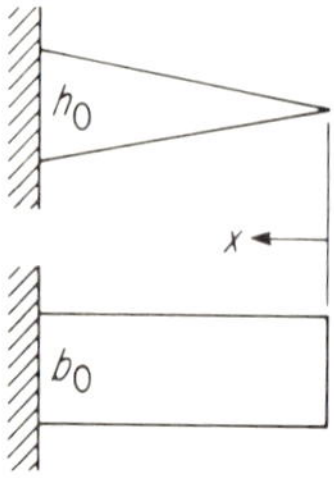	$h = h_0 \dfrac{x}{L}$ $b = b_0 = \text{const.}$	17.09	48.89	96.57		
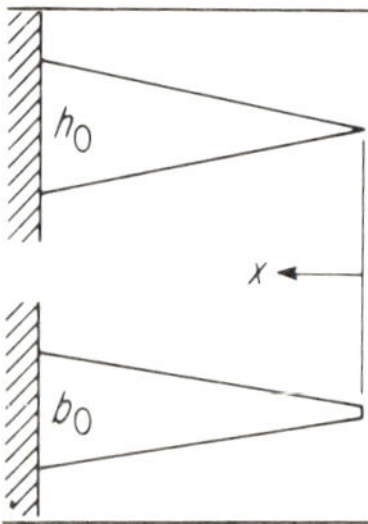	$h = h_0 \dfrac{x}{L}$ $b = b_0 \dfrac{x}{L}$	26.08	68.08	123.64		
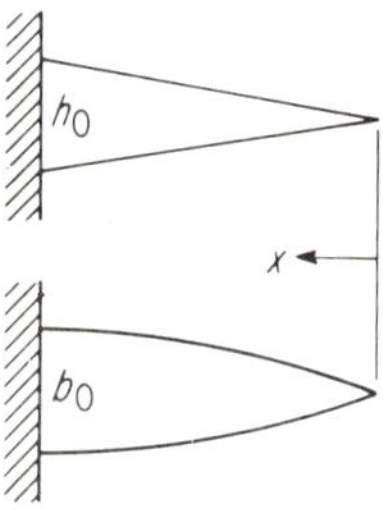	$h = h_0 \dfrac{x}{L}$ $b = b_0 \sqrt{\dfrac{x}{L}}$	22.30	58.18	109.90		
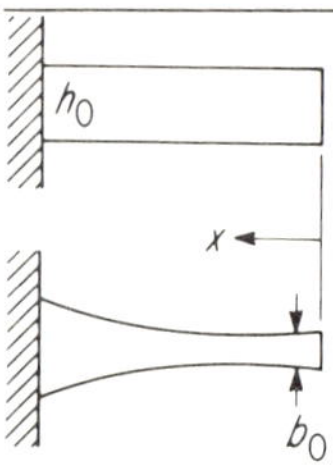	$h = h_0$ $b = b_0 e^{x/L}$	15.23	77.78	206.07		

Condition	Number of Spans	n				
		1	2	3	4	5
	2	31.73	49.59	126.94	160.66	285.61
	3	31.73	40.52	59.56	126.94	143.98
	4	31.73	37.02	49.59	63.99	126.94
	5	31.73	34.99	44.19	55.29	66.72

TABLE 11-1 (continued)

Condition		K				
		$n = 1$	$n = 2$	$n = 3$	$n = 4$	$n = 5$
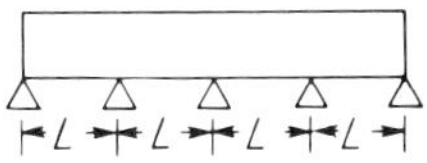	6	31.73	34.32	40.52	49.59	59.56
	7	31.73	33.67	38.40	45.70	53.63
	8	31.73	33.02	37.02	42.70	49.59
	9	31.73	33.02	35.66	40.52	46.46
	10	31.73	33.02	34.99	39.10	44.19
Ends simply supported	11	31.73	32.37	34.32	37.70	41.97
	12	31.73	32.37	34.32	37.02	40.52
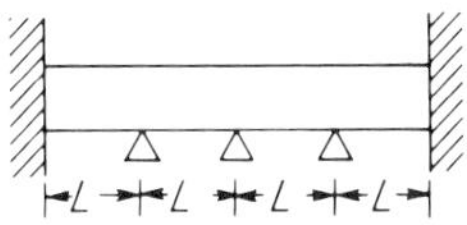	2	49.59	72.36	160.66	198.34	335.20
	3	40.52	59.56	72.36	143.98	178.25
	4	37.02	49.59	63.99	72.36	137.30
	5	34.99	44.19	55.29	66.72	72.36
	6	34.32	40.52	49.59	59.56	67.65
	7	33.67	38.40	45.70	53.63	62.20
	8	33.02	37.02	42.70	49.59	59.98
	9	33.02	35.66	40.52	46.46	52.81
Ends clamped	10	33.02	34.99	39.10	44.19	49.59
	11	32.37	34.32	37.70	41.97	47.23
	12	32.37	34.32	37.02	40.52	44.94
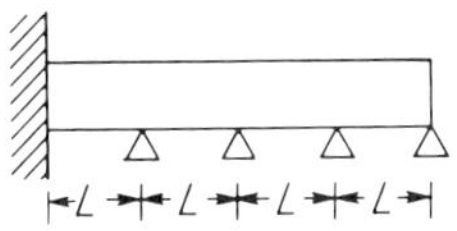	2	37.02	63.99	137.30	185.85	301.05
	3	34.32	49.59	67.65	132.07	160.66
	4	33.02	42.70	56.98	69.51	129.49
	5	33.02	39.10	49.59	61.31	70.45
	6	32.37	37.02	44.94	54.46	63.99
	7	32.37	35.66	41.97	49.59	57.84
	8	32.37	34.99	39.81	45.70	53.63
	9	31.73	34.32	38.40	43.44	49.59
Ends clamped-supported	10	31.73	33.67	37.02	41.24	46.46
	11	31.73	33.67	36.33	39.81	44.19
	12	31.73	33.02	35.66	39.10	42.70

Example 9

In this example we shall use the Rayleigh-Ritz method to study the vibration of a rectangular membrane supported at the boundary (see Fig. 11-12). A harmonic series solution which satisfies the boundary conditions in this case can be given as

$$Z = Z_{xy} \sin \omega t$$

where

$$Z_{xy} = \sum_{i=1}^{n} \sum_{j=1}^{n} A_{ij} \sin \frac{i\pi x}{a} \sin \frac{j\pi y}{b}$$

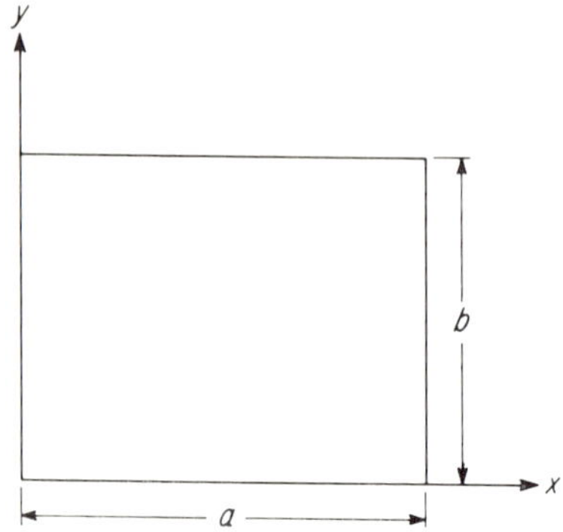

FIG. 11-12

The maximum kinetic energy in the membrane can be given by

$$T = \frac{1}{2}\omega^2\rho \int_0^a \int_0^b Z_{xy}^2 \, dxdy$$

where ρ is the mass per unit area. The expression for the strain energy at the maximum deflected position is

$$U = \frac{1}{2} P \int_0^b \int_0^a [(\partial Z_{xy}/\partial_x)^2 + (\partial Z_{xy}/\partial_y)^2] \, dxdy$$

where P is the uniform tension per unit length of the membrane and the natural frequencies are given by the values of ω which make the determinant of coefficients vanish for the set of equations:

$$\frac{\partial(T - U)}{\partial A_{ij}} = 0$$

Example 10

The frequencies for a rectangular plate simply supported at the boundary, Fig. 11-13, can be obtained in a manner similar to Example 9.

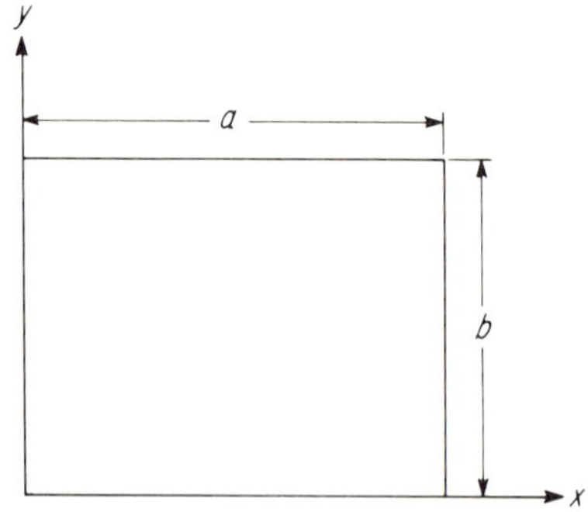

FIG. 11-13

In this case also the geometrical boundary can be satisfied by a harmonic series

$$Z_{xy} = \sum_j \sum_i A_{ij} \sin \frac{i\pi x}{a} \sin \frac{i\pi y}{b}$$

The maximum K.E. is given by

$$T = \frac{1}{2} \omega^2 \rho \int_0^a \int_0^b Z_{xy}^2 \, dxdy \tag{11-28}$$

and the strain energy in the maximum deflected position is obtained (by theory of plates) as

$$U = \frac{1}{2} D \int_0^a \int_0^b [(\partial^2 Z_{xy}/\partial x^2)^2 + (\partial^2 Z_{xy}/\partial y^2)^2$$
$$+ 2\nu(\partial^2 Z_{xy}/\partial x^2)(\partial^2 Z_{xy}/\partial y^2)$$
$$+ 2(1 - \nu)(\partial^2 Z_{xy}/\partial x \partial y)^2] \, dx \, dy \tag{11-29}$$

where

$$D = \frac{Eh^3}{12(1 - \nu^2)} = \text{flexural rigidity of the plate}$$

The natural frequencies are similarly obtained from the set of equations

$$\frac{\partial(T - U)}{\partial A_{ij}} = 0$$

11-6. GENERAL PROCEDURE FOR THE APPLICATION OF THE RITZ METHOD TO THE VIBRATION OF PLATES OF DIFFERENT SHAPES AND BOUNDARY CONDITIONS

In this section a power-series solution is described by Rao and Pickett (Ref. 5) to obtain numerical solutions for the lowest natural frequencies of plates with different shapes and various boundary conditions. This procedure is also applicable for plates with holes. The procedure is as follows.

The deflection is assumed as

$$Z = Z_{xy} \sin \omega t$$

If the rectangular coordinates are used and no holes are present, then Z_{xy} can be given by

$$Z_{xy} = f(x, y) h \Sigma A_{mn} (x/a)^m (x/b)^n$$

where $f(x, y)$ is an appropriate function that provides the necessary geometrical requirements, h is the plate thickness, a, b are characteristic dimensions, and the A's are arbitrary parameters. If the plate is clamped

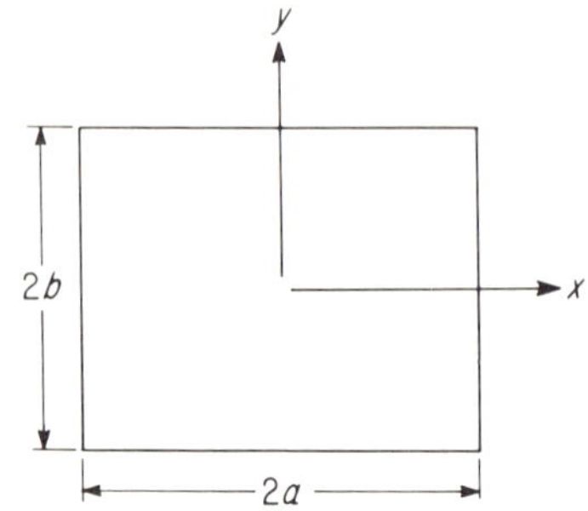

FIG. 11-14

at the boundaries

$$f(x,y) = [1 - (x/a)^2][1 - (y/b)^2]^2$$

If the plate is simply supported at $x = \pm a, y = \pm b$,

$$f(x,y) = [1 - (x/a)^2]^2[1 - (y/b)^2]$$

If the plate is clamped at $x = 0$ but otherwise is free of support,

$$f(x,y) = x^2/a^2$$

If free of support entirely,

$$f(x,y) = 1$$

In this case the first parameter A_{00} if present, is determined from the requirement

$$\int\int Z_{xy}\,dxdy = 0$$

and eliminated by substitution in terms of the other A's. The maximum strain energy U is found from Eq. 11-28 and the maximum kinetic energy from Eq. 11-29 with the correct integral limits. The natural frequency is found by trial such as to make the determinant of coefficient of the set of simultaneous equations $[\partial(T - U)/\partial A_{mn} = 0]$ vanish. For certain modes of vibration it will be obvious in some cases that not all the A's will appear. The work is simplified if these are omitted at the beginning. The number of significant A's required for less than 1 percent error for different boundary conditions is as follows:

	A's Used
Boundary clamped	2
Simple support	3
Free	4

The clamped plate requires the fewest number of terms because the variational procedure only has to approximate the equation of equilibrium. The simple support condition requires more terms to satisfy the requirement of no boundary moments. The free plate requires additional terms to satisfy the requirement of no boundary reaction. Some tabulation of results for different cases is given in Tables 11-2 through 11-7. [Refs. 5, 6, 9, and 10.]

TABLE 11-2
Natural Frequencies of Square Plates

$$f_n = K_n \frac{h}{a^2} \times 10^4 \times \frac{c}{c_{\text{steel}}}$$

where f_n = nth natural frequency, cps
K_n = frequency constant listed in table
h = plate thickness, inches
a = plate-edge length, inches
c = $\sqrt{E/\rho}$

Boundary conditions F = free C = clamped S = simply supported	n					
	1	2	3	4	5	6
top F; left F, right F; bottom C	3.40	8.32	20.86	26.71	30.32	
top F; left C, right F; bottom C	6.77	23.43	26.07	46.75	61.44	
top F; left F, right F; bottom F	13.72	19.99	23.26	34.98	59.93	63.47
top S; left S, right S; bottom S	19.20	48.00	76.83	96.01	124.82	163.25
top S; left S, right S; bottom C	23.01	50.28	57.06	83.79	97.58	110.13
top S; left C, right C; bottom S	28.16	53.26	67.44	92.02	99.43	125.60
top C; left C, right C; bottom C	35.01	71.42	105.36	128.03	128.71	160.72

TABLE 11-3
Natural Frequencies of Cantilever Plates

$$f_n = K_n \frac{h}{a^2} \times 10^4 \times \frac{c}{c_{\text{steel}}}$$

where f_n = nth natural frequency, cps
K_n = frequency constant listed in table
h = plate thickness, inches
a = plate dimension, as shown, inches

Boundary Conditions F = free C = clamped S = simply supported		n				
		1	2	3	4	5
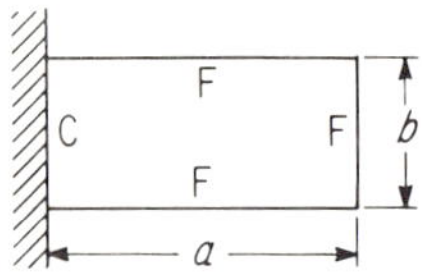	$a/b = 1/2$	3.41	5.23	9.98	21.36	24.18
	$a/b = 1$	3.40	8.32	26.71	20.86	30.32
	$a/b = 2$	3.38	14.52	91.92	21.02	47.39
	$a/b = 5$	3.36	33.79	548.60	20.94	103.03
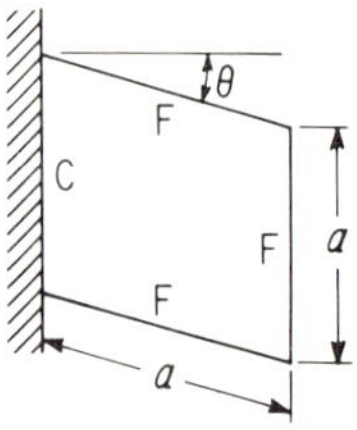	$\theta = 15°$	3.50	8.63			
	$\theta = 30°$	3.85	9.91			
	$\theta = 45°$	4.69	13.38			
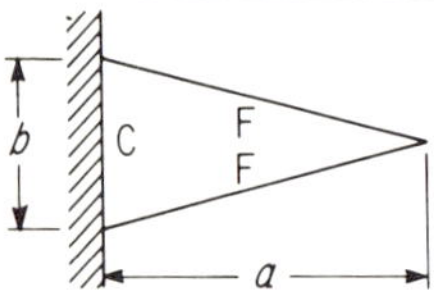	$a/b = 2$	6.7	28.6	56.6	137	
	$a/b = 4$	6.6	28.5	83.5	240	
	$a/b = 8$	6.6	28.4	146.1	457	
	$a/b = 14$	6.6	28.4	246	790	
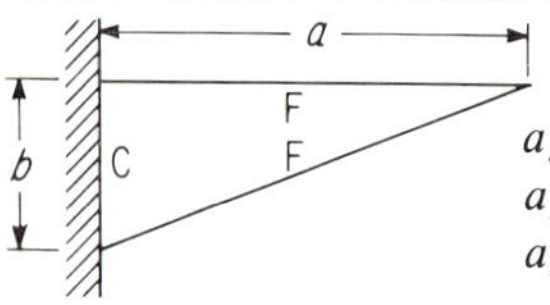	$a/b = 2$	5.5	23.6			
	$a/b = 4$	6.2	26.7			
	$a/b = 7$	6.4	28.1			

11-7. HERTZIAN IMPACT

The problem of two elastic solids impacting along the common normal to their curved surfaces is generally known as *Hertzian impact* and is treated in books on theory of elasticity. This type of impact will be illustrated here by the case of radial impact of two spheres shown in

TABLE 11-4
Natural Frequencies of Rectangular Plates

$$f = K\frac{h}{a^2} \times 10^4 \times \frac{c}{c_{\text{steel}}}$$

where f = lowest natural frequency, cps
K = frequency constant given in table
a = plate-edge length, inches, as shown
h = plate thickness, inches

Boundary conditions S = simply supported C = clamped	b/a					
	1.0	1.5	2.0	2.5	3.0	∞
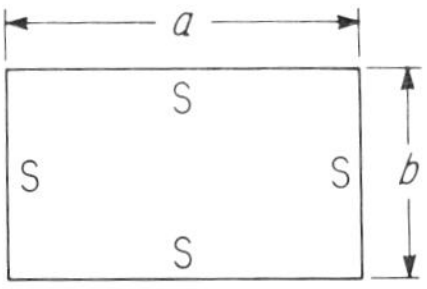	19.20	13.87	12.00	11.14	10.67	9.60
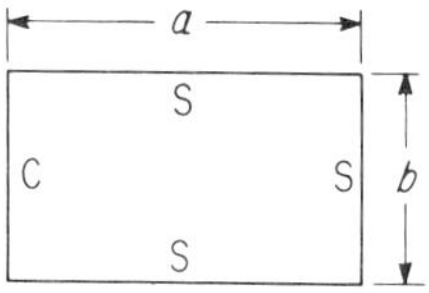	23.01	18.39	16.86	16.18	15.82	15.01
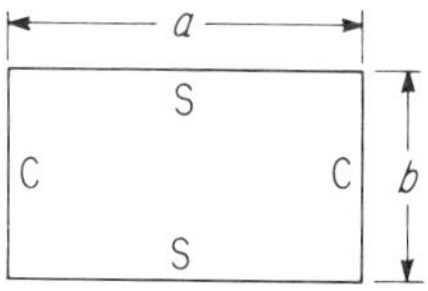	28.16	24.37	23.17	22.64	22.37	21.76

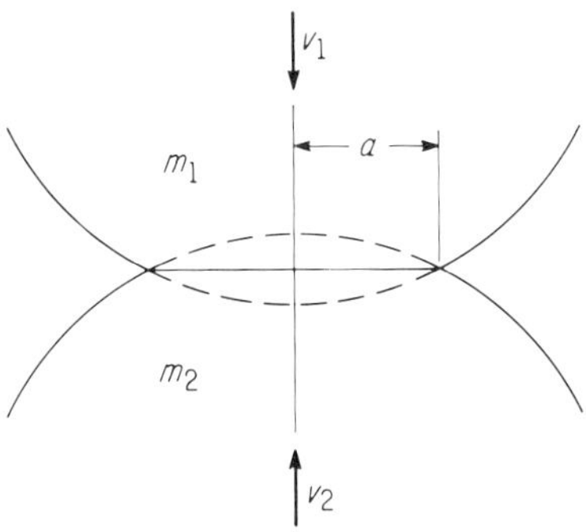

FIG. 11-15

Fig. 11-15. Assuming P to be the common impact force at any instant, therefore

$$-P = m_1\frac{dv_1}{dt}, \qquad -P = m_2\frac{dv_2}{dt}$$

TABLE 11-5
Natural Frequencies of Rectangular Plates

$$f_n = K_n \frac{h}{a^2} \times 10^4 \times \frac{c}{c_{\text{steel}}}$$

where f_n = nth natural frequency, cps
K_n = frequency constant listed in these tables
h = plate thickness, inches
a = plate-edge length, inches

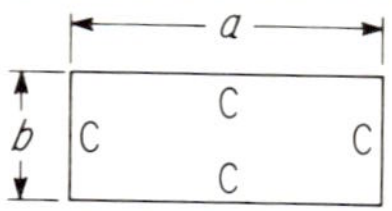	b/a									
	1.0	0.9	0.8	0.7	0.6	0.5	0.4	0.3	0.2	0.1
n 1	35.0	31.8	29.1	26.9	25.1	23.9	23.0	22.4	22.0	21.8
2	71.4	60.6	51.1	42.9	36.2	30.9	27.1	24.4	22.7	22.0
3	105.3	68.8	66.6	64.8	55.4	43.6	33.5	28.0	24.2	22.3
4	128.0	95.4	86.7	69.8	63.5	61.6	45.4	33.6	26.3	22.8

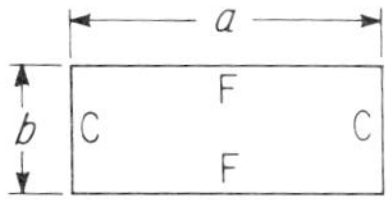

n	b/a = 1.0	0.9	0.8	0.7	0.6	0.5	0.4	0.3	0.2	0.1
1	21.5	21.4	21.4	21.4	21.3	21.3	21.3	21.2	21	21
2	25.6	26.7	27.6	29.2	31.5	34.8	40.4	50.0	57	57
3	42.3	47.4	54.2	59.2	59.0	59.0	58.6	58.5	70	
4	59.4	59.5	59.2	64.4	74.2	79.7	89.3	106.5	115	

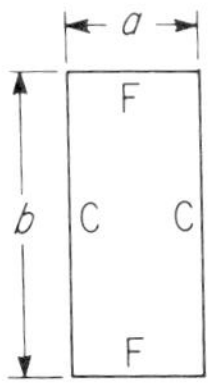

n	a/b = 1.0	0.9	0.8	0.7	0.6	0.5	0.4	0.3	0.2	0.1
1	21.5	21.5	21.5	21.5	21.5	21.5	21.6	21.6	21.6	21.6
2	25.6	24.9	24.2	23.6	23.1	22.7	22.3	22.0	21.8	21.8
3	42.3	38.3	34.8	21.6	28.9	26.7	24.8	23.4	22.4	22.0
4	59.4	59.4	56.5	47.9	40.6	34.5	29.7	26.0	23.5	22.4

TABLE 11-6
Natural Frequencies of Circular Plates

$$f = K\frac{h}{a^2} \times 10^4 \times \frac{c}{c_{\text{steel}}}$$

where f = natural frequency, cps
K = frequency constant listed in table
h = plate thickness, inches
a = plate radius, inches

Boundary conditions	Number of nodal circles	Number of nodal diameters			
		0	1	2	3
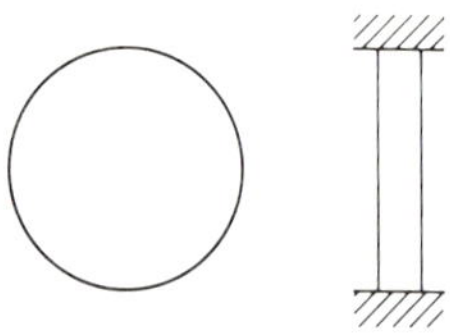	0	9.936	20.651	33.906	
	1	38.713			
	2	86.516			
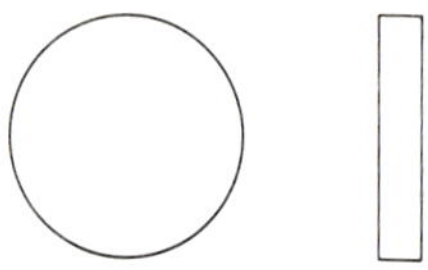	0			5.110	11.902
	1	8.832	19.970	34.295	51.491
	2	37.487	58.255		
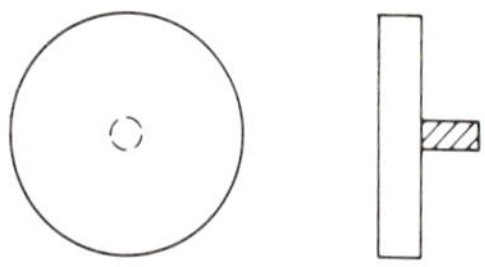	0	3.649			
	1	20.349			
	2	59.053			
	3	116.490			
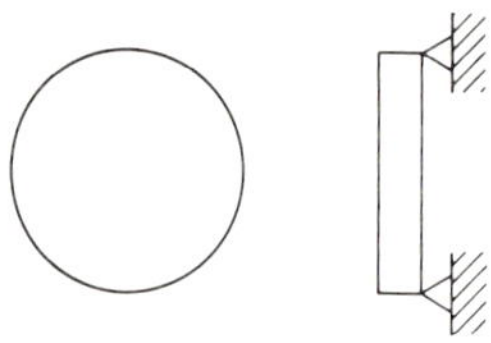	0	5.0			

where m_1, m_2 = masses of the spheres
v_1, v_2 = velocities of the two spheres at any instant during their impact

and the relative velocity α of the center of mass can be written as

$$\dot{\alpha} = v_1 + v_2$$

$$\ddot{\alpha} = -\left(\frac{P}{m_1} + \frac{P}{m_2}\right)$$

$$= -P\left(\frac{1}{m_1} + \frac{1}{m_2}\right) \tag{11-30}$$

TABLE 11-7
First Natural Frequency of Circular Plates with Circular Holes

$$f = \beta^2 \frac{1}{a^2} \sqrt{\frac{g}{\gamma} \frac{Eh^3}{12(1 - \nu^2)}}$$

where f = first natural frequency
a = outer radius of plate
b = radius of hole
h = thickness
E = modulus of elasticity
γ = weight per cubic inch
ν = Poisson's ratio = 1/3
g = gravitational acceleration = 386 in/sec^2
β = values tabulated below

b/a	β		
	Clamped	Supported	Free
0	3.195	2.233	2.292
0.1	3.188	2.212	2.281
0.2	3.221	2.175	2.248
0.3	3.363	2.158	2.196
0.4	3.680	2.180	2.137
0.5	4.195	2.245	2.050
0.6	5.028	2.380	1.966
0.7	6.520	2.620	1.876
0.8	9.242	3.071	1.790
0.9	≈ 15	4.190	1.702

The expression for the relative displacement α of the centers of the spheres under a force P is evaluated from Hertz theory of elastic bodies in contact as

$$\alpha = \sqrt[3]{P^2 \frac{9\pi^2}{16} \frac{(K_1 + K_2)^2(R_1 + R_2)}{R_1 R_2}} \qquad (11\text{-}31)$$

or

$$P = \{(4/3\pi)[\sqrt{R_1 R_2/(R_1 + R_2)}]/(K_1 + K_2)\}\alpha^{3/2}$$

where R_1, R_2 are the radii of the spheres

$$K_1 = \frac{1 - \nu_1^2}{\pi E_1}, \qquad K_2 = \frac{1 - \nu_2^2}{\pi E_2}$$

where ν, E are the Poisson ratio and modulus of elasticity respectively.

Eliminating P, Eqs. 11-30 and 11-31 therefore yields

$$\ddot{\alpha} = -C\alpha^{3/2} \qquad (11\text{-}32)$$

where

$$C = \frac{4}{3\pi}\left(\frac{m_1 + m_2}{m_1 m_2}\right) \frac{\sqrt{R_1 R_2/(R_1 + R_2)}}{K_1 + K_2}$$

Equation 11-30 can be rewritten as

$$\frac{1}{2}\frac{d(\dot{\alpha})^2}{d\alpha} = -Ca^{3/2}$$

By integration, therefore

$$\frac{1}{2}\dot{\alpha}^2 = -\frac{2}{5}C\alpha^{5/2} + \text{const.}$$

At $t = 0, \quad \alpha = 0, \quad \dot{\alpha} = v_0$; therefore

$$\tfrac{1}{2} v_0^2 = \text{const.}$$

where v_0 is the approach velocity before impact. The solution of Eq. 11-30 is therefore

$$\frac{1}{2}(\dot{\alpha}^2 - v_0^2) = -\frac{2}{5}C\alpha^{5/2} \qquad (11\text{-}33)$$

$$\dot{\alpha} = \sqrt{v_0^2 - \frac{4}{5}C\alpha^{5/2}} \qquad (11\text{-}34)$$

The maximum impact force and consequently the maximum approach $\dot{\alpha}_{\max}$ occur when $\alpha = 0$. This condition gives

$$\alpha_{\max} = \left(\frac{4}{5}\frac{v_0^2}{C}\right)^{2/5} \qquad (11\text{-}35)$$

from which

$$P_{\max} = \frac{4}{3\pi}\frac{\sqrt{R_1R_2/(R_1 + R_2)}}{(K_1 + K_2)}\left(\frac{4}{5}\frac{v_0^2}{C}\right)^{6/10}$$

$$= \frac{4}{3\pi}\frac{\sqrt{R_1R_2/(R_1 + R_2)}}{(K_1 + K_2)}\left(\frac{4}{5C}\right)^{0.6} v_0^{1.2}$$

$$= C\left(\frac{m_1m_2}{m_1 + m_2}\right)\left(\frac{4}{5C}\right)^{0.6} v_0^{1.2}$$

$$= \left(\frac{4}{5}\right)^{0.6}\frac{m_1m_2}{m_1 + m_2}C^{0.4}v_0^{1.2} \qquad (11\text{-}36)$$

The duration of impact τ can be calculated by recasting Eq. 11-34 as

$$dt = \frac{d\alpha}{\sqrt{v_0^2 - \frac{4}{5}C\alpha^{5/2}}}$$

from which integration yields

$$\tau = 2\int_0^{\alpha_{max}} \frac{d\alpha}{\sqrt{v_0^2 - \frac{4}{5}C\alpha^{5/2}}}$$

$$= 2\int_0^{\alpha_{max}} \frac{d\alpha}{v_0\sqrt{1 - \left(\frac{\alpha}{\alpha_{max}}\right)^{5/2}}}$$

$$= 2.9432\frac{\alpha_{max}}{v_0} \tag{11-37}$$

11-8. SUMMARY

A brief discussion of the problems associated with impact and vibration of continuous systems is presented in this chapter. The relationship between the wave motion and the natural period of oscillation is given for simple examples of such systems. Series solutions has been discussed for beam and plate problems. The procedure for calculation of the pulse resulting from impact of spheres is presented. Design data for the natural frequencies of beams and plates with different configurations and boundary conditions are also given.

chapter 12

Sound

12-1. INTRODUCTION

An important phenomenon associated with continuous systems is sound. Sound is a mechanical wave produced by an excitation or disturbance in a vibratory system and transmitted through a solid or fluid continuum. It is a form of oscillation which has a most notable significance and influence on man's daily activities.

Sound waves are classified according to their frequencies as follows:

Below 20 cps. . . . . Infrasonic range
20 cps to 20,000 cps . . . . . . Audible range
Above 20,000 cps. . . . . . Ultrasonic range

While a mechanical wave is generally characterized by its amplitude, frequency and waveform, sound waves are described by their intensity, pitch, and quality. These characteristics are respectively correspondent to each other in the case of harmonic waves. For a complex sound wave like that associated with the human voice or a musical instrument, the relations between amplitude and intensity, frequency, and pitch are no longer simple.

Because of the growing interest of acoustics as related to mechanical systems, this chapter will be devoted to a brief introduction to the phenomenon of sound.

12-2. PRODUCTION OF SOUND

It has been shown in the previous chapters that many mechanical systems oscillate upon subjection to a disturbance. As these systems vibrate they transmit energy to the surrounding medium and consequently undergo a decay of their motion. The transmitted energy to the surroundings manifests itself as sound waves with the same frequencies as those of the vibrating system. If the mechanical system is a linear single-degree-of-freedom system, the corresponding sound waves will be monotonic. On the other hand, if the system is oscillating with many frequencies (as in lumped-parameter or continuous systems) the sound will be multiharmonic.

In the case of a string subjected to a general disturbance the different vibrational frequencies are multiples of a fundamental frequency. The sound wave corresponding to the fundamental frequency is called the *fundamental tone*. The waves corresponding to the higher-order natural frequencies are called the *first overtone*, *second overtone*, etc.

12-3. SOUND-GENERATING ELEMENTS

All vibratory systems are sources of sound at the frequency of their vibration. The following is a tabulation of the natural frequencies f_n of some vibrations of continuous systems which are generally associated with sound generation.

1. Lateral vibration of strings

$$f_n = \frac{n\sqrt{T/m}}{2L}$$

where f_n = nth natural frequency
n = order of vibration mode (1, 2, 3, ...)
T = string tension
m = mass per unit length of the string
L = length of the string

2. Longitudinal vibration of rods

$$f_n = \frac{n\sqrt{E/\rho}}{2L}$$

where $\sqrt{E/\rho}$ = wave-propagation velocity
E = modulus of elasticity
ρ = mass density
L = length of rod

3. Lateral vibration of beams

$$f_n = K\frac{r}{L^2} \times 10^4 \frac{c}{c_s}$$

and can be calculated for the different modes by utilizing Table 11-1 for beams with different end conditions and variable cross sections.

4. Vibration of fluid columns
   (a) One end open and the other closed

$$f_n = \frac{(2n - 1)\sqrt{K/\rho_0}}{4L}$$

where $\sqrt{K/\rho_0}$ = wave propagation velocity in fluid

K = bulk modulus

ρ_0 = equilibrium density

L = length of column

n = order of vibrational mode

(b) Columns with both ends open

$$f_n = \frac{n\sqrt{K/\rho_0}}{2L}$$

where n = order of vibrational mode

Table 12-1 shows the natural frequencies and modal shapes for the first four modes of vibration

TABLE 12-1
Natural Frequencies in Fluid Columns

Condition	f_n			
	$n = 1$	$n = 2$	$n = 3$	$n = 4$
One open end and one closed end	$\frac{\sqrt{K/\rho_0}}{4l}$ 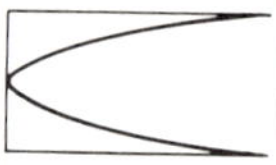	$\frac{3\sqrt{K/\rho_0}}{4l}$ 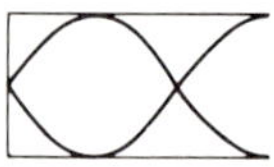	$\frac{5\sqrt{K/\rho_0}}{4l}$ 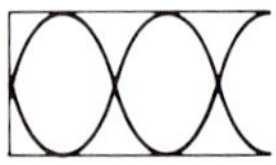	$\frac{7\sqrt{K/\rho_0}}{4l}$ 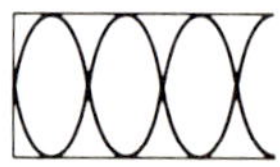
Two open ends	$\frac{\sqrt{K/\rho_0}}{2l}$ 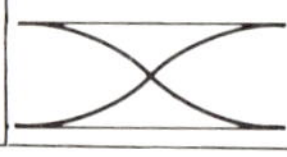	$\frac{\sqrt{K/\rho_0}}{l}$ 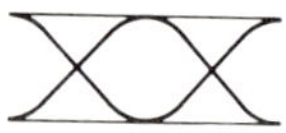	$\frac{3\sqrt{K/\rho_0}}{2l}$ 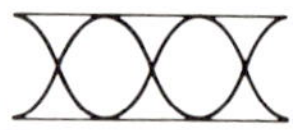	$\frac{2\sqrt{K/\rho_0}}{l}$ 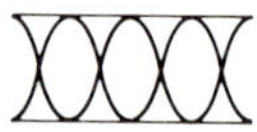

5. Vibration of circular membranes

$$f_n = \frac{K}{r}\frac{T}{h}\frac{\sqrt{E/\rho}}{\sqrt{(E/\rho)_{\text{steel}}}}$$

where f_n = natural frequency for nth mode, cps

r = radius of membrane

T = tension at the circumference (lb/in.)

h = thickness in inches

E = modulus of elasticity

ρ = mass density

Values of K for different modes of vibration of the membrane are given in Table 12-2. The mode shapes are also illustrated for each case. The shaded areas represent motions in the opposite direction of the blank areas.

TABLE 12-2
Vibration Modes of Circular Membranes

Number of nodal circles	Number of Nodal Diameters		
	0	1	2
1	14.09 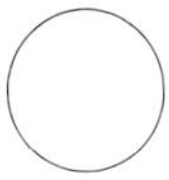	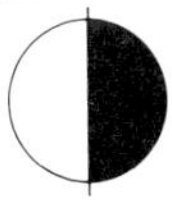 22.49	 30.12
2	32.41 	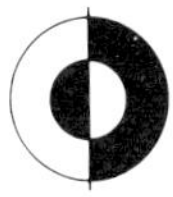 41.22	 49.44
3	50.79 	 59.71	 64.94

6. Vibrations of plates

The general equation for the natural frequencies for the different vibrational modes for several conditions of plates are given in Tables 11-2 through 11-7.

12-4. INTENSITY OF SOUND

Sound intensity is the power transferred through a unit area normal to the direction of sound propagation. The unit usually used for measurement of sound intensity is watts per square meters.

The equation for calculation of sound intensity can be derived as follows. Assuming that a harmonic sound wave with an amplitude A and frequency ω (rad/sec) is traveling in a medium with density ρ, the sound energy per unit volume E can be calculated from

$$E = \int_0^A (\text{mass} \times \text{acceleration})\, dx$$

$$= \int_0^A \rho_0 \omega^2 x\, dx$$

$$= \tfrac{1}{2} \rho_0 \omega^2 A^2$$

where ρ_0 is the equilibrium mass density of the medium. Since the wave is traveling with a propagation velocity $c = \sqrt{K/\rho_0}$, the power per unit area (or the intensity) is therefore

$$I = \tfrac{1}{2} \rho_0 \omega^2 A^2 c$$

Substituting $f = \omega/2$ cps in this equation gives

$$I = 2\pi^2 f^2 A^2 c$$

The loudness of a monotonic sound as detected by the ear increases with increasing intensity. The ratio of intensity for sound that is barely audible to that which is painful may be as low as 10^{-12}. The international standard for the threshold of audibility is $I = 10^{-12}$ watt/m^2, while the threshold of pain is in the neighborhood of $I = 1.0$ watt/m^2. It is, therefore, convenient to use a logarithmic scale for intensity evaluation. These scales are usually selected in units of *bels* or *decibels*. The bels represent units of the logarithm to the base 10 of the intensity ratio and the decibels are $^1/_{10}$ of the bels. Accordingly, the intensity level in bels and decibels of a sound with intensity I watts/m^2 relative to a reference intensity I_0 is given as

$$\text{Intensity level} = \log_{10} \frac{I}{I_0} \text{ bels}$$

$$= 10 \log_{10} \frac{I}{I_0} \text{ decibels (or db)}$$

Table 12-3 gives examples of intensity levels for some familiar sounds of different levels for illustration.

A plot of the auditory area between threshold of hearing and threshold of pain over the audible frequency range is given in Fig. 12-1.

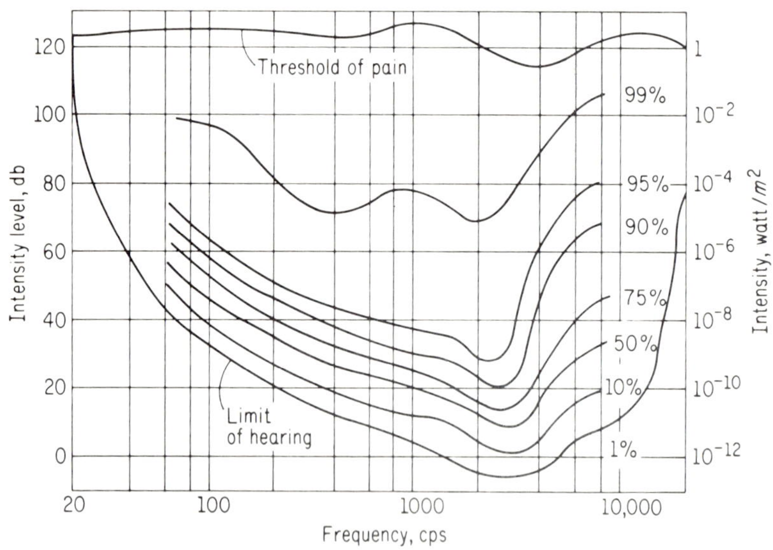

FIG. 12-1

TABLE 12-3

Sound	Intensity watt/m²	Intensity level, db
Threshold of hearing	10^{-12}	0
Whisper	10^{-10}	20
Busy traffic	10^{-5}	70
Elevated train	10^{-4}	90
Threshold of pain	1	120

12-5. INTERFERENCE, REFLECTION, REFRACTION, AND DIFFRACTION OF SOUND

Sound waves, like any other wave phenomena, are subject to interference, reflection, refraction, and diffraction. Interference of waves can be evaluated by superposition as described in Chapter 2. Beats will occur in sound when two waves have slightly different frequencies.

Reflection occurs due to impedance change at the interface where the change takes place. *Echo* is a familiar example of reflection.

Refraction of sound (as in the refraction of light) occurs when it passes the interface between two media with different propagation velocities. A well-recognized example of refraction of sound is the acoustic mirages that occur under conditions of variable temperatures and wind. Sound travels faster in warm air than in cold air, and faster with the wind than against it. As a result under adverse conditions of wind and temperature, sound may appear to come from a quite different direction than it would through a homogeneous medium.

Diffraction is one of the most striking properties of sound and explains its ability to bend around corners and obstacles. When the obstacle diffracting the sound wave is relatively small, as in the case of fog droplets, the bending makes the sound turn in all directions from the original one and causes scattering of the sound.

12-6. DOPPLER EFFECT

Christian Johann Doppler (1803–1853) observed that when a wave-generating source is moving with respect to the receiving station (or vice versa) a change of frequency in the received wave takes place. This change is influenced by both the speed of the source and the receiver.

This phenomenon, known as the *Doppler effect*, is easily observed by the change in pitch of an approaching sound source. Many useful devices for detection of velocity of moving objects are based on this effect. This is done by comparing the frequency of a transmitted wave to its frequency when received back after its reflection from the moving object. Radar is an example of such a device utilizing radio waves.

Let us assume for example that a source is generating sound at a frequency f_S and is moving with a velocity v_S toward a receiving station. The wavelength λ_0 of the received sound at the station will therefore undergo a change due to the change in the distance over which the waves are received. This can be calculated from

$$\lambda_0 = \frac{c - v_S}{f_s}$$

where c is the wave propagation velocity.

Similarly, if the receiving station is also moving with a velocity v_R toward the source, the receiver gathers away more waves in a unit time. The received frequency f_R will accordingly be

$$f_R = \frac{c + v_R}{\lambda_0}$$

The general expression for the frequency of the received wave when both the source and the receiver are moving is obtained by combining the previous equations. Accordingly,

$$f_R = \frac{c + v_R}{c - v_S} f_S$$

In this equation v_S and v_R are substituted with a positive or negative sign, depending on whether either the source or the receiver is moving toward or away from the other respectively.

12-7. NOISE

The quality of sound is characterized by the harmonic composition of its waveform. When the amplitude variations of the sound wave are random and do not follow a predictable pattern the sound is classified as noise.

Noise is generally defined as undesirable sounds. The degree of "undesirability" is a psychological question and is open to subjective interpretations. In the analytical sense, noise is usually defined as a random wave resulting from a random excitation. The analysis of this type of motion is discussed in the following chapter.

12-8. SUMMARY

This chapter gives a brief introduction to the analysis of sound. The objective is to familiarize the reader with the basic definitions and terminology. Since sound is a wave phenomenon, its analysis follows the laws of waves in continuous media. Similarly, the analysis of noise falls in the domain of random wave analysis which is considered in the following chapter.

REFERENCES

1. S. Timoshenko and J. Goodier, *Theory of Elasticity*. 2d ed. New York: McGraw-Hill, 1951.
2. D. Wallace, "Optimum Design of Prismatic Bars Subjected to Longitudinal Impact," M.S. thesis, U. of Wisconsin, Madison, Wis., 1968.
3. H. Lamb, "Wave in an Elastic Plate," *Proc. Roy. Soc.*, Vol. 93 (1917), p. 114.
4. S. Timoshenko, "On the Transverse Vibration of Bars of Uniform Cross Section," *Phil. Mag.*, Vol. 43, Series 6 (1922), p. 125.
5. C. V. Joga Roa and G. Pickett, "Vibrations of Plates of Irregular Shapes and Plates with Holes," *J. Aeron. Soc. of India*, Vol. 13, No. 3 (1961).
6. J. N. McDuff and R. P. Felgar, "Vibration Design Charts," *Trans. ASME*, Vol. 79 (1957), pp. 1459–1475.
7. S. Timoshenko, *Vibration Problems in Engineering*. 3d ed. Princeton, N. J.: Van Nostrand, 1955.
8. M. V. Barton, "Vibration of Rectangular and Skew Cantilever Plates," *Trans. ASME, J. Appl. Mech.*, Vol. 18 (1951), pp. 129–134.
9. B. W. Anderson, "Vibration of Triangular Cantilever Plates by the Ritz Method," *Trans. ASME, J. Appl. Mech.*, Vol. 21 (1954), pp. 365–370.
10. H. A. Rothbart (ed.), *Mechanical Design and Systems Handbook*, New York: McGraw-Hill, 1964.
11. S. H. Crandall, *Engineering Analysis*. New York: McGraw-Hill, 1956.
12. R. F. Rissone and J. J. Williams, "Vibration of Nonuniform Cantilever Beams," *Engineer* (London), Vol. 220 (Sept. 24, 1965), pp. 497–506.
13. W. Goldsmith, *Impact*. London: Arnold, 1960.
14. J. N. Brennan (ed.), "Bibliography on Shock and Shock-Excited Vibration," Eng. Res. Bull. No. 69, Pennsylvania State Univ., University Park, Pa., 1957.
15. A. E. Love, *A Treatise on the Mathematical Theory of Elasticity*. New York: Dover, 1944.
16. H. Kolsky, *Stress Waves in Solids*. London: Oxford U. P., 1953.
17. J. W. Strutt, Lord Rayleigh, *Theory of Sound*. 2d ed. New York: Dover, 1937.
18. L. L. Beranek, *Noise Reduction*. New York: McGraw-Hill, 1960.
19. G. H. Shortley and D. E. Williams, *Elements of Physics*. 4th ed. Englewood Cliffs, N. J., 1965.
20. J. T. Broch, *Acoustic Noise Measurements*. Denmark: Bruel and Kjaer, 1967; also B & K Instruments Co., Cleveland, Ohio.
21. J. L. Hunter, *Acoustics*. Englewood Cliffs, N. J.: Prentice-Hall, 1957.
22. C. M. Harris, *Handbook of Noise Control*. New York: McGraw-Hill, 1957.

part V

Systems Subjected to Random Excitations

I regard the University as the noblest school of mathematical education in the world. And, in spite of the injury that is done by a too mechanical use of the analysis it is the most exact. There is no accuracy of teaching like that of Cambridge, and in the applied sciences that are really taken up there, of which I would especially mention Astronomy, the education is complete and excellent. But with this, there is an excessively strong tendency to those mathematical subjects which are pursued in the closet, without the effort of looking into the scientific world to see what is wanted there.

G. B. Airy (1868), in a letter to G. G. Stokes

chapter 13

Systems Subjected to Random Excitations

13-1. INTRODUCTION

In the previous chapter, we have been concerned with systems and disturbances which are known deterministically. It has been shown that exact solutions for such systems become progressively difficult as the number of degrees of freedom increases. In the case of a continuous system subjected to a continually changing disturbance, the evaluation of system response by these classical techniques can be prohibitive.

It is not unusual in many practical cases that the system parameters and the geometrical boundaries may not be known in a precise manner or may not easily lend themselves to exact analysis. Moreover, the disturbance on the system may also be highly erratic and unpredictable. Structures which are expected to withstand natural forces such as wind, waves, gusts, and earthquakes are continually being designed. Public safety in many instances depends on the accuracy of prediction of their response to these forces. Some familiar examples are aircraft and spacecraft structures. These are known to undergo considerable stress due to the acoustic energy of their engines as well as the aerodynamic effects during flight. The description of the behavior of these systems by classical methods is all but impossible.

In the cases where the system parameters or the disturbances are only known probabilistically, statistical methods in vibration analysis provide predictions of system response which are also probabilistic. These statistical methods are usually very helpful in physical interpretation of system behavior and in development of general design guides.

Statistical concepts are often used in many branches of the physical sciences. It has been used for many years in design of electrical and acoustical systems. Its usefulness in mechanical and structural systems became evident in recent years in studying the structural response of sea, air, and space craft to shock and vibration.

This chapter provides a brief introduction to some statistical concepts and their applications in vibration problems.

13-2. THE RANDOM PROCESS

A process is categorized as random when different results can be obtained even though all the controllable conditions are unchanged. In such cases, one set of data cannot satisfactorily describe the process. Neither can it be predicted by mathematical means. Any description for such a process has to be sought within the laws of statistics.

In general, a set of discrete data can be described by means of some average value (which is a measure of a central tendency) and a dispersion which indicate a deviation from the average value.

A common measure of central tendency is the arithmetic mean. If $x_1, x_2, x_3, \ldots, x_i, \ldots, x_n$ represent a set of discrete data, the arithmetic mean $\bar{x}$ is therefore

$$\begin{aligned}\bar{x} &= \frac{x_1 + x_2 + x_3 + \cdots + x_n}{n} \\ &= \frac{1}{n} \sum_{i=1}^{n} x_i \end{aligned} \tag{13-1}$$

The deviation d_i of any value x_i from the mean $\bar{x}$ is

$$d_i = x_i - \bar{x} \tag{13-2}$$

A common measure of the dispersion known as the standard deviation σ is defined as

$$\begin{aligned}\sigma &= \sqrt{\frac{\Sigma (d_i)^2}{n - 1}} \\ &= \sqrt{\frac{(x_1 - \bar{x})^2 + (x_2 - \bar{x})^2 + \cdots + (x_n - \bar{x})^2}{n - 1}}\end{aligned} \tag{13-3}$$

The square of the standard deviation is known as the *variance* and is given by

$$\sigma^2 = \frac{\Sigma (d_i)^2}{n - 1} \tag{13-4}$$

In the case of zero mean, Eqs. 13-1 to 13-4 become

$$\begin{aligned}\bar{x} &= 0 \\ d_i &= x_i \\ \sigma &= \sqrt{\frac{\Sigma (x_i)^2}{n - 1}} \\ \sigma^2 &= \frac{\Sigma (x_i)^2}{n - 1}\end{aligned}$$

The standard deviation and the variance in this case represent the root mean square and the mean square respectively.

Stationary Random Process

A random process is called *stationary* if its statistical properties are not affected by a shift in the time origin.

Probability Distribution

If the total number of data of a discrete set is n, the fraction $P = \nu/n$ of the total number for which x_i will be less than a specified value x provides a representation of the probability distribution of the set. ν in this case represents the frequency of occurrence of $x_i < x$. The fraction of the total number for which x_i is less than ∞ is 1, since all values of the set should be less than ∞.

The probability of a value less than $-\infty$ is obviously zero. Accordingly, the fraction having a value less than some arbitrary value of x (between $-\infty$ and ∞) should fall between zero and 1. This can be written as

$$P(\infty) = 1$$

$$P(-\infty) = 0$$

and

$$0 \leq P(x) \leq 1$$

Figure 13-1 represents this probability distribution P and is known as the *first probability distribution.*

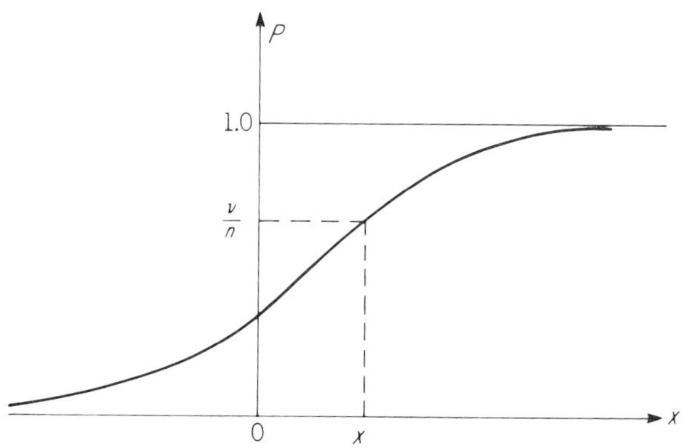

FIG. 13-1

Probability Density

The first derivative of the probability distribution is known as the *probability density p.* By definition, the probability density is

$$p(x) = \frac{d}{dx} P(x) \tag{13-5}$$

It follows that

$$P(x) = \int_{-\infty}^{x} p(x)\,dx \tag{13-6}$$

and

$$P(\infty) = \int_{-\infty}^{\infty} p(x)\,dx = 1$$

Figure 13-2 describes the probability density for the distribution given in Fig. 13-1. By virtue of Eqs. 13-5 and 13-6, the area under the

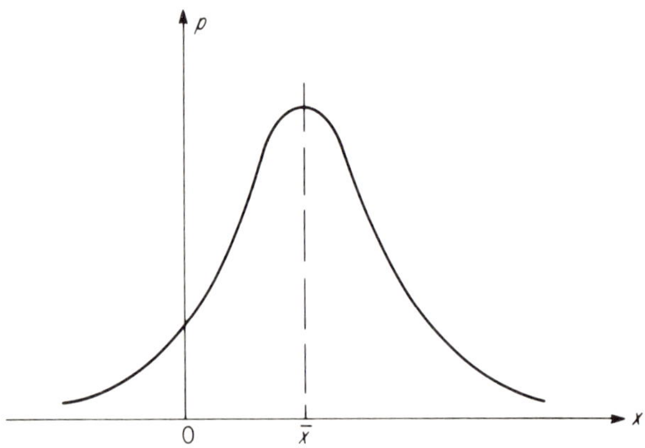

FIG. 13-2

probability-density curve between any two values of x gives the fraction of the total number having a value between these two limits. The area under the curve between x and ∞ represents the fraction of the total number having a value greater than x.

The probability-density curve therefore can provide all necessary information concerning the distribution of a random process. One of the most common forms of probability-distribution functions is the Gaussian or normal distribution shown in Fig. 13-3. The curve is described by the equation

$$p(x) = \frac{1}{\sigma\sqrt{2\pi}}\, e^{-\frac{(x-\bar{x})^2}{2\sigma^2}} \tag{13-7}$$

where $p(x)$ = probability density at any value x
$\bar{x}$ = arithmetic average defined by Eq. 13-1
σ = standard deviation defined by Eq. 13-2
e = logarithmic constant = 2.7183

The Gaussian distribution curves have the following characteristics:

1. The most frequently occurring value is the arithmetic avg. $\bar{x}$

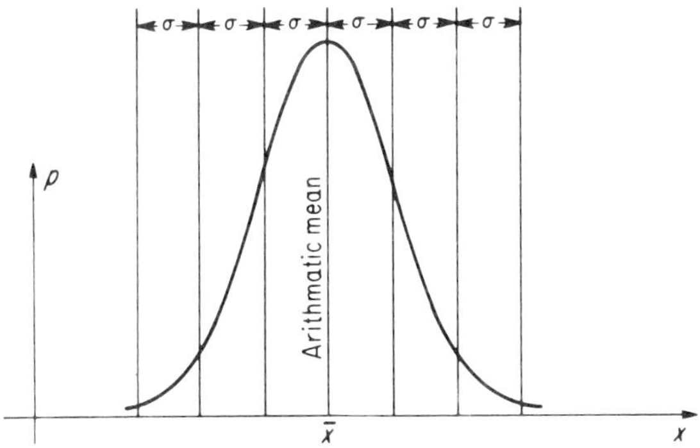

FIG. 13-3. Gaussian or normal probability density curve.

2. The curve is symmetrical about $\bar{x}$
3. The dispersion is measured by the standard deviation
4. The probability distribution as indicated by the area under the curve is as follows:

Value limits	Percentage of total events
$\bar{x} \pm \sigma$	68.26
$\bar{x} \pm 2\sigma$	95.46
$\bar{x} \pm 3\sigma$	99.73
$\bar{x} \pm \infty$	100.00

This shows that in a normal distribution, more than two-thirds of the total falls within one standard deviation from the arithmetic average. It also shows that 99.73 percent of the total falls within 3σ from the average value. Although many random processes are distributed in the Gaussian form, it is not unusual that other forms of distribution occur in practice. See Fig. 13-4.

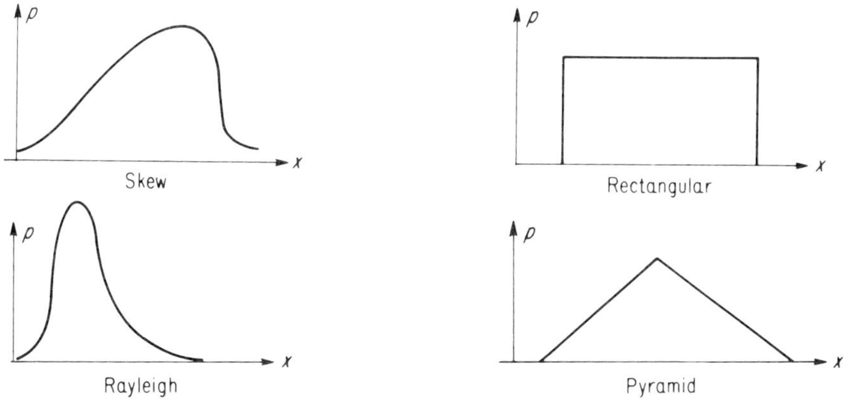

FIG. 13-4

13-3. RANDOM MOTION

Power Spectra

We have shown in Part II that a complex wave $x(t)$, if periodic, can be analyzed into an average value and different harmonic components by the Fourier analysis. When the wave has a zero mean, it can be described as

$$x(t) = \sum_{i=1}^{\infty} A_i \sin(\omega_i t + \phi_i)$$

where A_i = amplitude of each harmonic
ω_i = corresponding frequency
ϕ_i = corresponding phase angle

Since the mean-square value of a sinusoidal wave is equal to half the square of its amplitude,

$$\frac{1}{2}\int_0^{2\pi} A_i^2 \sin^2(\omega_i t + \phi_i)\,d\omega t = \frac{A_i^2}{2}$$

Therefore, the mean-square value of a complex wave with zero mean is half the sum of the squares of the amplitudes of its different harmonic components. This is written as

$$\begin{aligned} X_{\text{ms}} &= \text{mean-square value of the wave} \\ &= \sum_{i=1}^{\infty} \frac{A_i^2}{2} \end{aligned} \tag{13-8}$$

and consequently

$$\begin{aligned} X_{\text{rms}} &= \text{root-mean-square value of the wave} \\ &= \left[\sum_{i=1}^{\infty} \frac{A_i^2}{2}\right]^{1/2} \end{aligned}$$

Since the mean square of a harmonic wave is an indication of the power associated with it, the mean square of a complex wave can therefore be considered as proportional to the total power associated with the wave.

When a complex wave is fed into a narrow-band filter with a center frequency ω_0 and bandwidth $\Delta\omega$, the filter will only transmit those frequencies within the passband ($\omega_0 \pm \frac{1}{2}\Delta\omega$). If the output is then fed into a mean-square meter (which contains squaring and averaging devices), the output will therefore represent the mean-square amplitude $(\overline{\Delta x})^2$ associated with this particular band. When this is divided by the bandwidth $\Delta\omega$, the result gives the value of the mean-square spectral density $W(\omega_0)$ at the center frequency ω_0.

$$W(\omega_0) = (\overline{\Delta x})^2/\Delta\omega \tag{13-9}$$

The power spectral density can therefore be obtained over a wide frequency range and plotted versus the frequency as shown in Fig. 13-5.

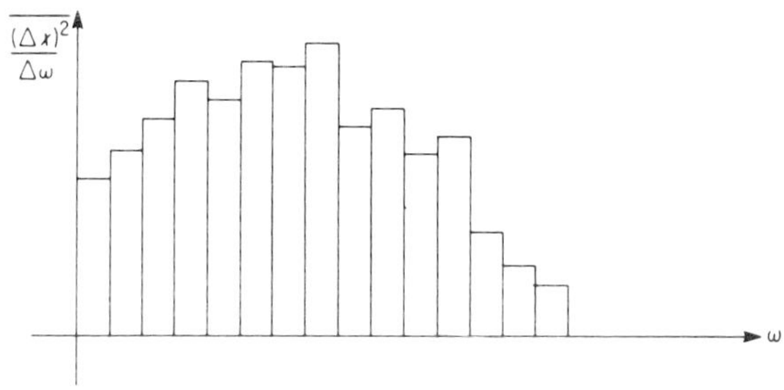

FIG. 13-5

When $\Delta\omega$ approaches zero, Eq. 13-9 takes the general form

$$W(\omega) = \frac{d\bar{x}^2(\omega)}{d\omega}$$

and the power spectral density plot approaches a smooth curve as shown in Fig. 13-6. The mean-square value of the random wave is given by the area under this curve.

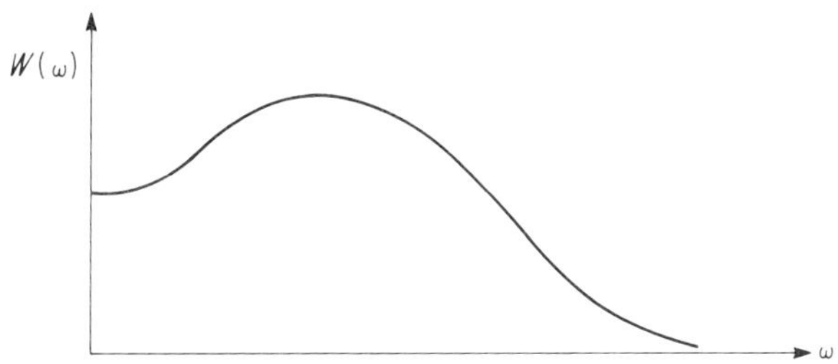

FIG. 13-6

$$\begin{aligned}\bar{x}^2 &= \text{mean-square value of the random wave}\\ &= \int_0^\infty W(\omega)\,d\omega \end{aligned} \tag{13-10}$$

The lower limit in the above integral is taken as zero, since there is no physical significance to negative frequencies. It should be noted here that when the passband width approaches zero, the filtered random wave approaches a sinusoidal vibration having the same frequency as the center frequency of the filter. The response of a structure to this filtered signal is therefore the same as its response to a sinusoidal excitation.

Common Types of Spectral Density

The following are some of the forms of spectral-density plots which represent idealizations of some of the common occurring random processes:

1. Wide-Band Random Process—"White Noise"

In a wide-band process, the mean-square spectral density (or power spectral density) $W(\omega)$ has significant value over a wide range of frequencies. Such wave is illustrated in Fig. 13-7 by a time history and a spectral density plot.

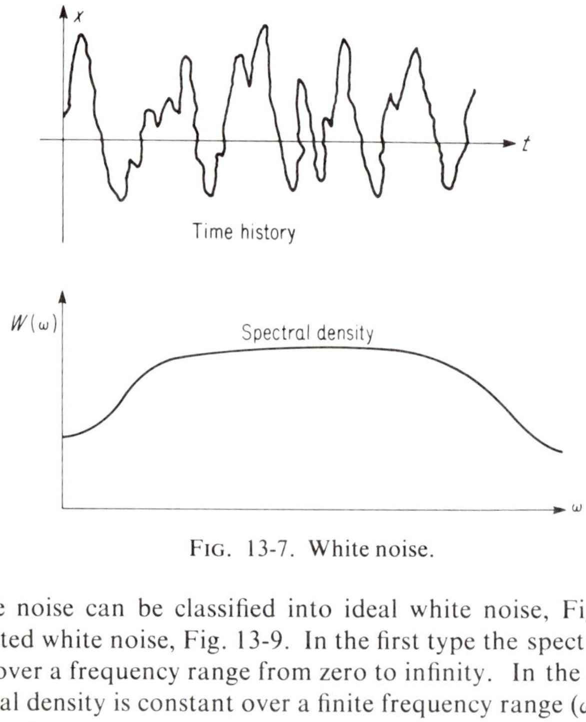

FIG. 13-7. White noise.

White noise can be classified into ideal white noise, Fig. 13-8, and band-limited white noise, Fig. 13-9. In the first type the spectral density is constant over a frequency range from zero to infinity. In the second type, the spectral density is constant over a finite frequency range $(\omega_2 - \omega_1)$ and is zero elsewhere.

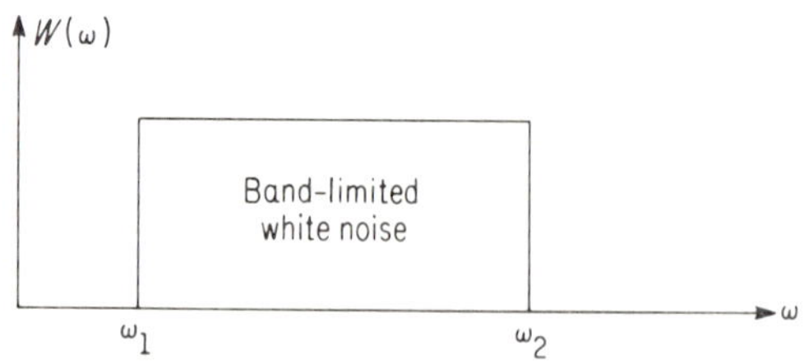

FIG. 13-8

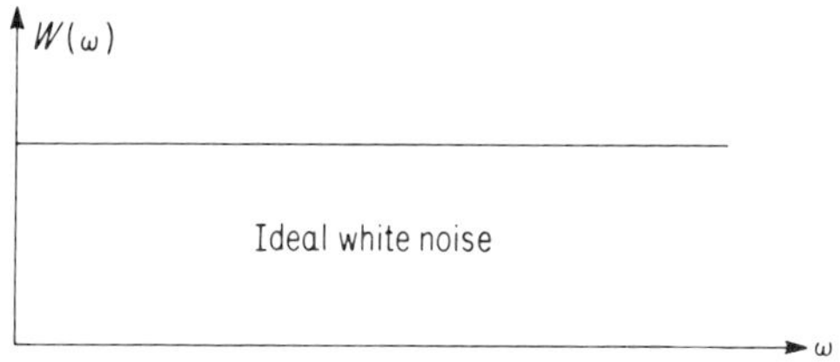

FIG. 13-9

2. Narrow-Band Process

In a narrow-band random process, the mean-square spectral density has significant values only in a narrow range of frequencies. They can be encountered, for example, as response in strongly resonant vibratory systems when the excitation is a wide-band process. Figure 13-10 shows a representation of the time history and spectral density of a narrow-band process.

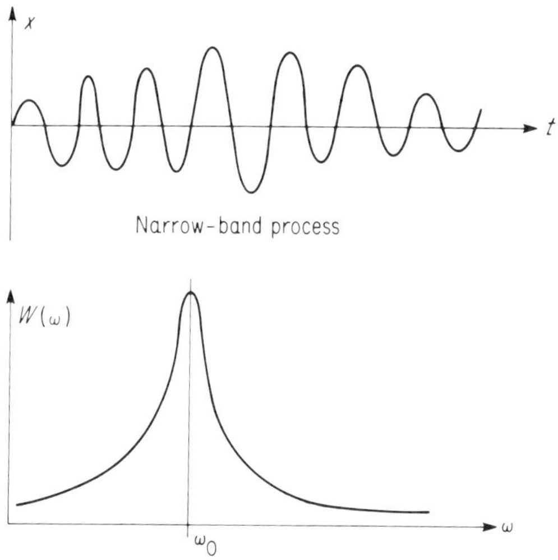

FIG. 13-10

13-4. RESPONSE OF SYSTEMS TO RANDOM EXCITATIONS

We have previously found that a filtered random vibration approaches a sinusoidal wave as the band width of the filter approaches zero. Since the mean-square value of this sinusoidal wave is known (which is the same as half the square of the amplitude), the response of a system (with known characteristics) to it can be readily obtained by applying the proper transfer function for the particular system at the particular frequency. By applying the principle of superposition, the response of a linear system to the power spectrum can theoretically be obtained by integration (or summation) of the responses to sinusoidal waves over the frequency range of the power spectrum.

This will be illustrated by considering the following examples.

A. Response of a Single-Degree-of-Freedom System to Random Excitation

We have discussed in Part II the relationship between the excitation and response (or transfer function) for the linear system represented in

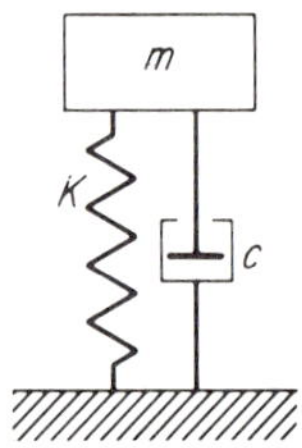

FIG. 13-11

Fig. 13-11. We found in Chapter 6 that for the case of a sinusoidal excitation

$$A_r = T(\omega)A_e \tag{13-11}$$

where A_r = amplitude of response
A_e = amplitude of excitation
$T(\omega)$ = transfer function relating the excitation and the response—a function of the frequency of the excitation and the physical characteristics of the system whose dimensions depend on the dimensions of the excitation and response

The following is a tabulation of the transfer function $T(\omega)$ for three different types of excitations. These results can be easily obtained from Chapter 6. Using the notation

ω = excitation frequency
p = natural frequency = $\sqrt{K/m}$
K = spring stiffness
m = mass
c = damping coefficient
ξ = damping factor = c/c_c
c_c = critical damping coefficient = $2\sqrt{Km}$

Condition	Excitation	Response	$T(\omega)$
1	Force excitation on the mass	Spring deflection	$\dfrac{1}{K\sqrt{\left[1-\left(\dfrac{\omega}{p}\right)^2\right]^2+\left[2\xi\dfrac{\omega}{p}\right]^2}}$
2	Acceleration of foundation	Spring deflection	$\dfrac{1}{p^2\sqrt{\left[1-\left(\dfrac{\omega}{p}\right)^2\right]^2+\left[2\xi\dfrac{\omega}{p}\right]^2}}$
3	Displacement of foundation	Spring deflection	$\dfrac{(\omega/p)^2}{\sqrt{\left[1-\left(\dfrac{\omega}{p}\right)^2\right]^2+\left[2\xi\dfrac{\omega}{p}\right]^2}}$

By squaring both sides of Eq. 13-11 we obtain the relationship between the square of the amplitudes of the response and excitation

$$A_r^2 = T^2(\omega)A_e^2(\omega) \tag{13-12}$$

Since the mean-square value is half the square of the amplitude, Eq. 13-12 can be written in terms of power spectral density as

$$W_r(\omega) = T^2(\omega)\,W_e(\omega) \tag{13-13}$$

where $W_r(\omega)$ = mean-square or power spectral density of the response

$W_e(\omega)$ = mean-square or power spectral density of excitation

The mean-square value of the response can therefore be obtained from Eq. 13-10,

$$\bar{x}^2 = \int_0^\infty W_r(\omega)\,d\omega$$

by substitution for $W_r(\omega)$ from Eq. 13-12

$$\bar{x}^2 = \int_0^\infty T^2(\omega)\,W_e(\omega)d\omega \tag{13-14}$$

A single-degree-of-freedom system with relatively low damping will have a relatively narrow response spectrum. The system will therefore receive energy only in the neighborhood of its natural frequency. In the case of a wide-band white-noise input, the spectral density of the excitation $W_e(\omega)$ is approximately constant in the neighborhood of the natural frequency of the system. Therefore $W_e(\omega) = W_e(p)$ can be considered constant and Eq. 13-14 takes the form

$$\bar{x}^2 = W_e(p)\int_0^\infty T^2(\omega)d\omega \tag{13-15}$$

Results of the integral of the previous equation for different transfer functions are given in Ref. 10.

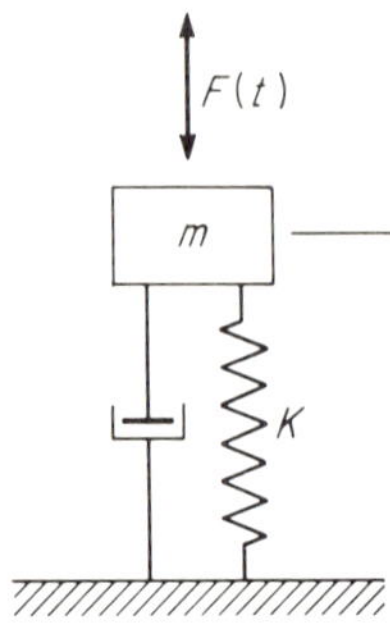

FIG. 13-12

Case 1

When the excitation is random force applied to the mass m, and the required response is the spring deflection, then

$$T(\omega) = \frac{1}{K\sqrt{\left[1-\left(\frac{\omega}{p}\right)^2\right]^2 + \left[2\xi\frac{\omega}{p}\right]^2}}$$

and the integral of the square of the transfer function in this case (from Ref. 10)

$$\int_0^\infty T^2(\omega)d\omega = \frac{\pi}{4K^2}\frac{p}{\xi} \tag{13-16}$$

The mean-square of the spring deflection, Eq. 13-16, becomes

$$\bar{x}^2 = W_F(p)\frac{\pi}{4K^2}\frac{p}{\xi} \tag{13-17}$$

where $W_F(p)$ = mean-square spectral density of the excitation (in the neighborhood of $\omega = p$ with dimensions of $\text{lb}^2/(\text{rad/sec})$

K = spring stiffness, lb/in.

p = resonance frequency, rad/sec = $\sqrt{K/m}$

ξ = damping factor (dimensionless) = c/c_c

$\bar{x}^2$ = mean-square displacement, in.^2

Case 2

In many practical situations, the excitation is usually given in terms of the power spectral density of the foundation acceleration and the required response is the spring deflection. The transfer function in this case is

$$T(\omega) = \frac{1}{p^2\sqrt{\left[1-\left(\frac{\omega}{p}\right)^2\right]^2 + \left[2\xi\frac{\omega}{p}\right]^2}}$$

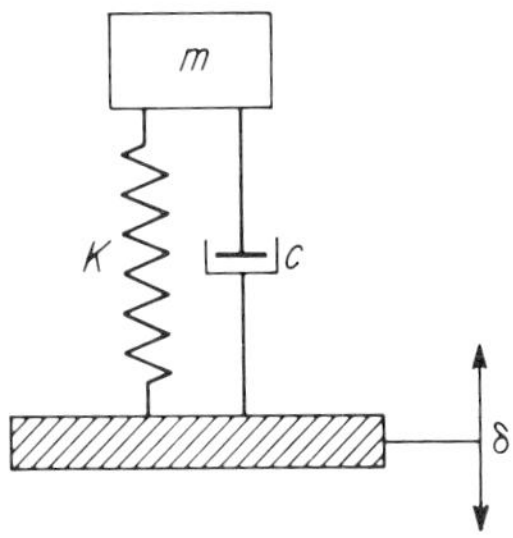

FIG. 13-13

and from Eq. 13-16,

$$\int_0^\infty T^2(\omega)d\omega = \frac{\pi}{4p^4}\frac{p}{\xi}$$

from which the mean-square of the spring deflection, Eq. 13-15 becomes

$$\bar{x}^2 = W_{\ddot{\delta}}(p)\frac{\pi}{4p^4}\frac{p}{\xi} = W_{\ddot{\delta}}(p)\frac{\pi}{4\xi p^3} \tag{13-18}$$

where $W_{\ddot{\delta}}(p)$ = mean-square acceleration spectral density of foundation (in the neighborhood of $\omega = p$) with dimensions of $(\text{in./sec}^2)^2/(\text{rad/sec})$

p = resonant frequency, rad/sec

ξ = damping factor (dimensionless)

$\bar{x}^2$ = mean-square value of spring deflection, in.^{2}

Approximate Solution for System with Two Degrees of Freedom

The response of a two-degree-of-freedom system to random excitation is basically similar in nature to that of a single-degree-of-freedom system which has been discussed. The problem, however, is complicated due to the various modes of the system and the degree of their participation in the response at any particular frequency. The problem would further be complicated when the damping is considered. A rigorous analysis of this case will require complex computations (Ref. 1). When the problem involves several degrees of freedom, the formulation may be prohibitive. It is intended here to present a simple approach to this problem which can be extended to include systems with a finite number of degrees of freedom as well as continuous systems.

Considering the two-degree-of-freedom system of Fig. 13-14 and using the notation

$(Y_1)_1, (Y_1)_2$ = displacement of the first mass relative to the platform due to the first and second modes respectively

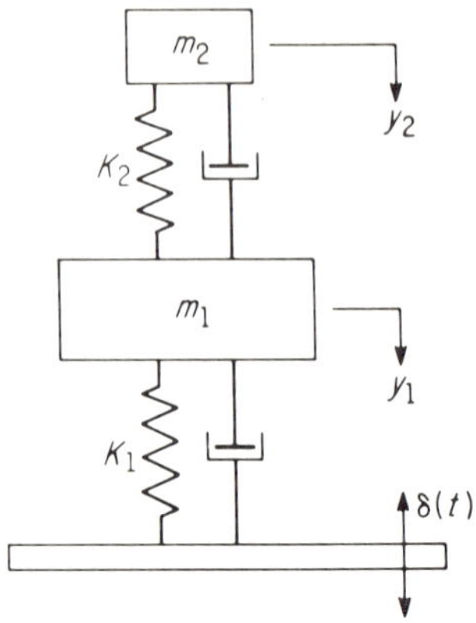

FIG. 13-14

$(Y_2)_1, (Y_2)_2$ = displacement of the second mass relative to the platform due to the first and second modes respectively

If the damping is neglected at this stage of analysis, the two amplitude ratios $(Y_1/Y_2)_1$ for the first mode and $(Y_1/Y_2)_2$ for the second mode, as well as the corresponding natural frequencies p_1, p_2 can be obtained from a free-vibration analysis of the system (Part III). Using the notation

$$(\theta_i)_k = (Y_i/Y_1)_k \tag{13-19}$$

to represent the normal-mode functions where

$(Y_i)_k$ = amplitude at mass i relative to the foundation for the k mode

$(Y_1)_k$ = the reference amplitude at the same mode

Therefore, for the two-mass system under consideration,

$$(\theta_1)_1 = (Y_1/Y_1)_1 = 1 \qquad (\theta_2)_1 = (Y_2/Y_1)_1$$
$$(\theta_1)_2 = (Y_1/Y_1)_2 = 1 \qquad (\theta_2)_2 = (Y_2/Y_1)_2$$

It has been shown in Part III that each of the two equations describing the behavior of the undamped system under sinusoidal excitation include the two independent variables y_1 and y_2. The response of the system is obtained by the simultaneous solution of the two equations. The undamped system, however, can be described as explained in Chapter 10 in the normal-mode from where each equation represents one mode of vibration and includes one independent variable representing the magnitude of vibration due to the particular mode. The equations in this case will be

$$\ddot{\eta}_1 + p_1^2 \eta_1 + a_1 \ddot{\delta}_{(t)} = 0 \tag{13-20}$$
$$\ddot{\eta}_2 + p_1^2 \eta_2 + a_2 \ddot{\delta}_{(t)} = 0 \tag{13-21}$$

where η_1 and η_2 = magnitude of the response for the two modes respectively

p_1, p_2 = natural frequencies corresponding to each mode

$\ddot{\delta}(t)$ = foundation acceleration function (or excitation function) assumed sinusoidal

a_1, a_2 = "mode-participating factors" representing intensity of excitation for each mode separately

These factors are defined as

$$a_1 = \frac{\sum_{i=1}^{i=2} m_i(\theta_i)_1}{\sum_{i=1}^{i=2} m_i(\theta_i)_1^2} \tag{13-22}$$

$$a_2 = \frac{\sum_{i=1}^{i=2} m_i(\theta_i)_2}{\sum_{i=1}^{i=2} m_i(\theta_i)_2^2} \tag{13-23}$$

Now the solution for η_1, η_2 for the two modes from Eqs. 13-20 and 13-21 is a simple matter, since we are dealing with each equation independently. It can be easily seen that the solution is

$$\eta_1 = \frac{a_1 \ddot{\delta}(t)}{p_1^2 \left[1 - \left(\frac{\omega}{p_1}\right)^2\right]}$$

$$\eta_2 = \frac{a_2 \ddot{\delta}(t)}{p_2^2 \left[1 - \left(\frac{\omega}{p_2}\right)^2\right]}$$

and by introducing in the solution, at this stage, an equivalent damping factor η_k to represent the modal damping. The solution for systems with Rayleigh damping can also be written as

$$\eta_1 = \frac{a_1 \ddot{\delta}(t)}{p_1^2 \sqrt{\left[1 - \left(\frac{\omega}{p_1}\right)^2\right]^2 + \left[2\xi_1 \frac{\omega}{p_1}\right]^2}} \tag{13-24}$$

$$\eta_2 = \frac{a_2 \ddot{\delta}(t)}{p_2^2 \sqrt{\left[1 - \left(\frac{\omega}{p_2}\right)^2\right]^2 + \left[2\xi_2 \frac{\omega}{p_2}\right]^2}} \tag{13-25}$$

The modal-damping factors ξ_1 and ξ_2 are either estimated or deduced from analysis of the response of the system to sinusoidal vibration.

After the determination of η_1 and η_2 the amplitude of the response of the two masses is calculated from

$$(Y_1)_1 = \eta_1(\theta_1)_1 = \eta_1$$
$$(Y_2)_1 = \eta_1(\theta_2)_1$$
$$(Y_1)_2 = \eta_2(\theta_1)_2 = \eta_2$$
$$(Y_2)_2 = \eta_2(\theta_2)_2$$

and in a general notation

$$(Y_i)_k = \eta_k(\theta_i)_k \tag{13-26}$$

where i represents the mass and k the mode. By substituting the value of η_k (Eqs. 13-14 and 13-15) we get

$$(Y_i)_k = \frac{a_k(\theta_i)_k\ddot{\delta}(t)}{p_k^2\sqrt{\left[1-\left(\dfrac{\omega}{p_k}\right)^2\right]^2 + \left[2\xi_k\,\dfrac{\omega}{p_k}\right]^2}} \tag{13-27}$$

The transfer function $T_{ik}(\omega)$ will be in this case

$$T_{ik}(\omega) = \frac{a_k(\theta_i)_k}{p_k^2\sqrt{\left[1-\left(\dfrac{\omega}{p_k}\right)^2\right]^2 + \left[2\xi_k\,\dfrac{\omega}{p_k}\right]^2}} \tag{13-28}$$

And the mean-square response can be obtained from Eq. 13-14 as

$$\overline{Y_{ik}^2} = \int_0^\infty [T_{ik}(\omega)]^2[W_{\ddot{\delta}}(p_k)]\,d\omega$$

from which

$$\overline{Y_{ik}^2} = a_k^2[(\theta_i)_k]^2[W_{\ddot{\delta}}(p_k)]\frac{\pi}{4\xi_k p_k^3} \tag{13-29}$$

where Y_{ik}^2 = mean-square displacement of mass m_i corresponding to the k mode relative to the foundation (inch^2)

a_k = mode-participation factor as defined by Eq. 13-34 (dimensionless)

$(\theta_i)_k$ = modal function or ratio of amplitude of mass m_i to the reference amplitude at mass m_i as obtained from free vibration analysis (dimensionless)

p_k = natural frequency corresponding to mode k (rad/sec)

$W_{\ddot{\delta}}(p_k)$ = power spectral density which is assumed constant in the neighborhood of the natural frequency p_k, $(\text{in./sec}^2)^2/(\text{rad/sec})$

ξ_k = effective damping factor for the k mode (dimensionless)

The total mean-square response at mass m_i is then obtained by summation of the different modal values

$$\overline{Y_i^2} = \sum_{k=1}^{n} \overline{Y_{ik}^2} \tag{13-30}$$

and in the case of the two-degrees-of-freedom system under consideration,

$$\overline{Y_1^2} = \overline{Y_{11}^2} = \overline{Y_{12}^2} \tag{13-31}$$

$$\overline{Y_2^2} = \overline{Y_{21}^2} = \overline{Y_{22}^2} \tag{13-32}$$

13-5. SYSTEMS WITH SEVERAL DEGREES OF FREEDOM

It is not difficult to see that the previous formulation and method of solution can be extended to deal with systems having several degrees of freedom. In this case

$$i = 1 \rightarrow n$$

and

$$k = 1 \rightarrow n$$

where n is the number of masses in the system (or degrees of freedom).

The general modal differential equation can be written as

$$\ddot{\eta}_k + p_k^2 \eta_k + a_k \ddot{\delta}(t) = 0 \tag{13-33}$$

where the mode participation factor a_k is given by

$$a_k = \frac{\sum_{i=1}^{i=n_1} (m_i)(\theta_i)_k}{\sum_{i=1}^{n} (m_i)(\theta_i)_k^2} \tag{13-34}$$

and by following the same procedure as for the two-mass system we obtain for the mean-square modal response:

$$\overline{Y_{ik}^2} = a_k^2 [(\theta_i)_k]^2 [W_{\ddot{\delta}}(p_k)] \frac{\pi}{4\xi_k p_k^3}$$

from which the total mean-square response at mass m_i becomes

$$\overline{Y_i^2} = \sum_{k=1}^{k=n} \overline{Y_{ik}^2}$$

Continuous Systems Subjected to Random Excitation at the Support

An extension of the previous method is possible for the case of continuous systems where the displacement at any instant can be described

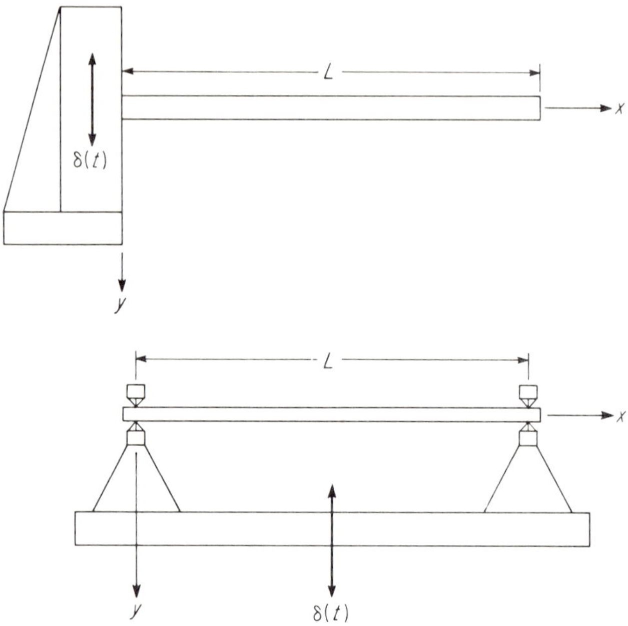

FIG. 13-15

by a continuous curve $y = f(x)$. The cantilever and the simple supported beam shown in Fig. 13-15 are examples of such systems.

In this case the system has infinite degrees of freedom and the modal functions $(\theta_i)_k$ are continuous curves $\theta_k(x)$. The modal participation factors a_k defined by Eq. 13-34 for the lumped-parameter systems take the form

$$a_k = \frac{\int_0^L \rho\phi_k(x)\,dx}{\int_0^L \rho\phi_k(x)\,dx}$$

where ρ = mass per unit length of beam

$\phi_k(x)$ = modal function corresponding to k mode, such functions can be obtained as shown in Part IV; they are also tabulated for beams with different boundary conditions in Ref. 15.

The differential equation in the modal form is the same as Eq. 13-33.

$$\ddot{\eta}_k + p_k^2\eta_k + a_k\ddot{\delta}(t) = 0$$

where η_k represents magnitude of vibration due to k mode.

p_k is the natural frequency corresponding to the mode. By proceeding in the same manner as for the case of n degrees of freedom, we obtain the same results (Eq. 13-29) for the mean-square deflection function corresponding to the k mode:

$$\overline{Y_k^2}(x) = a_k^2 \phi_k^2(x)[W_{\ddot{\delta}}(p_k)] \frac{\pi}{4\xi_k p_k^3} \tag{13-35}$$

where $\overline{Y_k^2}(x)$ = mean-square deflection of the beam with respect to support, in.^{2}

$\phi_k(x)$ = modal function (dimensionless)

p_k = natural frequency corresponding to mode, rad/sec

a_k = mode participation factor (dimensionless)

$W_{\ddot{\delta}}(p_k)$ = power spectral density of the support excitation, (in./sec)^2 /(rad/sec)

ξ_k = modal-damping factor (dimensionless)

And the total mean-square response for any number of modes n (from $1 \rightarrow \infty$) is obtained by summation of the individual modal responses

$$\overline{Y^2}(x) = \sum_{k=1}^{n} \overline{Y_k^2}(x)$$

13-6. APPROXIMATE NORMAL-MODE METHOD

The approximate normal-mode method explained in Chapter 10 can be conveniently used to analyze the response of a damped lumped-parameter system to random inputs. The response of the general case of a damped linear system to white-noise random input can be evaluated in a similar manner to the case of systems with Rayleigh damping (Ref. 3). In this case however, the fictitious damping ratios ξ_{ik} are used instead of the modal damping ratios ξ_k. These fictitious ratios are either estimated or evaluated from the response of the system to sinusoidal excitations as explained in Chapter 10.

The mean-square response of the system to white-noise random-force excitation on the first mass, for example, becomes:

$$\overline{X_i^2} = W_{F_1} \sum_k \frac{\pi \omega_k a_k}{4\xi_{ik} K_1^2} \tag{13-36}$$

where $\overline{X_i^2}$ = mean-square displacement of ith mass

W_{F_1} = mean-square spectral density, lb^2/(rad/sec)

K_1 = spring stiffness, lb/in.

ξ_{ik} = fictitious damping ratios

ω_k = natural frequencies

a_k = modal-participation factors

Similarly, when subjected to white-noise ground excitation, the response takes the form

$$\bar{X}_i^2 = W_{\ddot{x}_0} \sum_k \frac{\pi \phi_{ik}^2 a_k^2}{4\xi_{ik}\omega_k^3} \tag{13-37}$$

where $W_{\ddot{x}_0}$ = mean-square spectral density of excitation with dimensions, $(\text{in./sec}^2)^2/(\text{rad/sec})$

ϕ_{ik} = eigenvectors

Example

The damped-absorber system of Fig. 13-16 is assumed to be subjected

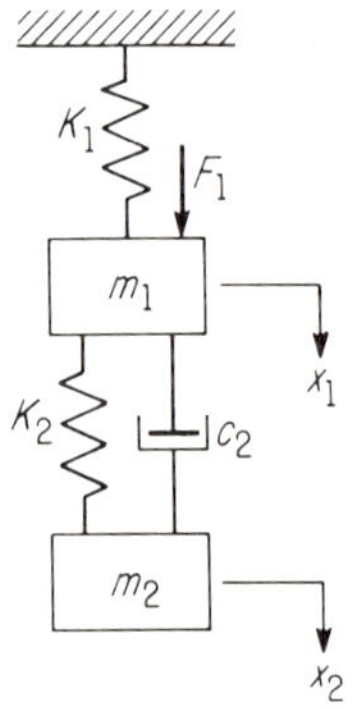

FIG. 13-16

to white-noise force excitation. Equation 13-36 for this case becomes

$$\bar{X}_1^2 = W_{F_1}\left(\frac{\pi\omega_1 a_1^2}{4\xi_{11}K_1^2} + \frac{\pi\omega_2 a_2^2}{4\xi_{12}K_1}\right) \tag{13-38}$$

$$\bar{X}_2^2 = W_{F_1}\left(\frac{\pi\omega_1 \phi_{21}^2 a_1^2}{4\xi_{21}K_1^2} + \frac{\pi\omega_2 \phi_{22}^2 a_2^2}{4\xi_{22}K_1^2}\right) \tag{13-39}$$

The mean-square value of system response can be obtained from the classical solution of the transfer functions and substituting in Eq. 13-14. This gives

$$\frac{\bar{X}_1^2/\omega_{11}}{\pi W_{F_1}/K_1^2} = \frac{\dfrac{1}{4\xi_2}[1 - f^2(2+\mu) + f^4(1+\mu)^2] + \xi_2 f^2(1+\mu)}{\mu f} \tag{13-40a}$$

$$\frac{\bar{X}_2^2/\omega_{11}}{\pi W_{F_1}/K_1^2} = \frac{f}{4\xi_2\mu}[\mu + f^2(1+\mu)] + \frac{2\xi_2(1+\mu)}{\mu} \tag{13-40b}$$

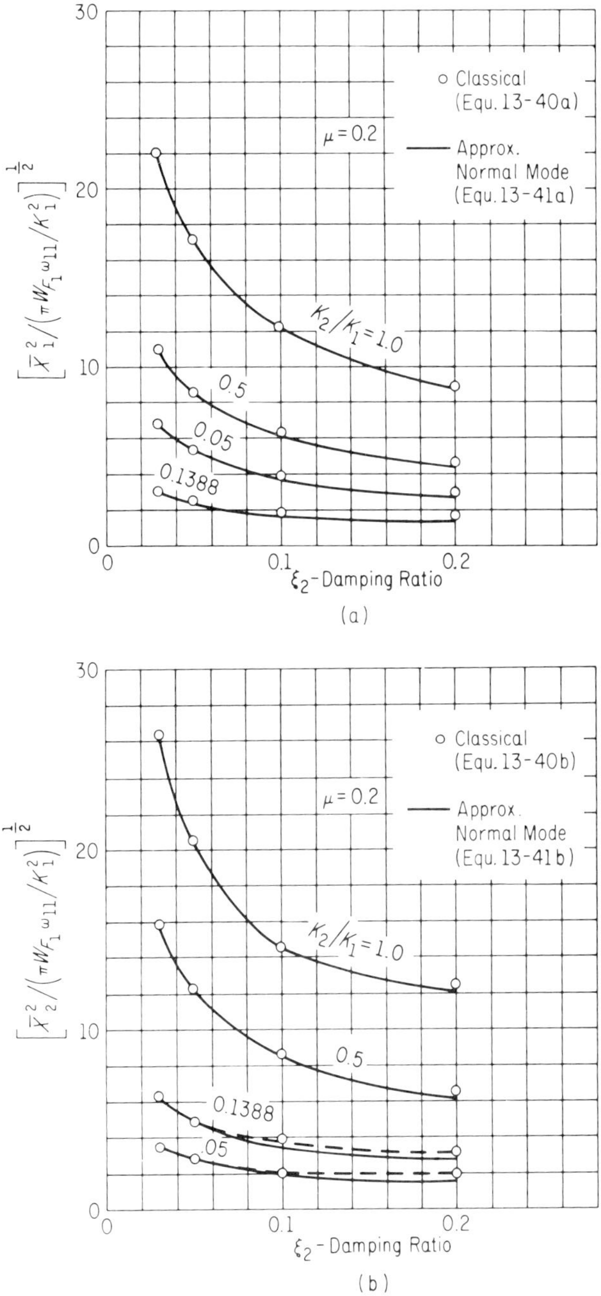
Classical (Equ. 13-40a)
Approx. Normal Mode (Equ. 13-41a)
$\mu = 0.2$
$K_2/K_1 = 1.0$
0.5
0.05
0.1388
$\left[\bar{X}_1^2/\left(\pi W_{F_1}\omega_{11}/K_1^2\right)\right]^{\frac{1}{2}}$
ξ_2-Damping Ratio
(a)
Classical (Equ. 13-40b)
Approx. Normal Mode (Equ. 13-41b)
$\mu = 0.2$
$K_2/K_1 = 1.0$
0.5
0.1388
.05
$\left[\bar{X}_2^2/\left(\pi W_{F_1}\omega_{11}/K_1^2\right)\right]^{\frac{1}{2}}$
ξ_2-Damping Ratio
(b)

FIG. 13-17

where $\mu = m_2/m_1$

$f = \sqrt{K_2 m_1/K_1 m_2}$

In order to compare system response as calculated by the approximate normal-mode method to that obtained by classical solution, Eq. 13-38 is rearranged to give

$$\frac{\overline{X_1^2}/\omega_{11}}{\pi W_{F_1}/K_1^2} = \frac{1}{4\xi_{11}(1 + \mu\phi_{21}^2)\dfrac{\omega_1^3}{\omega_{11}^3}} + \frac{1}{4\xi_{12}(1 + \mu\phi_{22}^2)\dfrac{\omega_2^3}{\omega_{11}^3}} \tag{13-41a}$$

Similarly, Eq. 13-39 gives

$$\frac{\overline{X_2^2}/\omega_{22}}{\pi W_{F_1}/K_1^2} = \frac{\phi_{21}^2}{4\xi_{21}(1 + \mu\phi_{21}^2)\dfrac{\omega_1^3}{\omega_{11}^3}} + \frac{\phi_{22}^2}{4\xi_{22}(1 + \mu\phi_{22}^2)^2\dfrac{\omega_2^3}{\omega_{11}^3}} \tag{13-41b}$$

Figure 13-17a shows the results from numerical comparisons made between Eqs. 13-40 and 13-41 for systems with different parameters. It can be seen that for all systems, the approximate normal-mode method agrees closely with the classical method when ξ_2 is equal to or less than 0.1. The classical solution consistently gives larger values. As a general rule, the agreement is good for the main mass response. The secondary-system response also shows excellent agreement for high response values, while significant errors result for low mean-square response values.

Optimum Parameters for Damped Absorber

The requirements for optimization of the system, Fig. 13-16, when subjected to sinusoidal excitation are shown by Den Hartog (Ref. 11, p. 93) to be

$$f = \text{frequency ratio } (\omega_{22}/\omega_{11}) = 1/1 + \mu$$

and the damping ratio

$$\xi_2 = [3\mu/8(1 + \mu)^3]^{1/2}$$

The latter equation was obtained by Brock (Ref. 12). The maximum response of the main mass of the optimized system is

$$\frac{X_1}{F_1/K_1} = \sqrt{1 + (2/\mu)}$$

In this study, Eq. 13-40 is utilized to optimize system performance under white-noise random excitation. Computer optimization techniques are used for selected values of mass ratios.

The requirements for minimum mean-square response for the main

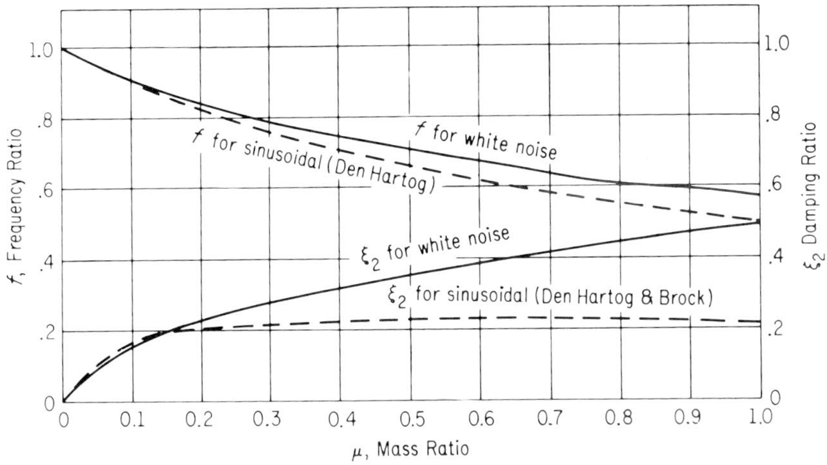

FIG. 13-18

mass are found to fit closely the following relationships

$$f \approx \frac{1}{\sqrt{1 + 2\mu}}$$

$$\xi_2 \approx \frac{\sqrt{\mu}}{2}$$

$$\frac{\bar{X}_1^2/\omega_{11}}{\pi W_{F_1}/K_1^2} \approx \frac{0.96}{\sqrt{\mu}}$$

These relationships are plotted in Figs. 13-18 and 13-19 and compared to the corresponding values for the case of sinusoidal excitation.

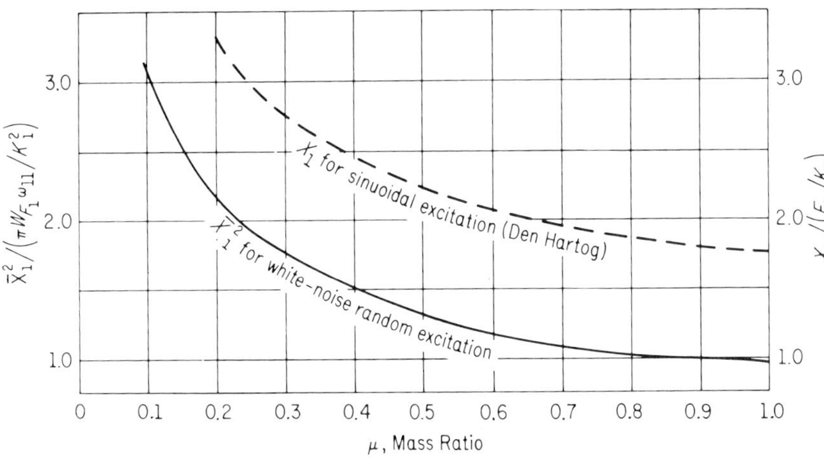

FIG. 13-19

13-7. AN INTRODUCTION TO CORRELATION

Correlation analysis is a convenient technique for determining the spectral characteristics of a waveform or the similarity of waveforms.

Autocorrelation

The autocorrelation of a waveform $f(t)$ is the plot of the mean of the product of the function and a delayed replica of itself $f(t - \tau)$ as a function of the time shift τ. This can be expressed mathematically as

$$R_f(\tau) = \lim_{T \to \infty} \frac{1}{T} \int_{-T/2}^{+T/2} f(t)f(t - \tau)dt$$

where $R_f(\tau)$ = autocorrelation function

f = waveform

τ = delay

T = selected period for summation of the product

When $f(t)$ is only defined in a finite interval, a finite autocorrelation can be obtained as

$$Y_f(\tau) = \frac{1}{T} \int_{-T/2}^{+T/2+\tau} f(t)f(t - \tau)dt$$

For $\tau = 0$ then $Y_f(0)$ represents the rms value of the function. The autocorrelation may be obtained by means of the instrumentation circuit shown in Fig. 13-20.

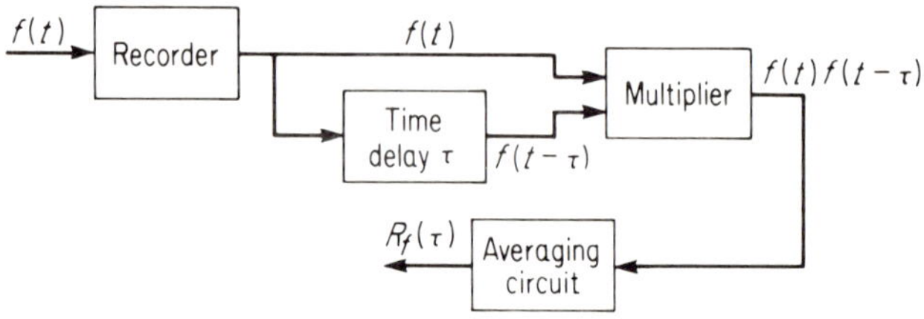

FIG. 13-20

Considering a sine wave for example, when the time shift τ is an integral number of periods, it will become identical to itself and as shown in Fig. 13-21, the autocorrelation function will itself be sinusoidal with the same frequency.

On the other hand in the case of a wide-band random noise, a very small time shift may be sufficient to destroy the similarity and waveform never repeats itself. Accordingly, as shown in Fig. 13-22, the autocorrelation function takes the shape of a sharp impulse decaying quickly from a

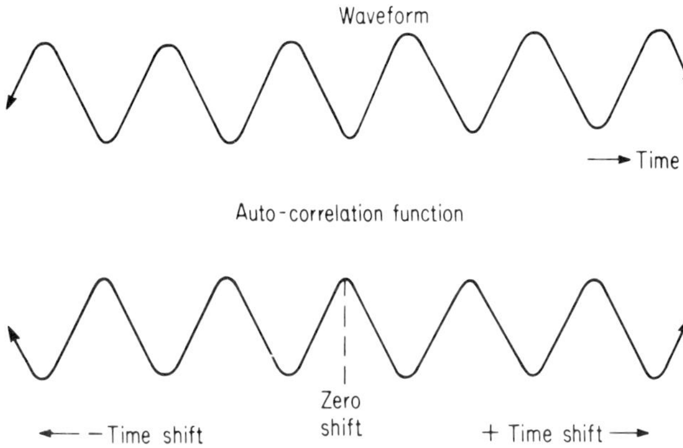

FIG. 13-21

maximum value at zero shift to very low values as the time shift increases. The rate of decay of the autocorrelation function increases as the bandwidth increases.

The autocorrelation function of a waveform depends not on the actual waveform but on its content. The autocorrelation in fact contains the same information as the power spectrum which is merely the Fourier transform of the autocorrelation. This can be written as

$$W(\omega) = \frac{1}{2\pi} \int_{-\infty}^{\infty} R_f(\tau) e^{-i\omega\tau} \, d\tau$$

where $W(\omega)$ is the mean-square spectral density.

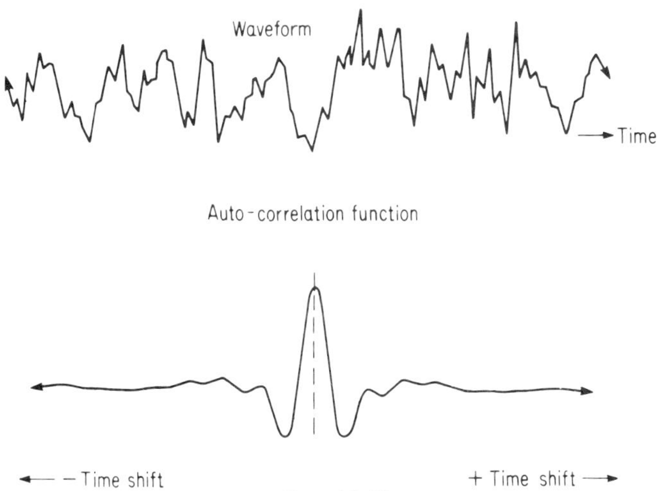

FIG. 13-22

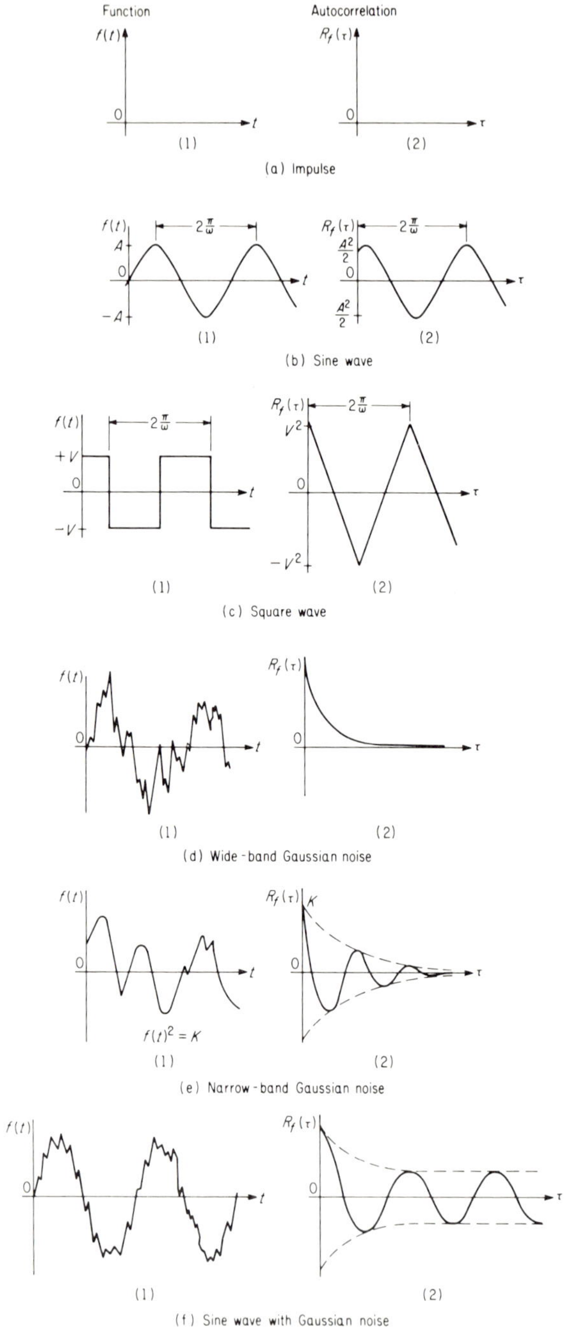
Function
Autocorrelation
$f(t)$
$R_f(\tau)$
0
t
τ
(1)
(2)
(a) Impulse
$2\frac{\pi}{\omega}$
A
$-A$
$\frac{A^2}{2}$
(b) Sine wave
$+V$
$-V$
V^2
$-V^2$
(c) Square wave
(d) Wide-band Gaussian noise
$f(t)^2 = K$
K
(e) Narrow-band Gaussian noise
(f) Sine wave with Gaussian noise

FIG. 13-23

Examples of autocorrelation functions for some common functions are shown in Fig. 13-23.

Cross-Correlation

While the autocorrelation function of a waveform is concerned with averaging the product of the wave and a time-shifted version of itself, the cross-correlation is concerned with averaging the product of two different waveforms one of which is time-shifted. The averaging process for the two waves is performed over a specified interval called the *window*. The time T is known as the *width* of the window.

Accordingly, the finite cross-correlation function of two waveforms $f(t)$ and $g(t)$ within a window of width T as shown in Fig. 13-24 can be

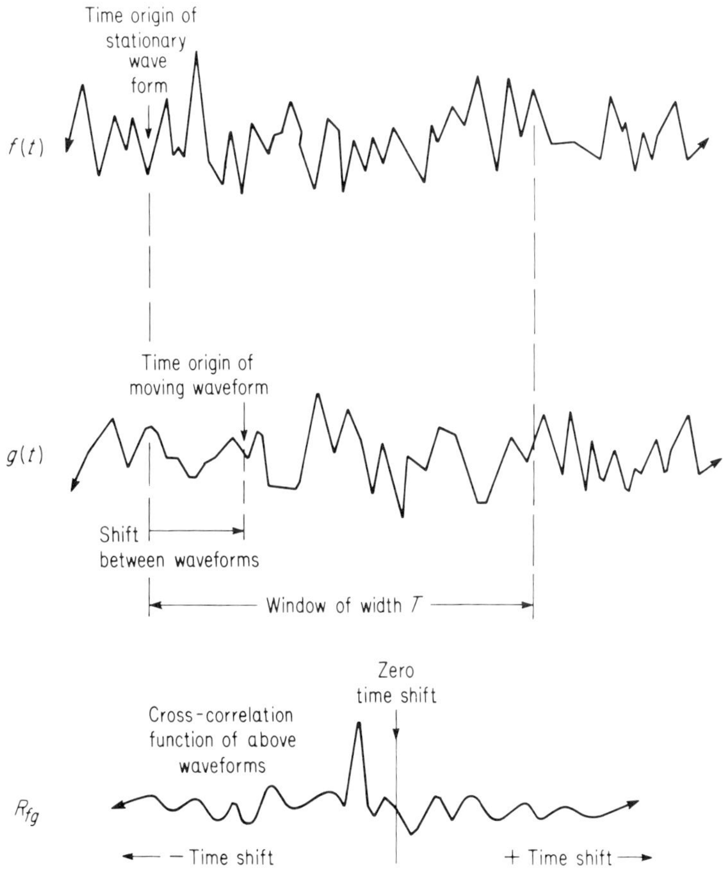

FIG. 13-24

written as

$$R_{fg}(\tau) = \frac{1}{T}\int_{-T/2}^{+T/2} f(t)g(t - \tau)dt$$

It can also be obtained from an instrumentation circuit similar to that of autocorrelation as shown in Fig. 13-25.

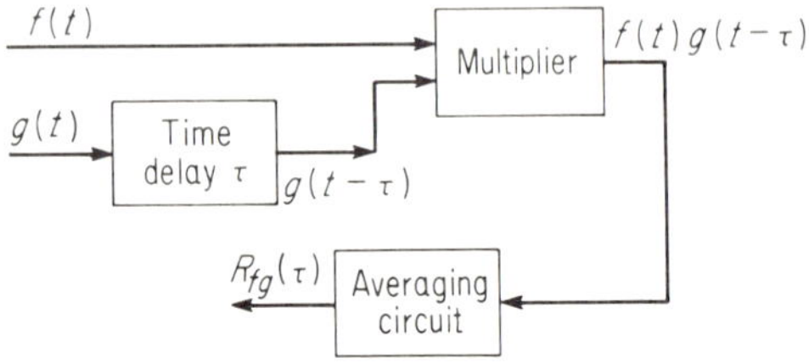

FIG. 13-25

An important practical aspect of cross-correlation is that the cross-correlation function of two waveforms contains only those frequencies common to both waveforms. For example, if a complex wave is cross-correlated with a sine wave, the output will contain only those components in the complex wave which are very close to the frequency of the sine wave. The correlation therefore produces a sharp-peaked filter at the frequency of the sine wave.

13-8. SUMMARY

A simplified and compact treatment of the response of mechanical systems to random excitations has been presented. The use of normal-mode techniques for undamped as well as damped systems is discussed. Autocorrelation and cross-correlation concepts have been briefly introduced, since they provide powerful tools of analysis for random signals.

REFERENCES

1. S. H. Crandall and W. D. Mark, *Random Vibrations in Mechanical Systems*. New York: Academic, 1963.
2. S. H. Crandall (ed.), *Random Vibrations*. 2 vols. Cambridge, Mass.: M.I.T. Press, 1963.
3. A. Seireg and L. Howard, "An Approximate Normal-Mode Method for Damped Lumped-Parameter Systems," *Trans. ASME*, Vol. 89, Ser. B, No. 4 (November 1967).
4. I. Dyer, "Statistical Vibration Analysis," *Int. Sci. and Tech.*, No. 20 (August 1963).

5. E. Ungar, "Statistical Energy Analysis of Vibrating Systems, *Trans. ASME*, Vol. 89, Ser. B, No. 4 (November 1967).
6. A. C. Eringen, "Response of Beams and Plates to Random Loads, *Trans. ASME, J. Appl. Mech.*, Vol. 24 (1957), pp. 46–52.
8. H. Press and J. Houboldt, "Some Applications of Generalized Harmonic Analysis to Gust Loads on Airplanes," *J. Aeron. Sci.*, Vol. 22 (1955).
9. S. O. Rice, "The Mathematical Analysis of Random Noise," *Bell System Tech. J.*, Vol. 23 (1944), p. 282, and Vol. 24 (1945), p. 46
10. H. M. James et al., *Theory of Servomechanisms*. M.I.T. Radiation Laboratory Series, Vol. 25. New York: McGraw-Hill, 1947.
11. J. P. Den Hartog, *Mechanical Vibrations*. 4th ed. New York: McGraw-Hill, 1956.
12. J. E. Brock, "A Note on the Damped Vibration Absorber, *Trans. ASME, J. Appl. Mech.*, Vol. 68 (1946), p. A-284.
13. Y. J. Lubkin, *An Introduction to Correlation*, Monograph No. 1, Federal Scientific Corp., New York, 1966.
14. B. LuBow, "Correlation Entering New Fields with Real-Time Signal Analysis," *Electronics*, Oct. 31, 1966.
15. R. Bishop and D. Johnson, "*The Mechanics of Vibration*," New York: Cambridge University Press, 1960.

part VI

Experimental Investigation of Dynamic Systems

In 1787 E. F. F. Chladni published his celebrated treatise (Entdeckungen über die Theorie des Klanges) *in which he described his method of using sand sprinkled on vibrating plates to show the nodal lines. His figures were very beautiful and in a general way could be accounted for by considerations similar to those relating to vibrating strings. The exact forms however defied analysis for many years even after the publication of Chladni's classic work* (Die Akustike) *in 1802. Napoleon provided for the Institute of France a prize of 3000 francs to be awarded for a satisfactory mathematical theory of the vibration of plates. The prize was awarded in 1815 to Mlle. Sophie Germain, who gave the correct fourth-order differential equation. Her choice of boundary conditions proved, however, to be incorrect.*

R. B. Lindsay, historical introduction to Lord Rayleigh's Theory of Sound, *Dover, New York, 1937.*

chapter 14

Experimental Investigation of Dynamic Systems

14-1. INTRODUCTION

In all areas of physical science, a close interdependence exists between the analytical and experimental methods. The analytical method, whenever applicable, possesses the elegance and generality. The experimental method, on the other hand, is invaluable in situations where the system parameters and the boundary conditions are not known in proper detail or where a mathematical analysis of the system is unduly complicated and intractable.

In many instances the most fruitful results are obtained from a combination of both. The mathematical model provides the guide and general plan, while the physical model describes more accurately the material properties and boundary conditions.

In this chapter the reader is introduced to some of the tools necessary for experimental investigations. A brief description is given of some types of dynamic tests, vibration generators, transducers, and instrumentation systems, as well as physical models that are utilized for dynamic studies.

14-2. TYPES OF DYNAMIC TESTS

Dynamic tests are usually classified according to the type of disturbance on the system. Most tests fall in one of the following categories.

1. Steady-state forced-vibration tests, where the system is subjected to a periodic disturbance.

2. Transient vibration tests, where the disturbance is in the form of a change in position or velocity, or when the system is subjected to a disturbance of finite duration.

3. Impact, which is a special case of the transient disturbance of a continuum resulting in a propagation, attenuation, and reflection of the disturbing wave on the boundaries.

4. Random tests, in which the disturbance is unpredictable in a precise manner and analytical methods give only probabilistic results.

14-3. VIBRATION AND SHOCK GENERATORS

Vibration and shock exciters utilized in dynamic tests can be classified into three main categories.

Sinusoidal Exciters

These are useful in studying steady-state response of systems. The following are different varieties of this type.

1. Constant Displacement Exciters

This consists basically of a variable-speed motor which is rigidly connected to the driving shaft of a crank or an eccentric hub (Fig. 14-1). A

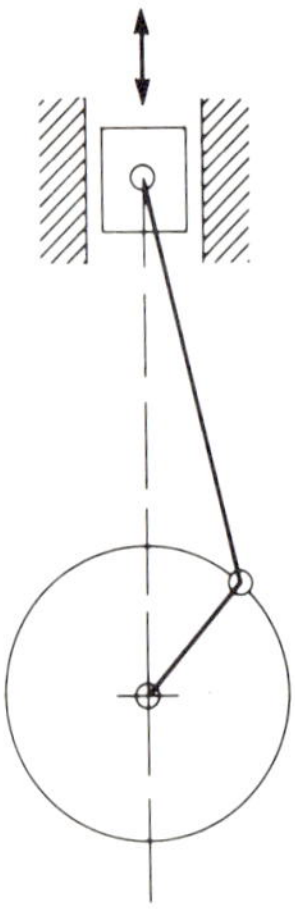

FIG. 14-1

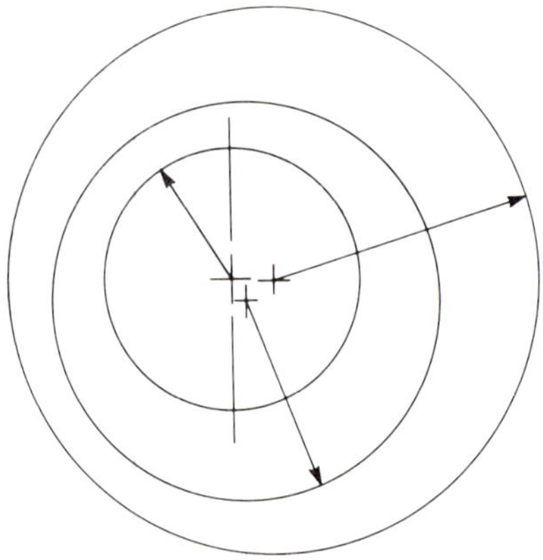

FIG. 14-2

connecting rod converts the rotary motion to a reciprocating one. Crossheads or sliders may be used to support force components perpendicular to the direction of the oscillation. Large "connecting rod/crank" ratios are essential in order to insure a sinusoidal excitation.

A double eccentric (Fig. 14-2) can be used in cases where different degrees of eccentricities are needed. The inner eccentric hub is keyed to the motor shaft while the outer one can be rotated on the inner to obtain the required eccentricity. This type of exciter is particularly useful when excitations of large magnitudes at low frequencies are required.

2. Constant Force Exciters

This type provides exciting forces with constant amplitudes at any desired frequency.

(a) Electro-dynamic shakers (Fig. 14-3). This class of exciters has a

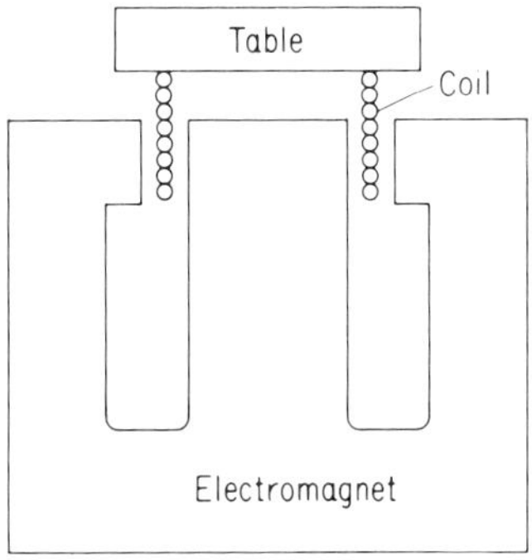

FIG. 14-3

driven coil connected rigidly to the shake table. The coil is free to move in the magnetic field of an intense electromagnet. An alternating voltage of known frequency is supplied to the coil by a signal generator and a power amplifier. The intensity of the force acting on the coil (and, consequently, on the table) is proportional to the current flowing in the coil. Electrodynamic shakers are available commercially in sizes up to 30,000 lb and ± 1/4-in. stroke with frequencies up to 2000 cps.

(b) Hydraulic shakers. This type is capable of sinusoidal motion at extremely high forces and large strokes. These shakers can be obtained commercially with a capacity of 60,000 lb, a stroke of 4 in. and a frequency range of 1 to 150 cps. There are five basic components in hydraulic shakers: the driving head, hydraulic accummulator, a high-pressure

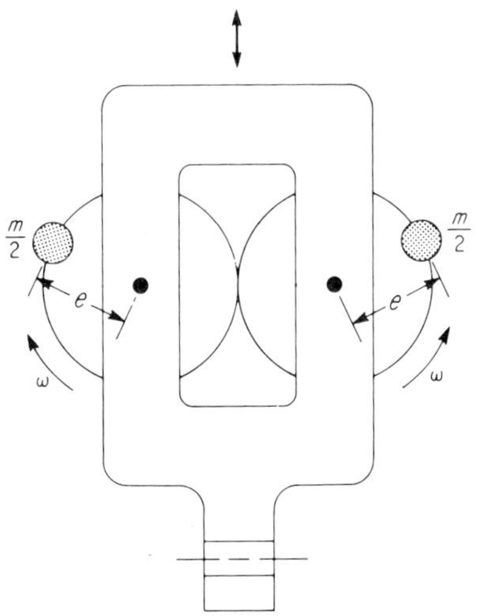

FIG. 14-4

hydraulic pump, electronically controlled servo valve, and an electronic control system. The limits of force, stroke, and frequency depend on the capacities of the servo valve actuating cylinders and hydraulic pump.

3. Unbalanced Rotating-Weight Type Exciters (Fig. 14-4)

Such exciters are composed of two disks rotating at the same speed and in opposite directions by means of a pair of gears. Each disk carries a mass $m/2$ at an eccentricity e. The resultant effect on the frame is a sinusoidal force as shown. The magnitude of this force is proportional to the square of the rotating frequency ω:

$$F = me\omega^2 \sin \omega t$$

This relationship makes this type of exciters useful in many practical applications especially at relatively high frequencies.

Impact Exciters

Exciters of this type apply impulse loading for shock testing. These loads are characterized by high magnitudes and short duration. There are many types of shock-applying systems, among which are the drop weight, inclined plane, ballistic pendulums, mechanical hammers, air guns, and explosive guns for high-speed impacts. The application of explosive layers, strips or slabs directly on the surface is an effective means of producing controlled shock loads. The detonation technique is very critical in this case.

Random Vibration Exciters

The electrodynamic shakers described previously are also utilized for random-vibration tests. A random-noise generator is used in place of the signal generator. Filters are also used between the noise generator and the power amplifier to adjust the frequency characteristics, to limit the frequency band, and to achieve as nearly as possible a selected overall frequency characteristic. Limiting or protective features are usually utilized in random-vibration testing to limit the peak acceleration and to avoid high impacts that may result if the table hits the stops restricting the table amplitude.

In many cases of random testing, a recorded signal is used as the noise source to reproduce a particular type of expected excitation.

14-4. TRANSDUCERS AND RECORDING EQUIPMENT

The fundamental quantities to be measured in dynamic tests are the time history of the system configuration together with the disturbance and their phase relation. In general, time recording of force (or stress), displacement (or strain), velocity, and acceleration are necessary to study a dynamic system. In cases where the vibration is known to be periodic and sinusoidal, the amplitudes of any of the motion parameters (displacement, velocity, or acceleration) and the frequency are all that is necessary to define the motion of the system. In the general case if an acceleration record is obtained, it will be sufficient to describe the motion since the velocity and displacement histories can be obtained by integration while the inverse differentiation cannot be carried out with sufficient accuracy.

Transducers and measuring techniques are available in numerous varieties ranging from simple do-it-yourself devices to very delicate and complex instruments and associated equipment.

Vibration-measuring devices can be broadly classified as visual, optical, mechanical, or electrical. The latter are the most common due to the availability and sophistication of electrical systems for detection, transmission, amplification, recording, and data analysis.

Visual or Optical Methods

1. Penumbra Vibrometer

It is a very simple technique utilizing sensory perception. It is only useful for measuring relatively large amplitudes at low frequencies. A triangle as shown in Fig. 14-5 is drawn on a card attached to the vibrating

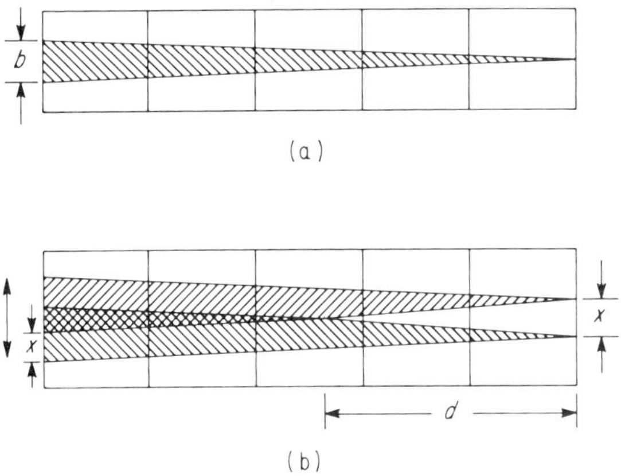

FIG. 14-5

surface in such a way that the base b is in the direction of the vibration. The triangle will appear as shown in Fig. 14-5 and the amplitude is represented by "x". The amplitude can be more accurately read by scaling the distance d perpendicular to the direction of vibration. A variation of this vibrometer is used for torsional oscillations and is shown in Fig. 14-6.

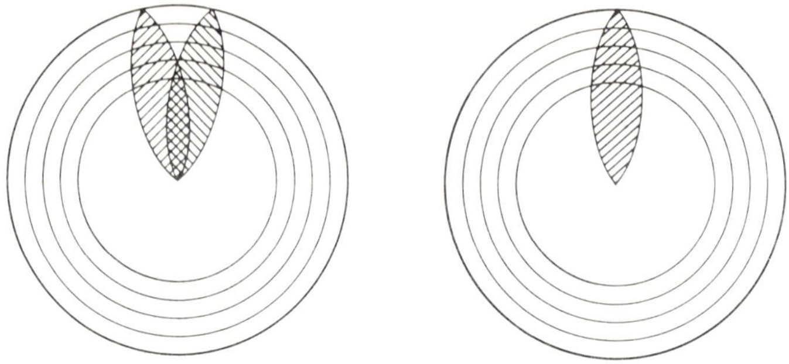

FIG. 14-6

2. Stroboscope

A stroboscope is a light source capable of providing intermittent flashes at a desired rate. When the frequency of the stroboscope flashes is the same as that of the vibrating object, it will appear to be at a standstill. Allowing a small difference in the frequency, the object will appear to vibrate at a very slow frequency. Its amplitude can therefore be easily determined by comparison to a fixed reference. The stroboscope can also be calibrated to give the frequency of vibration.

3. Microscope

The microscope can be used for measurement of small amplitude. A light spot or a pinhole is marked on the vibrating surface for optical contrast. The microscope is then focused on it. Upon vibration the point will appear as a straight line whose length L can be measured by means of a micrometer slide. The amplitude of vibration can be calculated from

$$a = \frac{L - d}{2}$$

where d = diameter of pinhole.

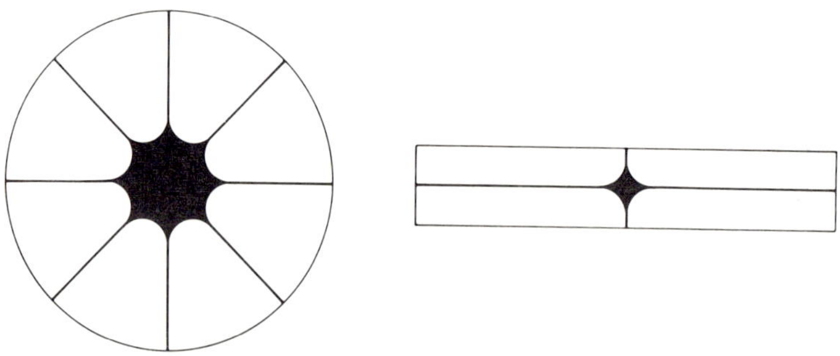

FIG. 14-7

4. Chladni Nodal Patterns

This technique which was developed by Ernest F. Chladni (1802) gives a simple method for observing the modes of vibration of plates. By sprinkling sand or fine powder on the plate it is possible to show the existence of nodal lines for various modes of its vibration and to determine the corresponding frequencies. Figure 14-7 shows the nodal pattern for the bending vibration of a circular plate and torsional vibration of a rectangular plate.

5. High-Speed Photography

High-speed photography provides a reliable and direct method of investigation for both transient and steady-state dynamic phenomena. Extremely high speeds can be attained in smear cameras and those employing streak photography or rotating-drum techniques. Speeds up to several millions of frames per second are possible at the present time.

6. Photoelastic Techniques

The photoelastic method consists of examining by optical means a model of the structure to be investigated. The model being a transparent plastic having special optical properties and is examined under a load system precisely similar to that applied to the structure. By studying the optical behavior of the model, the stress pattern in the model (and consequently in the structure) can be readily obtained.

With the development of high-speed photography, photoelasticity is becoming an important tool in studying dynamic phenomena, especially those involving stress-wave propagation.

Another phase of photoelasticity is the photoelastic coating technique. In this method the photoelastic material is coated directly on the actual part under investigation whether this part is metal, wood, glass, rubber, or concrete. The basic concepts of the photoelastic method are summarized as follows.

(a) The stresses in any loaded element can be studied by observing the optical behavior of a similar transparent model loaded under similar conditions. They may also be studied by observing the behavior of a photoelastic coating on the actual model.

(b) The direction of the principal axes of stress at any point in the element can be found from the isoclinic lines obtained by studying the model in the plane polariscope with a white-light source.

(c) The difference between the principal stresses at any point (or the magnitude of the maximum shear stress at the point) can be obtained by observing the model in the circular polariscope with a monochromatic light source.

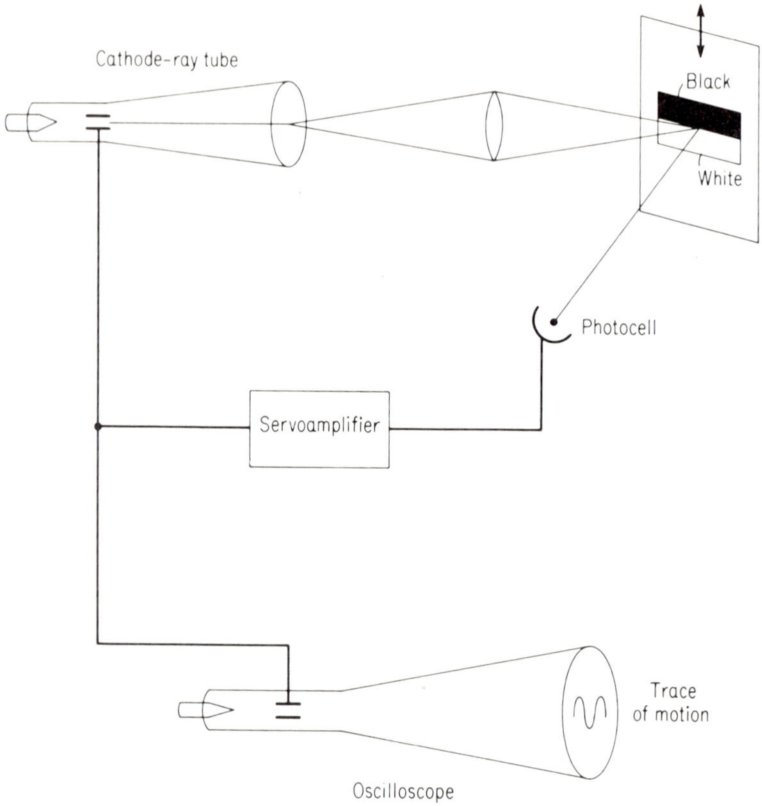

FIG. 14-8

7. The Optron Displacement Follower (Fig. 14-8)

This transducer is capable of measuring displacement without requiring contact with the vibrating object. The intense beam from a cathode-ray tube is focused on an edge of the vibrating object or a demarcation line between black and white regions painted on the observed surface. The principle of operation of the system is to force this beam to follow the motion of the edge or demarcation line and consequently the motion of the object. Since the beam is focused on a reflecting surface, it will reflect back and is guided by a mirror to a photocell whose output is fed to the vertical plates of the scope. The beam will then move upward and will pass over the edge without reflection. The output from the photocell will consequently vanish and the beam will then drop to the original position. In this sense the intense beam will always follow the edge or the demarcation line between a reflective and absorbing surface. The trace of the motion can be seen on an oscilloscope whose vertical plates are connected in parallel to those of the cathode-ray tube.

Microwave and Ultrasonic Transducers

There are transducers which employ the reflection of microwaves from the vibrating surface such as the waveguide system and the microwave-beam reflecting system. Others employ reflected ultrasonic energy from the vibrating surface to detect displacements as small as 10^{-5} in. with frequencies ranging from 5 to 100,000 cps.

Mechanical Methods

Mechanical devices are mainly suitable in the low-frequency range due to their inertia. Dial indicators can be used to measure amplitudes ranging from 0.0001 in. with practically no lower limit for the frequency. There are numerous types of mechanical systems ranging from scratch gages and other types of mechanical strain indicators to spring-mounted devices as in the different varieties of vibrographs.

Electrical Methods

As mentioned previously, electrical methods of detecting vibrations are by far the most widely utilized due to the rapid development in electronics and solid-state components and circuitry. In the following, some types of electrical transducers are described with a brief analysis of the principles of their operation.

1. Variable Resistance Transducers

(a) Slider Type (Fig. 14-9). The slider of a fixed variable resistance is rigidly connected to the vibrating object. The change in resistance is

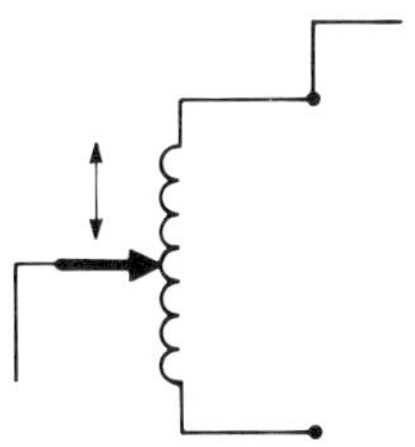

FIG. 14-9

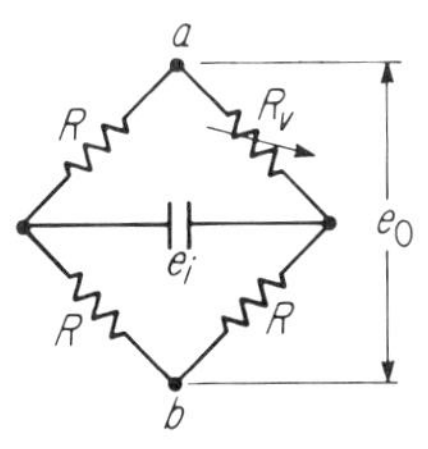

FIG. 14-10

therefore indicative of the mechanical motion. This change in resistance with time can be measured by several techniques.

The Wheatstone bridge (Fig. 14-10) is one of the more accurate and useful methods. This device is formed of four identical sides one of which is the transducer resistor. When the amplitude of vibration is zero the bridge will be balanced and no potential difference will exist between a and b. In this case $e_0 = 0$. Under conditions of vibration the motion of the slider is converted into a potential difference e_0 which can be calibrated to give the actual mechanical displacement.

(b) Strain gage type. The principle employed here is that the electrical resistance of a very thin and long wire will undergo a change due to a change in its length. When this wire is connected to the vibrating object in such a way as to follow its motion, the change in wire resistance is indicative of the motion. It can therefore be measured or recorded by converting it to an emf by employing the Wheatstone bridge.

There are several types of resistance strain gages. Some are of bonded type where they are cemented to the vibrating object. Others are of the unbonded type where only the two ends of the wire are cemented to measure the strain between two points. This type of transducer is very sensitive to displacement and is capable of measuring motions in the order of microinches. It is also very sensitive to temperature variations and careful attention should be given to temperature compensation.

2. Variable Inductance (Fig. 14-11)

One of the most common transducers employing the variable inductance principle is the differential transformer type, which consists of a coil

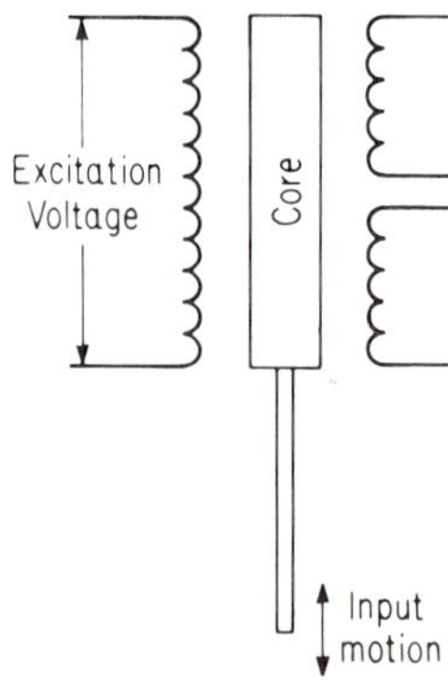

FIG. 14-11

assembly and a movable magnetic core. The coil assembly has three coaxial coils. The cylindrical high permeability core is positioned within the coil assembly without making mechanical contact. Normally the primary coil is equi-distant from the two balanced secondary coils. When the primary is energized with alternating current, voltages are induced in the two secondary windings which are connected differentially. When the core is moved from the null position, voltage increases in one coil and decreases in the other. The distance and direction of the movement is determined by reading the electrical output.

This type of transducer is capable of measuring relative displacements as small as 10^{-6} in. between the core and coil assembly.

3. Variable Capacitance (Fig. 14-12)

In the variable-capacitance transducer the mechanical motion is detected by measuring the change in capacitance due to the relative motion

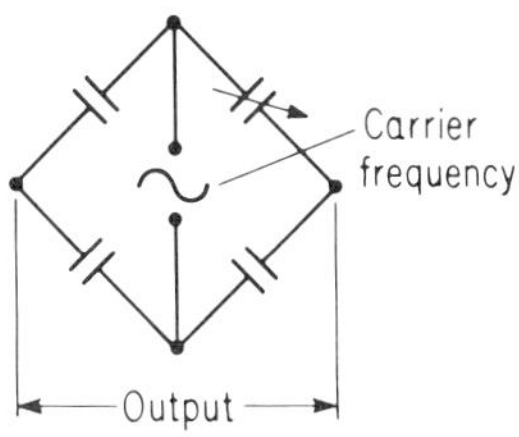

FIG. 14-12

between the capacitor plates. Since capacitance C is inversely proportional to the distance d between the plates,

$$C \propto \frac{A}{d}$$

where A is the area of plate. The impedance $1/\omega C \propto d/\omega A$ is therefore proportional to the distance d. As in the case of the variable-resistor transducer, a Wheatstone bridge can be used. In this case, the frequency of the excitation voltage (or carrier frequency) should be much greater than the frequency of the measured vibration.

4. Electrodynamic Transducers (Fig. 14-13)

In this type the relative motion between a coil and a permanent magnet induces a voltage in the coil with magnitude

$$e_{\text{ind}} = HL \frac{dx}{dt}$$

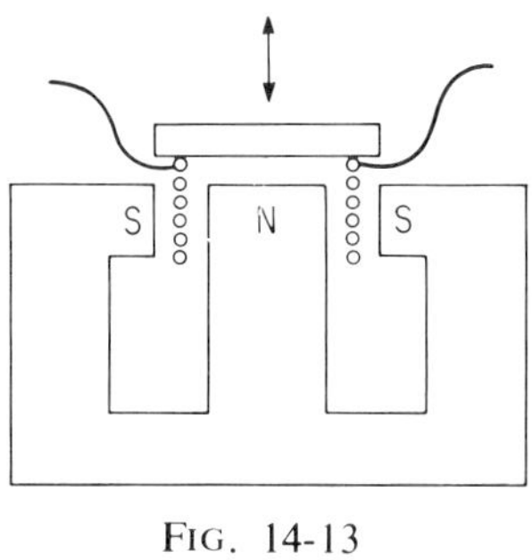

FIG. 14-13

where H = average field strength
L = length of coil wire
x = relative displacement

This type of transducer is therefore a velocity pickup.

5. Electromagnetic Type (Fig. 14-14)

In this type a coil is wound concentric with a permanent magnet. A plate made of a magnetic material is placed at a distance d from the

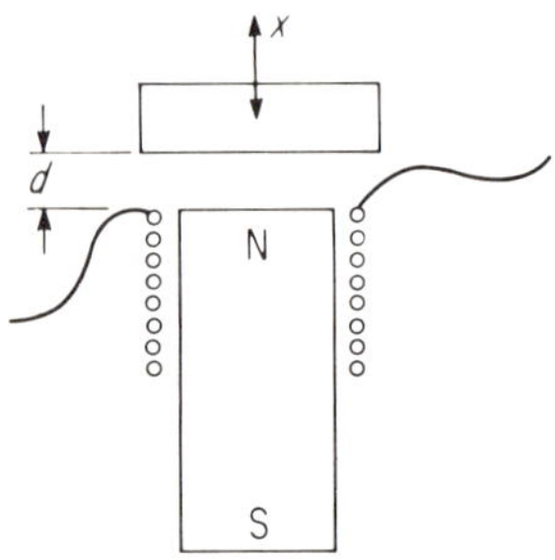

FIG. 14-14

magnet. A voltage (e_{ind}) will be induced in the coil which will be linearly proportional to the relative velocity dx/dt between the plate and the magnet as long as $x < d$.

6. Piezoelectric Transducers

Certain crystals such as quartz and rochelle salt, tourmaline, barium titanate, and lead metaniobate, etc., contain domains which are oriented in the same direction, either by nature or by polarization during manufacturing processes. When stress is applied, the minute relative displacement of the crystal domains generates a charge. A full-scale crystal deflection of three microinches may produce as much as 100 volts.

An expression for the induced emf can be given by

$$e_{ind} = \left(\frac{k}{C}\, e^{-t/RC}\right) x$$

where x = deflection
R = resistance of crystal
C = capacitance of crystal
t = time
k = a constant

The voltage is therefore time-dependent and consequently frequency dependent. It is, however, practically linear with displacement above a certain frequency until mechanical resonance of the crystal disturbs this linearity.

7. The Transistor Transducer

Transistors have been produced which can be used as pressure sensitive components as well as point-force ones. The Pitran PT2 and PT3 types (Ref. 3) are examples. These particular transducers are silicon *n-p-n* planar versions with the emitter-base junction mechanically coupled to a diaphragm at the top of the sealing can. When a pressure or point force is applied to the diaphragm a large reversible change in the transistor's characteristics is produced.

This principle has made it possible to produce sensitivities of the order of 1 volt per 0.25 gm force with linearities of better than 1 percent over 1-volt range. The Pitran is operated as a conventional transistor with the addition that a mechanical variable can be produced to modulate the output. Due to its size the device is useful for acceleration sensing devices. The basic circuit for the Pitran is shown in Fig. 14-15.

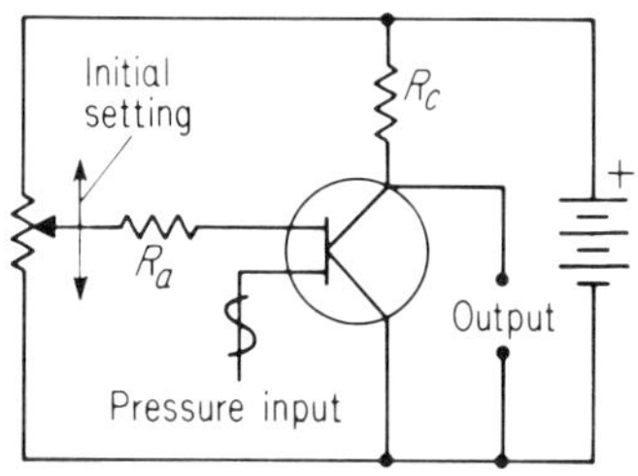

FIG. 14-15

Seismic Instruments

All the previous transducers depend on the relative displacement or relative velocity between two of their elements in order to function. In many instances it is rather difficult to mount the transducer in such a way as one of its elements is fixed. The use of the seismic principle is therefore necessary in order to measure a vibration without the use of a fixed reference.

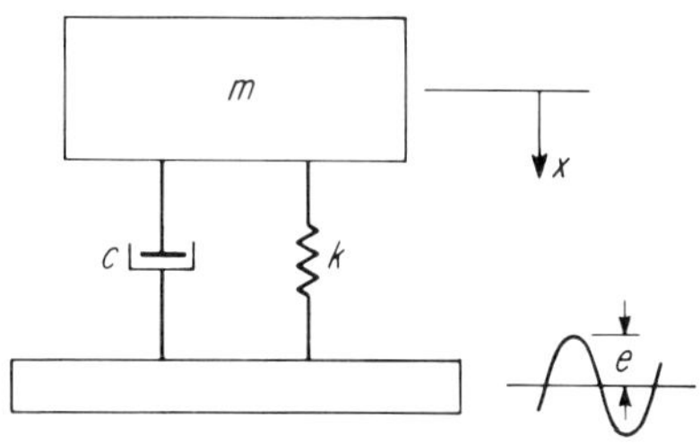

FIG. 14-16

The theory of the seismic unit has been discussed briefly with regard to foundation disturbance of a viscous-damped spring-mounted mass (Fig. 14-16). The relative motion between the mass and base is given in Chapter 6 as

$$X_s = \frac{e\left(\dfrac{\omega}{\omega_n}\right)^2}{\sqrt{\left[1 - \left(\dfrac{\omega}{\omega_n}\right)^2\right]^2 + \left(2\zeta\dfrac{\omega}{\omega_n}\right)^2}}$$

$$\tan\phi = \left[2\zeta\left(\frac{\omega}{\omega_n}\right)\right] \Big/ \left[1 - \left(\frac{\omega}{\omega_n}\right)^2\right]$$

where e = amplitude of foundation excitation
$\omega_n = \sqrt{k/m}$
$\zeta = c/c_c$
c = damping coefficient
c_c = critical damping coefficient $= 2\sqrt{km}$

IF $\omega_n \ll \omega$

$$X_s \approx e \quad \text{and } \phi = 180^\circ$$

and the relative motion is the same as the foundation displacement. In other words, the seismic mass will act as a fixed reference in space in this case.

IF $\omega_n \gg \omega$

$$X_s = \frac{\omega^2 e}{\omega_n^2\sqrt{\left[1 - \left(\dfrac{\omega}{\omega_n}\right)^2\right]^2 + \left[2\zeta\dfrac{\omega}{\omega_n}\right]^2}}$$

$$= \omega^2 e \times \text{const.}$$

In other words, the relative displacement will be proportional to the base acceleration in this case. The constant can be chosen to be approximately unity by the proper choice of ζ for a particular frequency range.

14-5. DIFFERENTIATING AND INTEGRATING CIRCUITS

It has been shown that one advantage of converting mechanical motions into electrical signals is the ability to integrate and differentiate these signals electrically to convert a measured acceleration to a velocity or displacement and vice versa. We shall discuss here the basic theory of differentiation and integrating circuits.

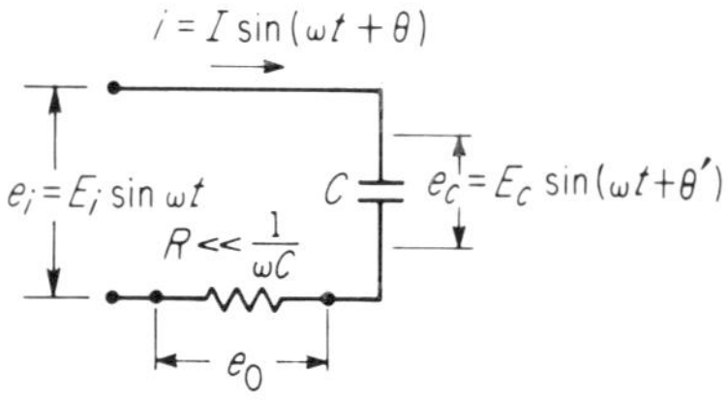

FIG. 14-17

Differentiating Circuits

An elementary differentiation circuit is shown in Fig. 14-17 and is composed of a resistor R and a capacitor C in series. It can be seen that

$$i = C \frac{de_c}{dt}$$

$$E_c = \frac{I}{\omega C}$$

where ω is the frequency of the input signal e_i

$$e_0 = iR = RC \frac{de_c}{dt}$$

If $$R \ll 1/\omega C$$
then $$IR \ll E_0$$
and $$e_c \approx e_i$$
Accordingly,

$$e_0 \approx RC \frac{de_i}{dt}$$

which shows that the voltage across the resistor in this case is the time rate of the input voltage.

Integrating Circuits (Fig. 14-18)

In this case

$$e_0 = \frac{1}{C} \int i \, dt$$

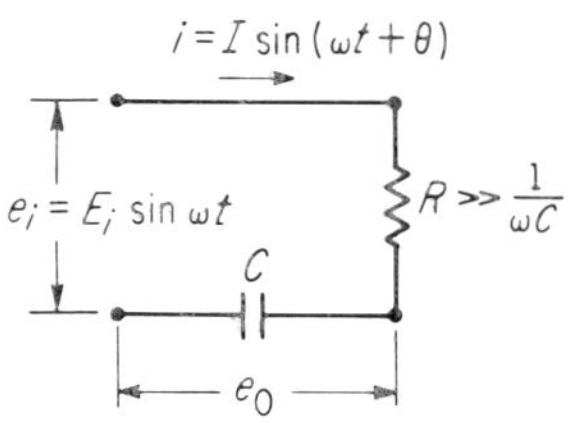

FIG. 14-18

$$I = \frac{E_i}{\sqrt{R^2 + \left(\frac{1}{\omega C}\right)^2}}$$

if

$$R \gg \frac{1}{\omega C}$$

$$I \approx \frac{E_i}{R}$$

and

$$e_0 = \frac{1}{RC} \int e_i dt$$

which indicates that the voltage across the capacitor is the time integral of the input signal.

14-6. FILTERS AND SIGNAL ANALYZERS

Transducers of the electrical type are designed to sense the motion of dynamic systems and transform it into a corresponding electrical signal. If the motion is substantially sinusoidal, the transducer output provides all the necessary information about the motion since the amplitude and frequency can be readily determined. When the motion contains many harmonic components or is random in nature, the transducer output by itself gives little information about the motion. Filters or signal analyzers are therefore necessary to give the frequency content and the power spectrum of the signal.

A filter is a device for isolating components of a signal on the basis of their frequency. It allows components in one or more frequency bands to pass relatively unattenuated and it attenuates components in other frequency bands. Filters may be classified as of the *low-pass, high-pass, band-pass,* or *band-rejection* type. A diagrammatic representation of the frequency characteristics of these different types is shown in Fig. 14-19.

An ideal low-pass filter allows frequencies below the cutoff frequency ω_c to pass unattenuated and blocks off all the higher frequencies. On the other hand, an ideal high-pass filter allows all frequencies above the cutoff frequency to pass unattenuated and blocks off all the lower frequencies. An ideal band-pass filter allows the frequencies within its specified band width to pass unattenuated and blocks off all other frequencies. The opposite occurs with the ideal band-rejection filter.

Band-pass filters may be of the *fixed-band* or *octave-band* type. The filter response in the first type is constant with respect to a particular number of cycles deviation from its center frequency. In the second type, the ratio between the high and low frequencies defining the band is ap-

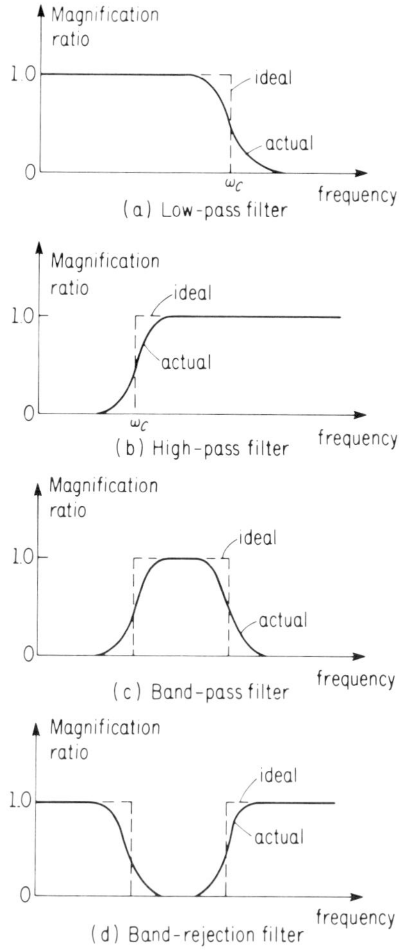

FIG. 14-19. Characteristics of different filters.

proximately equal to two. Example of octave bands are 18.75 to 37.5 cps 37.5 to 75 cps, 75 to 150 cps, etc.

Signal analyzers contain a filter system and a system for indicating the relative energy that is passed through the filter system. Fixed-band wave analyzers use fixed frequency filters in tunable systems similar to those of radio receivers. They are continuously tunable throughout the audio band. Octave-band analyzers contain sets of octave-band filters centered on particular sets of preferred frequencies. While fixed-band analyzers are more convenient to use, the octave-band type are more useful in noise analysis since they give a rapid evaluation of the signal.

14-7. INDICATION AND RECORDING INSTRUMENTS

Signals obtained from transducers should be fed in indicating or recording instruments. The following has to be kept in mind in this process.

1. The connecting leads and wires should be shielded to avoid interference from outside signals.
2. The signal may require amplification in order to meet the input power specifications of the indicating and recording equipment. The amplifiers used should be linear within the range of the test.
3. Filters may also be needed to stop the undesirable signals picked up from the surroundings or inherent in the carrier voltage.
4. The input as well as the output impedances of all components has to be matched for full utilization of the input signal and to avoid possible damage.

Moving-Coil Galvanometers and Recorders

The moving-coil galvanometers (Fig. 14-20) are an important part of measuring and recording instruments and a knowledge of their behavior is imperative to experimental studies of dynamic phenomena.

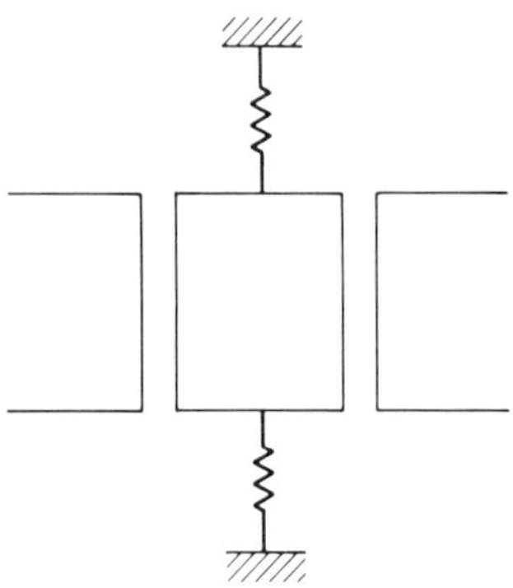

FIG. 14-20

The galvanometer can be represented by a coil whose motion is resisted by a restoring element in the form of a torsional spring. In essence a signal fed to the spring-suspended coil will cause it to move in a magnetic field in proportion to the input signal. The damping is electromagnetic and is very closely viscous. The damping can be increased by inducing a current in the coil which is accomplished by making the coil a part of a low-resistance circuit. Extremely high damping can therefore be obtained when the coil terminals are shorted. This is usually done whenever it is necessary to bring the galvanometer to an immediate stop.

A critically damped galvanometer is preferred when measuring steady currents. In this case the resistance which will just give critical damping is used and the final deflection of the galvanometer can be read without delay.

The underdamped galvanometer is useful for measurement of pulses. The galvanometer is calibrated with no damping and used with a known amount of damping. A correction should be made for the decreases in the maximum amplitude when the damped galvanometer is used for measuring a pulse.

The coil may carry a pointer for indicating on a graduated scale. It may also carry an ink pen, a heat pen, or a mirror reflecting a beam light for recording. Recording systems employing galvanometers can be utilized for signals up to several thousands cps. Systems with mechanical recording pens can only be used up to 150 cps.

Magnetic-Tape Recorders

This type of recorder has some definite advantages over other types. One such advantage is the ability to store the record indefinitely. The signal can also be played back almost immediately without any need for processing. Another advantage is the ability to use the record in a digital converter and feed the information directly into a digital computer for analysis. The tape recording can be done with amplitude modulation for up to 100,000 cps signal frequency with flat response. It can also be done with frequency modulation for signal frequencies up to 10,000 cps. The frequency modulation method is the most accurate method of recording signals up to 10,000 cps.

Cathode-Ray Oscilloscope (Fig. 14-21)

In cathode-ray oscilloscopes an electron gun provides a beam of electrons moving in a vacuum tube straight toward a fluorescent screen where

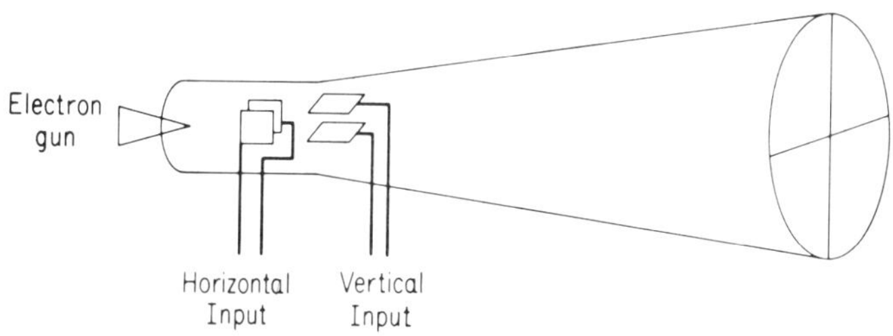

FIG. 14-21

it becomes visible as a light spot. When input signals are fed to the vertical input plates, the electrostatic field tends to deflect the beam according to the signal. If the horizontal input plates are connected to a sweep circuit, forcing the beam to move at a constant speed in the horizontal direction, the picture on the screen will show the input signal to a time base.

In many cases both the horizontal and vertical plates are used for signal inputs to form the well-known *Lissajous figures*. These figures can be used to check frequencies, amplitude ratios, and phase relations. One of the input signals may be a reference signal. Some of these figures are given in Fig. 14-22.

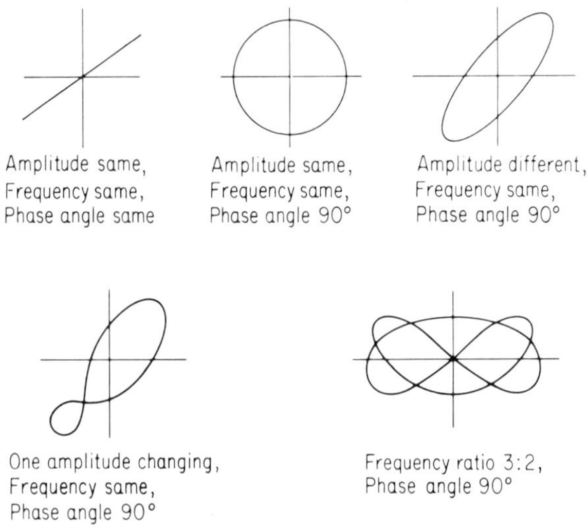

FIG. 14-22

Frequency Measurement

There are many methods of frequency measurement, among which is the oscilloscope shown previously in relation to the Lissajous figures. The stroboscope can also be used in cases where the amplitude is large enough for visual determination. Electronic counters can be effectively utilized for frequency measurements. The Reed gage is also an important means of frequency determination. Several reeds with different natural frequencies are placed on the vibrating object. The reed whose natural frequency corresponds to the frequency of vibration will be in resonance, and therefore will give a direct indication of the frequency.

14-8. PHYSICAL MODELS

The physical systems encountered in experimental investigations can be classified as follows.

Actual Systems or Structures

These systems may be buildings, dams, bridges, machines, ships, or structures for aircraft missiles and spacecraft, etc. The purpose of the test in such cases is to evaluate their vibratory characteristics (e.g., natural frequencies and modes, damping characteristics, etc.), or to determine their response under field excitations. Tests on models of this type require a great deal of ingenuity and experience. The planning of the test, obtaining the significant data, and evaluating these data usually represent a complex task.

Scale Models

Testing of scale models of the previous system is in many instances very desirable. It provides a guide for design before the actual construction is made. It can also be tested to destruction, if necessary, to provide information for limiting conditions and behavior under severe loading. Models of this type should be simple enough to retain correlation with mathematical theory yet realistic enough that the results can be related to the behavior of actual physical systems.

The design of scale models requires application of dimensional analysis. If dynamic similarity is required, the development of a criterion will be as follows. In any dynamic problem the variables are the force, mass, length, and time. From these variables one dimensionless factor (or π factor) can be formed. This factor is

$$\pi = \frac{Ft^2}{mL}$$

where F, m, L, t are the force, mass, length, and time respectively.

The dynamic similarity between the actual system and the model is satisfied if

$$\left(\frac{Ft^2}{mL}\right)_{\text{system}} = \left(\frac{Ft^2}{mL}\right)_{\text{model}}$$

Using the fundamental scale factors

$$\frac{m_{(\text{system})}}{m_{(\text{model})}} = S_1 \qquad \frac{L_{(\text{system})}}{L_{(\text{model})}} = S_2 \qquad \frac{t_{(\text{system})}}{t_{(\text{model})}} = S_3$$

Therefore for dynamic similarity

$$\frac{F_{(\text{system})}}{F_{(\text{model})}} = S_4 = \frac{S_1 S_2}{S_3^2}$$

which can be used as the design criterion for dynamic similarity.

In many instances dynamic similarity is impractical, since it requires complete geometrical similarity. Models are therefore designed based on other criteria such as system variables or system equations of motion. In the first case the variables are arranged in dimensionless groups. The model design conditions are based on the equality of these dimensionless parameters in the prototype and the model. For example, the grouping in the case of structure response to dynamic loading can be given by

$$\phi\left(\frac{\sigma L^2}{F}, \frac{F}{EL^2}, \frac{\rho g L}{E}, \frac{L\omega^2}{g}, \frac{a_n}{L}, \frac{\delta_n}{L}, c_n, \nu\right) = 0$$

where $F, c_1 F, c_2 F, \ldots, c_n F$ = loads
F = reference load
L = reference dimension
a_n = geometric dimensions of the structure
δ_n = deflections of the structure
ω = frequency of applied loads
g = gravitational acceleration
ρ = density
E = modulus of elasticity
σ = stress
ν = Poisson's ratio

In many cases the model analysis is based on the equation of motion and the boundary conditions. This is done by transforming all variables into dimensionless parameters. The equations of motion and the boundary conditions will accordingly be applicable to both the prototype and the model. For example, the equation of motion for the lateral vibration of a uniform beam fixed at one end and free at the other is given by

$$EI\,\frac{\partial^4 y}{\partial x^4} + \rho A\,\frac{\partial^2 y}{\partial t^2} = 0$$

where y = deflection
E = modulus of elasticity
I = moment of inertia of cross section
ρ = mass density
A = cross-sectional area

The boundary conditions are

$$\text{at } x = 0, \quad \frac{\partial^2 y}{\partial x^2} = 0, \quad \frac{\partial^3 y}{\partial x^3} = 0$$

$$\text{at } x = L, \quad y = 0, \quad \frac{\partial y}{\partial x} = 0$$

The dimensionless parameters can be taken as

$$\alpha = \frac{x}{L} \qquad \beta = \frac{y}{L} \qquad \tau = \frac{ct}{L} \qquad \zeta = \frac{I}{AL^2}$$

where $c = \sqrt{E/\rho}$ = the velocity of wave propagation. The equation now takes the form

$$\zeta \frac{\partial^4 \beta}{\partial \alpha^4} + \frac{\partial^2 \beta}{\partial \tau^2} = 0$$

from which it can be seen that the necessary condition for model design is that ζ remains the same in both the prototype and the model. Geometric similarity is therefore not necessary.

The system of Fig. 14-23 will be also considered. In this case the

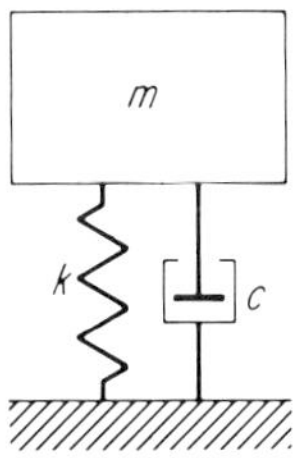

FIG. 14-23

equation of motion is

$$m\ddot{x} + c\dot{x} + kx = 0$$

from which

$$\ddot{x} + 2\zeta\omega_n\dot{x} + \omega_n^2 x = 0$$

Using the dimensionless parameters,

$$\alpha = \frac{x}{L}$$

$$\beta = \frac{t}{\tau}$$

where $\tau = \dfrac{2\pi}{\omega_n}$

L = a reference length

therefore the equation can be rewritten as

$$\frac{L}{\tau^2}\frac{d^2\alpha}{d\beta^2} + 2\zeta\frac{2\pi}{\tau}\frac{L}{\tau}\frac{d\alpha}{d\beta} + \left(\frac{2\pi}{\tau}\right)^2 L\alpha = 0$$

or

$$\frac{d^2\alpha}{d\beta^2} + 4\pi\zeta\frac{d\alpha}{d\beta} + 4\pi^2\alpha = 0$$

Accordingly, ζ is the only required dimensionless group to satisfy similarity.

This brief introduction to scale models gives the reader an appreciation of the problem. The reader who is interested in pursuing the matter further should refer to texts on theory of models.

Research Models

An important type of model is one which is designed for research and investigation purposes. This type provides an unending challenge to the experimentalist. It requires a great deal of skill and a thorough knowledge of the applicable theory and its limitations. Considerable caution must be observed in interpretation of data obtained from such tests to insure the validity of the measured results.

A device designed for investigation should provide real conceptual insight in the complexities of physical systems. With the continuous development of the experimental techniques, new concepts and more fruitful investigations on the frontiers of knowledge are always possible.

The area of collision of solids is an example of this challenge. The analysis of impact problems is as yet limited to very simple projectile-target systems involving primarily elastic deformations. In the case of a brittle projectile striking a harder target, it is known that the impactor will shatter even at moderate velocities. Other impact phenomena requiring conceptual experimental investigations are the spalling, dishing, perforation, and crater formation in solids subjected to impact.

Within the domain of nonlinear vibrations, random vibrations, and gyrodynamics, many experiments are necessary for deep understanding of associated phenomena.

Laboratory Demonstration Models

Laboratory demonstration models are usually designed to perform experiments illustrating phenomena which have been discussed theoretically. They should provide a visual aid to demonstrate these phenomena and to help the students in better understanding of the effect of the fundamental parameters of a dynamic system on its behavior under shock and vibration. Models of this type should be capable of producing both the simple and complicated dynamic situations to stimulate interest in the students to explore the complicated situation after thoroughly understanding the simple ones. In this sense they should be of the multipurpose variety. The equipment should produce meaningful data to introduce the student to experimental methods and data analysis; and to be utilized in testing the effectiveness and in calibration of new instruments or circuits designed by the students for measuring dynamic phenomena.

An example of a laboratory demonstration model (Ref. 5) will be described in the following together with some elementary tests showing its capabilities.

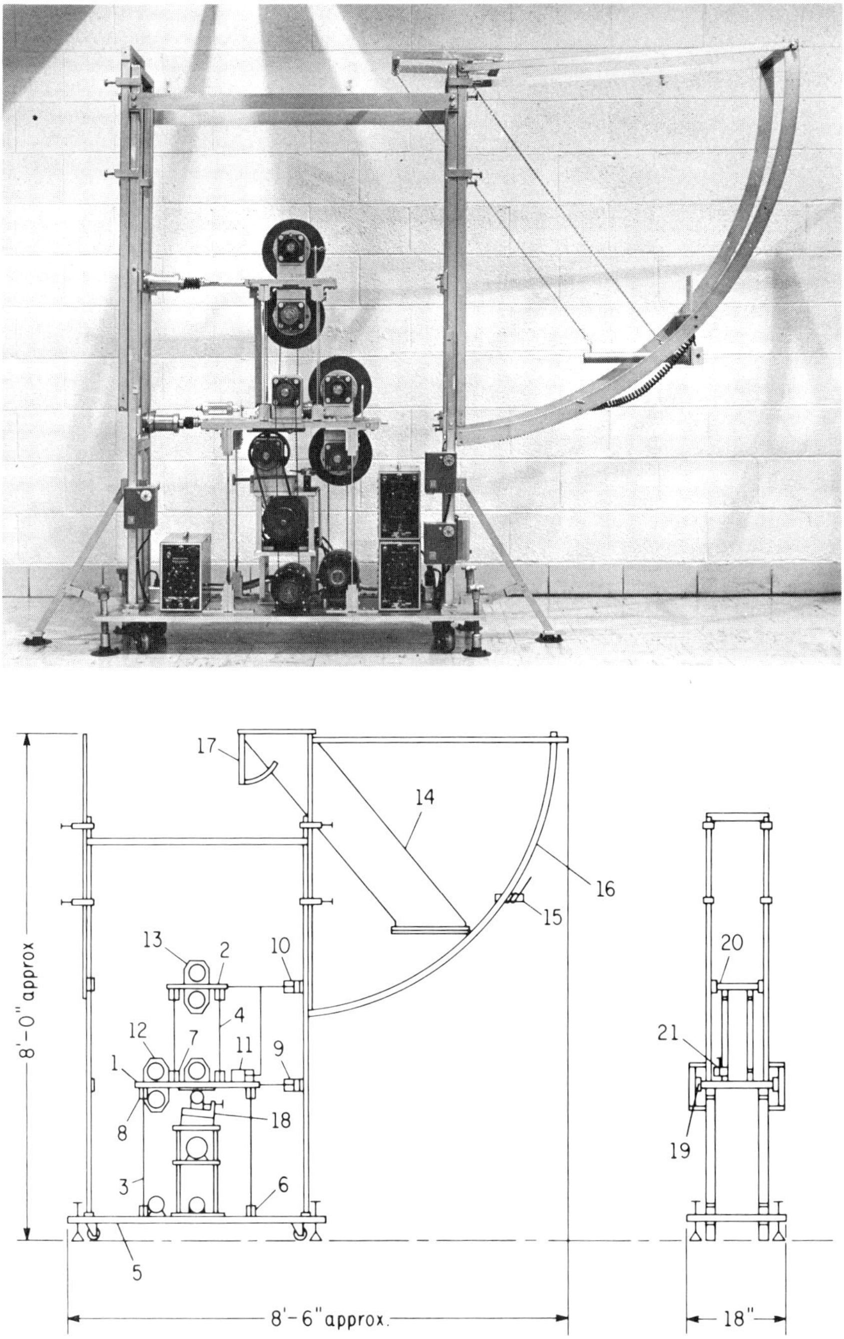

FIG. 14-24. (a) Tester. (b) Diagrammatic representation of tester.

The tester which is shown in Figs. 14-24a and 14-24b is a two-mass system with controllable elastic and damping characteristics. The system excitations are provided by two variable-speed exciters on the main and secondary mass respectively, a frictional wheel drive on the main mass and ballistic pendulums capable of providing different impulses on both masses. The principal elements are:

1. Main table plate.
2. Secondary table plate.
3. Four leaf springs for the main mass.
4. Four leaf springs for the secondary mass.
5. Base plate furnished with wheels for transportation and suction-cup mounts for temporary installation.
6. Steel angles for supporting the main springs.
7. Steel angles for supporting the secondary springs.
8. Exchangeable spacers with different lengths having straight sides for different values of linear stiffness and various types of profiles for nonlinear stiffness.

9, 10, 11. Dampers furnished with heaters and thermostatic controls providing different degrees of approximately viscous damping for the main mass, secondary mass, and their relative motion respectively.

12, 13. Variable-speed exciters on the main and secondary masses respectively with adjustable eccentricities capable of providing different excitations at frequencies from 0 to 150 cps. Each exciter is driven through a timing belt by a $^1/_4$-hp variable-speed motor. The two exciters can be also driven simultaneously at a particular frequency ratio and phase relationship by utilizing one of the two motors and a timing belt and pulley assembly connecting the two drives. Each exciter is composed of two shafts rotating in opposite directions at the same speed by means of a pair of gears made of synthetic material to eliminate the need for lubrication. The shafts rotate in identical bearing blocks fastened symetrically on both sides of the table plate. Each shaft carries a disk with an adjustable eccentric mass. The eccentric masses are always at equal and opposite positions with respect to the central plane of the plate. Their resultant effect is therefore a horizontal sinusoidal force. Constant force excitations as well as random excitations can be provided by utilizing a small electromagnetic shaker.

14. A chain-suspended ballistic pendulum with provisions for exchangeable hammers capable of applying impulses on the two masses.

15. An electric magnet for initial positioning and releasing of the hammer.

16. A calibrated arc to support the magnet and indicate the initial amplitude of the ballistic pendulum. Provisions are made for folding the arc when the set-up is moved.

17. A calibrated arm for indicating the rebound of the pendulum.

18. A frictional wheel assembly including a variable-speed motor (0 to 1,500 rpm) and a wedged table to provide different frictional forces between the wheel and frictional plates fastened to the main mass.

19, 20. Crossbars fastened to the frame to support adjustable frictional plates for applying Coulomb friction of different intensities on both sides of the main and secondary masses respectively.

21. An adjustable frictional plate which can replace the viscous damper to provide Coulomb damping of different intensities for the relative motion between the two masses.

Basic Instrumentation

The fundamental quantities to be measured in the setup are the displacement, velocity, and acceleration of both the main and secondary masses.

Strain gages are mounted on the leaf springs and calibrated to measure the displacement of the main mass and the relative displacement between the two masses. U-shaped strain-gage displacement transducers or differential transformer-type displacement transducers can be used to measure the displacement of either mass. Velocity and acceleration transducers can be mounted directly on the masses. The output from the transducers can be recorded simultaneously after the proper amplification by means of a six-channel recorder. The phase relationship between the excitation and the response waves can be easily obtained by superimposing on the particular response signal an electric pulse indicating the position of the peak of the exciting wave. The pulse is produced by allowing a notch in the rim of the eccentricity disk to excite the field of a permanent magnet and induce a pulsating emf in the enclosing coil to indicate the position of maximum excitation. (See Fig. 14-25).

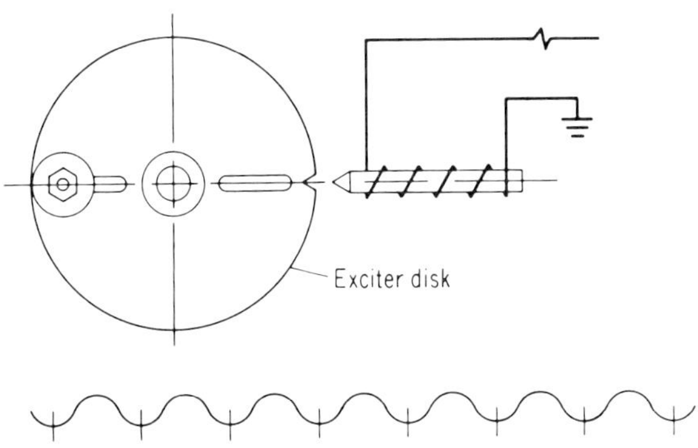

FIG. 14-25. Phase-angle measurement.

Capabilities of the Tester

The tester can be used for demonstrations and tests on the following:

1. Steady-state response and phase relationships for single- or two-mass systems to sinusoidal excitations at the main mass or secondary mass or both. The system can have linear or nonlinear stiffness and different magnitudes of viscous or Coulomb damping at different locations.
2. The principles of shock and vibration isolation for systems with linear as well as nonlinear characteristics.
3. The principle of dynamic damper with different parameters.
4. Response of the linear or nonlinear systems to impulses of different characteristics.
5. Approximate measurement of the force-time relationship during impacts of different hammers on systems with different characteristics.
6. Self-excited vibrations. The frictional wheel assembly can be used for studies on self-excited vibrations.
7. The vibration of a single or two-mass system under conditions of Coulomb damping.

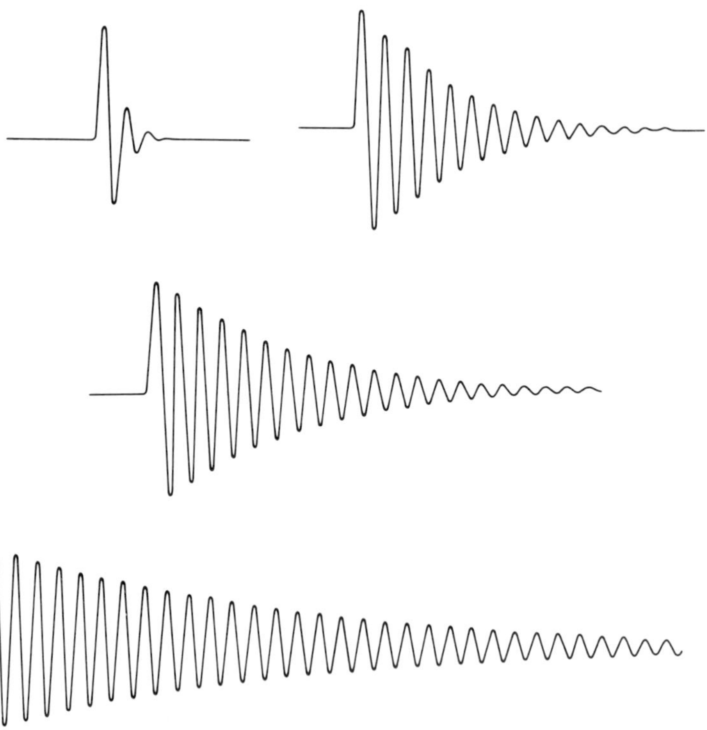

FIG. 14-26. Examples of vibration decay.

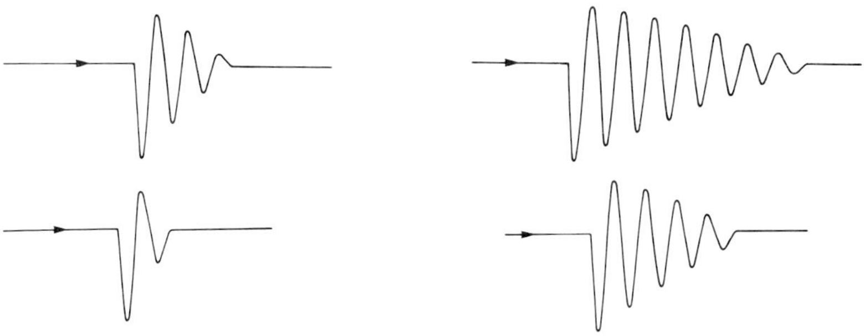

FIG. 14-27. Vibration decay with coulomb damping.

8. Response of single or two-mass systems of linear or nonlinear characteristics to different combinations of impacts, sinusoidal excitations and a constant frictional force acting simultaneously.

9. The equipment can be utilized as a mechanical model to simulate particular vibration problems and to study the influence of changing the different parameters. This is particularly useful in systems with nonlinear characteristics or when certain mechanical or material properties are involved. The mathematical analysis in these instances are rather difficult and the electrical analogy methods are only approximations.

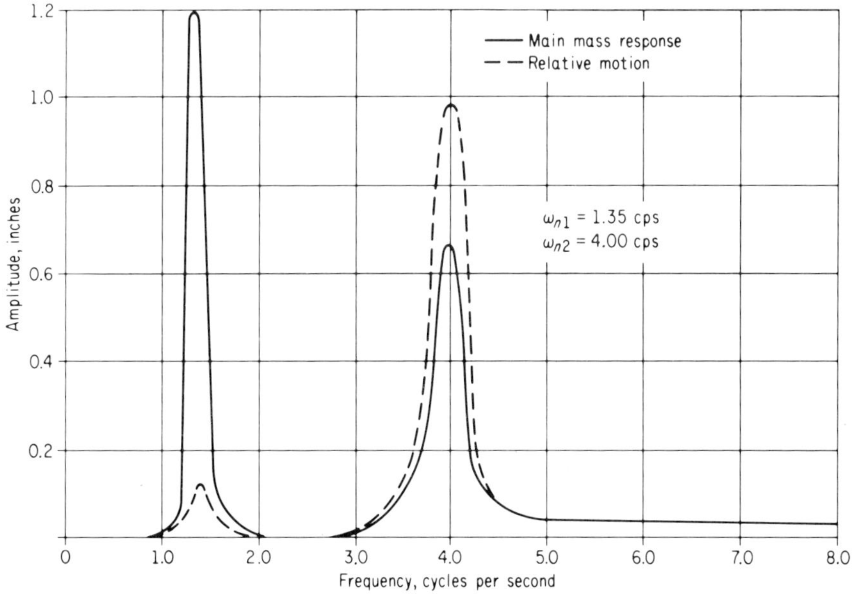

FIG. 14-28. Response of two-mass system to excitation on main mass.

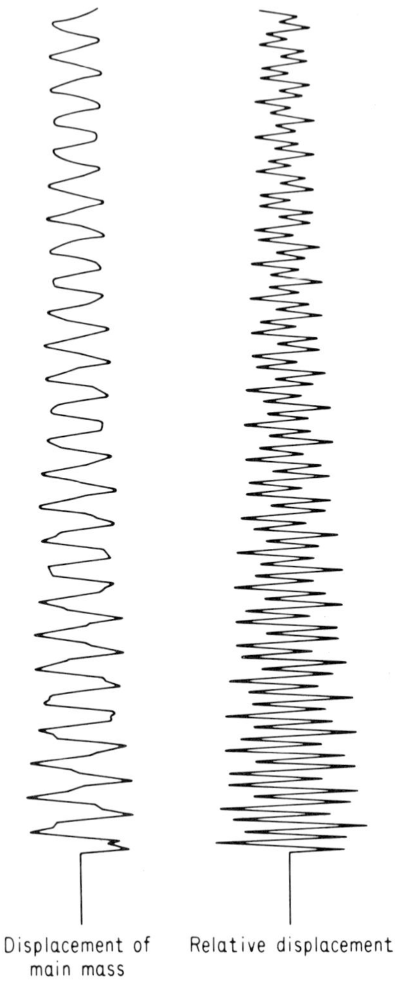

FIG. 14-29. Response of system to impulse on main mass.

10. It can also be utilized as a shake table to study systems of smaller dimensions or to calibrate vibration instruments or damping devices.

11. The setup is readily adaptable to any modifications or changes which may seem advantageous based on calibration or experimentation.

Examples of Some Basic Demonstrations and Tests

Sample of damped-free vibration curves with a petroleum jelly (Vaseline) as the damping medium are shown in Fig. 14-26. The decay curves for a temperature range between room temperature and 200° F show that Vaseline gives an approximately viscous damper within the described temperature range.

Figure 14-27 shows the vibration decay with Coulomb damping.

The response of an undamped two-mass system to sinusoidal excitation and an impulse at the main mass is shown in Figs. 14-28 and 14-29 respectively.

The response of single-mass system with nonlinear spring characteristics to sinusoidal excitation is illustrated in Fig. 14-30. The main mass

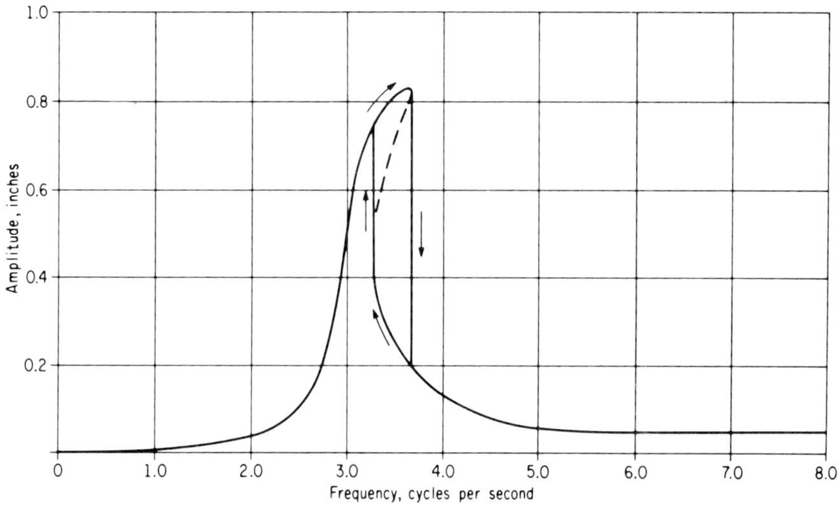

FIG. 14-30. Response of secondary system with nonlinear springs.

is connected rigidly to the frame in this test. The support spacers for the secondary spring system are changed to provide a decreasing support length with increasing deflection. The spring constant therefore increases with deflection. It can be easily seen that, as the frequency increases, at a particular value of the frequency a sudden drop in the amplitude of vibration occurs. Also a sudden rise in the amplitude will occur at a lower frequency when the frequency decreases. This illustrates the "jump phenomenon" which is characteristic of systems with nonlinear springs.

Data on self-excited vibration of the main system are shown in Fig. 14-31. The bearing block of the frictional wheel is moved on the supporting wedge a predetermined amount to produce a particular value of static friction. The motion of the block is indicated by means of a dial gage. The driving motor is then run at different speeds and the resulting

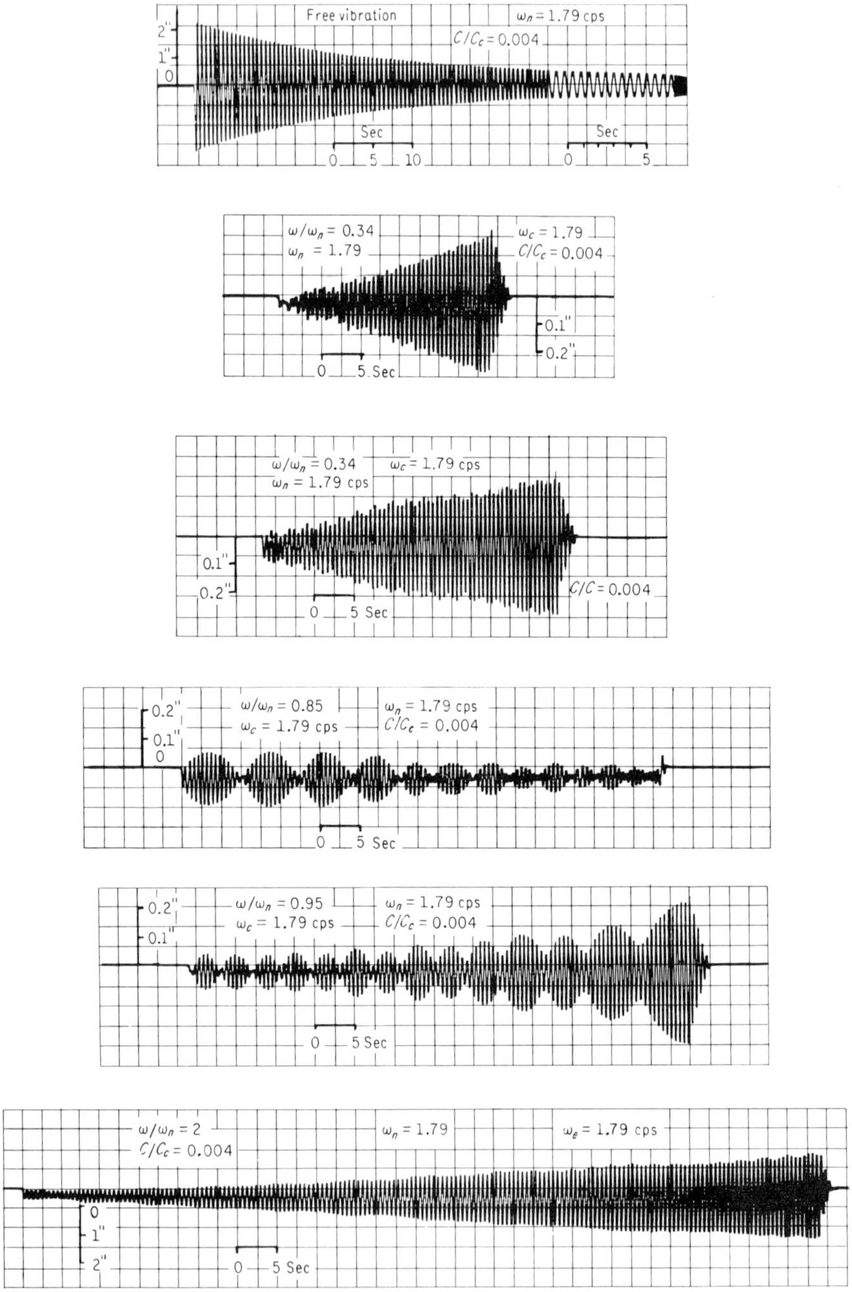

FIG. 14-31. Examples of self-excited vibration.

vibration of the system is recorded. It should be noted that the frequency of the motion is the same as the natural frequency of the system and is independent of the motor speed.

14-9. BIOLOGICAL SYSTEMS

Considerable attention has been given to the effect of shock and vibration on man over the past four decades. The development of faster means of travel on land, sea, air, and space accelerated these investigations. From the very beginning of man, he has always been subjected to vibrations of one form or another. The effect of these excitations on him,

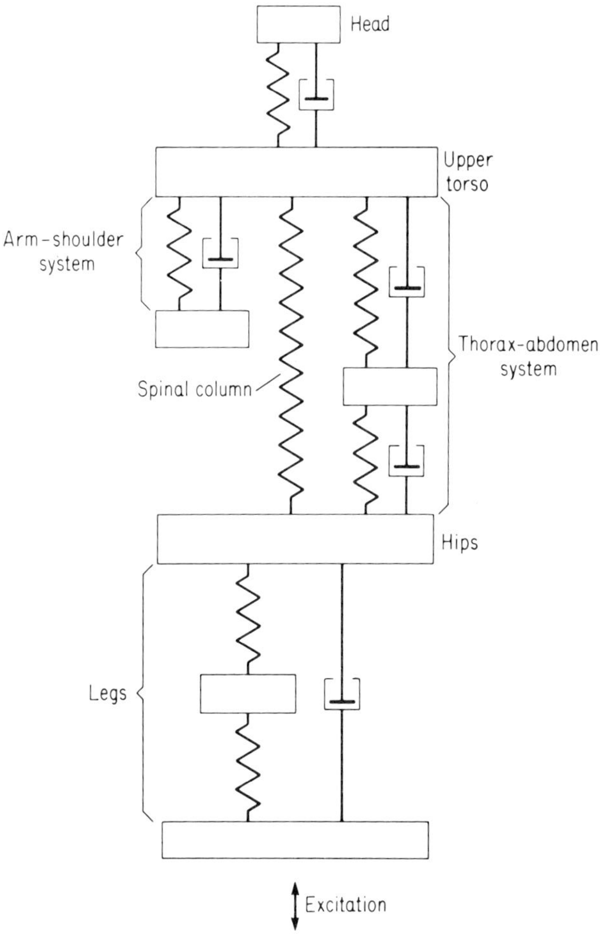

FIG. 14-32. (After Coermann *et al.*, 1960.)

as well as other biological systems ranges from pleasant sensation to unbearable or injurious. There is no question as to the complexity of living organism and the variety of means by which excitations are transmitted to them. There is, for example, a marked difference in human response to vibrations, depending on position relative to the direction of vibration. High-frequency vibrations are known to be generally perceived only through the medium of the ear. The interrelationship between the physical, psychological and physiological responses makes experimental investigation almost necessarily an interdisciplinary activity. As in the case of other highly nonlinear physical systems, the dynamic characteristics of living organisms depend on the response of other parts of the system to vibration. Furthermore, these characteristics are dependent on the psychological reaction to the test or test equipment. Due to the variation in the physical and psychological characteristics of the subjects, the design and evaluation of tests has to bear heavily on statistical analysis. The transducers and instrumentation for sensing human responses has to be chosen in a way as to produce the least possible effects on the subject. Noncontact methods such as photographic and X-ray techniques are desirable in detecting physical responses. Also, as in many complex structures, acceleration measurements are preferable to displacement or velocity. Other physiological and psychological measurements include blood pressure, respiratory rates or depth, heart potential, brain potentials, visual activity, manual steadiness, balancing senses, metabolic rate, and physical weariness or depression.

In order to study physical responses of human systems in regions where fear of injury to the subjects limits the investigation, simplified mechanical and mathematical models are attempted. Figure 14-32 shows a simplified system representing human body standing on a platform vibrating in the vertical direction. Figure 14-33 represents a simplified electrical analog circuit simulating the respiratory system. An example of

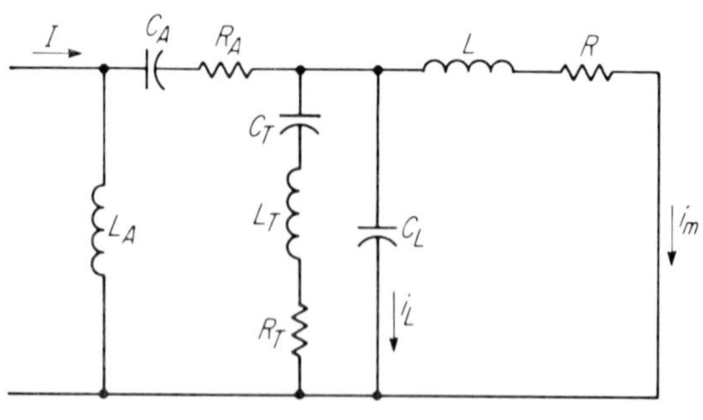

FIG. 14-33

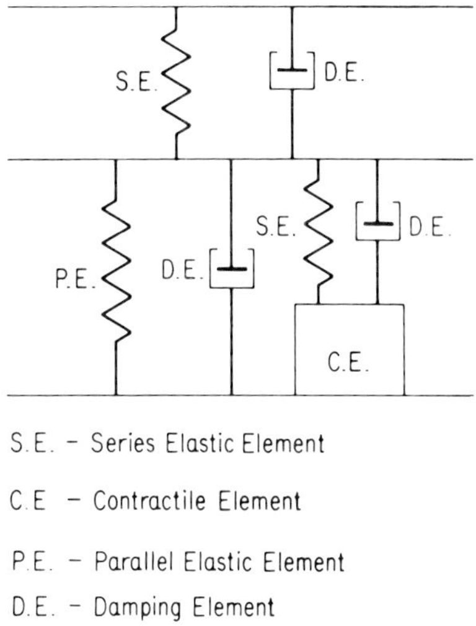

FIG. 14-34. Model for simulating muscle dynamics. (Ref. 17)

a mechanical analog of animal muscles is shown in Fig. 14-34. A model for investing whiplash injuries is given in Fig. 14-35. The analysis of the dynamics of some common human activities is described in Reference (16) and a sample of the results obtained for the case of a man jumping off a platform is illustrated in Fig. 14-36.

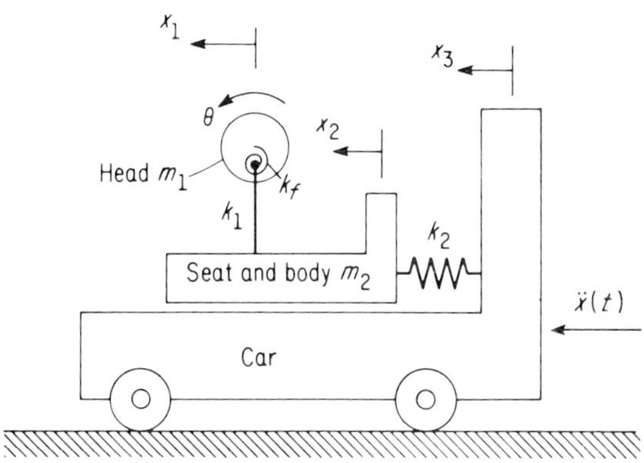

FIG. 14-35. Model for study of whiplash injuries. (Ref. 13)

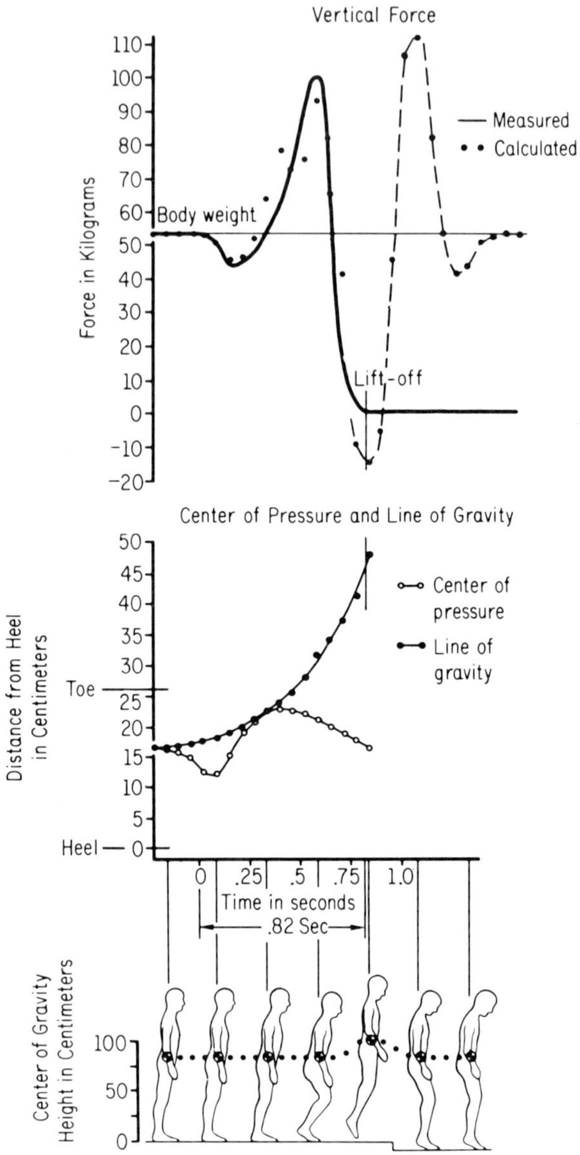

FIG. 14-36. Normal male, jumping forward off of force platform. Upper graph shows vertical force patterns measured from platform and force values calculated from photographic records of displacements of mass centers of body segments. Calculated values after the lift-off phase of jumping are connected by dashed line. Middle graph shows fore-aft excursions of center of pressure and line of gravity on the supporting surface. Lower graph shows vertical trajectory of the center of gravity. (Ref. 16)

14-10. SUMMARY

This chapter is intended as an introduction to the experimental investigation of dynamic systems.

The objective is merely to familiarize the reader with some of the numerous methods and techniques of experimental analysis and scale modeling; the material presented is far short of that necessary to develop proficiency in this area. The interested reader should consult texts and publications dealing primarily in experimental techniques and measurement systems.

REFERENCES

1. C. M. Harris and C. E. Crede, *Shock and Vibration Handbook.* New York: McGraw-Hill, 1961.
2. E. Doebelin, *Measurement Systems Application and Design.* New York: McGraw-Hill, 1966.
3. "The Transistor as a Pressure and Force Transducer," *Biomed. Eng.*, (May 1968).
4. V. J. Skoglund, *Similitude: Theory and Applications.* Scranton, Pa.: International Textbook, 1967.
5. A. Seireg, "A Multipurpose Vibratory Model," *J. Eng. Educ.*, Vol. 54, No. 9 (May 1964).
6. I. Vigness, "Navy High-Impact Shock Machines for Lightweight and Medium-Weight Equipment," Naval Research Laboratory Report No. 5618, January 1961.
7. L. S. Pechter et al., "Technique for Determining the Frequency Response of Linear Mechanical Systems," *J. Acous. Soc. Amer.*, Vol. 37, No. 1 (January 1965).
8. C. E. Crede, *Vibration and Shock Isolation.* New York: Wiley, 1951.
9. *Trans. ASME, J. Eng. Ind.*, Special Vibration Issue, Vol. 89, Ser. B, No. 4 (November 1967).
10. M. V. Barton (ed.), "Shock and Structural Response," *Colloquium on Shock, ASME*, New York (1960).
11. W. J. Worley (ed.), "Experimental Techniques in Shock and Vibration," *ASME*, New York (1962).
12. P. Peterson and E. Gross Jr., "Handbook of Noise Measurement," 5th ed., General Radio Co., West Concord, Mass., 1963.
13. J. Martinez, J. Wickstrom, and B. Barcelo, "The Whiplash Injury—A Study of Head-Neck Action and Injuries in Animals," *ASME* publication No. 65-WA/HUF-6, 1965.
14. D. Goldman and H. von Gierke, "Effect of Shock and Vibration on Man," *Shock and Vibration Handbook*, New York; McGraw-Hill Co., 1961, Vol. 3, Chapter 44.
15. V. Roberts, E. Stech, and C. Terry, "Review of Mathematical Models which

Describe Human Response to Acceleration," *ASME* publication No. 66-WA/BHF-13, 1966.
16. P. Murray, A. Seireg, and R. Scholz, "Center of Gravity, Center of Pressure, and Supportive Forces During Human Activities," *J. of App. Physiology*, Vol. 23, No. 6, Dec. 1967.
17. W. Parmley and E. Sonnenblick, "Series Elasticity in Heart Muscle," *Circulation Research*, Vol. 20, January 1967.

part VII

Material Behavior Under Dynamic Loading

Our experience would seem to prove that, even with the greatest care in manufacturing, these axles are subject to a rapid deterioration owing to the vibration and jar which operates with increased severity, on account of their peculiar form. So certain and regular is the fracture at the corner of the crank from this cause, that we can almost predict in some classes of engines the number of miles that can be run before signs of fracture are visible. . . .

James E. McConnell, *Proc. Inst. Mech. Engineers*,
London (1849)

chapter 15

Material Behavior Under Dynamic Loading

15-1. INTRODUCTION

In the previous chapters, different analytical and experimental techniques were discussed with the purpose of evaluating the dynamic loading on different components of structures or mechanical systems. A very important question remains to be answered: What load can the element sustain without failure for a certain predetermined life?

Because of the demands of modern design problems, the engineer is confronted with material selection for applications involving applied loads which are continually increasing in severity and frequency. Many situations also require designs for loads which are transient or random. During recent years there has been considerable effort toward better understanding of material behavior under dynamic loads. An extensive account of this subject is beyond the scope of this book, however, a general discussion of it is necessary for an integrated treatment of mechanical systems analysis. This chapter presents a brief review of the fundamental aspects of fatigue and fracture.

15-2. FATIGUE

The weakness of metals under repeated loading is a very familiar phenomenon to engineers handling the design of machine parts or their operation. This phenomenon of decreased resistance of a material to varying stresses is called *fatigue.*

Fatigue is undoubtedly one of the most serious of all causes of breakdowns of machines and structure. Whereas wear, corrosion, and creep produce progressive deteriorations which at certain stages make it necessary to replace some parts or components, fatigue results in sudden failure with sometimes complete destruction as a whole. Uses of metallic materials in which fatigue is not to be feared are very few. In the past, the classical method of engineering design was based on the static strength of materials when safety factors on stress were large enough to cover all the unknowns. Where repeated loading was involved, the assumptions of static design were generally adequate as the allowable stresses were of such magnitude that the life of the part was almost infinite. At that stage of development a knowledge of the fatigue limit was the only additional requirement for dynamic design, but in spite of the intention of keeping the stresses below the safe limit, the ever-present local stress raisers were causing difficulties, and in almost every industry numerous unpleasant and even calamitous failures have been recorded.

With the increasing emphasis on weight and speed, use of static-strength criteria became inadequate and a better understanding of the behavior of metals under different loading histories assumed great importance. The more refined analytical and experimental stress-evaluating techniques can only be utilized effectively by more refinements in predicting a safe allowable stress for any required life of the member.

The Mechanism of Fatigue Failure

Fatigue failures can be very easily distinguished from static failures because they look entirely different. Two zones are usually observed in a fatigue fracture—one due to the gradual development of a crack, the other due to a sudden fracture which occurs as soon as the crack develops far enough to weaken the member below its static strength. The first zone can be easily recognized by a rubbed or ground area indicating the nucleus of failure and "oyster shell" markings indicating the propagation of the crack. The second zone resembles the fracture of a tensile-test specimen and is of the brittle type such as occurs in cast iron even if the material is ductile.

The mechanism of fatigue failure is not as yet fully understood. It is well agreed, however, that it develops as follows:

1. Formation of dislocations on the atomic level.
2. Formation and growth of submicroscopic cracks.
3. Growth of visible cracks.

In general, fatigue cracks initiate in the exposed surface of a metal part unless unusual circumstances or processing treatments such as nitriding develop surface layers of substantially higher strength than the inner zones of metal.

Apparently, the exposed surfaces are inherently weaker than the interior grains that are confined by bonds with adjoining grains. That is why shot-peening, cold-working, or polishing provides a marked increase in fatigue strength because it strengthens the surface layers.

Fatigue may also initiate inside the material due to the presence of discontinuities in the material; that is, internal defects and different weakened regions of a structure. Textural stresses on a microscale can be developed at interfaces between the various constituents of the metal because of their difference in physical and mechanical characteristics.

Factors Influencing Fatigue Failures

Some of the most significant factors influencing the fatigue failure are:

1. The material, its heat treatment, its making and shaping.

2. The application of stress, frequency of stress cycles, simultaneous action of static and variable stresses of different kinds (tension, bending, torsion), prestresses resulting from mechanical or thermal treatment.

3. Temperature, chemical action, fretting corrosion in press-fitted assemblies.

4. Notch effects produced by changes of section, fillet radii, grooves, holes, threads, keyways, and—on a microgeometrical scale—machining effects.

5. The size effect. A small part is known to have longer life than a geometrically similar large part.

Of all the above enumerated factors, notch effects must be considered as most important.

It is not difficult to visualize that abrupt changes in the cross section of a loaded member give rise to great irregularities in the stress distribution. Accordingly at certain points, the stress value is far above the nominal (or engineering) stress, and under the action of reversal of stresses progressive cracks are likely to start from such points. These points of high stress values are of great practical importance and are conventionally known as points of stress concentration.

Under static loading, regions of high stress concentration do not appreciably reduce the strength of a component *unless the metal is very brittle.* Such regions usually represent a small portion of the whole cross section, so that if the metal possesses only slight ductility, a small amount of yielding at the high stress region will relieve the stress concentration considerably. Because of this, the stress distribution becomes more and more uniform as the load increases. This explains why, with ductile materials, notches do not lower the ultimate strength when the notched piece is tested statically. (The word *notch* is used to include all discontinuities.)

Under reversed loading, much less plastic deformation occurs during each cycle of loading, and the influence of stress concentration is of greater importance. This dynamic plastic strain per cycle varies during the course of the fatigue test for most of the steels used in practice.

The pattern of variation is different for different materials and even for the same material at different temperatures. It should be noted that the ability of materials to deform plastically under cyclic stressing results in reduced sensitivity to stress concentration. This explains why lower fatigue stress-concentration factors are obtained for notched parts made of mild steel than for parts made of harder steels. The ability of mild steel to deform plastically during each cycle of the fatigue tests so alters the stress distribution that the maximum stress at the notch is appreciably reduced. This is generally described by a notch-sensitivity factor q:

$$q = \frac{K_f - 1}{K_t - 1}$$

where K_f = fatigue stress concentration factor
K_t = theoretical stress concentration factor

Examples of values of q for shafts with fillets are shown in Fig. 15-1.

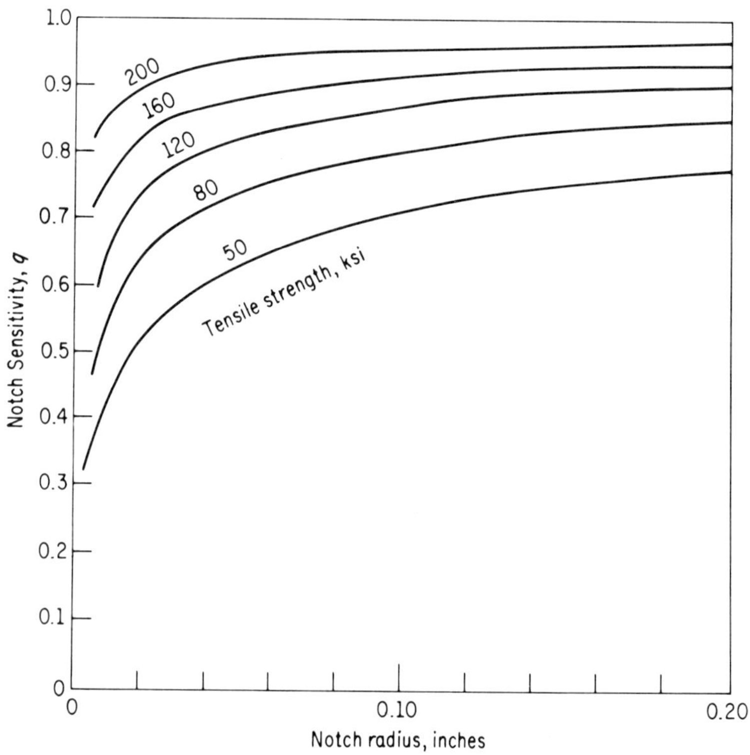

FIG. 15-1. Notch sensitivity for heat-treated steel.

It is well recognized that the size of the part affects its fatigue characteristics. It was found from many tests that under fatigue loading there appears to be a marked difference in behavior between small-scale test pieces and those of larger sizes. The decrease of fatigue strength with the increase of size is generally attributed on one hand to the growth of number of internal defects and different weakened regions of the part, and on the other hand to the change that takes place in the stress condition of the material in the surface layer due to the change of stress gradient.

Heat treating of steels, as would be expected, increases the fatigue strength due to the increase in hardness, provided that no quenching cracks or decarburization of the surface occurs in the process. However, this increase in hardness is usually accompanied by an increase in the notch sensitivity of the part because of the decrease in ductility.

Surface treatment also has an important influence on the fatigue behavior of the part. The machining and finishing processes as well as other mechanical treating methods (such as cold-working, shot-peening, grinding, polishing, etc.) will have a significant influence on the fatigue life, partly because of the residual stresses they induce and partly because of their effect on the surface roughness. Chemical treatments such as carburizing, nitriding, and cyaniding also increase the fatigue strength. The quantitative effect of these different surface treatments depends to a good extent on the material, its geometry, and the type of stress.

The temperature effect on fatigue strength is of considerable importance. A decrease of the fatigue limit of unnotched specimens is easy to visualize. A reduction in the notch sensitivity for conditions of severe stress concentration usually results from increasing temperature. Fatigue fractures can also result from temperature fluctuations even if the load stress is very low. The latter is known as *thermal fatigue*.

It is not difficult to visualize that lower fatigue limits would be expected if the part is loaded in a chemically active atmosphere or under a condition of chemical corrosion.

Another type of corrosion of considerable influence on fatigue behavior is *fretting corrosion*. This usually occurs where relative motion exists such as in close fitted or clamped assemblies that are supposed to be rigid but actually have a very small amount of play. The occurrence of fretting is commonly experienced in press fits where it is usually responsible for cutting the endurance strength of the shaft down to one-half. Increasing the hardness of such shafts does not necessarily increase its endurance limit, and surface rolling (cold rolling) appears to be the most effective method of combating the fretting effects.

15-3. ANALYSIS OF FATIGUE DATA

Stress Life Diagrams and Endurance Limit

Starting with a number of identical specimens, loading them at different stress levels, and recording the number of cycles they withstand until failure, the *S-N* curve or the stress life curve can be obtained (see Fig. 15-2). It is noticed that at the beginning the stress value decreases rapidly

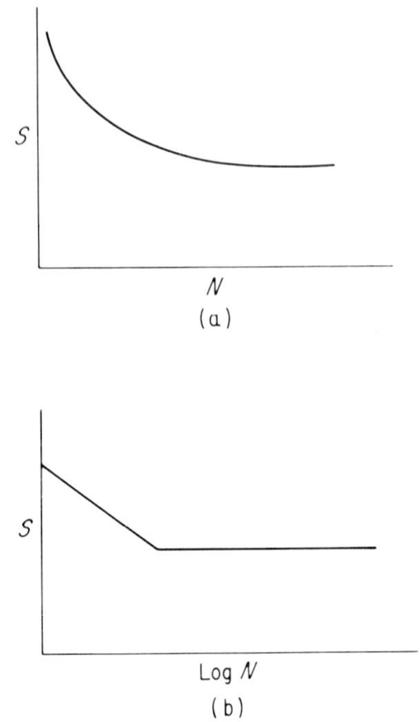

FIG. 15-2. Stress-life curves.

as the number of cycles increases. Then the curve approaches asymptotically a horizontal line. The stress corresponding to such an asymptote is called the *endurance limit* of the tested specimens.

It is a common practice in fatigue tests to plot the curve S against log N. In this manner a definite mark of discontinuity in the curve defining the magnitude of the endurance limit is usually observed.

Statistical Nature of Fatigue

It is a well-known fact that the results obtained from fatigue tests represent a scattered pattern rather than fit a smooth uniform curve. Recent studies indicate that most of the scatter observed in finite fatigue-life testing of metals is an inherent characteristic of the material and does not necessarily indicate poorly adjusted machines or improper testing techniques.

The distribution of the fatigue life of groups of similar specimens is quite varied. Therefore one cannot legitimately speak of the fatigue limit in terms of a single value but must regard the S-N curve more as a family of curves, each of which indicates a definite probability of failure (Fig. 15-3). However, it is seldom economically feasible to run enough experi-

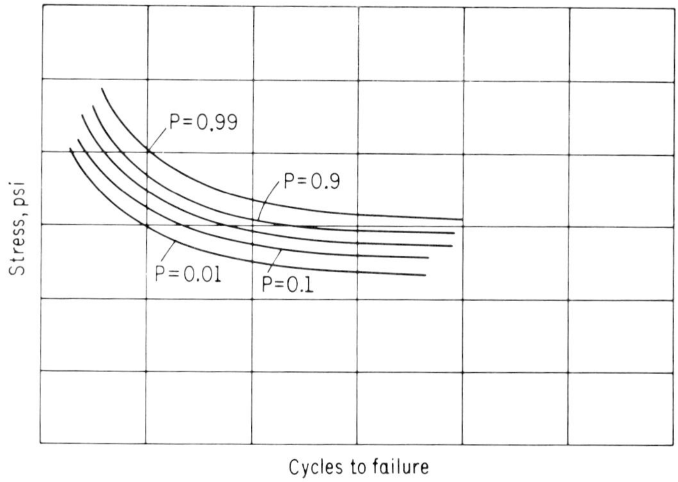

FIG. 15-3. Composite *S-N* curves for various probabilities of failure, *P*.

ments to obtain the statistical information necessary to design a part for limited life with a predictable degree of certainty. Furthermore, consideration of the many variables and uncertainties that occur in practical service conditions makes it impossible to predict a precise fatigue life for a member in service.

The presence of slag, oxides, sulfides, or other inclusions in steel is frequently suspected as being harmful and often directly blamed for premature fatigue failures. Unfavorable residual stress patterns resulting from metalworking or heat-treating also contribute to the uncertainty of the fatigue strength of a part.

Endurance Limits for Different Stress Cycles—The Goodman Diagram

In the previous discussions, the endurance limit for the case of complete reversal of stresses was considered. In actual service a great majority of parts are subjected to various combinations of fluctuating and steady stress of the type shown in Fig. 15-4.

There is a general agreement that the relationship between strength and such stresses is given by the modified Goodman diagram of which three forms are shown in Fig. 15-5. These diagrams can be plotted provided that the ultimate strength, yield point, and endurance limit under complete reversal of stress are known.

In these diagrams, the heavy lines are considered as the limiting allowable stress lines. If for a practical part, the minimum stress and the maximum stress (or fluctuating stress) are known and when plotted on the

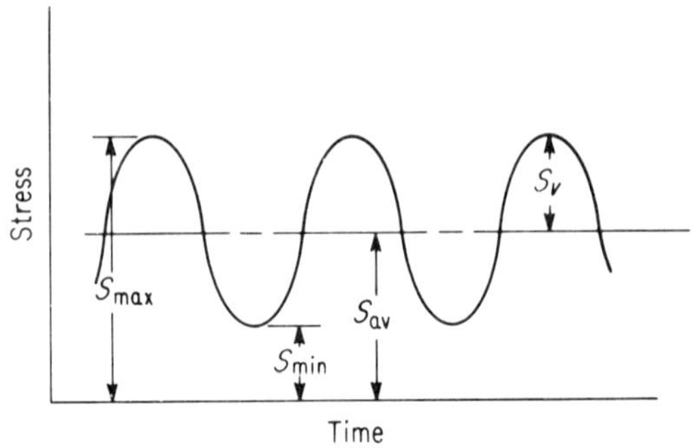

FIG. 15-4. Stress cycle.

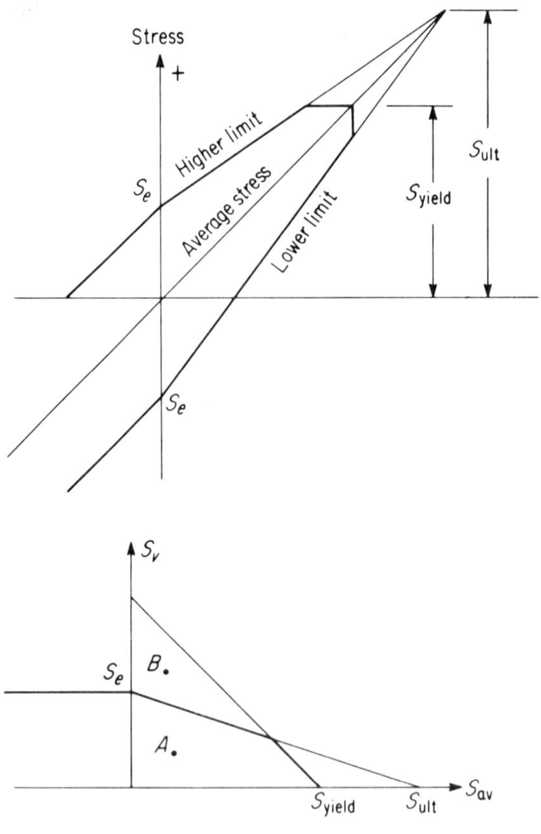

FIG. 15-5. Modified Goodman diagrams.

diagram of Fig. 15-5 fall within the heavy lines such as point A, the part is not expected to fail. Conversely, if the plotted point falls outside the allowable stress lines such as at point B, failure is expected.

When the average stress is compressive, it will have little effect on fatigue failure. In this case fracture will occur only when the fluctuating component exceeds the endurance limit. For any combination of stresses, the maximum and minimum combined stresses can be obtained by the Mohr circle construction and by plotting them in Fig. 15-5 failure or safety can be predicted.

Soderberg Criterion for Factor of Safety

A simplified design criterion which allows an evaluation of a factor of safety for conditions where the stress cycles having an average tensile value is due to C. R. Soderberg. It is assumed that the safe limit for any material is determined by a straight line connecting the endurance stress S_e and the yield stress S_y as shown in Fig. 15-6. It will also be assumed

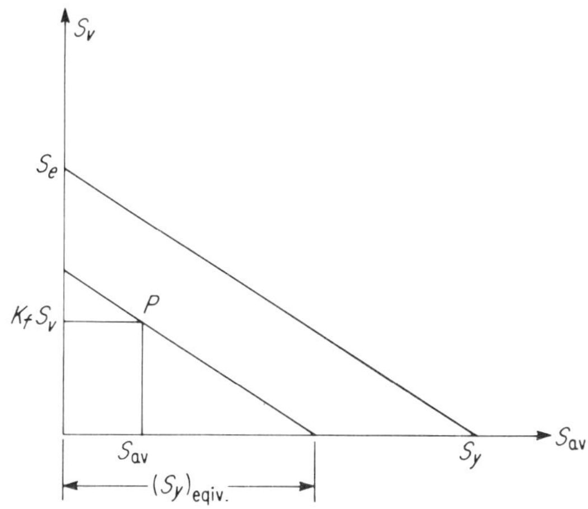

FIG. 15-6. Soderberg diagram.

that the stress concentration has no effect on the average stress value for ductile materials. The fluctuating stress is multiplied by the fatigue stress-concentration factor k_f corresponding to the notch and the material under consideration. The state of stress can therefore be represented by point P in the figure. The factor of safety can therefore be considered as

$$\text{F.S.} = \frac{S_y}{(S_y)_{\text{equiv}}}$$

where $(S_y)_{\text{equiv}}$ = an equivalent static stress

$$= \left[S_{\text{av}} + K_f S_v \left(\frac{S_y}{S_e}\right)\right] \text{ for ductile materials}$$

Values of ultimate strength, yield strength, and endurance limit for some common materials are given in Table 15-1.

TABLE 15-1
Mechanical Characteristics of Some Common Materials

Material	Ultimate Strength, psi	Yield Strength, psi	Endurance Limit, psi
Cast iron			
B.H.N. 180	26,000	—	12,000
Steel			
B.H.N. 140	70,000	30,000	30,000
B.H.N. 200	100,000	68,000	45,000
B.H.N. 400	200,000	170,000	85,000
Ultrahigh-strength steel			
H 11, HT	310,000	245,000	132,000
Copper			
Annealed	32,000	11,000	10,000
Cold-rolled	52,000	45,000	16,000
Aluminum alloy			
A-13 as cast	39,000	21,000	19,000
7075 HT(T6)	83,000	73,000	23,000
Nodular cast iron			
60-45-10 annealed	70,000	55,000	30,000
100-70-03 heat-treated	110,000	80,000	44,000

Cumulative Fatigue

Numerous machines and structures are subjected to variable loading conditions where the stress cycle does not remain the same during the operation of the machine. This can be easily seen in cases such as land vehicles, ships, or aircraft. Several theories are available to predict the endurance life of a member under varying load conditions. Two of the best-known theories of cumulative fatigue under sinusoidal loads of changing amplitudes are explained briefly in the following.

Miner's Theory

This theory suggests that failure occurs when

$$\Sigma \frac{n_i}{N_i} = 1$$

where n_i = number of cycles at a certain stress level
N_i = number of cycles to failure at same stress level obtained from S-N curves for the material under consideration

This rule is very simple to apply but it makes no statement as to the chronological sequence of stress cycles of varying magnitude, although it is a well-known fact that the effect of varying sequence is by no means negligible. Because of its simplicity it is widely used in practice.

Corten-Dolan Hypothesis

A relationship similar to that of Miner for calculations of life under cumulative fluctuating loads is proposed by Corten and Dolan as

$$N_g = \frac{N_1}{a_1 + a_2(S_2/S_1)^d + \cdots + a_i(S_i/S_1)^d}$$

where N_g = total number of cycles of stress for failure with fluctuating stress history
N_1 = number of cycles of stress S_1 for failure
S_1 = highest stress level
$S_2, \ldots, S_i$ = different stress levels
$a_1, a_2, \ldots, a_i$ = ratios of total number of cycles for stress $S_1, S_2, \ldots, S_i$
d = constant for material describing slope of log S versus log N curve, Fig. 15-7.

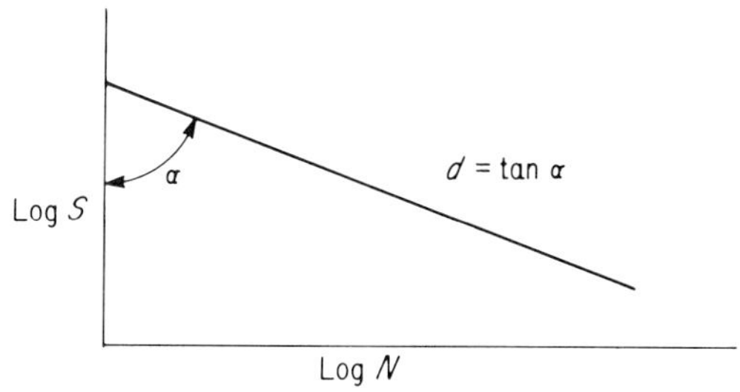

FIG. 15-7

Cumulative Damage Criterion

A hypothesis which takes into account the chronological sequence is also proposed by Corten and Dolan based on a cumulative damage criterion. Fig. 15-8 illustrates this criterion. The figure shows the plot of damage percent versus number of cycles at different stress levels. The 100 percent damage characterizes the appearance of a detectable crack. The diagram shows the total damage resulting from N_1, N_2, N_4, N_3, cycles at stress levels S_1, S_2, S_4, and S_3 in this order.

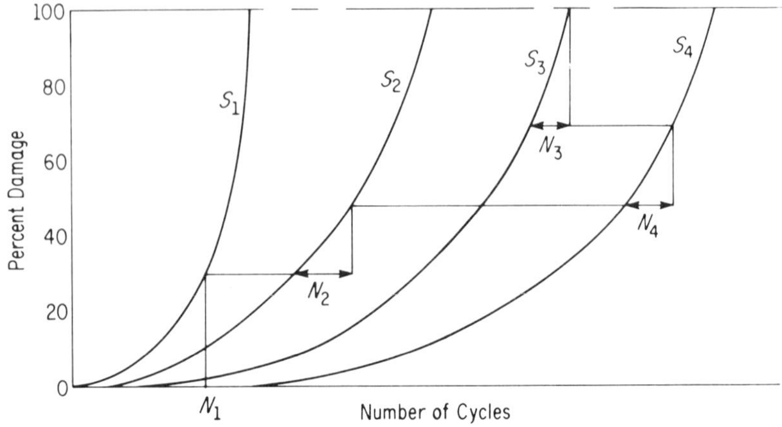

FIG. 15-8. The Corten-Dolan hypothesis.

15-4. PLASTIC STRAIN AND ENERGY CONCEPT FOR FATIGUE

Better correlation of fatigue data is attained when the plastic strain rather than the stress amplitude is used as the criterion for failure. An empirical formula describing the relationship between plastic strain amplitude and life (generally known as the *Coffin-Manson criterion*) is given by

$$\epsilon_p = \epsilon_f(2N_f)^{-C} \tag{15-1}$$

where ϵ_p = plastic strain amplitude as indicated in the hysteresis loop of Fig. 15-9
N_f = number of cycles required to cause failure
ϵ_f = ductility of the material (constant for a particular material)
C = constant which was reported by Tavernelli and Coffin (Ref. 1) to be equal to 0.5 regardless of the material, the temperature of testing (below the creep range), or the manner of testing

It was also found to apply for static testing ($N = \frac{1}{4}$) and for thermal fatigue. The value of $C = 0.58$ is suggested by Halford (Ref. 4).

Accumulated plastic hysteresis energy has been suggested as a criterion for fatigue failure by Feltner and Morrow (Ref. 2), Martin (Ref. 3), Halford (Ref. 4) and Pimbley and Ku (Ref. 5). Halford presented an analysis that agrees with the Coffin-Manson relationship for axial stress in the low and medium-cycle region. The equation for prediction of failure is given by

$$W_f = N_f \Delta W \tag{15-2}$$

where W_f = "fatigue toughness," which is accumulated plastic hysteresis energy at which failure will occur

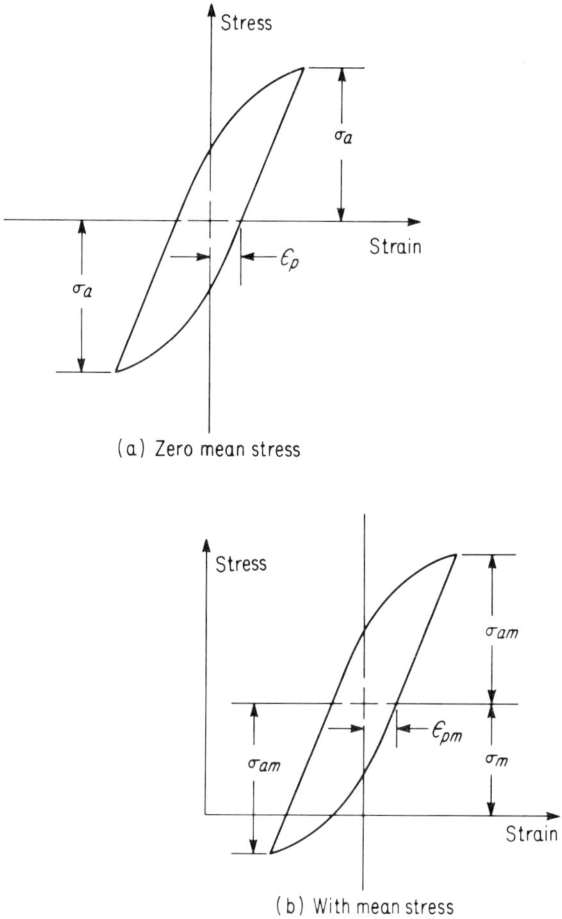

FIG. 15-9. Hysteresis loops.

N_f = number of cycles to failure
ΔW = plastic hysteresis energy per cycle represented by area of hysteresis loop

This equation assumes that the area of the hysteresis loop remains constant throughout the test. This is known to be untrue for the first phase of cycling where there is a change in the loop area as well as in the slope of the stress-strain curve due to strain hardening or softening. It is therefore reasonable to consider ΔW in the previous equation as the plastic hysteresis energy per cycle at half the life of the sample.

The Coffin-Mason criterion can be rewritten in terms of stress by utilizing the stress-strain relationship for the reproducible loop. This is given by

$$\left(\frac{\epsilon_p}{\epsilon_f}\right)^{n'} = \left(\frac{\sigma_a}{\sigma_f}\right) \tag{15-3}$$

where σ_a = stress amplitude
σ_f = fracture stress for material
ϵ_f = ductility of the material
ϵ_p = plastic strain amplitude
n' = a material constant representing a cycling strain-hardening exponent

By substituting in Eq. 15-1, we obtain

$$\sigma_a = \sigma_f (2N_f)^{-Cn'} \tag{15-4}$$

It can be easily seen that Eqs. 15-1, 15-2, and 15-4 predict an exponential increase of life as the stress value decreases. They do not describe situations where an endurance limit exists.

Generalized plastic hysteresis energy criteria are proposed by Pimbley and Ku (Ref. 5) which extend the previous theories to allow for the existence of an endurance limit. It is assumed that if an endurance hysteresis energy ΔW_0 exists, this portion of the plastic hysteresis energy during each cycle is not effective in contributing to fatigue. Equation (2) can therefore be modified as

$$W_f = N_f(\Delta W - \Delta W_0) \tag{15-5}$$

Using the expression derived by Halford (Ref. 4) for the plastic energy,

$$\Delta W = \frac{4(1 - n')}{(1 + n')} \sigma_a \epsilon_p \tag{15-6}$$

an expression for the fatigue toughness can be obtained from Eq. 15-2, 15-3, 15-4, and 15-6 as

$$W_f = 2\frac{(1 - n')}{(1 + n')} \sigma_f \epsilon_f (2N_f)^{1 - C(1 + n')} \tag{15-7}$$

This equation accounts for increased fatigue toughness with number of cycles.

Now by substituting in Eq. 15-5 from Eq. 15-3 and 15-7, a relationship between the applied stress and number of cycles to failure can be derived as

$$\left(\frac{\sigma_a}{\sigma_f}\right)^{(1+n')/n'} - \left(\frac{\sigma_0}{\sigma_f}\right)^{(1+n')/n'} = (2N_f)^{-C(1+n')} \tag{15-8}$$

where σ_0 = endurance limit. This expression approaches the Coffin-Manson relationship for low-cycle fatigue and gives $\sigma_a = \sigma_0$ when N_f is very large.

The effect of mean stress on fatigue can also be included by considering it also to influence the plastic-strain amplitude. Equation 15-3 can be modified for this case as

$$\left(\frac{\epsilon_{p_m}}{\epsilon_f}\right)^{n'} = \left(\frac{\sigma_{a_m}}{\sigma_f - \sigma_m}\right) \tag{15-9}$$

where σ_m = mean stress.

The equation indicates that a tensile mean stress $(+\sigma_m)$ increases ϵ_p and the hysteresis, while a compressive stress $(-\sigma_m)$ reduces both. An expression for including the influence of the mean stress can therefore be written as

$$\sigma_{am} = (\sigma_a)\left(1 - \frac{\sigma_m}{\sigma_f}\right)^{1/(1+n')} \tag{15-10}$$

where σ_{am} = stress amplitude with mean stress
σ_a = stress amplitude causing failure in the same number of cycles N_f with zero mean stress

The term σ_m is considered to have a positive sign when tensile and negative when compressive.

15-5. FATIGUE DUE TO RANDOM LOADS

Since the load in many mechanical systems is random in nature, it is of interest to the system analyst and designer to have some understanding of methods of life prediction for the different components under such conditions. It has been shown in Part V that the response of a mechanical system to random inputs can be evaluated as a mean-square value. The square root of this value can be used to calculate a root-mean-square stress ($\sqrt{\sigma^2}$) for the critical sections of the system components. The design of these components or the prediction of their life can be simplified considerably if this mean-square value of stress can be transformed to an equivalent sinusoidal stress. Such transformation has been developed by Miles (Ref. 29) and is given by

$$\frac{\sigma_r}{\sqrt{\overline{\sigma^2}}} = (\sqrt{2\pi}/2)^{1/\lambda d}\,\sqrt{\lambda d/e}$$

where σ_r = equivalent stress at which failure will occur in sinusoidal vibration after a life equal to the life under the random load

$\sqrt{\overline{\sigma^2}}$ = root-mean-square stress calculated from the system response to random load

d = material fatigue index given by the slope of the log S versus log N curve from sinusoidal stress tests (Fig. 15-7)

e = 2.71828

λ = a load parameter which characterizes the statistical nature of the response (or its power spectrum) which is evaluated from experimental data or field experience

The experimental evaluation of λ can be accomplished by comparing data from a fatigue test with a programmed random load with those from tests with sinusoidal loads on similar specimens.

15-6. DESIGN CONSIDERATIONS

From the previous presentation it can be appreciated why our knowledge of the fatigue phenomena is rather limited. Fatigue tests are usually expensive and time-consuming and the factors involved are too numerous to be easily controlled.

It is fortunate that small decreases in the peak stresses developed in a member under a given set of conditions will result in large improvement in the fatigue life. Thus an effective engineering solution to the problem is not impossible.

Many of the chronic cases of fatigue failure in service are traceable to poor design details or to processing mistakes or service-induced defects that result in conditions well known to be poor from the fatigue viewpoint.

Even though exact procedures of a mathematical nature for proportioning parts are not available, the elimination of severe stress raisers and care in processing operations will do much to prolong fatigue life and give satisfactory service.

As long as the metal is sound and free from inherent defects, selection of material within wide limits is not as important in achieving fatigue life as is the case in design, fabrication, and maintenance. For members containing severe stress raisers, it is sometimes found that only *minor* improvements can be expected by changing to a "higher-strength" metal, except under unique conditions such as those involved in operations at extremely high temperatures.

Quite often the elimination of harmful conditions is found to be a more economical and effective means of improving fatigue strength. In designing a machine part, the following should be borne in mind.

1. Eliminate severe stress raisers by providing smooth gradual transitions of stress.

2. Avoid sharp surface tears from rough machining, punching, stamping, shearing, etc.

3. Prevent surface decarburization during processing or heat treatment.
4. Control or afford protection against corrosion, erosion, or chemical attack in service.
5. Alter design, if possible, to eliminate press fits, dovetails, or other mating parts that develop a microscopic chafing action in transferring the load that seriously reduces fatigue strength.

15-7. FAILURE UNDER IMPULSIVE LOAD

Impulsive stresses are developed in an elastic body when it collides with another body. Stress waves are generated at the impact location due to velocity differences and are propagated throughout the body as dilatational, transverse, and shear waves. These waves are subjected to reflection, refraction, and defraction at the boundaries and undergo attenuation due to the internal friction in the materials. Reflections from the boundaries can give rise to numerous waves which may interfere with one another as they propagate throughout the body. As has been shown in Part IV, transverse and dilatational waves travel at different velocities and their reflection at the different boundaries can produce complex stress patterns which change with time.

The problem is further complicated by the fact that stress waves change the sign of stress upon reflection from free boundaries while they maintain the same sign when reflected from fixed boundaries. Dilatational waves upon reflection from a free surface may also produce transverse waves as well (Ref. 24). Impulsive loads which usually produce extremely high stresses at very rapid rates may change significantly the mechanical properties of the material. This change is more pronounced when plastic deformations or cracks develop.

The particle velocity at which the ultimate stress is reached and fracture occurs is called the *critical velocity* and can be calculated from

$$V_u = \int_0^{\epsilon_u} \sqrt{\frac{E_\epsilon}{\rho}}\, d\epsilon = \int_0^{\epsilon_u} \sqrt{\frac{(d\sigma/d\epsilon)}{\rho}}\, d\epsilon \qquad (15\text{-}11)$$

where E_ϵ = slope of the stress strain curve
ϵ_u = ultimate strain

Similarly, the velocity that will produce a plastic strain is calculated from

$$V_p = \int_0^{\epsilon_p} \sqrt{\frac{(d\sigma/d\epsilon)}{\rho}}\, d\epsilon$$

where ϵ_p is the strain which will produce plastic deformation.

It can be seen from Eq. 15-11 that the critical velocity of impact in the case of a tensile bar is a material property and is independent of the cross-sectional area of the bar or the mass of the striker. Values of the critical velocity for some common materials are given in Table 15-2.

TABLE 15-2
Dynamic Properties of Materials*

Material	Ultimate strength, psi		Ultimate strain, in./in.		Critical velocity, fps	
	Static	Dynamic	Static	Dynamic	Experimental	Theoretical
SAE 1015						
Annealed	50,600	63,500	0.28	0.30	200	—
SAE 1022						
Cold-rolled	84,000	105,000	0.06	0.15	100	95
SAE 1040						
Annealed	78,050	91,800	0.204	0.207	200	—
SAE 1045						
Quenched and tempered	142,900	169,000	0.057	0.092	190	88
SAE 2345						
Quenched and tempered	145,250	175,250	0.084	0.141	200	161
SAE 4140	134,250	151,000	0.085	0.147	175	132
SAE 5150	139,000	148,100	0.085	0.133	170	159
Type 302 stainless steel	93,300	110,800	0.585	0.466	200	490
Copper						
Annealed	29,900	36,700	0.327	0.438	200	231
Cold-rolled	45,000	60,000	0.025	0.107	50	42
2S Aluminum, annealed	11,600	15,400	0.23	0.30	200	176
24S-T aluminum alloy	65,150	68,600	0.113	0.135	200	290
Magnesium alloy	43,750	51,360	0.096	0.109	200	303

*From D. S. Clark, "The Behavior of Metals Under Dynamic Loading," *Metal Progress*, November 1953, p. 67.

15-8. FRACTURE DUE TO IMPACT

The peak stresses producing failure under impulsive loads are influenced by factors which are different from those affecting the stress distribution under static or alternating loads. The most important difference is that under conventional loads each part of the loaded body has an influence on the stresses and deformation at all other locations in the body. On the other hand, under impact, stresses and deformations can occur in one part of the body independent of the conditions in the rest of the body. Furthermore, the density and elastic modulus have a marked influence on the stress patterns resulting from impulsive loads but do not affect the stress distribution under conventional loads. Random failures which must be treated statistically occur more frequently under static loading than under impulsive loading.

Some of the common modes of failure under impulsive loading are explained in the following.

Scabbing

This is a type of fracture that occurs near the bottom of a plate when it is subjected to an impact on its surface. Interference between the incident and reflected waves can cause the tensile stress to be built up a short distance from the bottom of the plate to values which are high enough to cause fracture. This type of failure was first observed by Hopkinson and is sometimes called *Hopkinson fracture*.

When the level of stress within the wave is very high (more than double the critical normal fracture stress of the material) multiple scabbing may occur.

Corner Fracturing

This is one of the most common types of failure under shock loads. Fracture occurs due to the high concentration of stress near the intersection of two free boundaries when a compressive wave with high stress value reaches it.

Fracturing Associated with Trajectories of Maximum Shear

Many examples of fractures have been observed to develop along the shear trajectories. Most cracks when formed propagate at velocities well below the wave-propagation velocities. The maximum velocity of crack propagation under impulsive loads is approximately 5000 fps.

15-9. FATIGUE DUE TO REPEATED IMPACT

Experimental data reported by Davidenkev and Belyaeva (Ref. 28) on test of alloy steels subjected to 100,000 repeated bending impacts showed that the fatigue strength was 5 to 8 percent higher than that under conventional loading when the specimens were tempered at 600°C. It was also found to be 8 to 12 percent lower when the specimens were tempered at 200°F. The peak stresses were measured by resistance-type strain gages.

15-10. DYNAMIC BEHAVIOR OF RUBBER AND SYNTHETIC MATERIALS

In spite of the increased utilization of rubber and synthetic materials in many components of mechanical systems, there is rather limited experimental data on the dynamic characteristics of these materials. A comprehensive discussion of the dynamic properties and structural changes of rubber with load can be found in Ref. 35. It is known that rubber undergoes a gradual reduction in modulus of elasticity with number of load cycles which is a form of stress relaxation. The fatigue life up to failure is known to decrease rapidly with increases in static deformation.

In fatigue testing of rubber and plastics, considerable difficulties have been encountered with the high damping capacity of these materials and the resulting problems of specimen-temperature control. A testing machine developed by Pringle and Harker (Ref. 36) applies vibratory fluid loading to multiple cantilever specimens. The fluid loading facilitates temperature control and makes possible high testing frequencies.

Data on the dynamic behavior of rubber and synthetic materials can be found in Refs. 35 to 42.

15-11. MECHANICAL CHARACTERISTICS OF BIOLOGICAL MATERIALS

There has been considerable interest in recent years in the static and dynamic behavior of biological materials. References 43 to 52 represent some of the recent contributions to the literature on the mechanical characteristics of bone, muscle, skin and ligaments.

Tests conducted on samples taken from various individual bones to determine their mechanical characteristics show that these characteristics depend on the age of the individual, bone form, and whether the bone is fresh or embalmed. For compact bones the average values of tensile strength, compressive strength, and modulus of elasticity are 12,000 psi, 18,000 psi, and 2.25×10^6 psi respectively. (See Table 15-3). Dry samples have linear stress-strain relationship to rupture. Wet samples exhibit a softening stress-strain curve which is indicative of better energy absorption.

Static and dynamic tests of complete long bones show that bones are weaker in tension than in shear. Accordingly, a torsional load produces a spiral fracture as frequently seen in skiing accidents.

Deformation of skulls under blows was the subject of extensive studies by H. R. Lissner and coworkers at Wayne State University. Most of the tests utilized cadavers or dry skulls. Results from about 1200 tests show that when the skull is subjected to a blow some regions bend inward and others outward simultaneously. A blow to the side of the head well

TABLE 15-3
Passive Mechanical Characteristics of Bone and Soft Tissue

	Soft Tissue	Compact Bone	
		Fresh	Embalmed
Density lb/in^3	0.036–0.043	0.07–0.072	0.0713
Modulus of Elasticity lb/in^2	1.1 Low Loads	3×10^6 Low Loads	2.25×10^6
Bulk Modulus lb/in^2	3.8×10^5 Low Loads		1.9×10^6
Shear Modulus lb/in^2	0.36 Low Loads		1.0×10^6
Tensile Strength lb/in^2	47 soft tissue 630 skin	1.2×10^4	1.3×10^4
Shear Viscosity dynes-sec/cm^2	1.5×10^2		

below the ear would produce a fracture in the temporal region on the same side in 85 percent of the cases. Fractures are detected by the opening of the circuit of the 4-R strain gages cemented on the skull. A complete Atlas has been developed and used by radiologists in examining X-rays for fractures when the site of a blow was known. Also from the legal point of view the site of a blow can be inferred for medical legal purposes.

Tests on cadavers show that the energy required to cause a small linear fracture is absorbed in 0.001 sec and varied between 200–1000 in-lb. with an average of 500 in-lb. Very little additional energy beyond this would produce a very severe multiple fracture and depression. These results are in good agreement with tests on dry skulls.

High speed impact tests are carried out by dropping steel balls of different weights down an elevator shaft on the skull. For impact velocities smaller than 45 ft/sec linear fractures occur. At 90 ft/sec clean round punctures result. Hammer blows at intermediate velocities produce depression under the head of the hammer with some linear fracture surrounding the depressions. In order to find the validity of tests made on cadavers and dry skulls stress coat and strain gages are used on anaesthetized dogs and monkeys. The crack pattern is obtained after delivering a hammer blow to the skull and then the animal is killed. Several days later the old lacquer coating is removed and the identical impact repeated. Finally the skull is dried out and the same tests repeated. A comparison of the crack pattern indicates that while the extent of the deformation differs the patterns obtained in all three tests are similar. The mechanical characteristics of bone and its constituents (Collagen and Apatite) have been investigated by Kroon, Sweeney, and Byers (Ref. 57).

Most soft tissue in compression resemble fluids in their mechanical properties but in shear they approximate stiff non-linear gels with internal losses. Studies on the resistance of skin to puncture by Gadd et al (Ref. 46) show that pigskin and human skin appear to have the same mechanical characteristics in tension compression and shear. The rheological properties of ligaments are investigated by Haut and Little (Ref. 49). Other studies also show viscoelastic behavior for bones, muscle, cartilage blood, brain tissue . . . etc.

The dynamic response of bone and muscle tissue to impacts of varying duration is studied by Sweeney and Byars (Ref. 45). The behavior of bone and muscle "in-vivo" under cyclic loading show both degenerative and regenerative effects. While increased muscle capacity can result from exercise, chronic overstressing causes them to fatigue and to stop contracting. Also, if they are not used, they tend to shrink and waste away. Muscles dissipate considerable energy during their load cycle and are therefore an important source of heat.

Tests by Seireg and Kempke (Ref. 52) on the tibia of the live rat indicate that the regenerative process due to cyclic load may be influenced by psychological and physiological as well as diet effects. These tests also show that a well defined endurance limit exists at approximately 42 percent of the static fracture load. The fatigue fractures of the tested bones occur quite suddenly and are of the brittle type.

15-12. SUMMARY

In this chapter a brief discussion of material behavior under dynamic loads has been presented. The objective is to introduce the reader to some of the fundamental concepts necessary for material selection for systems subjected to loads which are periodic, random, or impulsive in nature.

REFERENCES

1. J. F. Travernelli and L. F. Coffin, "A Compilation and Interpretation of Cyclic Strain Fatigue Tests on Metals," *Trans. ASM* Vol. 51 (1959), p. 438.
2. C. E. Feltner and J. D. Morrow, "Microplastics Strain Hysteresis Energy as a Criterion for Fatigue Fracture," *Trans. ASME, J. Basic Eng.*, Vol. 83, Ser. D (1961), pp. 15–22.
3. D. E. Martin, "An Energy Criterion for Low-Cycle Fatigue," *Trans. ASME, J. Basic Eng.*, Vol. 83, Ser. D (1961), pp. 565–571.
4. G. R. Halford, "The Energy Required for Fatigue," *J. Matl.*, Vol. 1, No. 1 (March 1966), pp. 3–18.
5. W. T. Pimbley and T. C. Ku, "Design for Fatigue Based on Energy Concept," *ASME* Paper No. 68-DE-17, 1968.

6. L. E. Coffin, Jr., "Low Cycle Fatigue: A Review," *J. Appl. Matl. Res.*, Vol. 1, No. 3 (October 1962), pp. 129–141.
7. S. S. Manson, "Fatigue, A Complex Subject—Some Simple Approximations," *Exptl. Mech.*, Vol. 5, No. 7 (July 1965), pp. 193–226.
8. D. E. Martin and J. Brinn, "Some Observations on the Plastic Work Required to Fracture Stainless Steel Under Cyclic Loading," *Proc. ASTM*, Vol. 59 (1959), pp. 677–690.
9. C. S. Chang et al., "An Analysis of Metal Fatigue Based on Hysteresis Energy," *Experimental Mechanics*, Vol. 8, No. 3 (1968).
10. T. J. Dolan and C. S. Yen, "Some Aspects of the Effect of Metallurgical Structure on Fatigue Strength and Notch-Sensitivity of Steel," *Proc. ASTM*, Vol. 48, No. 0 (1948), pp. 664-695.
11. J. Marin, *Mechanical Behavior of Engineering Materials.* Englewood Cliffs, N.J.: Prentice-Hall, 1962, pp. 142–144.
12. J. Marin, "Interpretation of Fatigue Strengths for Combined Stresses," *Proc. Int. Conference on Fatigue of Metals*, London Inst. of Mech. Engrs., 1956, pp. 184–194.
13. R. E. Peterson, "Torsion and Tension Relations for Slip and Fatigue," *Proc. Colloquium on Fatigue, Stockholm, May 25–27, 1955.* Berlin: Springer, 1956, pp. 186–196.
14. H. Neuber, *Theory of Notch Stresses.* Ann Arbor, Mich.: Edwards, 1946.
15. P. Kuhn and H. F. Hardrath, "An Engineering Method for Estimating Notch-Size Effect in Fatigue Tests on Steel," NACA Technical Note 2805, National Advisory Committee for Aeronautics, Washington, D.C., 1952.
16. P. Kuhn, "Influence des dimensions sur la fatigue des pièces entaillées," *Rev. Met.*, Vol. 9 (1958), pp. 860–864.
17. E. Z. Stowell, "Stress and Strain Concentration at a Circular Hole in an Infinite Plate," NACA Technical Note 2073, National Advisory Committee for Aeronautics, Washington, D.C., 1950.
18. K. Gunn, "Effect of Yielding on the Fatigue Properties of Test Pieces Containing Stress Concentrations," *Aeron. Quart.* (1955), pp. 277–294.
19. W. N. Findley, "A Theory for the Effect of Mean Stress on Fatigue of Metals under Combined Torsion and Axial Load or Bending," *Trans. ASME, J. Eng. Ind.*, Vol. 81 (1959).
20. R. R. Gatts, "Cumulative Fatigue Damage with Random Loading," *Trans. ASME, J. Basic Eng.*, Vol. 83D, No. 4 (1961).
21. W. B. Bishop, "Use of Failure Statistics in Design," *Electro-Technol.* (July 1963), pp. 97–99.
22. G. Kardos, "Design Criteria for Generalized High Strain," *Trans. ASME, J. Eng. Ind.*, Vol. 90B, No. 3 (1968).
23. A. Esin, "The Microplastic Strain Energy Criterion Applied to Fatigue," *Trans. ASME, J. Basic Eng.*, Vol. 90D, No. 1 (1968).
24. H. Kolsky, *Stress Waves in Solids.* London: Oxford U.P., 1953.
25. J. S. Rinehart and J. S. Pearson, *Behavior of Metals Under Impulsive Loads.* New York: Dover, 1954.
26. C. M. Zener and J. H. Hollomon, "Effect of Strain Rate upon Plastic Flow in Steel," *J. Appl. Phys.*, Vol. 15 (1944).

27. W. P. Welch, "Plates and Shells," Chap. 8 in O. J. Horger (ed.), *ASME Handbook—Metals Engineering: Design.* New York: McGraw-Hill, 1965.
28. N. N. Davidenkov and E. I. Belyaeva, "Study of Impact Fatigue Strength," *Metalloved i Obrabotka Metal,* No. 9 (September 1958), Brutcher trans. No. 4349.
29. J. W. Miles, "On Structural Fatigue under Random Loading," *J. Aeron. Sci.,* Vol. 21, (1952).
30. W. Starkey and S. Marco, "Effect of Complex Stress-Time Cycles on the Fatigue Properties of Metals," *Trans. ASME,* Vol. 79, (1957).
31. C. R. Soderberg, "Factors of Safety and Working Stress," *Trans. ASME* Vol. 52, No. 1 (1930).
32. R. E. Peterson, *Stress Concentration Design Factors*, New York: Wiley, 1953.
33. M. A. Miner, "Cumulative Damage in Fatigue," *J. App. Mech.* Vol. 12, A159–A164 (1954).
34. H. Corten and T. Dolan, "Cumulative Fatigue Damage," paper No. 2, Session 3, Int. Conf. on Fatigue of Metals, Inst. Mech Engrs., London (1956).
35. A. B. Davey and A. R. Payne, *Rubber in Engineering Practice.* New York: Palmerton Publishing Co., 1964.
36. O. A. Pringle and R. J. Harker, "Environmental Fatigue Testing of Molded Plastics for Prosthetic Heart Valves," *Experimental Mechanics,* Vol. 9, No. 4 (1969).
37. G. J. Lake, P. B. Lindley, "The Mechanical Fatigue Limit for Rubber," *J. App. Polymer Sci.,* Vol. 9, pp. 1233–1251 (1965).
38. D. Telfair, C. H. Adams and H. W. Mohrman, "Creep, Long-Time Tensile and Flextural Fatigue Properties of Melamine and Phenolic Plastic Materials," *Trans. ASME,* Vol. 69, pp. 789–793 (1947).
39. B. B. Pusey, "Flextural Fatigue Strengths of Reinforced Thermosetting Laminates," *Plas. Technol,* Vol. 3, No. 9 pp. 721–726 (1957).
40. E. Hagerup, "Flextural Fatigue Testing of Polyesters," *J. App. Polymer Sci.,* Vol. 7, No. 3 pp. 1093–1116 (1963).
41. R. H. Carey, "Fatigue Testing of Nonrigid Plastics," *ASTM Bulletin* No. 206, pp. 52–54 (1955).
42. K. H. Boller, "Resume of Fatigue Characteristics of Reinforced Plastic Laminates Subjected to Axial Loading," in *Fatigue—An Interdisciplinary Approach.* Syracuse, N. Y.: Syracuse Univ. Press, (1964), pp. 325–341.
43. H. R. Lissner, "*Biomechanics—What Is It?*" *Mechanical Engineering,* Vol. 85, No. 1, January 1963.
44. L. M. Prarick, "Biomechanics—A Multidisciplinary Field for Engineers," *Proceedings of the ASEE,* Vol. 71, 1963–1964.
45. J. H. McElhaney and E. F. Byars, "Dynamic Response of Biological Materials," *ASME* paper no. 65-WA/HUF-9 (1965).
46. C. W. Gadd, W. A. Lange and F. J. Peterson, "Strength of Skin and Its Measurement," ASME paper no. 65-WA/HUF-8 (1965).
47. R. P. Kroon, A. W. Sweeney and R. K. Byers, "Mechanical Characteristics of Bone and Its Constituents," ASME paper no. 65-WA/HUF-7 (1965).
48. D. D. Webster, "A Method of Measuring the Dynamic Characteristics of Muscle Rigidity, Strength, and Tremor in the Upper Extremity," *IRE Transactions on Medical Electronics* (1959), p. 159.

49. R. Haut and R. W. Little, "Rheological Properties of Canine Anterior Cruciate Ligament," ASME paper No. 69-BHF-9 (1969). Also Jour. of Biomechanics Vol. 2 (1969).
50. G. H. Bell, "Bone as a Mechanical Engineering Problem," in *The Biochemistry and Physiology of Bone,* New York: Academic Press, Inc. (1956).
51. F. G. Evans, *Stress and Strain in Bone,* Springfield, Illinois: Charles C. Thomas (1957).
52. A. Seireg and W. Kempke, "Behavior of 'In-vivo' Bone Under Cyclic Loading, "ASME publication No. 69-BHF-8 (1969). Also *Jour. of Biomechanics,* Vol. 2 (1969).

part VIII

Introduction to Optimum Design of Dynamic Systems

To know how to appreciate, in each case, at their true value, the considerations of convenience and economy which may present themselves; to know how to discuss the more important of those which are only accessories; to balance them properly against each other in order to obtain the best results by the simplest means; such should be the leading characteristics of the men called to direct and coordinate among themselves the labor of their comrades, to make them cooperate towards one useful end, of whatsoever sort it may be.

S. Carnot (1824)

chapter 16

Introduction to Optimum Design of Dynamic Systems

16-1. INTRODUCTION[1]

In the previous chapters a comprehensive treatment of mathematical modeling and analysis of dynamic systems has been presented. A brief review of the behavior of system materials under different types of stress regimes is also given. This provides the necessary information which governs the design of mechanical systems when subjected to known dynamic excitations.

The equations describing the behavior of such systems are usually complex in nature and the influence of the different parameters on the system dynamics is not generally expressible in an explicit manner. Accordingly, designers have to rely on intuitive or empirical guides for the selection of the main parameters. The system behavior is then checked to insure the feasibility of the design, and modifications are introduced whenever necessary.

The development of high speed computers and related mathematical sciences opened new horizons for the designers of mechanical systems. In many cases computer-aided design is limited to analytically evaluating, comparing and modifying different solutions which are obtained from experience and intuitive conception. Recent trends in computer utilization are moving toward completely automating the design process. The designer is expected to develop the procedure in which an optimal system is exercised by the designer through the establishment of the governing criteria. Many of the judgment aspects of the design can be quantified so that they can be handled by computers. Thus the utilization of the computer can be extended beyond the mere analytical function of predicting the behavior of trial designs. Procedures are devised to allow the computer to continually modify a trial design until the optimum solution is found.

[1] This chapter is based on a study by A. Seireg and B. Hamad at the University of Wisconsin 1966.

This chapter is intended as an introduction to the synthesis of vibratory systems. The design of torsional systems is considered as an example. A three-rotor system is selected for illustration since many power transmitting systems can be modeled as such. The rotors represent the prime mover, the connecting elements, and the driven machine respectively. Damped and undamped systems are investigated, where one of the masses is subjected to a pulse, a step load, and sinusoidal load. In the case of the damped systems, normal mode analysis is utilized, and Rayleigh damping is assumed in order to simplify the analysis. The objective is to develop the design methods for coupled systems with a minimum expenditure of computing time.

16-2. THE DESIGN STRATEGY

Rational design procedures require formulating the problem under consideration in mathematical terms and developing the logic and decision steps necessary to facilitate the search for optimum solutions. Many aspects of mechanical elements and systems are rather difficult to describe in a mathematical form. It is also difficult in many cases to express mathematically a general merit function by which a single optimum solution with the highest possible merit value can be attained. Mechanical systems in general can be classified as poorly structured systems where the search may not necessarily result in one optimum solution but in several good solutions. The examples discussed in this chapter utilize a unified plan of attack which is suitable for the systems under consideration. The plan can be briefly described in the following steps:

a. Defining the independent parameters of the problem.

b. Mathematically defining the constraints which are the limits imposed on each of the system parameters or any particular combination of them. These constraints generally define the available space, the material characteristics, and manufacturing limitations.

c. Stating the functional constraints which are the physical laws governing the behavior of the element or system under consideration.

d. Developing a suitable criterion for decision which mathematically expresses the objective of the designer. It is, therefore, possible that this criterion varies according to the goal set by the individual designer or his judgment.

e. Developing effective search techniques for systematic investigation of feasible solutions for the design with the highest possible merit value.

The first three steps deal with the description of the design domain which includes all feasible solutions. The boundaries of this domain are defined by the constraints. The last two steps define the objective and the search strategy to attain this objective. These steps will be followed in all given examples.

Example 1

Three-Rotor System Subjected to an Impulse (Fig. 16-1). In this example, the problem is to design the shafting connecting a prime mover

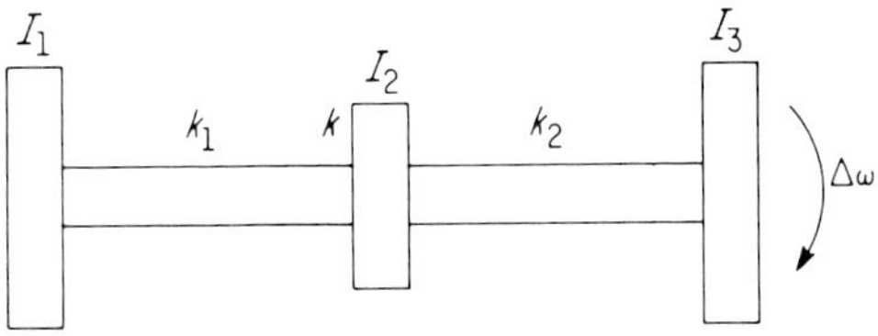

FIG. 16-1

(with moment of inertia I_1) to a machine (with moment of inertia I_3) through a coupling (with moment of inertia I_2). The machine in this case is subjected to an impulse causing a sudden change in velocity $\Delta\omega$. It is required to select the diameter of the two connecting shafts so that the total shaft volume is minimum. The allowable torsional stresses for the shafts are given as S_1 and S_2 psi respectively. Inertias may be added to the coupling (within a maximum value of $I_{2(\text{max})}$) if a considerable reduction in shaft size can be expected. It will be assumed that no limitations are imposed on the angular displacements of the different rotors.

The significant parameters can be tabulated as follows:

Given Inputs

$$I_1, I_3, I_{2\min}, I_{2\max}$$
$$L_1, L_2$$
$$S_1, S_2$$
$$\Delta\omega$$

Required Outputs

$$d_1, d_2, I_2$$

Parameter Constraints

$$d_1 > 0$$
$$d_2 > 0$$
$$I_{2\max} \geq I_2 \geq I_{2\min}$$

Functional Constraints

The problem is then formulated by first establishing the equations governing the system design. These equations for the case under consideration can be written as

1. The dynamic shaft torques:

$$T_1 = k_1[A_1(a_{21} - 1)\sin\omega_1 t + C_1(a_{22} - 1)\sin\omega_2 t]$$
$$T_2 = k_2[A_1(a_{31} - a_{21})\sin\omega_1 t + C_1(a_{32} - a_{22})\sin\omega_2 t]$$

The derivation of these equations is given in Chapter 8.

2. Maximum Torsional stress equation for both shafts:

$$\tau = \frac{T}{\frac{\pi}{16} d^3}$$

Objective Function

It is now necessary to select a *Merit Criterion* which describes the designer's objective and can be used to rate the different designs and guide the search for an optimal solution.

The designer's objectives in this example are assumed to be: (a) to minimize the total volume of shafting, (b) to bring the stresses in both shafts as close as possible to the allowable values, and (c) to add inertia within a given maximum limit if a considerable improvement may result from it.

A mathematical expression which describes this objective can be written as

$$\text{Minimize } U$$

where

$$U = [L_1 d_1^2 + L_2 d_2^2] \sqrt{\left(\frac{\tau_1}{S_1} - 1\right)^2 + \left(\frac{\tau_2}{S_2} - 1\right)^2} \; [I_2^n] \qquad (16\text{-}1)$$

This function serves as the merit criterion. The three brackets correspond to the a, b and c objectives respectively. The power n to which the second inertia is raised represents the relative emphasis placed on the reduction of volume and the inertia reduction.

It should be noted here that Eq. 16-1 is not the only form which satisfies the designer's objective. Many expressions can be constructed to accomplish this purpose. It is this step of the design procedure which requires the designer's greatest attention since it establishes the criterion which determines the final design. In many cases the objective function is individual and draws on the particular circumstances of the designer and his experience. An alternate expression which satisfies the design objective can be given by:

$$\text{Minimize } U = [L_1 d_1^2 + L_2 d_2^2] + W_1\left[\left|\frac{\tau_1}{S_1} - 1\right| + \left|\frac{\tau_2}{S_2} - 1\right|\right] + W_2[I_2]$$

In this case the summation of the sub-objectives is used instead of their multiplication. The constants W_1 and W_2 are *Weight Factors* which can be selected to describe the relative emphasis placed on the different sub-objectives. The middle brackets in both expressions of the objective function give alternate means of bringing the shaft stresses as close as possible to their allowable values.

The Search for Optimal Solution

The problem under consideration is now formulated and the merit function mathematically defined. It is therefore necessary for the designer to find the design parameters which give the lowest possible value for the merit function. The search for maxima and minima has long received considerable attention from the mathematical and physical scientists. Classical techniques such as differential calculus and calculus of variation methods can be used for single and multiple parameter problems respectively where the objective function is continuous and with a well-defined optimal value. Many design problems deal with design space (which is the multiple parameter space containing all feasible solutions) which is constrained in a complex manner and may have several disconnected regions. It is therefore necessary, for efficient search in design problems, to develop a systematic procedure which is well-suited to the problem at hand. The search program developed for this problem is shown schematically in Fig. 16-2. A starting point is selected for input to the program and the corresponding value of the objective function is calculated. A *Gradient Search* (or *Steepest Descent*) is then used to determine a new point. This type of search is illustrated in Fig. 16-3. This is done by selecting incremental changes in each design parameter which are proportional to the partial derivative of the objective function with respect to the parameter. This can be expressed mathematically as

$$(x_i)_{n+1} = (x_i)_n - \Delta \frac{\partial U}{\partial x_i}$$

where x_i represents the different design parameters and Δ is a selected increment which controls the step size.

The new point is then checked for constraints and the parameters adjusted if necessary to bring it into the design domain. If any of the parameter constraints are violated, the parameter in question is forced to be equal to the nearest constraint. This is continued until the merit rating ceases to improve. At such point a *Univariate Search* is started. In this type of search, the design parameters are changed one at a time in the direction of improving merit as shown in Fig. 16-4. In the given example, each parameter except for the diameters is varied in both the positive and negative directions in incremental steps and the new points checked against a secondary merit criterion $V_n = \tau_1 + \tau_2$ which is selected to guide the search towards lower values of shaft stresses. This is repeated until no further improvement is possible or a constraint is reached. The gradient search is then reactivated and so on until the optimum point with the lowest possible value of U is reached. This is usually accomplished when τ_1 and τ_2 reach the allowable values S_1 and S_2.

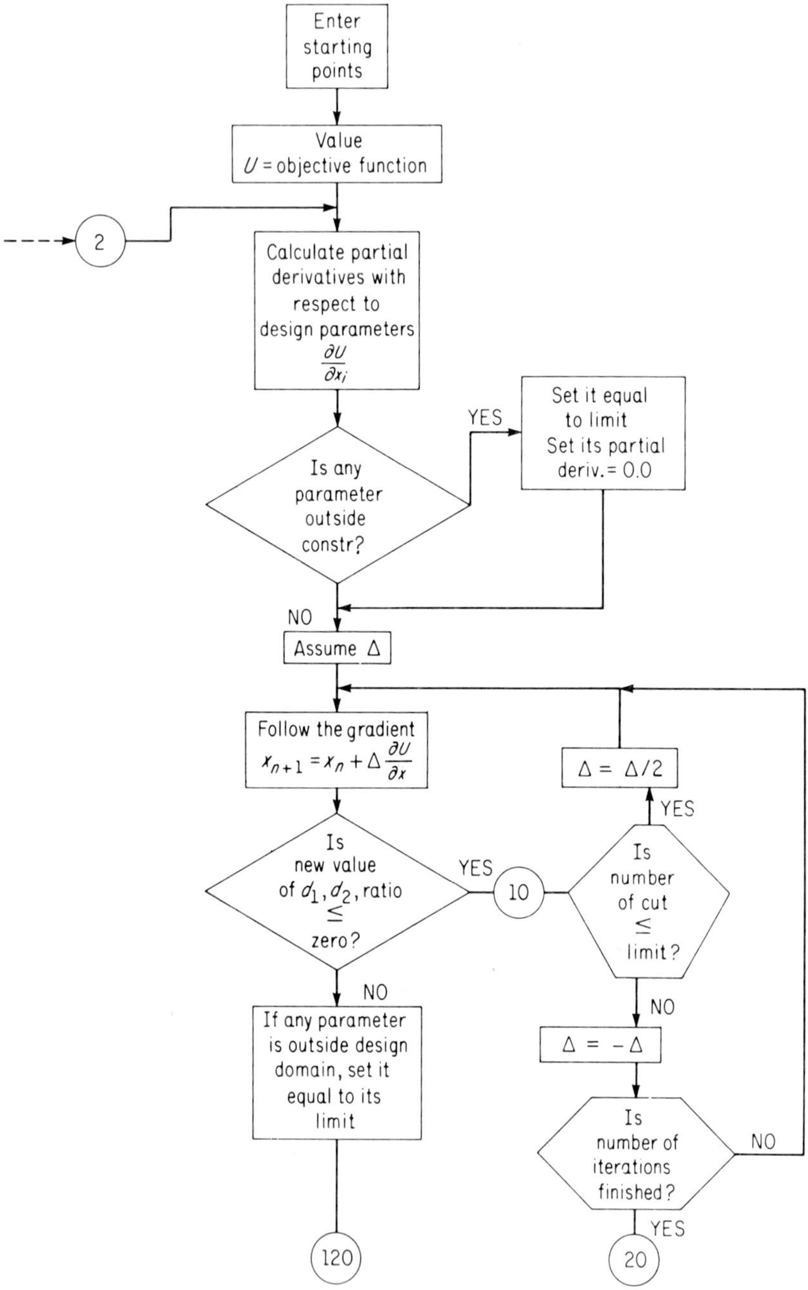

(a)

FIG. 16-2. Flow chart of gradient–univariate search.

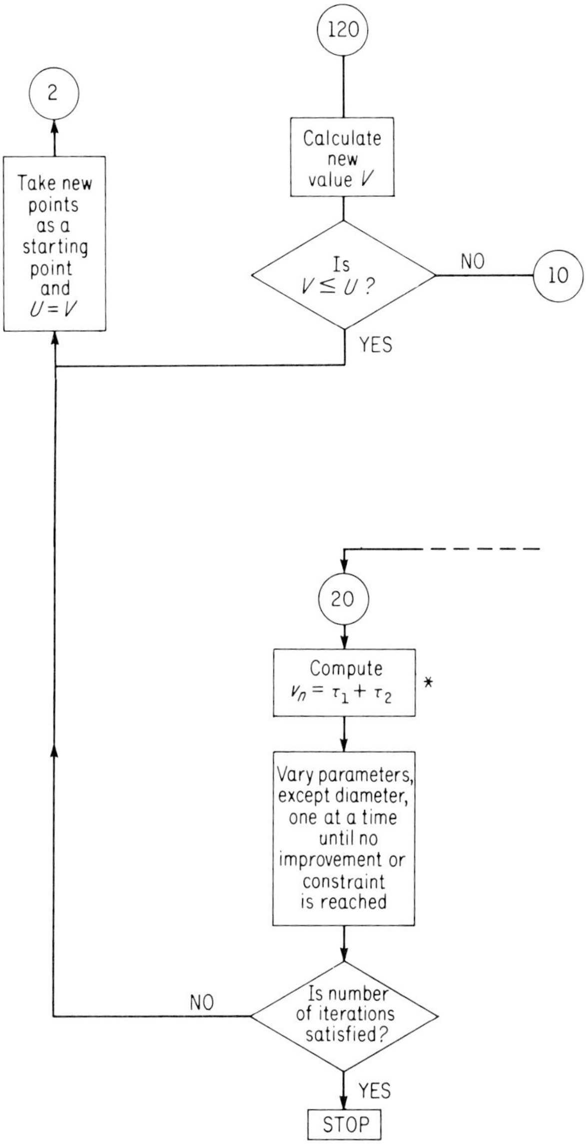

(b)

FIG. 16-2 (continued)

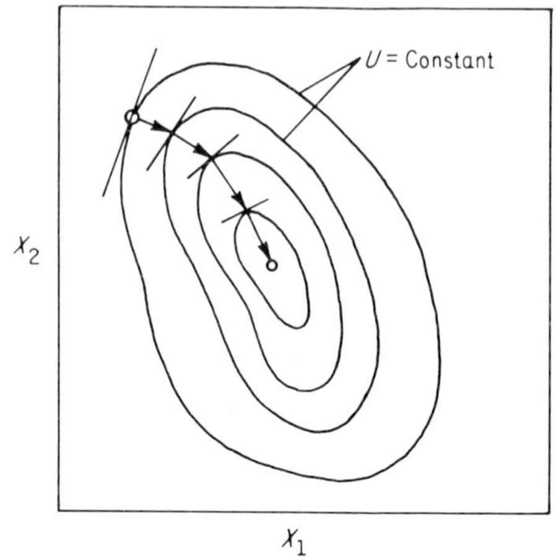

FIG. 16-3. Gradient search.

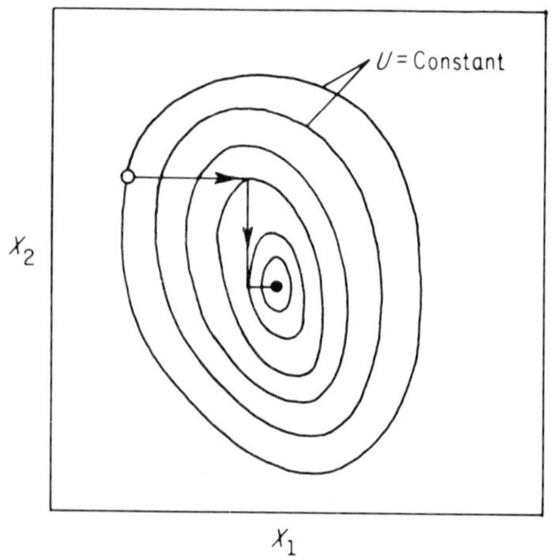

FIG. 16-4. Univariate search.

Sample problems are calculated to illustrate the previous program and the results are shown in Table 16-1 and Fig. 16-5. The input parameters in these examples are taken as

$$I_3 = 500 \text{ lb in. sec}^2$$
$$I_1/I_3 = 0.2, 0.4, 0.6, 0.8, 1.0$$
$$I_{2\,\min} = 10, 100 \text{ lb in. sec}^2$$
$$I_{2\,\max} = 200 \text{ lb in. sec}^2$$
$$L_1 = L_2 = 24 \text{ in.}$$
$$S_1 = S_2 = 50{,}000 \text{ psi}$$
$$\Delta\omega = 1.0 \text{ rad/sec}$$

TABLE 16-1
Optimum design for system subjected to an impulse producing a sudden change of velocity of I_3 equal to one radian per second

I_2	I_1/I_3	d_1	d_2	S_1	S_2	Shaft Volume
10	0.2	0.1706	0.1722	49770	50060	1.1
	0.4	0.2131	0.2141	49990	49970	1.71
	0.6	0.2398	0.2401	49930	50050	2.16
	0.8	0.2584	0.2585	49960	50050	2.5
	1.0	0.2723	0.2724	50010	50000	2.78
100	0.2	0.2328	0.2506	49990	49960	2.19
	0.4	0.2697	0.2781	49990	50040	2.81
	0.6	0.2936	0.2971	49750	50080	3.27
	0.8	0.3103	0.3129	50040	49730	3.64
	1.0	0.3225	0.3229	50050	49910	3.9
0	0.2	0.145	0.145	50000	50000	0.79
	0.4	0.189	0.189	50000	50000	1.335
	0.6	0.216	0.216	50000	50000	1.755
	0.8	0.235	0.235	50000	50000	2.06
	1.0	0.25	0.25	50000	50000	2.34

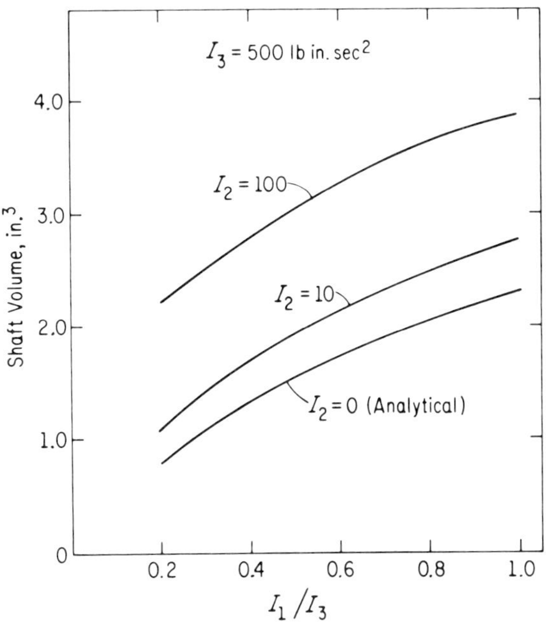

FIG. 16-5. Optimum shaft volume for impulse condition.

The closed-form analytical solution for the two mass system ($I_2 = 0$) is also given for comparison. The shaft diameter in this case can be calculated from the equation:

$$T_{max} = \text{Maximum dynamic torque}$$

$$= \Delta\omega \sqrt{k \frac{I_1 I_3}{I_1 + I_3}}$$

Therefore

$$\frac{\pi}{16} d^3(S_1) = \Delta\omega \sqrt{\frac{\pi}{32} d^4 \frac{G}{L_1 + L_2} \frac{I_1 I_3}{I_1 + I_3}}$$

by substitution and rearranging

$$d = \left[\frac{16}{\pi} \frac{\Delta\omega}{S_1} \sqrt{\frac{\pi}{32} \frac{G}{(L_1 + L_2)} \frac{I_1 I_3}{I_1 + I_3}}\right]^{1/3}$$

Example 2

Three-Rotor System Subjected to Step Load Input on One Mass (Fig. 16-6).

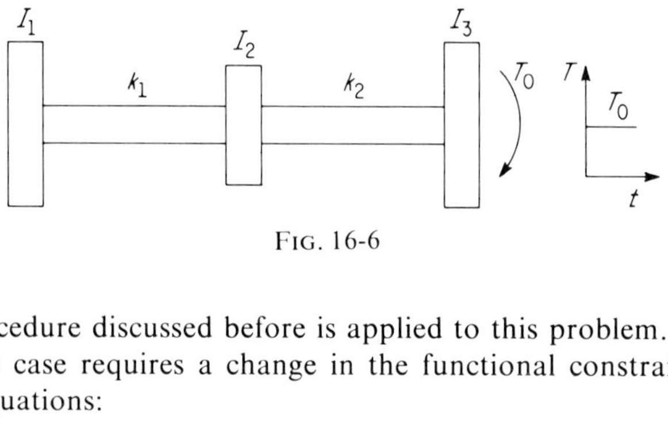

FIG. 16-6

The procedure discussed before is applied to this problem. The load input in this case requires a change in the functional constraints to the following equations:

$$T_1 = T_0 \frac{k_1}{I_3} \left[\frac{A_1(a_{21} - 1)}{\omega_1}(1 - \cos \omega_1 t) + \frac{C_1(a_{22} - 1)}{\omega_2}(1 - \cos \omega_2 t)\right]$$

$$T_2 = T_0 \frac{k_2}{I_3} \left[\frac{A_1(a_{31} - a_{21})}{\omega_1}(1 - \cos \omega_1 t) + \frac{C_1(a_{32} - a_{22})}{\omega_2}(1 - \cos \omega_2 t)\right]$$

The development of these equations is given in Chapter 8. The step load T_0 is taken equal to 5000 lb in. The other input parameters are kept the same as in Example 1. The results are shown in Table 16-2 and Fig. 16-7. The analytical solution for the two-rotor problem ($I_2 = 0$) is

TABLE 16-2
Optimum design for system subjected to a step torque = 5000 lb-in
(I_3 = 500 lb-in. sec^2)

I_2	I_1/I_3	d_1	d_2	S_1	S_2	Shaft Volume
0	0.2	0.55	0.55	50000	50000	11.4
	0.4	0.66	0.66	50000	50000	16.4
	0.6	0.72	0.72	50000	50000	19.9
	0.8	0.765	0.765	50000	50000	22.1
	1.0	0.795	0.795	50000	50000	23.8
10	0.2	0.5610	0.5649	49170	50000	11.96
	0.4	0.6630	0.6697	50010	49210	16.75
	0.6	0.7239	0.7296	50090	49250	19.95
	0.8	0.7639	0.7705	50390	49230	22.2
	1.0	0.7975	0.7947	49790	50280	23.9
100	0.2	0.6039	0.6621	50040	49200	15.14
	0.4	0.6940	0.7240	50150	49390	18.97
	0.6	0.7439	0.7667	51180	49290	21.52
	0.8	0.7839	0.7979	50820	49200	23.60
	1.0	0.8210	0.8189	49280	49650	25.40

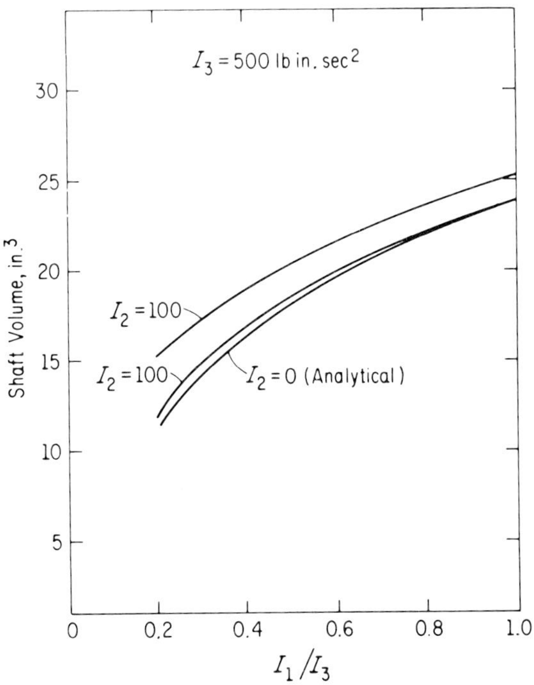

FIG. 16-7. Optimum shaft volume for a step torque.

is also given for comparison. This is calculated from

$$T_{max} = \text{maximum dynamic torque}$$
$$= \frac{2T_0}{I_3}\sqrt{\frac{kI_1I_3}{I_1+I_3}}\sqrt{\frac{k(I_1+I_3)}{I_1I_3}}$$
$$\frac{\pi}{16}d^3S_1 = \frac{2T_0}{1+(I_3/I_1)}$$

Therefore

$$d = \sqrt[3]{\frac{32T_0}{\pi S_1\left(1+\dfrac{I_3}{I_1}\right)}}$$

Example 3

Three-Mass System Subjected to a Sinusoidal Input on One of the Masses (Fig. 16-8).

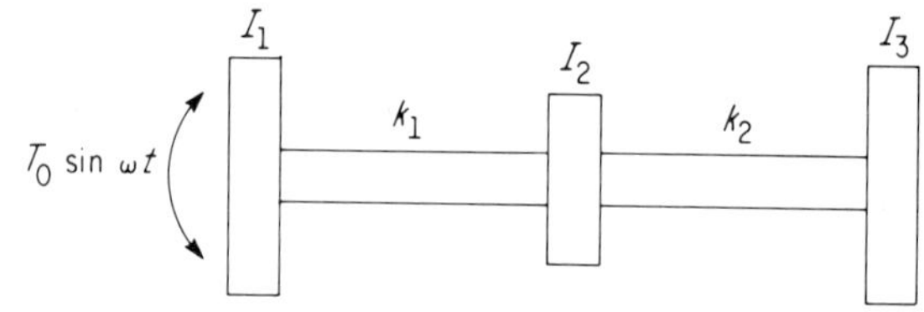

FIG. 16-8

In this case it is assumed that I_1 is subjected to a sinusoidal torque input $T_0 \sin \omega t$. The design procedure discussed in Example 1 also applies to this problem. The functional constraints, however, are changed due to the change of input torque. The equations for the dynamic shaft torques are obtained as discussed in Chapter 10 by utilizing the normal mode method for conservative systems. The normal mode coordinates are calculated from

$$\eta_1 = \frac{F_1}{\omega_1^2 M_1\left[1-\left(\dfrac{\omega}{\omega_1}\right)^2\right]}$$

and

$$\eta_2 = \frac{F_1}{\omega_2^2 M_2\left[1-\left(\dfrac{\omega}{\omega_2}\right)^2\right]}$$

The shaft torques are therefore given by

$$T_1 = k_1(\theta_2 - \theta_1) = k_1[(\phi_{21} - 1)\eta_1 + (\phi_{22} - 1)\eta_2]$$
$$T_2 = k_2(\theta_3 - \theta_2) = k_2[(\phi_{31} - \phi_{21})\eta_1 + (\phi_{32} - \phi_{22})\eta_2]$$

where

$$
\begin{aligned}
\eta_1, \eta_2 &= \text{normal mode coordinates} \\
\omega_1, \omega_2 &= \text{natural frequencies of the system} \\
\phi_{21}, \phi_{31} &= \text{eigen vectors for the first mode} \\
\phi_{22}, \phi_{32} &= \text{eigen vectors for the second mode} \\
M_1 &= I_1 + I_2\phi_{21}^2 + I_3\phi_{31}^2 \\
M_2 &= I_1 + I_2\phi_{22}^2 + I_3\phi_{32}^2
\end{aligned}
$$

The numerical inputs in this case are the same as for Example 1. The amplitude T_0 of the torque excitation is 10,000 lb in. and its frequency ω is 5.0 rad/sec.

The results are given in Table 16-3 and Fig. 16-9. The analytical

TABLE 16-3
Optimum design for system subjected to a sinusoidal excitation on the first mass of amplitude of 10,000 lb-in.
(ω = 5 rad/sec)

I_2	I_1/I_3	d_1	d_2	S_1	S_2	Shaft Volume
10	0.2	0.9762	0.9674	49580	50250	35.7
	0.4	0.9543	0.9500	50080	50070	34.25
	0.6	0.9386	0.9394	50320	49530	33.4
	0.8	0.9337	0.9232	49330	50390	32.6
	1.0	0.9181	0.9212	50480	49340	32.0
100	0.2	0.9794	0.9397	50050	49940	34.8
	0.4	0.9623	0.9298	50350	49320	33.8
	0.6	0.9498	0.9187	50550	495000	32.9
	0.8	0.9428	0.9090	50220	49800	32.2
	1.0	0.9422	0.8995	48940	50120	31.75
0	0.2	0.9800	0.9800	50000	50000	36.3
	0.4	0.9600	0.9600	50000	50000	34.8
	0.6	0.9450	0.9450	50000	50000	33.8
	0.8	0.9320	0.9320	50000	50000	32.9
	1.0	0.9240	0.9240	50000	50000	32.2

solution for the two mass system ($I_2 = 0$) is also given for comparison. This is derived as follows.

The equations of motion for a system with two masses (I_1, I_3) connected by a shaft with length L and diameter d, and subjected to a sinusoidal torque excitation $T_0 \sin \omega t$ on I_1 can be written as

$$
\begin{aligned}
-I_1\omega^2\theta_1 + k(\theta_1 - \theta_3) &= T_0 \\
k(\theta_3 - \theta_1) - I_3\omega^2\theta_3 &= 0
\end{aligned}
$$

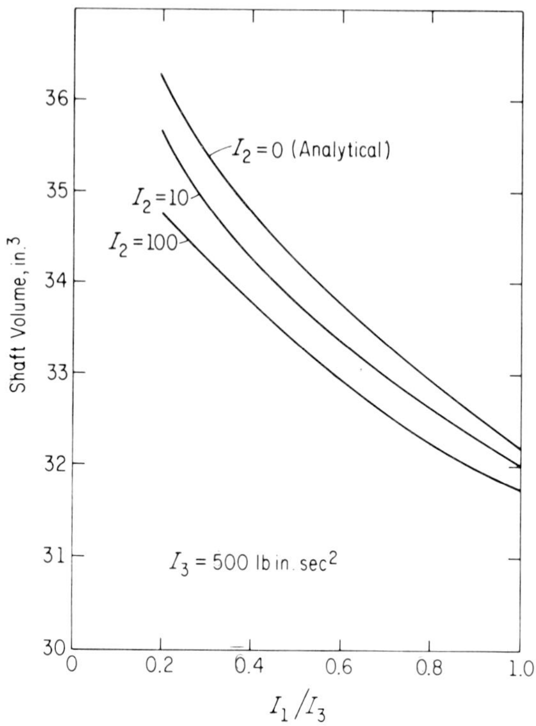

FIG. 16-9. Optimum shaft volume for sinusoidal excitation.

where

$$k = \frac{G\pi d^4}{32\,L}$$

from which

$$\theta_3 = \frac{k}{k - I_3\omega_2}$$

and

$$\theta_1 \left\{\frac{[I_1 I_3\omega^2 - k(I_1 + I_3)\omega^2]}{k - I_3\omega^2}\right\} = T_0$$

Dynamic torque is therefore

$$T_d = \pm T_0 + I_1\omega^2\theta_1$$
$$= \frac{\pi}{16} d^3 S$$

where S is the allowable torsional stress. By substitution and rearranging, the equation for evaluation of shaft diameter becomes

$$0.2\,S\omega^2 I_1 \pm 1.2 \times 10^6 \frac{d}{L} T_0 - 1.2 \times 10^6 \frac{d^4}{L}[1 + (I_2/I_3)] = 0 \qquad (16\text{-}2)$$

It should be noted here that due to the nature of the response curves under sinusoidal excitations, the possibility exists that several optimal designs may result depending on the choice of the starting point. Such solutions can be expected to the extent of $n + 1$ designs, where n is the number of system resonances. This can be easily seen for the case of a two mass system from Eq. 16-2. In this case the system has one natural frequency and therefore there is a possibility for two optimal solutions corresponding to the in-phase and out-of-phase responses. If, for example, the search for the case of $I_1/I_3 = 0.2$ is started with a large value of d, the optimum shaft diameter will be $d = 0.98$ in. which corresponds to the in-phase condition (positive sign in the equation). On the other hand, if the search is started with a very small value of the diameter, the optimum design will be $d = 0.1$ in. which corresponds to the out-of-phase condition (negative sign in the equation). The existence of multiple optimal designs for a three mass system where up to three solutions can be expected is illustrated in Fig. 16-10 and Table 16-4. In most engineering

TABLE 16-4
Multiple optimums for a 3-mass system subjected to sinusoidal excitation on the first mass of amplitude of 10,000 lb-in. and frequency $\omega = 5$ rad/sec

d_1	d_2	ω_1	ω_2	Volume
0.9762	0.9674	16.09	95.3	35.7
0.4649	0.0619	0.0894	15.88	4.15
0.05596	0.2649	0.0758	4.97	1.38

applications, it is usually desirable to select the solution where the forcing frequency is lower than the first natural frequency of the system. Although this design, according to the expressed objectives has the lowest merit rating of all possible optimal solutions yet it has the advantage of avoiding all resonances during starting and stopping. It also has the lowest values for the angular displacements of the different rotors. The designer, however, can exercise his judgment as to the choice between these different local optima. The optimum designs given in Fig. 16-9 and Table 16-4 correspond to the condition where the forcing frequency is lower than the first natural frequency.

Example 4

Three-Rotor System with Rayleigh Damping of the "Uniform Spring Damping" Type Under Sinusoidal Excitation (Fig. 16-11).

This case is considered to include damping in the design of vibratory systems. Classical damping as explained in Chapter 10 is considered to simplify the analytical solution and therefore reduce the computation

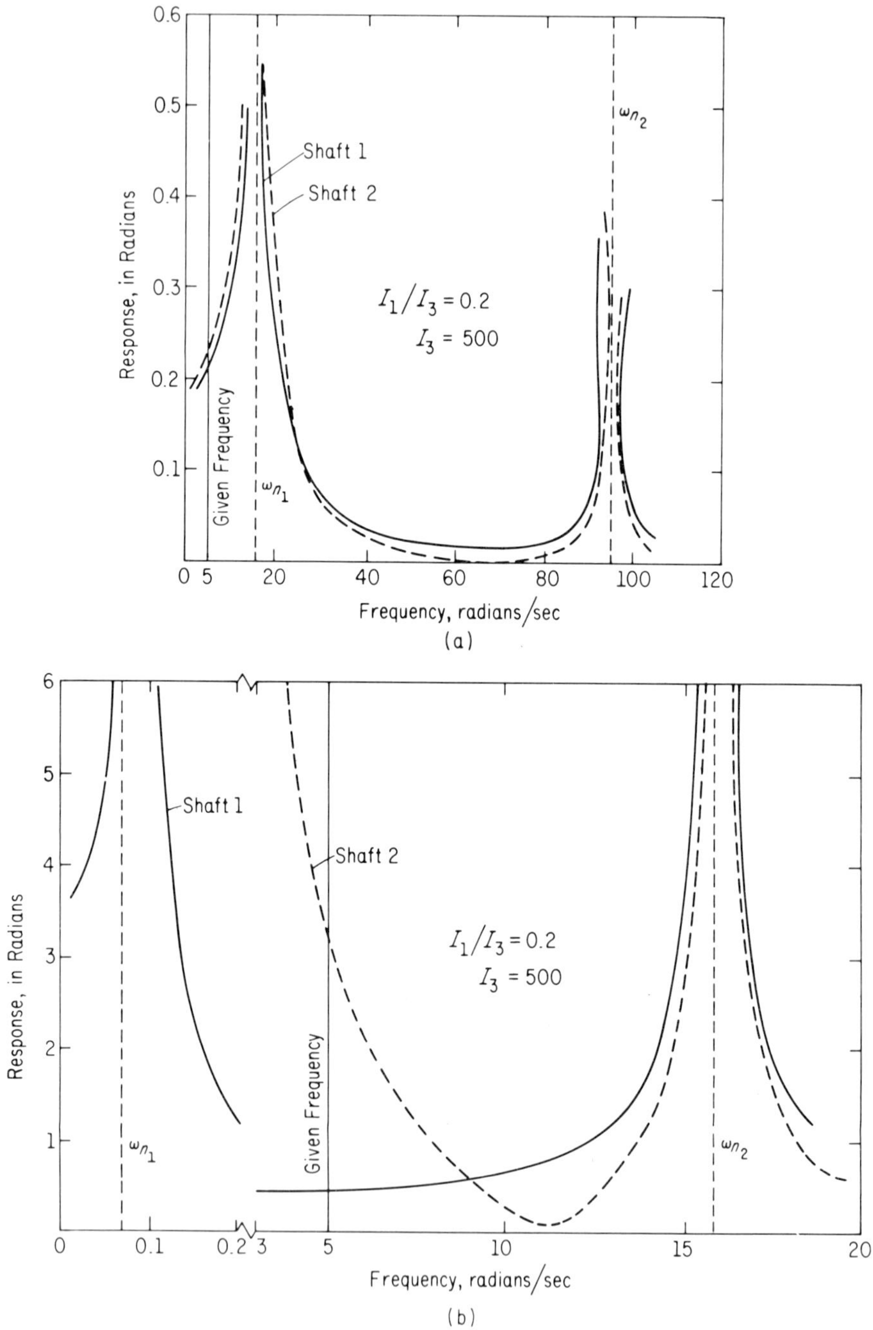

FIG. 16-10. (a) Plot of system response for first optimum conditions. (b) Plot of system response with second optimum conditions.

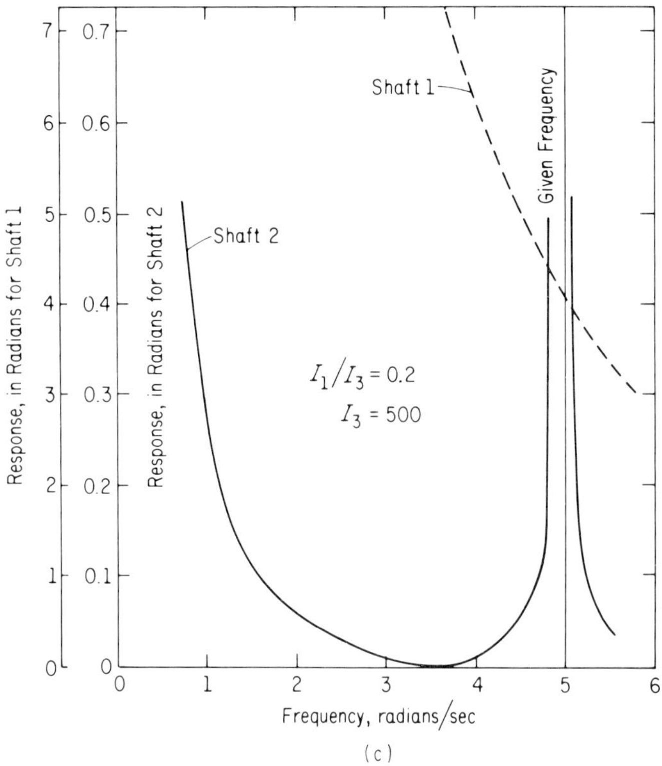

FIG. 16-10. (c) Plot of system response with third optimum conditions.

time. The damping in this example is taken to be proportional to the spring constant of the shafts (i.e., c/k = constant = R). As discussed in Chapter 10, the equation representing the kth mode of the linear lumped parameter system (Fig. 16-11) with uniform spring damping subjected to a general force excitation $F_i(t)$ is

$$\ddot{\eta}_k + R\omega_k^2\dot{\eta}_k + \omega_k^2\eta_k = \frac{\Sigma\phi_{ik}F_i(t)}{M_k}$$

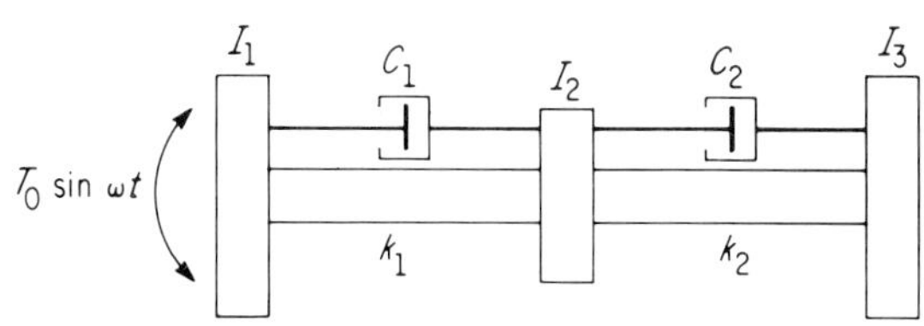

FIG. 16-11

where

$$
\begin{aligned}
\phi_{ik} &= \text{eigen vectors} \\
R\omega_k^2 &= 2\xi_k \omega_k \\
\xi_k &= \frac{R\omega_k}{2} = \text{mode damping ratio} \\
M_k &= \sum_i I_i \phi_{ik}^2
\end{aligned}
$$

The solution of the normal mode equation for sinusoidal excitation on the first mass is given by

$$
\eta_1 = \frac{F_1}{\omega_1^2 M_1 \sqrt{\left(1 - \left(\frac{\omega}{\omega_1}\right)^2\right)^2 + \left(2\xi_1\left(\frac{\omega}{\omega_2}\right)\right)^2}}; \quad \alpha_1 = \tan^{-1} \frac{2\xi_1\left(\frac{\omega}{\omega_1}\right)}{1 - \left(\frac{\omega}{\omega_1}\right)^2}
$$

$$
\eta_2 = \frac{F_1}{\omega_2^2 M_2 \sqrt{\left(1 - \left(\frac{\omega}{\omega_2}\right)^2\right)^2 + \left(2\xi_2\left(\frac{\omega}{\omega_2}\right)\right)^2}}; \quad \alpha_2 = \tan^{-1} \frac{2\xi_2\left(\frac{\omega}{\omega_2}\right)}{1 - \left(\frac{\omega}{\omega_2}\right)^2} \tag{16-3}
$$

which define the normal code coordinates (η_1, η_2) and phase angles (α_1, α_2) respectively.

The equations for the dynamic shaft torques in this case can therefore be written as

$$
\begin{aligned}
T_1 &= k_1(\theta_2 - \theta_1) = k_1[(\phi_{21} - 1)\bar{\eta}_1 + (\phi_{22} - 1)\bar{\eta}_2] \\
T_2 &= k_2(\theta_3 - \theta_2) = k_2[(\phi_{31} - \phi_{21})\bar{\eta}_1 + (\phi_{32} - \phi_{22})\bar{\eta}_2]
\end{aligned}
$$

The input parameters for this example are taken as

$$
\begin{aligned}
I_{1\max} = I_{2\max} = I_{3\max} &= 500 \text{ lb in. sec}^2 \\
I_{1\min} = I_{2\min} = I_{3\min} &= 200 \text{ lb in. sec}^2 \\
L_2/L_1 &= 1, 1.5, 3, 7, 15 \text{ respectively} \\
L_1 + L_2 &= 48 \text{ in.} \\
R_{\max} &= 0.1 \\
S_1 = S_2 &= 50{,}000 \text{ psi} \\
T_0 &= 10{,}000 \text{ lb in.} \\
\omega &= 1 \text{ rad/sec}
\end{aligned}
$$

The required outputs are: I_1, I_2, I_3, d_1, d_2 and R subject to the constraints

$$
\begin{aligned}
I_{\min} < I &< I_{\max} \\
d &\geq 0 \\
0 < R &< R_{\max}
\end{aligned}
$$

TABLE 16-5
Optimum Design for Systems Subjected to a Sinusoidal Excitation of Amplitude of 10,000 lb. at the First Mass.
$R = 0.1 \quad (\omega = 1 \text{ rad/sec})$

L_1	L_2	d_{1f}	d_{2f}	ω_1	ω_2	Shaft Volume	L_1/L_2
3	45	0.7650	0.6137	4.852	30.91	14.70	0.067
6	42	0.7667	0.6132	4.972	22.14	15.15	0.143
12	36	0.7664	0.6129	5.246	15.99	16.15	0.333
19.2	28.8	0.7678	0.6119	5.605	13.24	17.35	0.667
24	24	0.7678	0.6117	5.856	12.40	18.10	1.0
28.8	19.2	0.7695	0.6143	6.127	12.26	18.90	1.5
36	12	0.7672	0.6112	6.287	13.30	20.40	3.0
42	6	0.7684	0.6104	6.311	17.35	22.60	7.0
45	3	0.7687	0.6098	6.274	23.83	22.60	15.0

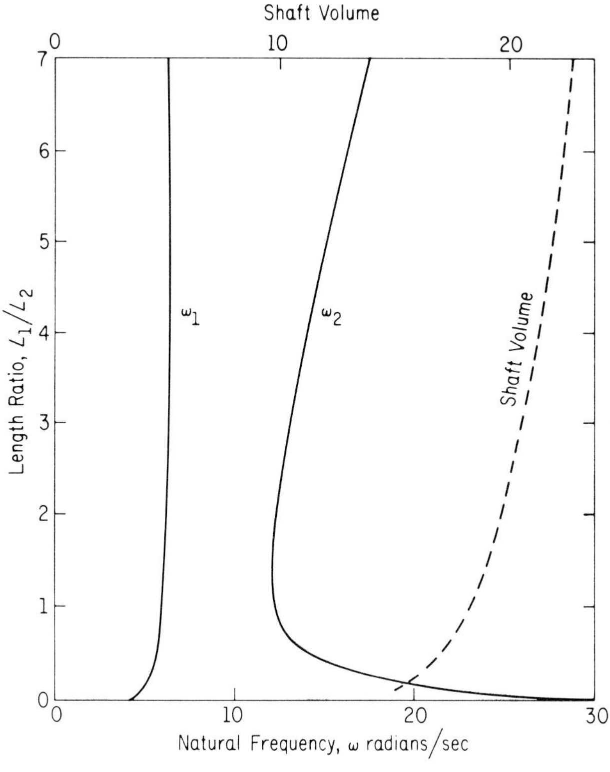

FIG. 16-12. Optimum shaft volume and the corresponding natural frequencies for different shaft length ratios. (ω = 1.0 radian/sec, R = 0.1)

The merit criterion in this case is expressed as:

$$\text{Minimize } U = (L_1 d_1^2 + L_2 d_2^2) + \sqrt{\left(\frac{\tau_1}{S_1} - 1\right)^2 + \left(\frac{\tau_2}{S_2} - 1\right)^2}$$

The search technique utilized here is the same as shown in Fig. 16-2. The optimum solutions corresponding to the condition when ω is smaller than the first natural frequency of the system are given in Table 16-5 and Fig. 16-12. The natural frequencies corresponding to these optimum designs are also given. It should be noted that the optimum diameters in this case are independent of the L_2/L_1 ratio. It is also noted that in all the performed calculations, the optimum values for I_1 and R are their upper limits while the optimum values for I_2 and I_3 are their lower limits.

16-3. SUMMARY

This chapter gives illustrative examples of the application of automated design method for evaluation of the optimum parameters for mechanical drives subjected to dynamic excitations. The procedure makes it possible to select the system parameters necessary to produce stress values as close as possible to the allowable stresses in all shafts under different types of dynamic loading. Traditional design of such systems has been usually based on intuitive and trial procedures which can only produce feasible solutions rather than optimum ones.

Although only two-degree-of-freedom systems were discussed in this introductory treatment, the technique can be used for systems with any number of degrees of freedom subjected to any type of excitation. Other design objectives and constraints can also be accommodated. The use of Hybrid computers can be invaluable for complex dynamic systems since it combines the advantages of digital logic for decision with the analog versatility for solution of the governing differential equations.

REFERENCES

1. G. Polya, *How to Solve It.* New York: Anchor, Doubleday, 1957.
2. G. Polya, *Mathematical Discovery.* New York: Wiley, Vol. 1, 1962, Vol. 2, 1965.
3. L. A. Schmit, "Automated Design," *International Science and Technology*, No. 53 (June 1966), pp. 66–78.
4. B. Hamad, "Optimum Design of Vibratory Systems," Ph.D Thesis, University of Wisconsin, 1968.
5. D. J. Wilde, *Optimum Seeking Methods.* Englewood Cliffs, N. J.: Prentice-Hall, Inc., 1964.
6. D. J. Wilde and C. S. Beightler, *Foundations of Optimization.* Englewood Cliffs, N. J.: Prentice-Hall, Inc., 1967.

7. D. Rudd and C. Watson, *Strategy of Process Engineering*. New York: Wiley, 1968.
8. T. L. Saaty, *Mathematical Methods in Operations Research*. New York: McGraw-Hill, 1959.
9. O. L. Mangasarian, *Non-Linear Programming*. New York: McGraw-Hill, 1969.
10. B. Carnahan and J. O. Wilkes, "Numerical Methods, Optimization Techniques and Process Simulation for Engineers," University of Michigan, Publication No. 6705-92, 1967.

appendix A

Laplace Transforms

LAPLACE TRANSFORMS FOR SOME COMMON FUNCTIONS

Transform	Function
1. $F(p) = Lf(t) = \int_0^\infty e^{-pt} f(t)\,dt$	$f(t)$
2. $aF(p) + bG(p)$	$af(t) + bg(t)$
3. $pF(p) - f(0)$	$\dfrac{df(t)}{dt}$
4. $p^2F(p) - pf(0) - \dfrac{df(0)}{dt}$	$\dfrac{d^2f(t)}{dt^2}$
5. $p^nF(p) - \sum_{k=1}^{n} p^{n-k}\dfrac{d^{k-1}f(0)}{dt^{k-1}}$	$\dfrac{d^nf(t)}{dt^n}$
6. $\dfrac{1}{p^n}F(p)$	$\overbrace{\int_0^t \cdots \int_0^t}^{n\text{ times}} f(t)\,\overbrace{dt \cdots dt}^{n\text{ times}}$
7. $(-1)^n \dfrac{d^nF(p)}{dp^n}$	$t^nf(t)$
8. $F(p-a)$	$e^{at}f(t)$
9. $e^{-ap}F(p)$	$\begin{cases} f(t-a) & \text{when } t > a \\ 0 & \text{when } t < a \end{cases}$
10. $F(p)G(p)$	$\int_0^t f(t-s)g(s)\,ds = \int_0^t f(s)g(t-s)\,ds$
11. $\dfrac{1}{p}$	1 (unit step function)
12. $\dfrac{1}{p+a}$	e^{-at}
13. $\dfrac{1}{(p+a)(p+b)}$	$\dfrac{1}{(a-b)}(e^{-bt} - e^{-at})$
14. $\dfrac{p}{(p+a)(p+b)}$	$\dfrac{1}{(b-a)}(be^{-bt} - ae^{-at})$

Transform	Function
15. $\dfrac{a}{p^2 + a^2}$	$\sin at$
16. $\dfrac{p}{p^2 + a^2}$	$\cos at$
17. $\dfrac{a}{p^2 - a^2}$	$\sinh at$
18. $\dfrac{p}{p^2 - a^2}$	$\cosh at$
19. $\dfrac{2ap}{(p^2 + a^2)^2}$	$t \sin at$
20. $\dfrac{p^2 - a^2}{(p^2 + a^2)^2}$	$t \cos at$
21. $\dfrac{a}{(p + b)^2 + a^2}$	$e^{-bt} \sin at$
22. $\dfrac{p + b}{(p + b)^2 + a^2}$	$e^{-bt} \cos at$
23. $\dfrac{2a^2 p}{p^4 + 4a^4}$	$\sin at \sinh at$
24. 1	$u'(t)$ (impulse)
25. e^{-pt_1}	$u'(t - t_1)$
26. p	$\mathfrak{D}(t)$ (impulsive couple)
27. pe^{-pt_1}	$\mathfrak{D}(t - t_1)$
28. $\dfrac{1}{p^n}$	$\dfrac{t^{n-1}}{(n-1)!}, n = 1,2,3,\ldots$
29. $\dfrac{n!}{p^{n+1}}$	$t^n, n = 0,1,2,\ldots$
30. $\dfrac{1}{(p + a)^n}$	$\dfrac{t^{n-1}e^{-at}}{(n-1)!}, n = 1,2,3,\ldots$
31. $\dfrac{p}{(p + a)^n}$	$\dfrac{(n-1) - at}{(n-1)!} t^{n-2} e^{-at}$
32. $\dfrac{1}{p^2 + 2\zeta\omega_0 p + \omega_0^2} = \dfrac{1}{(p + \omega_0)^2 + \omega_0^2(1 - \zeta^2)}$	$\dfrac{1}{\omega_0\sqrt{1 - \zeta^2}} e^{-\zeta\omega_0 t} \sin \omega_0 \sqrt{1 - \zeta^2}\, t$

LAPLACE TRANSFORMS FOR COMMON WAVESHAPES

No.	Waveshape	Function $f(t)$	$Lf(t) = F(p)$
1	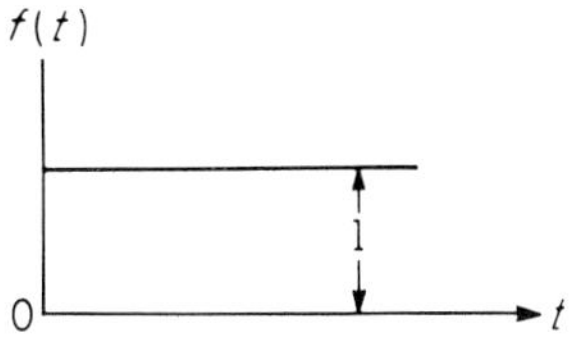	$u(t)$	$\frac{1}{p}$
2	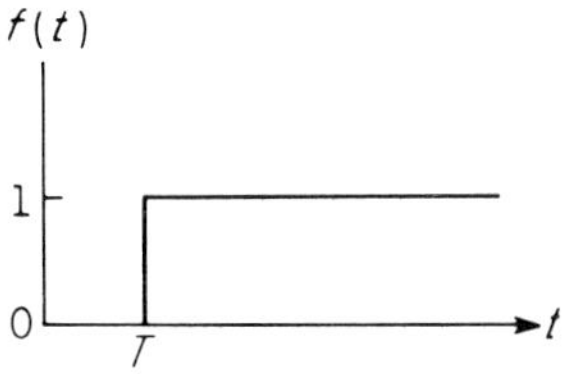	$u(t - T)$	$\frac{1}{p} e^{-pT}$
3	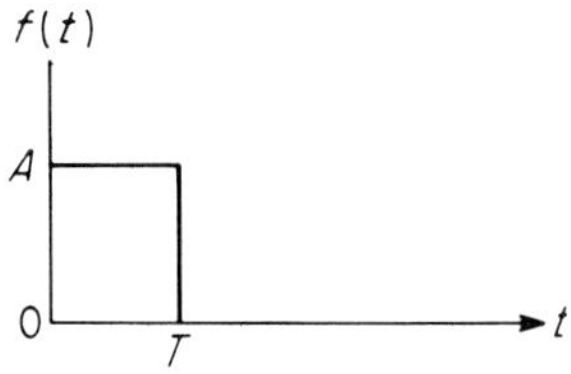	$A[u(t) - u(t - T)]$	$\frac{A}{p}(1 - e^{-pT})$
4	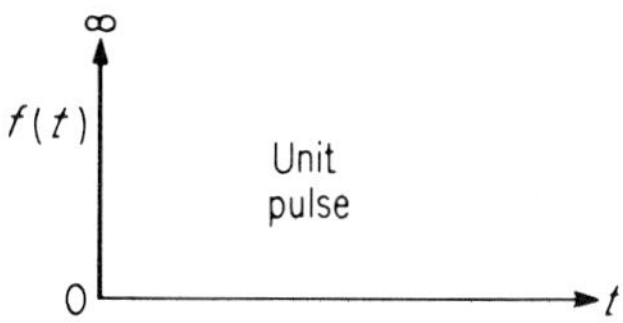	$u'(t)$	1
5	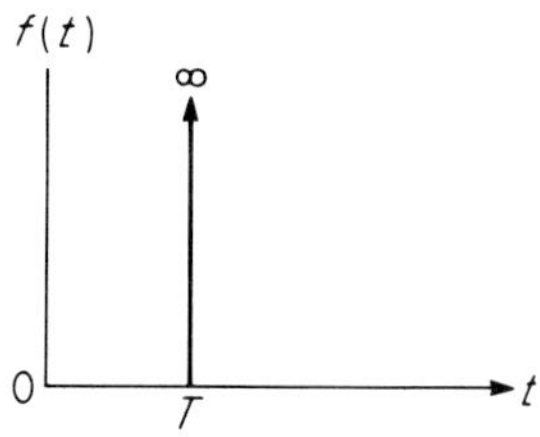	$u'(t - T)$	e^{-pT}
6	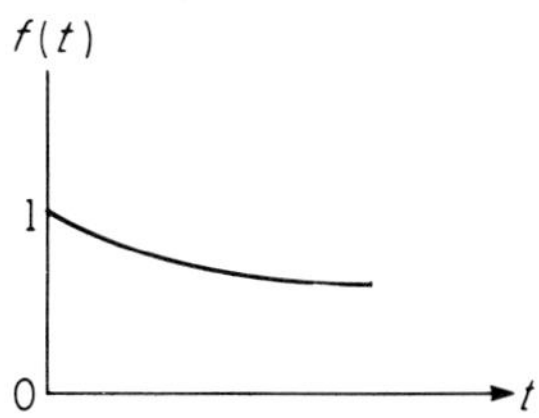	$e^{-at}u(t)(a \gtrless 0)$	$\frac{1}{p + a}$

LAPLACE TRANSFORMS FOR COMMON WAVESHAPES (*continued*)

No.	Waveshape	Function $f(t)$	$Lf(t) = F(p)$
7	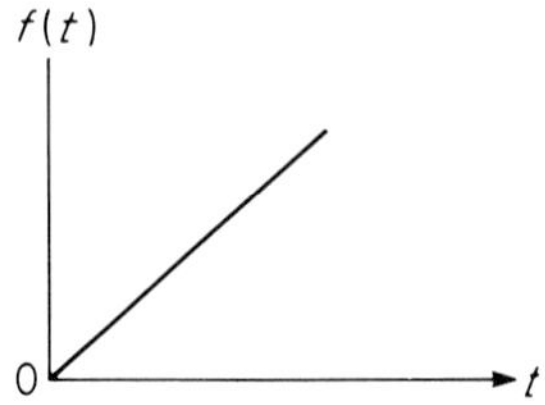	$tu(t)$	$\frac{1}{p^2}$
8	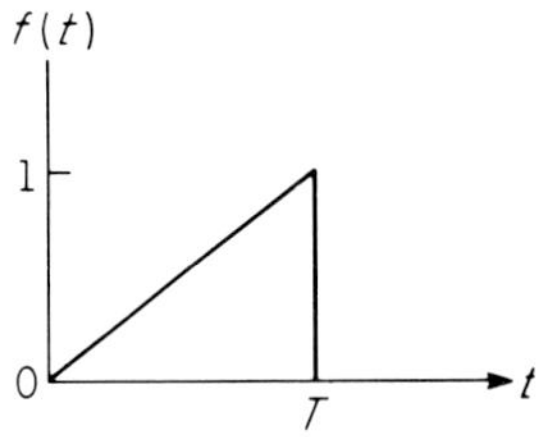	$\frac{t}{T}[u(t) - u(t - T)]$	$\frac{1}{Tp^2}(1 - e^{-pT}) - \frac{1}{p} e^{-pT}$
9	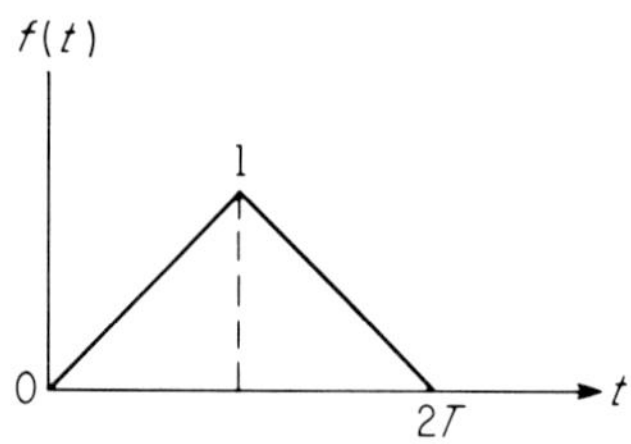	$\frac{1}{T}\{t[u(t) - u(t - T)] + (-t + 2T) \cdot (u(t - T) - u(t - 2T))\}$	$\frac{1}{T}\frac{1}{p^2}(1 - e^{-pT})^2$
10	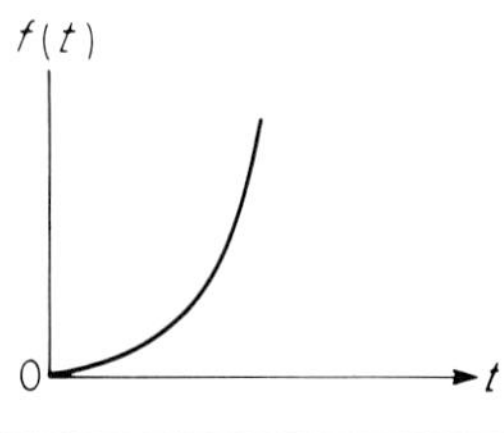	$t^n u(t)$ n positive and not an integer	$\frac{\Gamma(n + 1)}{p^{n+1}}$
11	Like No. 10	Like No. 10 for n positive integer	$\frac{n!}{p^{n+1}}$
12	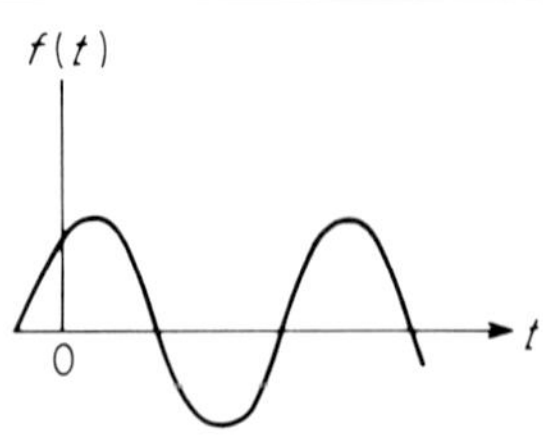	$[\sin(\omega_0 t + \phi)]u(t)$	$\frac{\omega_0 \cos\phi + p \sin\phi}{p^2 + \omega_0^2}$

LAPLACE TRANSFORMS FOR COMMON WAVESHAPES (*continued*)

No.	Waveshape	Function $f(t)$	$Lf(t) = F(p)$
13	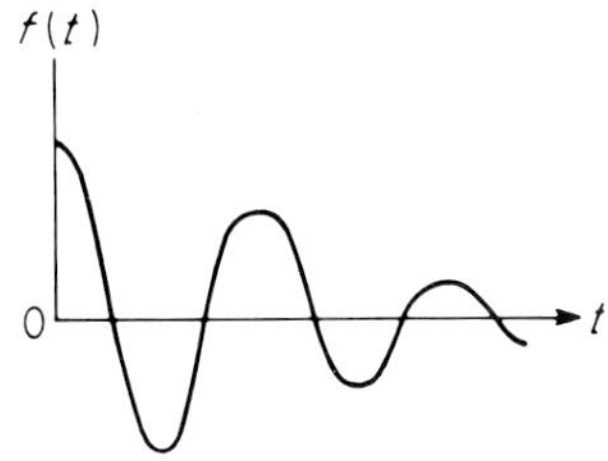	$[e^{-at} \cos \omega_0 t]\, u(t)$	$\dfrac{p + a}{(p + a)^2 + \omega_0^2}$
14	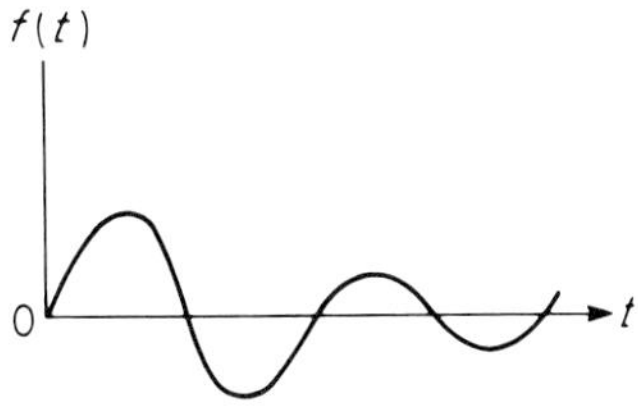	$[e^{-at} \sin \omega_0 t]\, u(t)$	$\dfrac{\omega_0}{(p + a)^2 + \omega_0^2}$
15	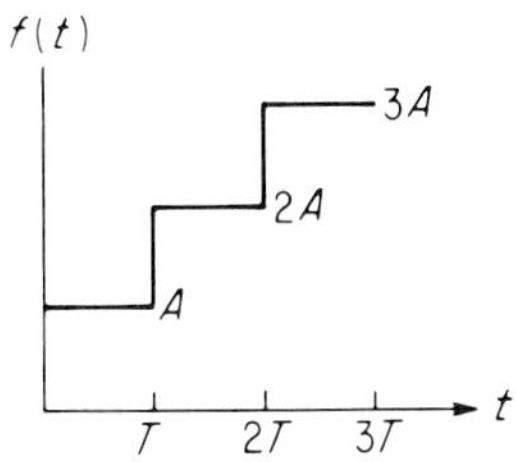	Stepping wave	$\dfrac{A}{p}(1 + e^{-pT} + e^{-2pT} + \cdots)$
16	Like No. 15	For infinite number of steps	$\dfrac{A}{p(1 - e^{-pT})}$
17	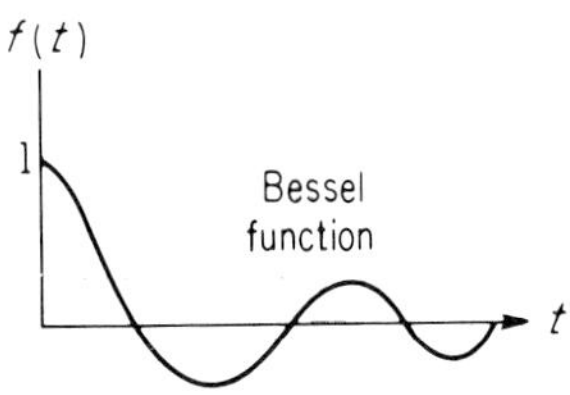	$J_0(t)\, u(t)$	$\dfrac{1}{p^2 + 1}$
18	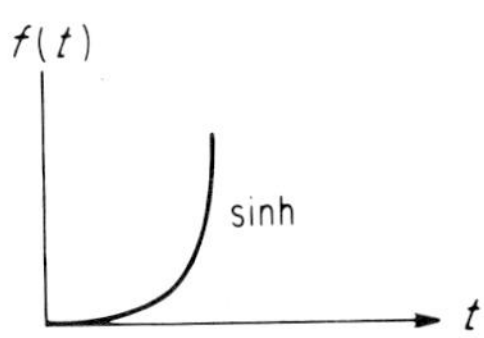	$(\sinh at)\, u(t)$	$\dfrac{a}{p^2 - a^2}\,(p/a)$

LAPLACE TRANSFORMS FOR COMMON WAVESHAPES (continued)

<table>
<tr><th>No.</th><th>Waveshape</th><th>Function $f(t)$</th><th>$Lf(t) = F(p)$</th></tr>
<tr><td>19</td><td>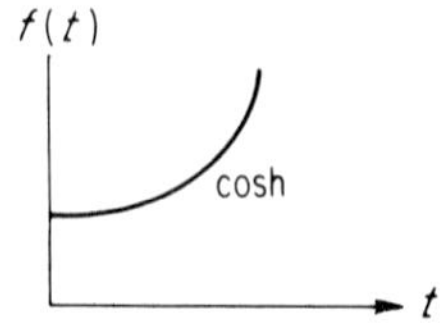</td><td>$(\cosh at)\,u(t)$</td><td>$\frac{p}{p^2 - a^2}\ (p/a)$</td></tr>
<tr><td>20</td><td>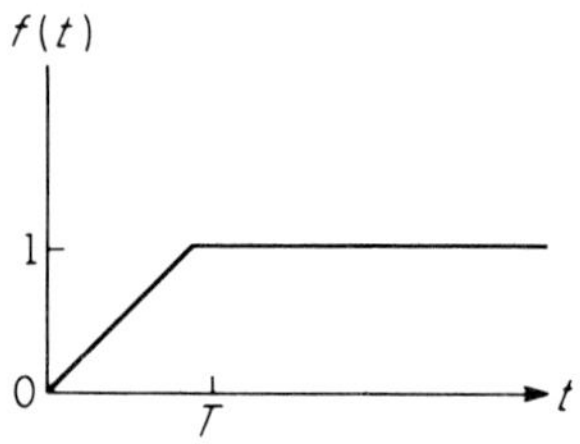</td><td>$\frac{t}{T}[u(t) - u(t - T)]$
$+ u(t - T)$</td><td>$\frac{1}{Tp^2}(1 - e^{-pT})$
$+ \frac{1}{p}e^{-pT}$</td></tr>
<tr><td>21</td><td>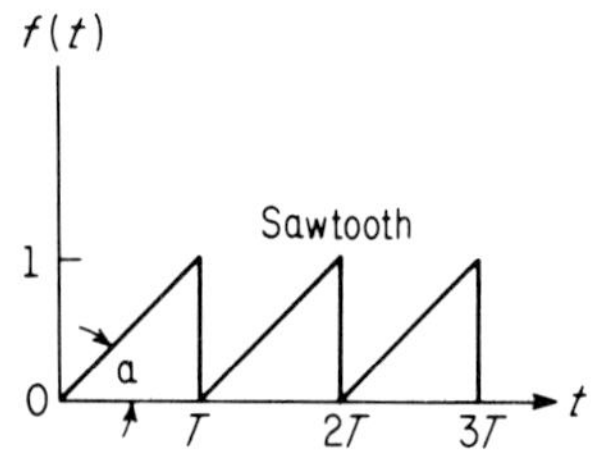</td><td></td><td>$\frac{a}{p^2} - \frac{aT}{2p}\left(\coth\frac{pT}{2} - 1\right)$</td></tr>
<tr><td>22</td><td>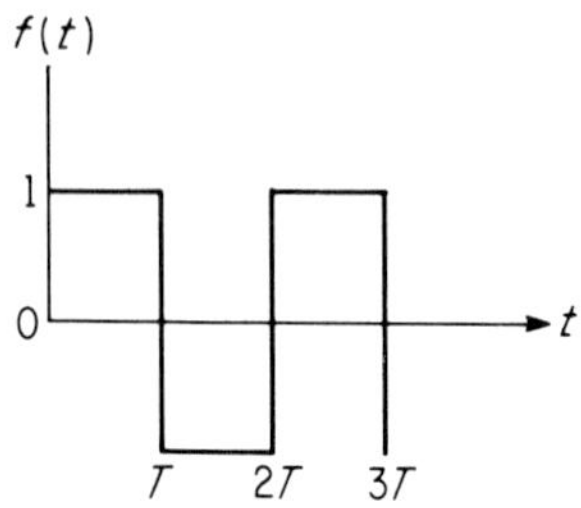</td><td>$u(t) - 2u(t - T)$
$+ 2u(t - 2T)\cdots$</td><td>$\frac{1}{p}.\tanh\left(\frac{pT}{2}\right)$</td></tr>
<tr><td>23</td><td>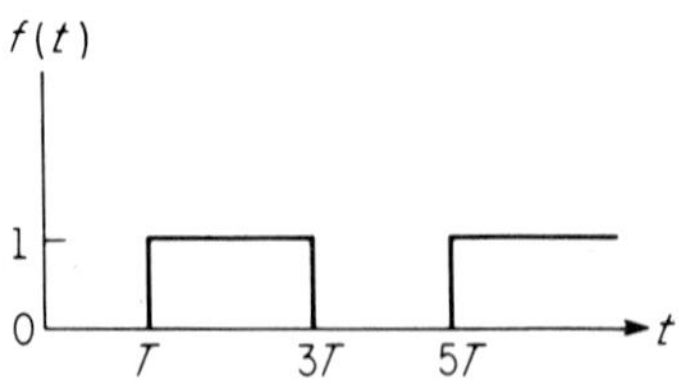</td><td></td><td>$\frac{1}{2p\cosh pT}$</td></tr>
</table>

LAPLACE TRANSFORMS FOR COMMON WAVESHAPES (*continued*)

No.	Waveshape	Function $f(t)$	$Lf(t) = F(p)$
24	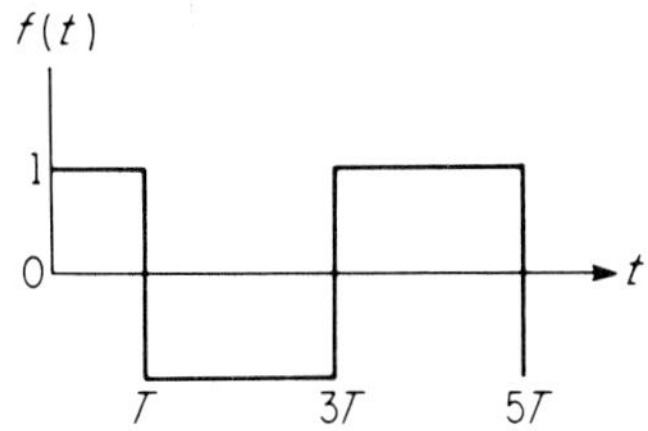		$\frac{1}{p}\left(1 - \frac{1}{\cosh pT}\right)$
25	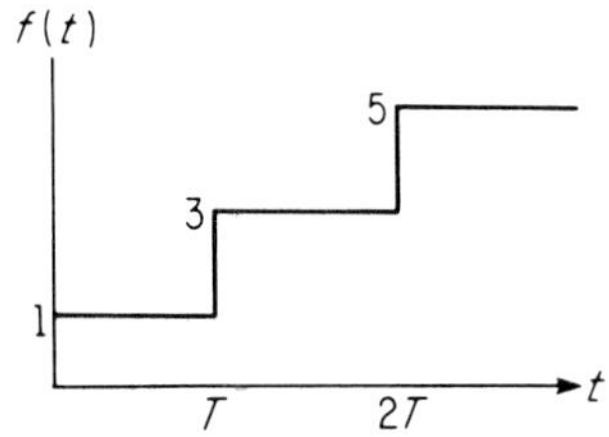		$\frac{1}{p}\coth\left(\frac{pT}{2}\right)$
26	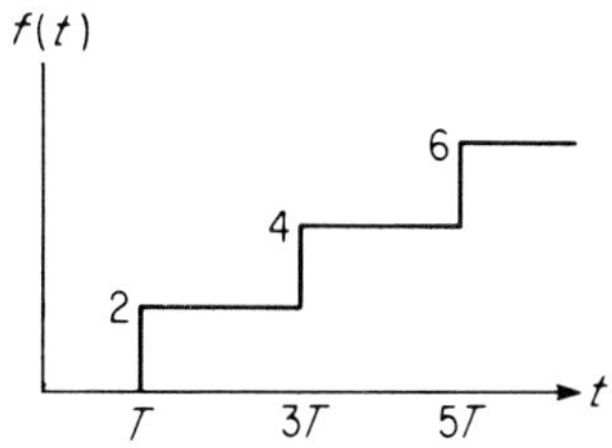		$\frac{1}{p \sinh pT}$

appendix B

Natural Frequencies for Two-Mass Torsional Systems with General Damping Condition

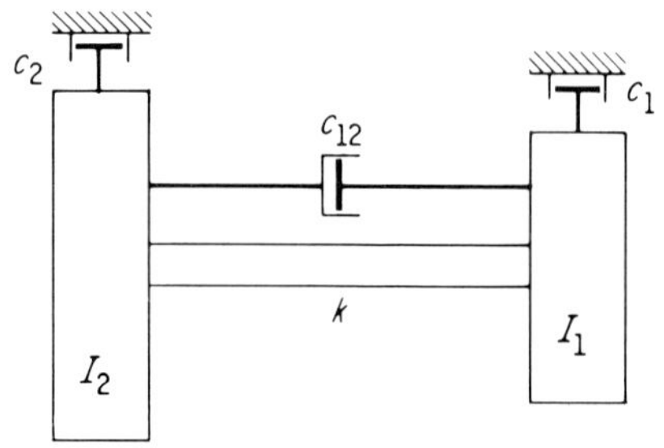

$$R = I_1/I_2$$

$$c = n\sqrt{k(I_1+I_2)}$$

b = damped natural frequency

$$\omega_0 = \sqrt{\frac{k(I_1+I_2)}{I_1 I_2}}$$

FIG. B-1. Two-mass torsional system with general damping condition.

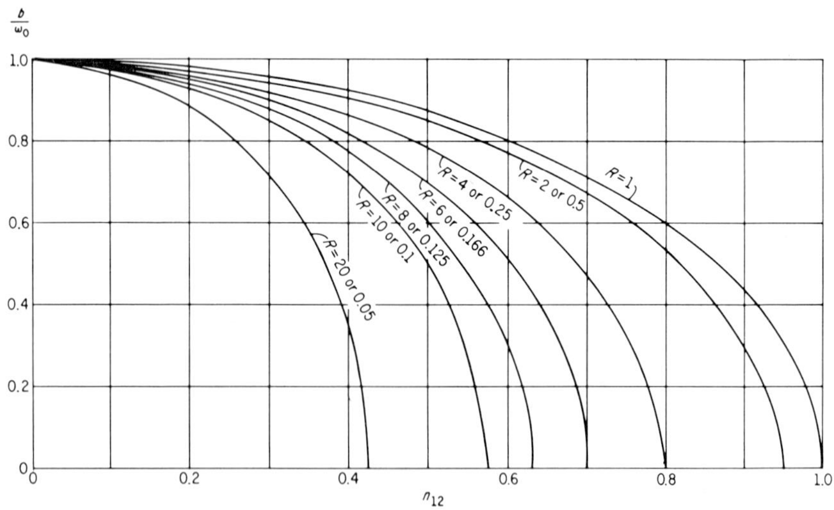

FIG. B-2. $n_1 = n_2 = 0$.

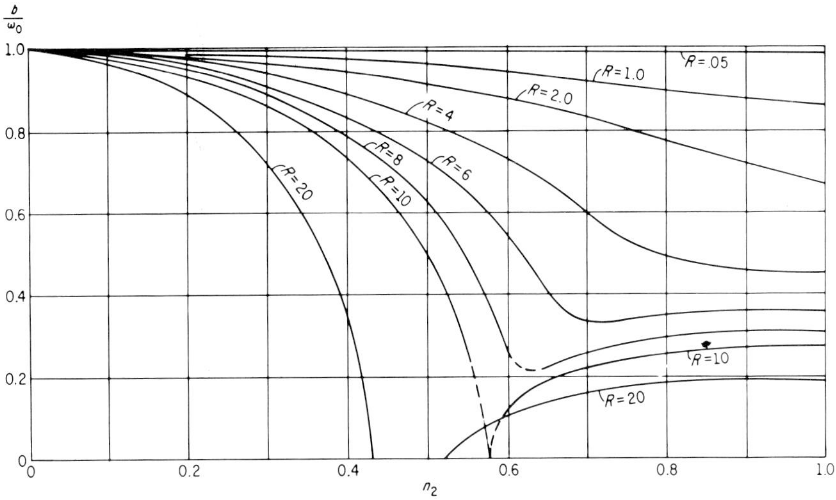

FIG. B-3. $n_1 = n_{12} = 0$.

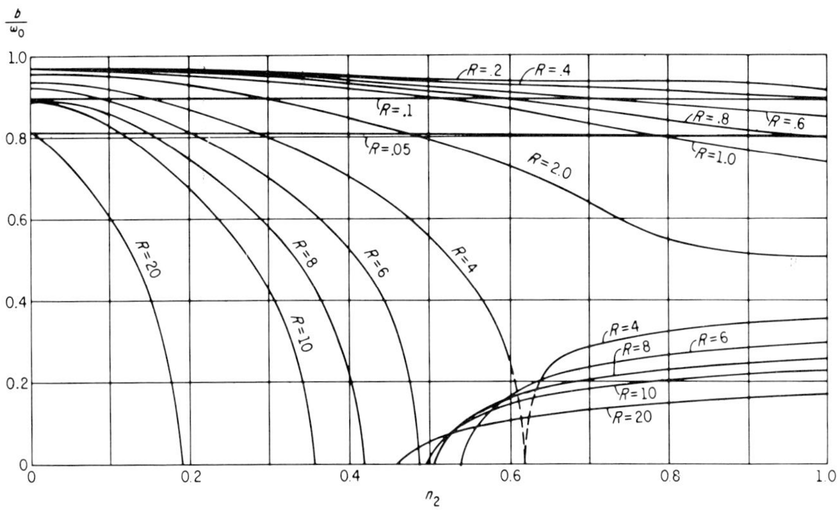

FIG. B-4. $n_1 = 0, n_{12} = 0.25$.

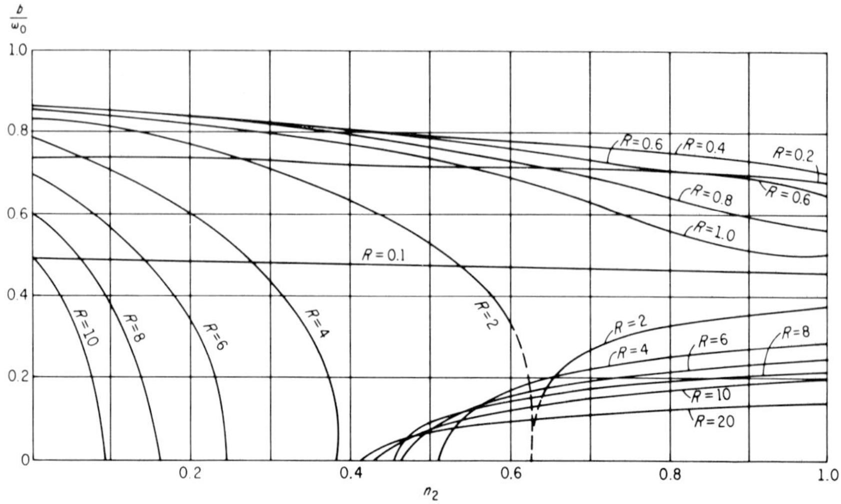

FIG. B-5. $n_1 = 0, n_{12} = 0.5$.

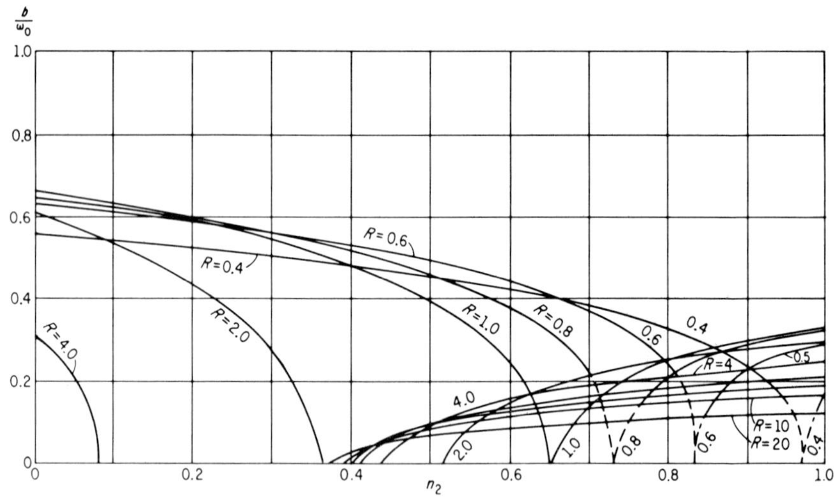

FIG. B-6. $n_1 = 0, n_{12} = 0.75$.

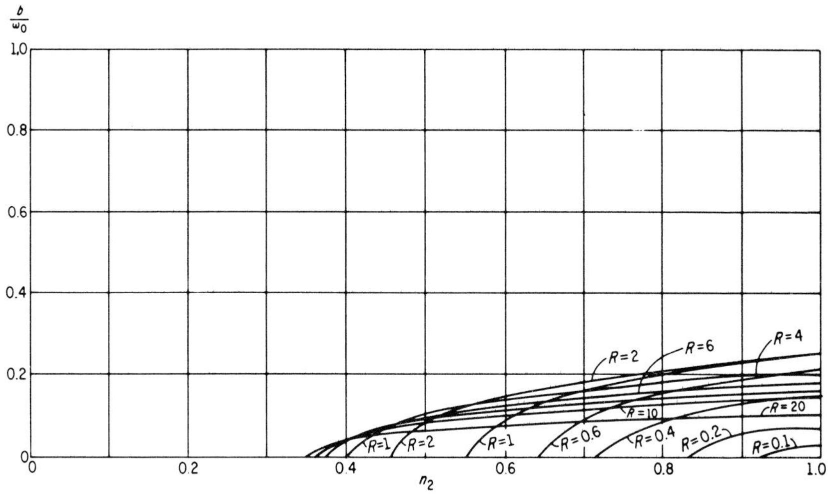

FIG. B-7. $n_1 = 0, n_{12} = 1.0$.

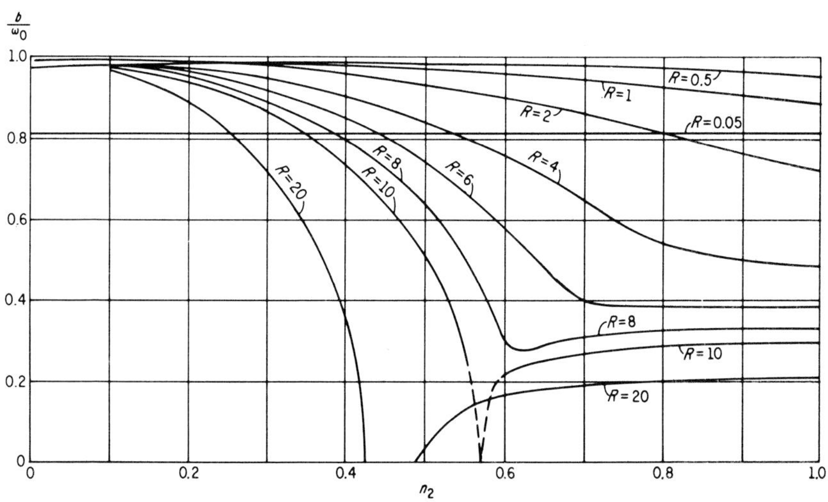

FIG. B-8. $n_{12} = 0, n_1 = 0.25$.

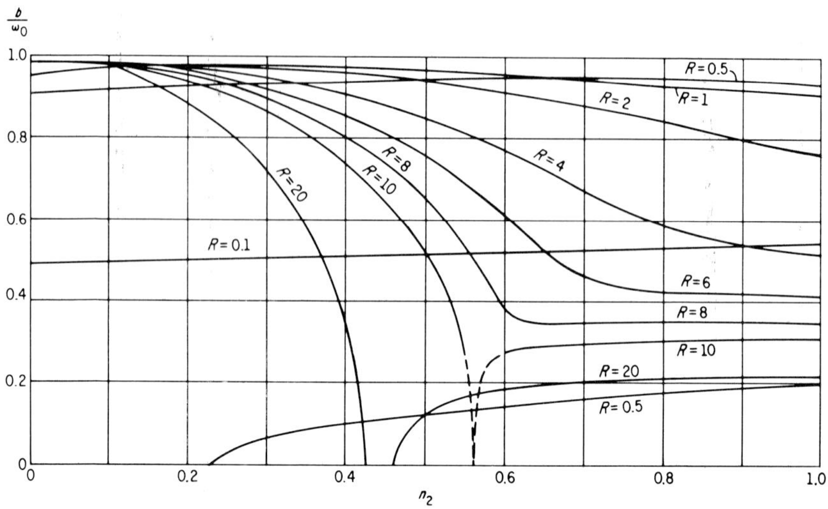

FIG. B-9. $n_{12} = 0, n_1 = 0.5$.

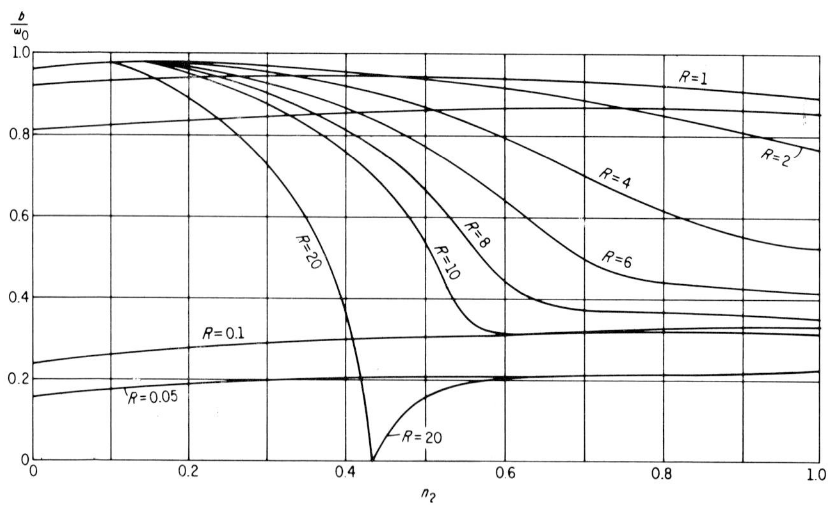

FIG. B-10. $n_{12} = 0, n_1 = 0.75$.

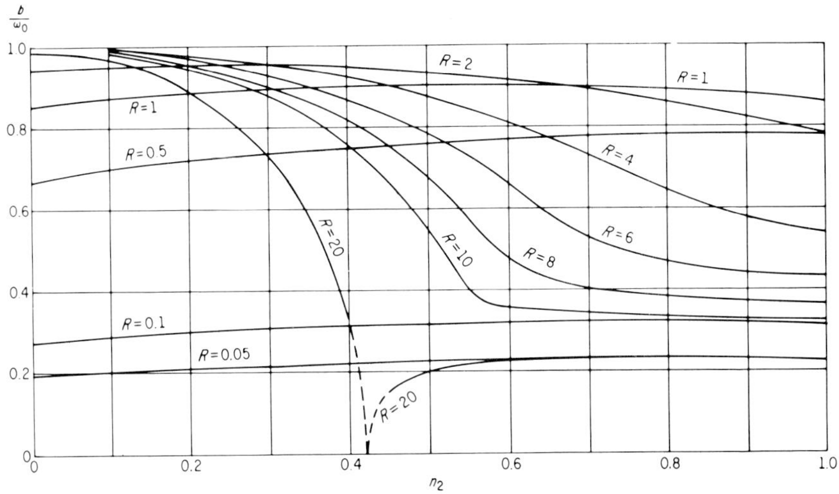

FIG. B-11. $n_{12} = 0, n_1 = 1.0$.

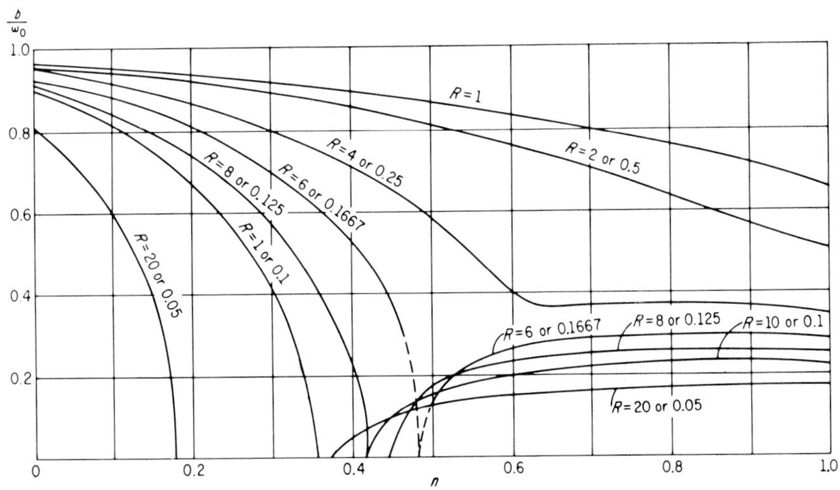

FIG. B-12. $n_1 = n_2 = n, n_{12} = 0.25$.

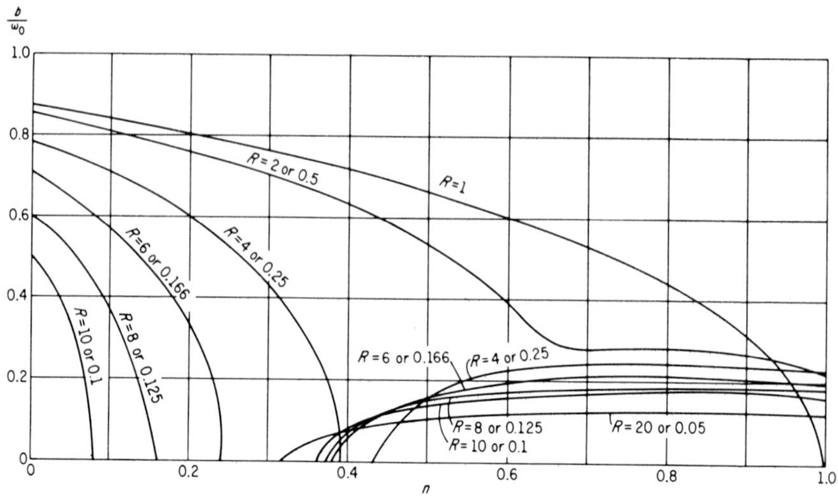

FIG. B-13. $n_1 = n_2 = n, n_{12} = 0.5$.

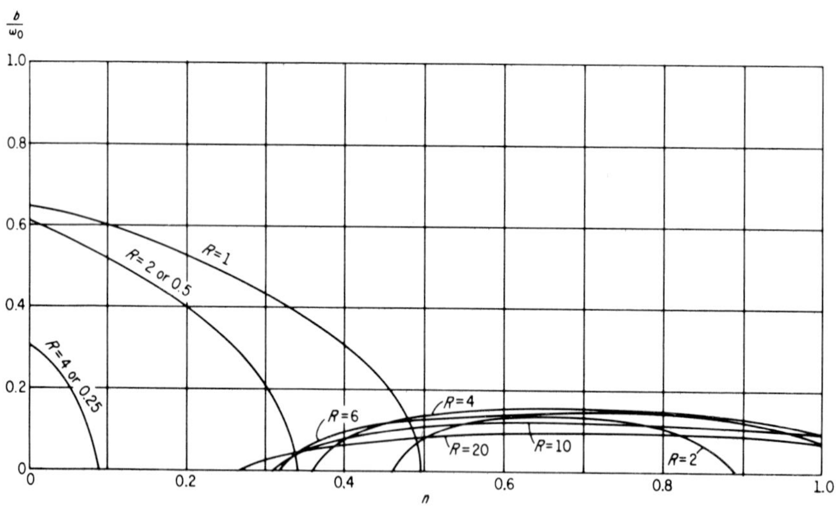

FIG. B-14. $n_1 = n_2 = n, n_{12} = 0.75$.

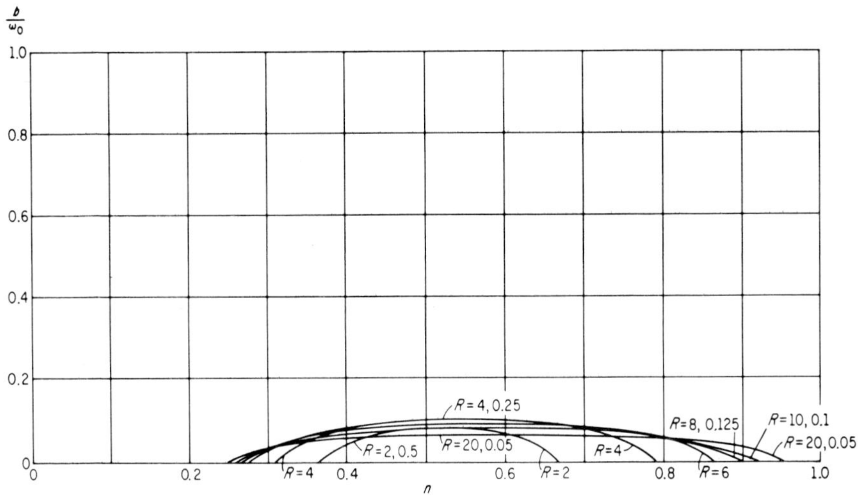

FIG. B-15. $n_1 = n_2 = n, n_{12} = 1.0.$

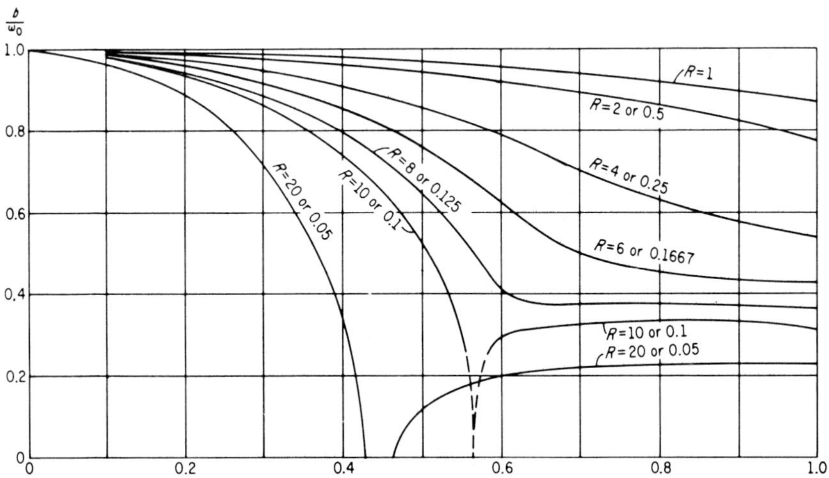

FIG. B-16. $n_1 = n_2 = n, n_{12} = 0.$

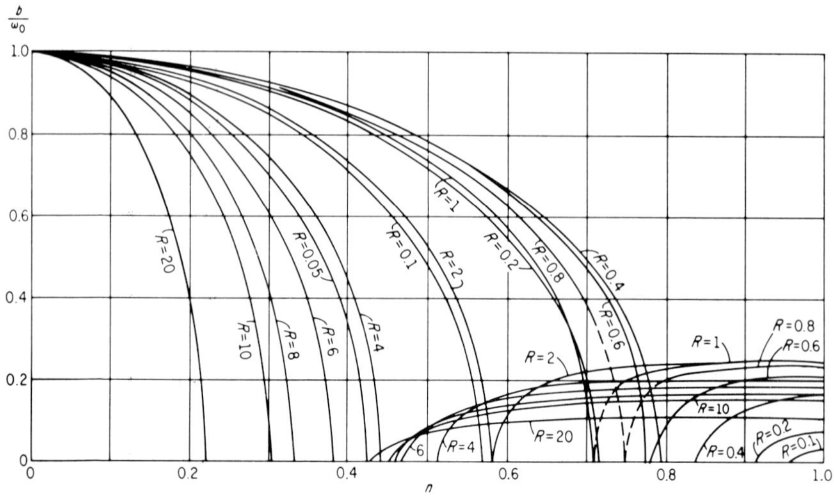

FIG. B-17. $n_2 = n_{12} = n, n_1 = 0.$

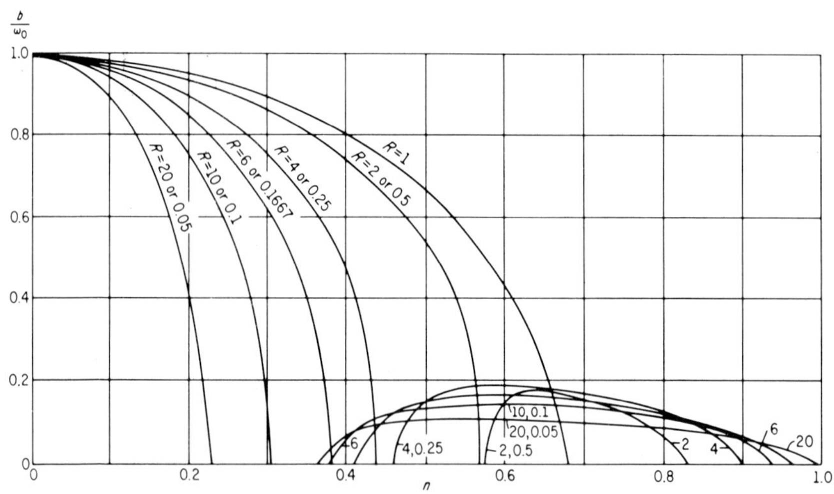

FIG. B-18. $n_1 = n_{12} = n_2 = n.$

appendix C

Design Data for Some Special Systems

C-1. SINGLE-DEGREE-OF-FREEDOM SYSTEMS ACCELERATING THROUGH CRITICAL SPEEDS

The magnification factors for such systems due to a harmonic excitattion are given in Fig. C-1.

ω_n = resonant frequency—cps

ζ = damping ratio

h = acceleration of system—cycles per sec^2.

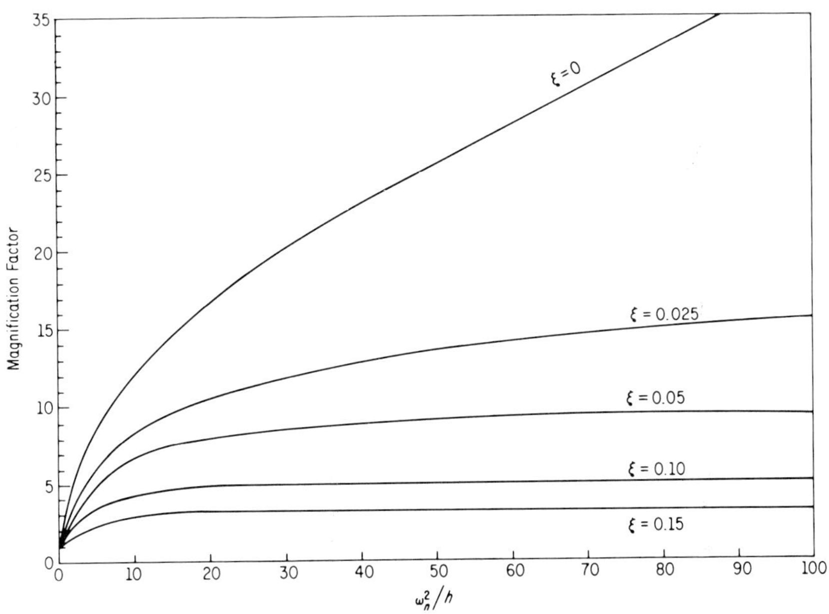

FIG. C-1

Reference

F. M. Lewis, "Vibration During Acceleration Through a Critical Speed," *Trans. ASME* Vol. 54, (1932) pp. 253–261.

Also see: "Mathematical-Machine Determination of the Vibration of Accelerated Unbalanced Rotor," by J. G. Baker, *J. App. Mech.* Vol. 6, *Trans ASME* Vol. 61 (1939) pp. A-145–A-150; and "Deceleration of an Unbalanced Rotor Through a Critical Speed," by W. K. Bodger, *Trans. ASME* Vol. 89-B No. 4 1967).

C-2. DYNAMIC FORCE IN A SPRING CONNECTING TWO MASSES MOVING WITH A CONSTANT VELOCITY V WHEN THE SYSTEM IMPACTS A STATIONARY SPRING

The peak force $F_{1(\max)}$ in the spring k_1 for such systems can be calculated from Fig. C-2 as

$$F_{1(\max)} = AVm_1\,\omega_{n2}$$

where

$$\omega_{n1} = \sqrt{k_1/m_1}, \quad \omega_{n2} = \sqrt{k_2/m_2}$$
$$\zeta_1 = c_1/2\sqrt{k_1 m_1}$$

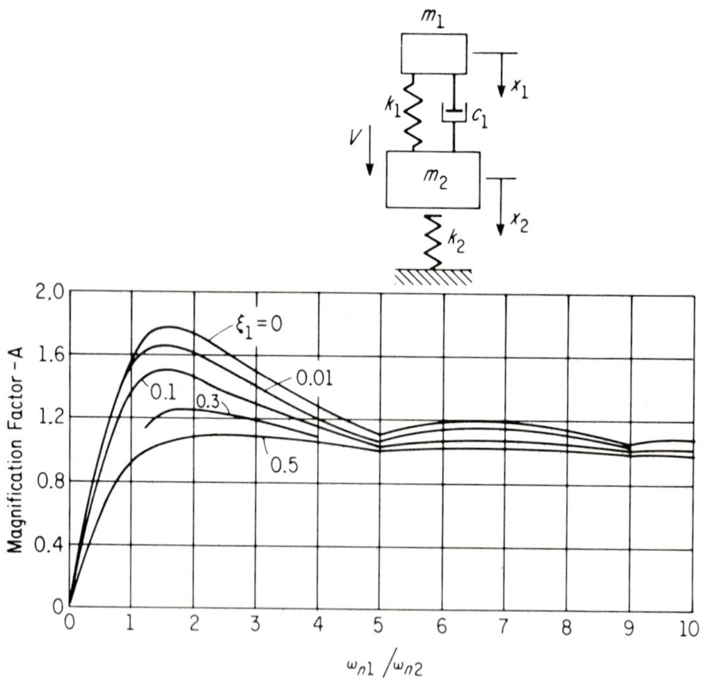

FIG. C-2

Reference

R. D. Mindlin et al, "Response of Damped Elastic Systems to Transient Disturbances," *Proc. Soc. for Exp. Stress Analysis*, Vol. V No. II (1948).

C-3. RESPONSE OF A TWO-ROTOR SYSTEM TO A RAMP-TYPE LOAD ON ONE ROTOR

The peak vibratory shaft torque T_s in this case can be obtained from Fig. C-3. The figure shows a plot of T_s/T_0 as a function of I_1/I_2 for different ratios of t_0/τ. The natural period τ is calculated from

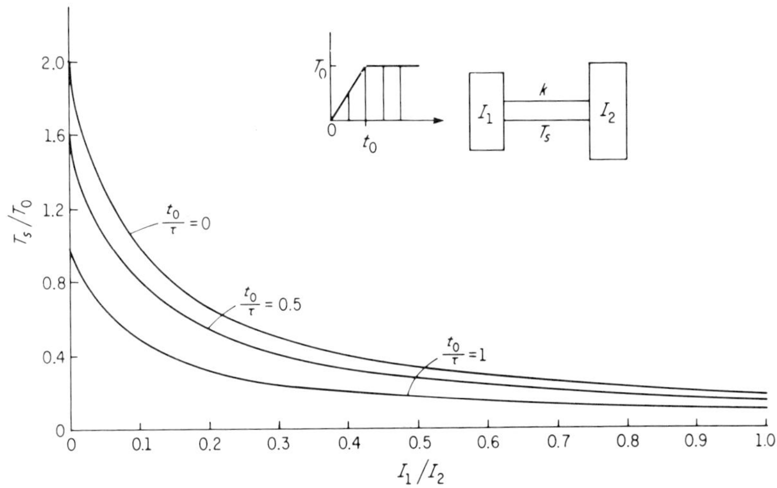

FIG. C-3

$$\tau = 2\pi/\sqrt{k(I_1 + I_2)/I_1 I_2}$$

C-4. WHIRLING OF SHAFTS IN GEARED SYSTEMS

The whirl frequency for a single-degree-of-freedom pinion shaft is affected by the gear restraint as given in Fig. C-4. In this case

Ω = whirl frequency for geared system
ω_n = whirl frequency for unconstrained pinion shaft
m_p = pinion mass
m_g = effective gear mass = I_G/R_G^2
I_G = moment of inertia of the gear
R_G = pitch radius of the gear

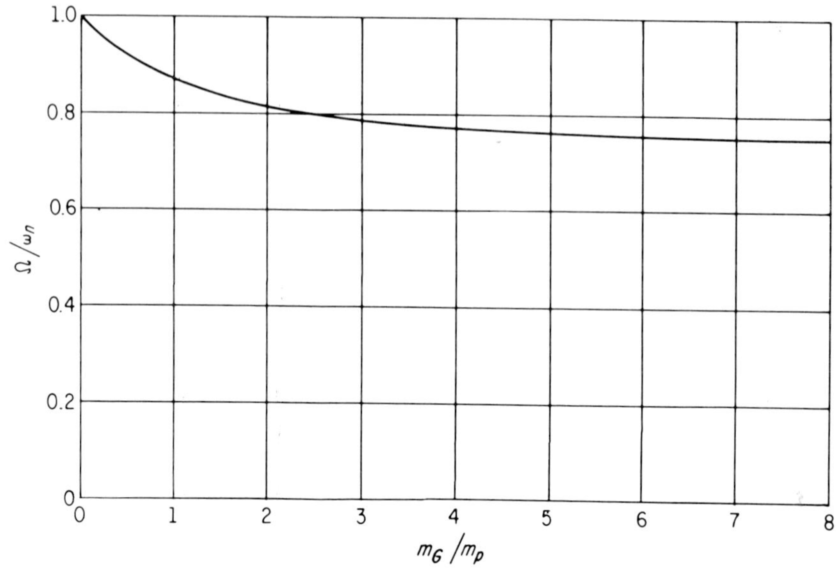

FIG. C-4

Reference

A. Seireg, "Whirling of Shafts in Geared Systems," *Trans ASME* Vol. 89-B No. 2 (1967).

For analysis of dynamic loads in geared systems and a comprehensive list of references on the subject see "Evaluation of Dynamic Factors For Spur and Helical Gears," by A. Seireg and D. Houser, to be published in *Trans ASME* Vol. 92-B (1970).

Index